ROCK SUPPORT AND REINFORCEMENT PRACTICE IN MINING

PROCEEDINGS OF THE INTERNATIONAL SYMPOSIUM ON GROUND SUPPORT
KALGOORLIE/WESTERN AUSTRALIA/15-17 MARCH 1999

Rock Support and Reinforcement Practice in Mining

Edited by

Ernesto Villaescusa
Western Australian School of Mines, Kalgoorlie, Western Australia

Christopher R. Windsor & Alan G. Thompson
Rock Technology, Perth, Western Australia

CRC Press
Taylor & Francis Group
Boca Raton London New York

CRC Press is an imprint of the
Taylor & Francis Group, an **informa** business
A BALKEMA BOOK

Supporting Organizations
Jetcrete Australia
Normandy Mining Ltd
WMC Resources Ltd
BHP Cannington
Sandvik/Tamrock
Rock Technology
ANI Arnall
Snowden Mining Consultants
Boral Shotcrete
Ingersoll Rand (Australia) Pty Ltd
Smorgon ARC/Steel

Organizing Committee
Dr Ernesto Villaescusa, WASM
Christopher Windsor, Rock Technology
Dr Alan Thompson, Rock Technology
David Finn, WMC Resources Ltd
Dr Peter Fuller, BFP Mining Consultants
Dr Tao Li, WMC Resources Ltd
Andrew Logan, BHP Cannington
Dr Peter Mikula, Kalgoorlie Consolidated Gold Mining
Mike Sandy, Australian Mining Consultants
Peter Teasdale, WMC Resources Ltd

Cover: Rock reinforcement of a highly stressed, blocky rockmass using combination cables during bench stoping operations at Mount Isa Mines (Provided by Dr E. Villaescusa)

Published by:
CRCPress/Balkema
P.O. Box 447, 2300 AK Leiden, The Netherlands
e-mail: Pub.NL@taylorandfrancis.com
www.crcpress.com – www.taylorandfrancis.com

ISBN 13: 978-90-5809-045-4 (hbk)

The texts of the various papers in this volume were set individually by typists under the supervision of each of the authors concerned.

Rock Support and Reinforcement Practice in Mining, Villaescusa, Windsor & Thompson (eds)
© 1999 Balkema, Rotterdam, ISBN 90 5809 045 0

Table of contents

2 *Mesh and membrane support systems*

3 *Shotcrete*

4 *Support and reinforcement in metalliferous mines*

 ISBN 90 5809 045 0

Foreword

The International Symposium on Rock Support and Reinforcement Practice in Mining was held at the Western Australian School of Mines, Kalgoorlie, Australia from March 15 to 17, 1999. The Symposium was the fourth in the series and follows on from the International Symposia held at Luleå, Sweden, 1983, Sudbury, Canada, 1992, and Lillehammer, Norway, 1997. The objective of the Symposium was to exchange experiences, knowledge and the lessons learnt in the areas of rock support and reinforcement with special attention being given to mining applications.

The Symposium dealt with seven main themes:

- Design, analysis, testing and monitoring applications;
- Mesh and membrane support systems;
- Shotcrete;
- Support and reinforcement in metalliferous mines;
- Support in burst-prone ground;
- Safety and training;
- Strata control in coal mines.

Over forty papers have been published in these Proceedings, including two Distinguished Lectures and four Keynote Addresses.

Distinguished lectures:

- E.T.Brown, Australia: The evolution of support and reinforcement philosophy and practice for underground mining excavations;
- E.Hoek, Canada: Support for very weak rock associated with faults and shear zones.

Keynote lectures:

- C.R.Windsor, Australia: Systematic design of reinforcement and support schemes for excavations in jointed rock;
- T.A.Melbye, Switzerland: International practices and trends in sprayed concrete;
- E.Villaescusa, Australia: The reinforcecd process in underground mining;
- P.G.Fuller, Australia: Roof strata reinforcement – Achievements and challenges.

The Organizing Committee wishes to thank Kalgoorlie Consolidated Gold Mines, Kundana Gold Mine, WMC Resources Ltd Kambalda Operations and Kanowna Belle Mine for their cooperation regarding field visits to mines. Sincere thanks are also given to all the authors for their valuable contributions.

E.Villaescusa
C.R.Windsor
A.G.Thompson

Distinguished lectures

Rock Support and Reinforcement Practice in Mining, Villaescusa, Windsor & Thompson (eds)
© 1999 Balkema, Rotterdam, ISBN 90 5809 045 0

The evolution of support and reinforcement philosophy and practice for underground mining excavations

E.T. Brown
The University of Queensland, Brisbane, Qld, Australia

ABSTRACT: This paper aims to provide some background to the subject of this conference. The purposes, terminology and principles of rock support and reinforcement in underground mining are outlined. The evolution of the use of rock and cable bolting, shotcrete and fill in underground mining in Australia and elsewhere is discussed. An example is given of the application of modern rock support and reinforcement to the Neves Corvo mine, Portugal. Finally, some observations are made about the recent application of support and reinforcement knowledge and technology in the Australian underground metalliferous mining industry.

1 INTRODUCTION

It is now possible to create underground excavations in rock, including mining excavations, of a size and in circumstances that would have been close to unthinkable 40 –50 years ago. That this is the case is largely the result of advances that have taken place over this period in rock mechanics understanding and in support and reinforcement techniques.

Despite the manifest advances that have been made in rock support and reinforcement technology much remains to be done particularly, in the author's opinion, in the application of existing knowledge and best practice (Brown 1998). Data provided to this conference by Lang (1999)[1] show that since 1980 rock falls have remained the largest single cause of fatalities and of lost time injury in underground mines in Western Australia. Lang's (1999) data also show that, except for occasional years over this period, Western Australia's rock fall fatalities per 1000 underground employees have been larger than those of South Africa and Ontario, Canada.

The purpose of this paper is to set the scene for the discussions of modern support and reinforcement practice in underground mining that will take place in this conference. The paper outlines the evolution of the terminology, philosophy and techniques of rock support and reinforcement in underground mining. The underlying principles of modern support and reinforcement practice will be discussed and a case-history example given of their practical application to outstanding effect. The emphasis of the paper will be on underground metalliferous mining. Natural support provided by pillars will not be considered.

2 PURPOSES

Brady & Brown (1985) have specified four common objectives for the performance of a mine structure as being

(a) to ensure the overall stability of the complete mine structure, defined by the main ore sources and mined voids, ore remnants and adjacent country rock;

(b) to protect the major service openings throughout their designed duty life;

(c) to provide secure access to safe working places in and around the centres of production; and

(d) to preserve the mineable condition of unmined ore reserves.

Support and reinforcement are major, but not the only, means of achieving these common objectives. The achievement of these objectives may also be aided by considerations of the shapes, sizes and orientations of excavations; the sequencing of

[1] References to papers presented at this conference are not included in the list of references given at the end of this paper.

mining operations; and drilling and blasting practice (eg Hoek & Brown 1980).

3 TERMINOLOGY

Historically, the terminology associated with rock support and reinforcement has been changeable, often confused and sometimes mistaken. Thankfully, today there appears to be general agreement about the "correct" usage of the terms support and reinforcement. The definitions widely, but not universally, adopted are those due to Windsor & Thompson (1993) and Windsor (1997):

"The words support and reinforcement are often used interchangeably. However, it is useful to consider the two terms as being explicitly different due to the method by which they stabilise the rock adjacent to an excavation. Essentially, **support** is the application of a reactive force at the face of the excavation, and includes techniques and devices, eg fill, timber, steel or concrete sets, shotcrete, etc. **Reinforcement** is considered to be an improvement of the overall rock mass properties from within the rock mass and will therefore include all techniques, and devices that act within the rock mass, eg rock bolts, cable bolts and ground anchors.

Pre-reinforcement is the application of reinforcement prior to the creation of the excavation. **Post-reinforcement** is the application of reinforcement at an appropriate time after the creation of the excavation." (Windsor 1997)

The classic mining text, Agricola's *De Re Metallica* (Agricola 1556) does not use the terms at all (admittedly in translation). It gives accounts of timbering in shafts, tunnels (see Fig. 1) and drifts but says little about its purpose other than to note that the lagging shown in Figure 1 is used "in case some portion of the body of the mountain may fall, and by its bulk impede passage or crush persons coming in and out". This is very similar to Brady & Brown's "to provide secure access to safe working places in

Figure 1. Timber tunnel support (A – posts, B – caps, C – sills, D – doors, E – lagging, F – drains) (Agricola 1556).

and around the centres of production".

In the early 20th century, the term **support** became widely used in underground mining. In Peele's famous *Mining Engineers' Handbook*, successive editions of which were published in 1918, 1927 and 1941, the term support is used much in the sense of Windsor's (1997) definition. It is interesting to note, however, that even in the third edition (Peele 1941), the emphasis is still on timber and dry fill as means of support in underground metalliferous mining.

One of the first consistent uses of the term **reinforcement** (in Windsor's sense) in a major text was in Obert & Duvall's (1967) account of rock bolting. By 1980, Hoek & Brown (1980) had introduced the term **pre-reinforcement** in its current sense. They did not use the general term reinforcement in a clear or explicit way even though the distinction between the roles of what we would now recognise as being either support or reinforcement was clearly demonstrated. In 1985, Brady & Brown (1985) distinguished between support and reinforcement in the now generally accepted sense. It should be noted, however, that this, or any, distinction between support and reinforcement is not always made, even by experts of high standing (eg Muir Wood 1993).

Hoek & Brown (1980) and Brady & Brown (1985) made the important terminological point that it was once the custom, especially in the civil engineering industry, to describe support as being either temporary or permanent. **Temporary support** was that support (or reinforcement) installed to ensure safe working conditions during mining. For centuries, such support consisted of some form of timbering (eg Agricola 1556, Beaumont 1903). If the excavation was required to remain open for an extended period of time, **permanent support** was installed subsequently. Quite often, the temporary support was partly or wholly removed to enable the permanent support to be installed (eg Peele 1941). As we now understand, this practice is dangerous and is counter to some of the principles of sound support and reinforcement practice.

Support and reinforcement may also be described as being primary or secondary although here shades of meaning exist. Brady & Brown (1985) say that **primary** support or reinforcement is applied during or immediately after excavation, to ensure safe working conditions during subsequent excavation, and to initiate the process of mobilising and conserving rock mass strength by controlling boundary displacements. This primary support or reinforcement will form part, and may form the whole, of the total support or reinforcement required. Any additional support or reinforcement applied at later stages is termed **secondary** or even **tertiary**.

In the author's experience, this usage is common in the civil engineering industry, especially in Europe. It differs in intent, although it may not differ in practice, from the approach taken by Windsor (1997) who writes:

"*Primary, secondary and tertiary reinforcement.* The preferred interpretation of these terms is associated with the role reinforcement plays in maintaining stability, mainly because the terms cannot be subsequently confused with the concept of permanent and temporary reinforcement. Primary reinforcement is used to maintain overall stability, secondary reinforcement is used for securing medium to large blocks or zones of rock between the primary reinforcement, and finally, tertiary reinforcement is used in conjunction with surface restraint to prevent surface loosening and degradation."

Support or reinforcement has also been classified as being either active or passive. **Active** support or reinforcement imposes a predetermined load to the rock surface at the time of installation. It can take the form of tensioned rock bolts or cables, hydraulic props, powered support for longwall faces or expanded concrete linings. **Passive** support or reinforcement is not installed with an applied loading, but develops its load as the rock mass deforms. Note that the terms "active" and "passive" as used here do not match the Windsor's (1997) definitions of "support" and "reinforcement".

Windsor and Thompson (1993) have included the concepts of tensioned (active) and untensioned (passive) reinforcement in a categorisation of reinforcement installation options (Fig. 2). In Figure 2, reinforcing elements are categorised according to the mechanism by which load is transferred to the rock – continuous mechanically coupled (CMC), continuous frictionally coupled (CFC) or discrete mechanically and frictionally coupled (DMFC).

4 PRINCIPLES AND MECHANISMS

4.1 *Support*

Until about 40 years ago, the concept of the mode of operation of support was the simplistic one of

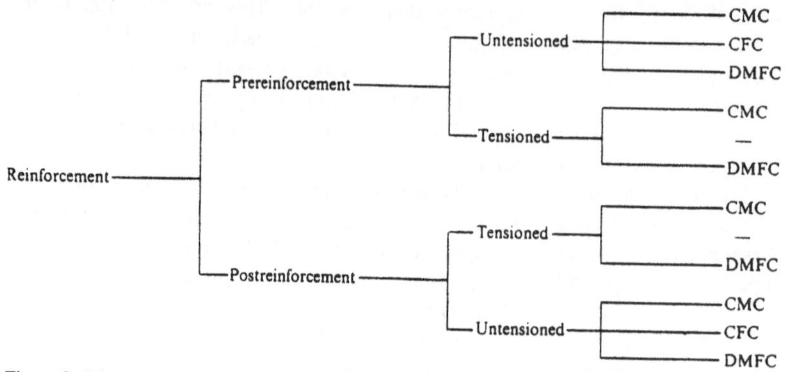

Figure 2. Categorisation of reinforcement and installation options (Windsor & Thompson 1993).

holding the rock around an excavation in place and resisting the gravity loading of loosened zones or blocks of rock. In Peele's *Mining Engineers' Handbook*, for example, there is next to no discussion of the mechanics of operation of the support methods described (mainly timbering). The ways in which they operated were "taken as read". This deficiency was remedied in Peele's successor, the *SME Mining Engineering Handbook* (Cummins & Given 1973).

Most of the early calculations carried out for the purposes of support design, particularly for steel sets and concrete tunnel or shaft linings, were concerned with the estimation of the loads to be carried by the support (eg Szechy 1970, Terzaghi 1946). These calculations, of which there were many types, usually attempted to estimate the loads applied by loosened zones above and around the tunnel (as in Fig. 3), a proportion of the overburden load or, in some cases, loads generated by swelling pressures.

An important advance in understanding came in the 1960s with the development of **ground-support interaction**, characteristic line or convergence-confinement concepts (eg Pacher 1964, Daemen 1977). Although initially expressed in terms of "support", these concepts, illustrated in a now well-known form in Figure 4, also apply to what we now refer to as reinforcement. The essential feature of this approach is that it shows how the support or reinforcement helps mobilise and conserve the inherent strength of the rock mass surrounding the excavation even when it is in a yielded or broken state. The support or reinforcement assists the rock to support itself. This approach illustrates clearly the importance of the timing of installation, stiffness and yield characteristics of support and reinforcing elements. Hoek (1999) gives a detailed practical example of the application of this approach in his paper to this conference.

Shotcrete is regarded as a form of support under Windsor's (1997) definition being adopted here. It provides support to the rock at the excavation

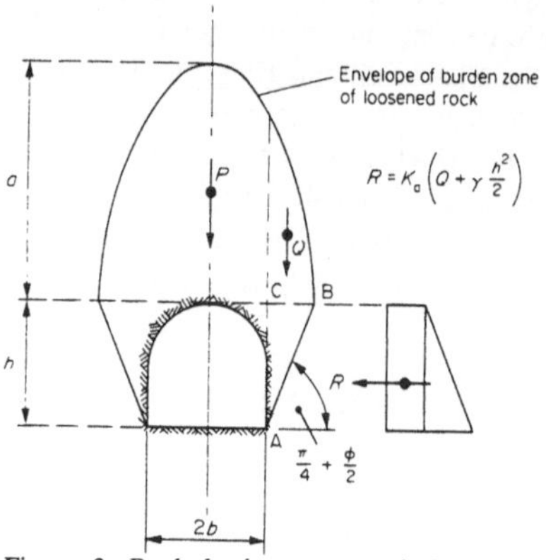

Figure 3. Rock load on a tunnel (Muir Wood 1993 after Terzaghi 1946).

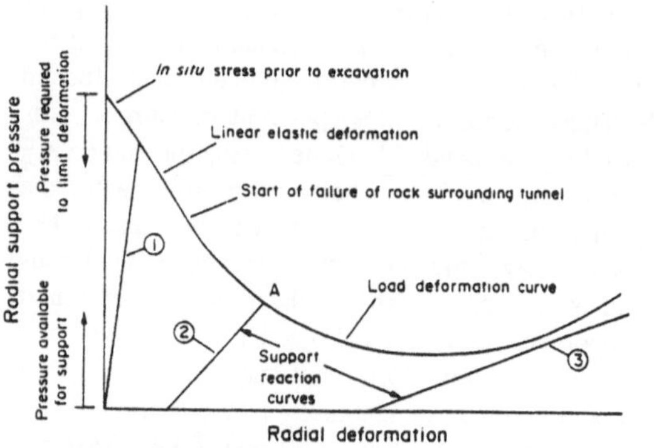

Support reaction curves
(= load induced in support by deformation of excavation)

1. Stiff support installed too early attracts excessive load.
2. Effective support at pressure A required to limit deformation = pressure available from support tunnel and support system in equilibrium.
3. Ineffective support not stiff enough and installed too late.

Figure 4. Simplified ground-support interaction curve (Douglas & Arthur 1983).

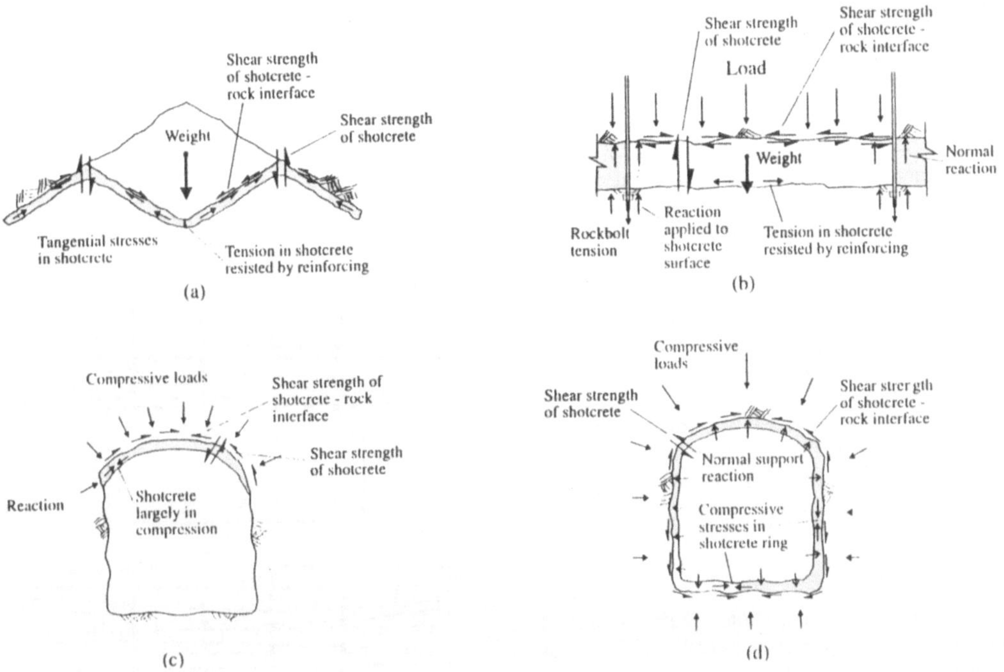

Figure 5. Some support mechanisms developed by shotcrete (a) single block, (b) beam anchored by bolts, (c) roof arch, (d) closed ring.

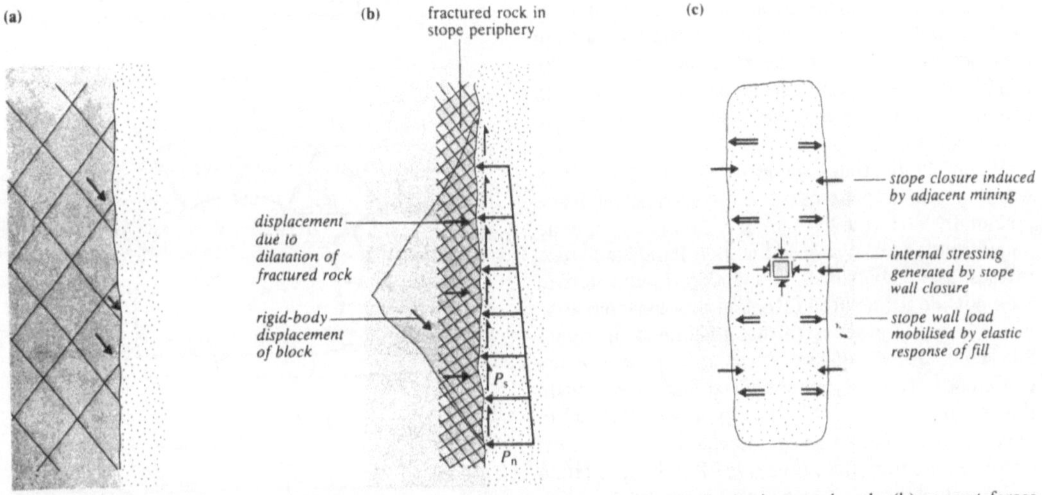

Figure 6. Modes of support of mine backfill: (a) kinematic constraint of surface blocks in destressed rock; (b) support forces mobilised locally by displacement of jointed and fractured rock; (c) global support due to compression of the fill by wall closure (Brady & Brown 1985).

periphery through a number of mechanisms, some of which are illustrated in Figure 5. These simplified diagrams illustrate the importance of the shear resistance at the shotcrete-rock interface, the shear strength of the shotcrete itself, the normal reactive support forces and the tensile strength of reinforced shotcrete. As suggested in Figures 5b-d, shotcrete may act in the ways considered in ground-support interaction or characteristic line analyses (Fig. 4). It is often used in conjunction with reinforcing elements (rock bolts, cables) to form a support and reinforcing system acting through a complex interaction of mechanisms.

Mesh, lacing and straps provide important parts of the support and reinforcement systems used in underground mining internationally. Essentially they act to provide surface restraint or containment to blocks, or more usually, loosened zones of rock. In their paper to this conference, Ortlepp et al (1999) provide examples of the use of containment support, including shotcrete, under large deformation static and dynamic loading conditions.

Fill provides an interesting example of a support mechanism. The ways in which fill acts to support the rock mass in both the near-field and the far-field of a stope have not always been agreed or well-understood. Brady & Brown (1985) show how fill may act to provide surficial, local and global components of support to a mine structure (Fig. 6). A recent example of the effective us of fill in an underground metalliferous mine is given in Section 6.

4.2 *Reinforcement*

Ever since the work of the United States Bureau of Mines in the 1950s (Obert & Duvall 1967), it has been recognised that reinforcing elements (initially rock bolts) act through a variety of mechanisms. The simplest mechanism is pinning individual blocks which may otherwise slip or fall under gravity loading, to more stable rock further removed from the face of the excavation. The reinforcing element may act in tension, in shear (at an interface), in combined tension and shear or, more rarely, in combined compression and shear (Windsor & Thompson 1993).

Figure 7 illustrates some of the more complex ways in which patterns of reinforcement stabilise potentially unstable rock masses around underground excavations. In non-laminated rock masses, long cables may be anchored in undisturbed rock outside a fractured or potentially unstable zone in a manner analogous to that illustrated in Figure 7a. Haile (1999) illustrates this in a paper to this conference. The very important arching mechanism illustrated in Figure 7c is discussed further in Section 5.1. The keying mechanism (Fig. 7d) expresses the important principle of seeking to retain the interlocking of blocks in discontinuous rock masses.

4.3 *Principles of good support and reinforcement practice*

Based on considerations of ground-support interaction mechanics (Fig. 4), Brady & Brown (1985) developed a set of principles to guide support and reinforcement practice. These principles were not meant to apply to the case of providing support for the self-weight of an individual block of rock, but to the general case in which yield of the rock mass surrounding the excavation is expected to occur. In this list of general principles the word reinforcement is used throughout. In this context it should be taken to refer to both support and reinforcement.

(a) Install the reinforcement close to the face

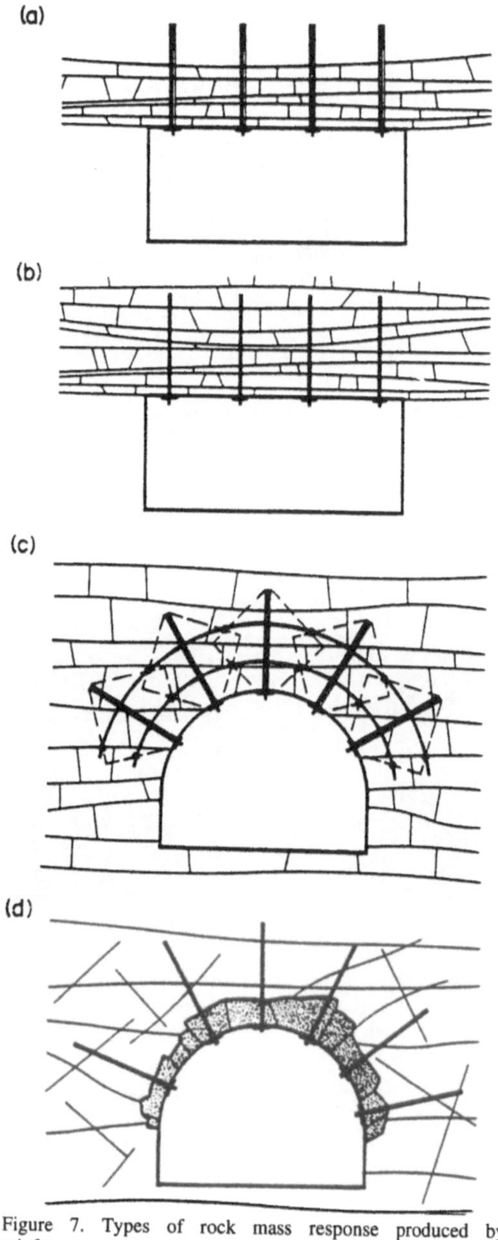

Figure 7. Types of rock mass response produced by reinforcement (a) suspension, (b) beam building, (c) arching and (d) keying (Windsor & Thompson 1993).

soon after excavation. (In some cases, it is possible, and advisable, to install the reinforcement before excavation. In others, usually involving high "squeezing" pressures, it may be advisable to permit some displacement to occur before the reinforcement is installed.)

(b) There should be good contact between the rock mass and the reinforcement.

(c) The deformability of the reinforcement should be such that it can conform to the displacements of the excavation surface.

(d) Ideally, the reinforcing system should help prevent deterioration of the mechanical properties of the rock mass with time due to weathering.

(e) Repeated removal and replacement of reinforcing elements should be avoided.

(f) The reinforcing system should be easily adaptable to changing rock mass conditions and excavation cross section.

(g) The reinforcing system should provide minimum obstruction of the working face.

(h) The rock mass surrounding the excavation should be disturbed as little as possible during the excavation process.

5 EVOLUTION OF TECHNIQUES

Table 1 lists some of the many support and reinforcement techniques referred to in papers presented to this conference. Windsor's (1997) definitions are used in this classification. This section will discuss the evolution of four of the most commonly used methods of support and reinforcement in underground metalliferous mining in Australia - rock bolts, cable bolts, fill and shotcrete. It is noteworthy that there are no significant discussions of fill included in Table 1.

5.1 *Rock bolts*

Gardner (1971) reports references in the literature to the use of rock bolting in some mines in the USA before 1900. He also reports that the St Joseph Lead Company in the USA used rock bolting successfully in the 1930s. After Conway (1948) described rock bolting experience by the Consolidated Coal Company in Illinois, experiments were conducted at the Erlington Colliery, NSW, in early 1949. The first full-scale use of rock bolting in collieries in Australia appears to have taken place in 1950 (McKensey 1953). The technique was soon introduced more widely into underground coal (Horseman 1954) and metalliferous (Yates & Holly 1956, Cawdle 1957) mines in Australia.

The use of rock bolts in Australia received its major impetus in civil engineering with the construction of the monumental Snowy Mountains Hydro-Electric Scheme in the period 1949-1969. Since 1999 represents the 50th anniversary of the commencement of construction of the Snowy Scheme, it is worth spending some time here to consider the contributions to rock bolting theory and practice made on the Snowy.

Although it had been used on a few projects in North America and France in the early 1950s, at the time of the design and construction of the major tunnels of the Snowy Scheme, rock bolting was not an established method of rock reinforcement (or support as it was called) in civil engineering, or in mining, in Australia or elsewhere. On the Snowy Scheme, some rock bolts were used in the Guthega project excavated in the period 1952-54, but the first major use of rock bolts in the Authority's works was in the Tumut 1 underground power station excavated in 1956-57 (Pender et al 1963). This represented a major and bold departure from the then traditional steel set and concrete support.

In the early to mid-1950s, it was generally held that the purpose of rock bolts was to pin surface rock (either individual blocks or bedded strata) to more stable rock some distance from the excavation boundary. At the time, mechanically anchored (slot and wedge or expansion shell) bolts were used. Rabcewicz (1957) had carried out model tests which suggested that a rock mass consisting entirely of blocks or fragments could be held stable by systematic bolting. The team of investigation and design engineers working on the Snowy, with the Assistant Commissioner T A Lang as the driving force, proved, developed and applied this concept with remarkable effect. Indeed, it has been suggested that the development and use of rock bolting for permanent "support" of underground excavations was probably the most significant of the many engineering developments made in the Snowy Scheme (McHugh 1988).

In the mid- to late 1950s, a detailed series of laboratory experiments was carried out by the Snowy Mountains Hydro-Electric Authority to investigate the action and effect of rock bolts. Simultaneously, experience was gained with the practical application of rock bolting, initially on the Tumut 1 project (Pinkerton et al 1961). The laboratory experiments described by Lang (1961) and by Alexander & Hosking (1971) included

- model tests of the arched roof of the Tumut 1 machine hall using gravel and regularly shaped perspex blocks to simulate the rock mass;
- cubical box models of various sizes in which the boxes were filled with model or prototype rock bolts and turned upside down with the open side facing downwards and often carrying an applied load;
- rod models;

Table 1. Some types of support and reinforcement referred to in papers to this conference.

Support		Reinforcement	
Type	Author(s)	Type	Author(s)
Steel sets	Hoek	Rock bolts	
Lattice girders	Hoek	Friction bolts (generic)	Schubert & Stewart
		Split Sets	Carr et al Chen & Collopy Li Rauert et al Sandy & Player Stacey & Ortlepp Thompson & Finn
Timber packs	Erasmus & Smit		
Yielding timber props	Glisson et al		
Yielding hydraulic props	Glisson et al	Swellex	Li & Håkansson Stacey & Ortlepp
Cellular lightweight concrete packs	Erasmus & Smit		
Shotcrete pillars	Hadjigeorgiou et al	Rebar (various types including cement and resin grouted	Haile Mikula Potvin et al Robinson & Tyler Stacey & Ortlepp
Shotcrete (plain or mesh reinforced)	Carr et al Chen & Collopy Clements Hoek Rauert et al Windsor & Thompson	Hollow groutable bolt (HGB)	Logan Sandy & Player
Steel fibre reinforce shotcrete/fibrecrete	Chen & Collopy Clements Ortlepp et al Ratcliffe Rauert et al Sandy & Player Spearing & Naismith Windsor & Thompson	Combi tube (CT)	Chen & Collopy Rauert et al Robinson & Tyler Sandy & Player Villaescusa & Wright
		Tubular groutable bolt (TGB)	Sandy & Player
		Universal	Chen & Collopy
		Yielding and smooth bar	Stacey & Ortlepp
Mesh	Chen & Collopy Haile Ortlepp et al Sandy & Player Thompson et al Villaescusa	Yielding cone bolts	Mikula
		Cables	
		Twin strand plain	Carr et al
		Destranded hoist rope	Stacey & Ortlepp
Lacing	Haile Ortlepp et al	Single strand bulbed Twin strand bulbed	Robinson & Tyler Logan Mikula Schubert & Stewart Villaescusa
Structural membrane	Finn et al Wojno & Toper	Rib-Roc cable dowels	Rauert et al
W straps	Sandy & Player	Grouted pipe forepoles	Hoek

- photo-elastic models using a solid plate of photo-elastic material or an assembly of blocks; and
- the famous inverted bucket model in which "a household bucket was filled with gravel and the surface layer was bolted with 1 inch (25.4mm) square bearing plates at both ends of the bolt. When the bucket was inverted, the lateral pressure developed on the sloping sides of the bucket was sufficient to support not only the gravel mass but also a central load of 50lb (0.22 kN)" (Alexander & Hosking 1971).

On the basis of the knowledge gained from these

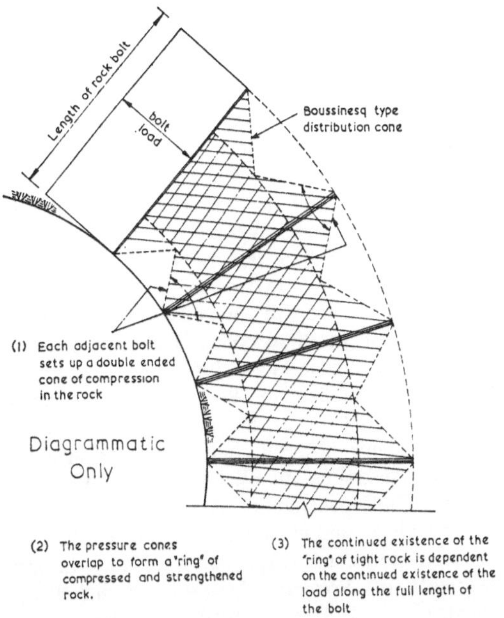

Figure 8. The action of rock bolts on the rock around an excavation (Pender et al 1963).

experiments and through field experience, the Snowy team developed an understanding of the way in which systematic rock bolting in a jointed rock mass forms a self-supporting compression zone within the rock mass (Fig. 8). In addition, Lang (1961) developed a set of design rules for pattern rock bolting which related bolt length and spacing to block size. These rules represented the state-of-the-art for many years subsequently (eg Hoek & Brown 1980). Lang (1961) also published pioneering analyses of the ways in which single rock bolts may prevent slip on single joints and single blocks of rock may be stabilised.

In the Tumut 1 machine hall, generally 4.6m long ungrouted mechanically anchored bolts on 1.4m centres were used in the walls. In the roof, 3.1m long bolts on 1.2m centres were used as "construction support" (primary support in our terms) between 1.2m square concrete ribs (Pinkerton et al 1961). The use of grouted rock bolts for permanent "support" was pioneered on the Tumut 2 underground power station constructed a few years later. Generally, 4.3m long cement grouted bolts on 1.2m centres were used. Some grouted bolts were 2.4m or 3.7m long (Pinkerton & Gibson 1963). Following its successful use on these large underground excavations, pattern rock bolting, often with grouting became an important feature of the several large tunnels constructed subsequently on the Snowy Scheme (Andrews et al 1964, Pender et al 1963).

The rock engineering expertise developed on the Snowy, including rock bolting technology, was soon transferred to the Australian mining industry. Rock bolts are now the most widely used reinforcing element in underground mines and civil construction. Many types of bolt including friction stabilisers and end- and fully-resin grouted bolts have since been developed (see Hoek et al 1995 and Windsor & Thompson 1993 for descriptions). New bolt types and installation techniques continue to evolve (eg Chen & Collopy 1999, Villaescusa & Wright 1999). Quality control and corrosion of rock bolts have become important issues in recent years and are addressed by a number of the papers presented to this conference (eg Logan 1999, Li & Lindblad 1999, Robinson & Tyler 1999, Villaescusa 1999).

5.2 *Cable bolts*

Hoek et al (1995) report that cable bolts were first used in underground mining in Canada (Marshall 1963) and South Africa (Thorn & Muller 1964). Cable bolt and dowel (untensioned) reinforcement was introduced into underground metalliferous mines in Australia in the early 1970s (Clifford 1974). Although there was simultaneous interest in the technique in Canada and in Scandinavia, Australian research and practice in cable bolting led the world. The initial applications were to cut and fill mining (Fuller 1981) but, subsequently, pre-placed cable bolt reinforcement was of fundamental importance to the introduction of large-scale open stoping methods of underground mining (Bywater & Fuller 1983, Thompson et al 1987).

Australian engineers have also made significant contributions to the understanding of the mechanics involved and the development of appropriate design methods (eg Fuller & Cox 1978, Windsor 1997). As in the case of rock bolts, advances continue to be made in cable bolt technology (eg Hutchins et al 1990, Hyett et al 1992). In recent years, cable bolting has been introduced into underground coal mining with some effect (O'Grady et al 1994, Kent et al 1997). For some time, cable bolting, often with yielding capability, has formed an important part of the strategies used to contain rock masses subject to high static and dynamic loading.

5.3 *Shotcrete*

Shotcrete's predecessor gunite (sprayed mortar) was used as a water- and weather-proofing agent, and to provide limited support, in both underground mining and civil engineering from the early part of this century. Peele (1941) gives several references to gunite being used to cover exposed rock and timber and steel supports for waterproofing, fireproofing and "to prevent air slacking of rock".

Shotcrete (sprayed concrete) was introduced into underground construction in Europe in the mid-1950s (Jaeger 1972). By the 1960s in Europe, shotcrete had become an integral part of the so-called New Austrian Tunnelling Method (NATM) (Rabcewicz 1964), so much so that some believed it to be the NATM. This is not correct. The NATM is an approach to underground construction which applies most of the sound principles and practices of ground control used elsewhere (Brown 1981).

Shotcrete technology using mainly the wet-mix process and involving the use, where appropriate, of accelerators, plastifiers, fibre reinforcement and other additives, is now widely used internationally in underground construction in hard and soft ground and in underground mining. In metalliferous mining, shotcrete is used not only in infrastructure and accesses but also to provide support around stopes and other production openings (eg Lourence et al 1996). The increased use of usually wet-mix steel fibre reinforced shotcrete has been one of the most remarkable changes in Australian underground metalliferous mining practice in recent years. Examples of its use in a variety of circumstances are given in papers to this conference by Clements (1999), Hoek (1999), Ortlepp et al (1999) and Sandy & Player (1999).

5.4 *Mining with fill*

According to Dickhout (1973), back fill was first used to control surface displacements above a mining area in 1864. Peele (1941) contains a large number of references to the filling systems in use in underground mines in the late nineteenth and early twentieth centuries. In the 1950s the dry fill formerly used was replaced with hydraulically placed fill in a number of Australian underground metalliferous mines (eg Cawdle 1957, Yates & Holly 1956). Mechanised cut and fill methods of mining were introduced for the Racecourse lead orebodies at Mount Isa in 1964 (Davies 1967) and adopted for the CSA Mine, Cobar in 1965 (Brady et al 1969). During the 1970s cut and fill was a primary underground metalliferous mining method at about 10 major Australian mines. The method was also being used in Canada and Scandinavia.

Although cut and fill mining continued to be used in some locations, especially in the Eastern Goldfields of Western Australia (Swindells & Szwedzicki 1991), economic imperatives subsequently caused a large number of Australian underground metalliferous mines to adopt open stoping methods, usually with backfill. These developments relied heavily on the mining geomechanics knowledge and expertise that had been developed in the "cut and fill era". The contributions of geomechanics to these

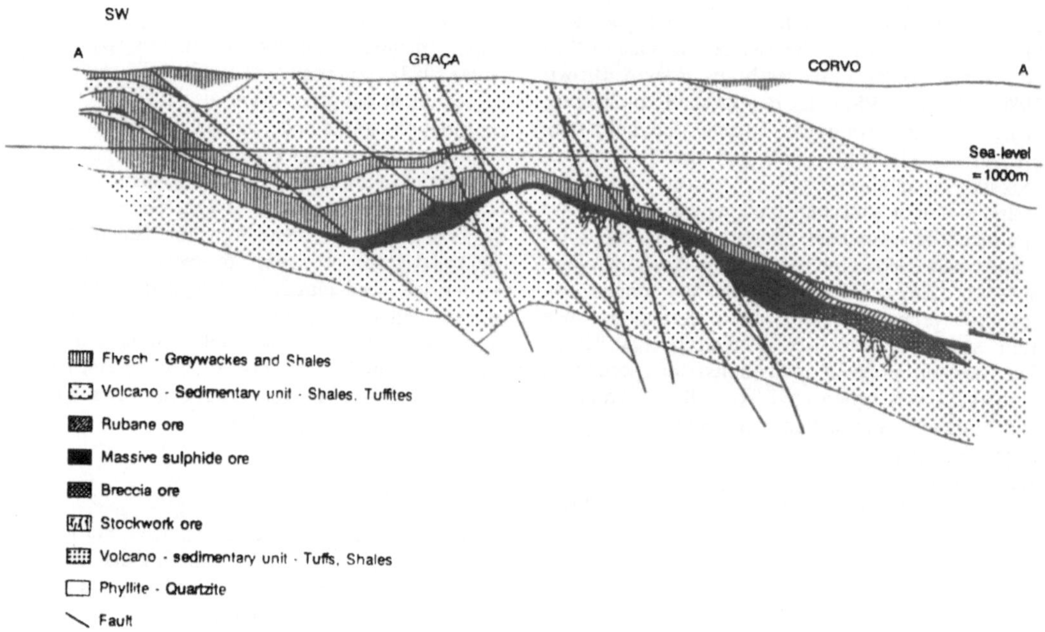

Figure 9. Transverse geological section through the Corvo and Graça orebodies, Neves Corvo mine, Portugal (Bailey & Hodson 1991.

developments are well illustrated by the development of open stoping in the 1100 orebody at Mount Isa (Watson 1987) which required the use of a specially developed low cost cemented rock fill (Williams 1977). More recent developments made in fill technology in other parts of the world include the use of fill to control convergence and improve safety in the deep level gold mines of South Africa (Jager *et al* 1987, Gürtunca 1997) and the development of paste fill in Canada (Landriault & Lidkea 1993).

Fill does not figure prominently in the papers presented to this conference. This may be because it is sometimes not thought of as a support or reinforcement technique and because recent cost-cutting measures have tended to limit its use in Australia. Nevertheless, fill remains an important support measure especially to ensure the global stability of a mine structure in the presence of large mined out voids (Brady & Brown 1985).

6 CASE HISTORY - THE NEVES CORVO MINE, PORTUGAL

The design, development and operation of the Neves Corvo copper-tin mine in the Iberian Pyrite Belt near the town of Almadovar in southern Portugal is a modern success story which owes much to the application of geomechanics knowledge and modern rock support and reinforcement philosophy and techniques. With annual production of 2.3 million tonnes of copper and tin ores it is one of the world's largest underground copper mines.

The discovery, project development and early mining of the Neves Corvo ore bodies are described by Bailey & Hodson (1991). Caupers et al (1993) provide an account of a trial stope experiment carried out in 1988-89 and early stope design and ground control measures. As Bailey & Hodson (1991) note, the choice of a stoping method for the Graça and Corvo orebodies was governed by a number of key factors:

- very high in situ copper grades which made maximum extraction and low dilution a high priority;
- the generally good geomechanical quality of the major orebodies and a strong hangingwall;
- flat orebody dips, averaging about 30° (see Fig. 9);
- a generally weak and sheared black shale footwall and footwall contact;
- the presence of several major faults which displace the ore; and
- the wide and rapid variation in ore thickness (see Fig. 9).

Some initial mining studies reached pessimistic

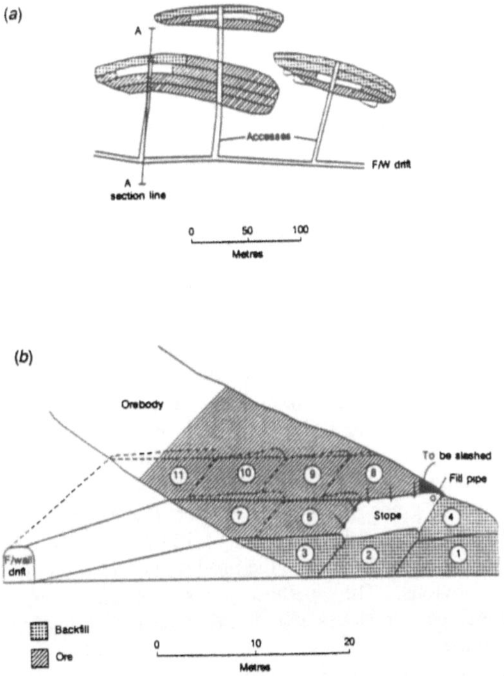

Figure 10. (a) Typical drift-and-fill layout plan, and (b) vertical section showing mining and filling sequence, Neves Corvo mine, Portugal (Bailey & Hodson 1991).

conclusions about stable opening spans, permissible overall mining spans and the possibility of achieving 100% extraction. Ultimately, a longitudinal drift-and-fill mining method was chosen and implemented with great success (Fig. 10). This method had several advantages (Bailey & Hodson 1991):

- the ability to develop initial drifts following the hangingwall contact minimised dilution and afforded mine geologists easy access to the orebody to define fault structures and contacts;
- the ability to allow considerable flexibility in stope layout and direction to take account of local variations in orebody characteristics;
- the continuous and full extraction concept avoids the creation of potentially troublesome highly stressed pillars;
- in wider parts of the orebody the technique could be modified as necessary to use transverse drift-and-fill from a central longitudinal drift; and
- the relatively large number of working faces would permit appropriate blending of the variable ore.

The quality of the cemented hydraulic fill and the achievement of tight placement against stope backs is of central importance to the success of the drift-and-fill mining method. Paste fill has been introduced recently (Caupers et al 1998a). Rock bolting (Swellex and resin grouted bolts), cable bolting and some shotcreting are used to ensure stability of access and drift backs. Tensioned and grouted cable bolting is used to help ensure the overall stability of the mine structure. Horizontal spans of the mined-out areas and filled areas can exceed 100m. Although by late 1998, some 16m tonnes of ore had been mined and 4 million m^3 of back fill had been placed since 1988, there were no indications of significant mine-scale deformations or instabilities.

Since 1993, the up to 20m high sill pillars left between stoping blocks have been mined successfully using a transverse cut-and-fill method allowing close to 100% extraction of the Corvo and Graça orebodies. These sill pillars together contained some 2.3 million tonnes of ore at an average grade of 11.7% copper (Caupers et al 1998b) compared to an average mine head grade of 6%. Their importance to the productivity of the mine is obvious. The method of sill pillar mining developed at Neves Corvo involves the following elements:

- pre-stoping development consists of cross-cuts from footwall accesses which connect with a longitudinal drift on the hangingwall contact (Fig. 11);
- all openings in the sill pillars are restricted to 4 to 5m in height and width;
- stoping is by transverse cut-and-fill retreating towards the cross-cuts;
- once the sill pillar is less than 20m thick it is pre-reinforced with cable bolts on 2m x 2m spacings;
- further cable bolts are installed as mining proceeds upwards;
- Swellex bolts are used for local reinforcement of the drift and stope backs;
- wet-mix shotcrete is used as necessary especially on the black shale footwall contact;
- ensuring tight fill against the drift and stope backs is important to the success of the mining method;
- the final lift of up to 8m is mined in two stages using a slinger truck to project cemented rock fill into place (Fig. 11); and
- displacements of the openings are monitored routinely.

Since 1996, mining of the better rock mass quality Neves North orebody (not shown in Fig. 9) to the north-west of Graça and Corvo has been by a bench-and-fill method similar to that used in the lead orebodies at Mount Isa (Villaescusa 1996). Stope heights vary between 10 and 43m, and the usual stope span is 16m with a few stopes being 22 or 24m in span. The average blast in the bench stoping area is about 8,000 tonnes. Up to 10m long grouted cables on 2m x 2m centres and Swellex bolts are used as reinforcement in drill drifts and stopes. In 1997, production by this method commenced in the Neves South orebody. In areas with poorer rock mass quality, a "mini" bench-and-fill method is being developed with smaller drift spans and stope widths and heights.

7 CONCLUDING REMARKS

It is clear from the outline given above that great

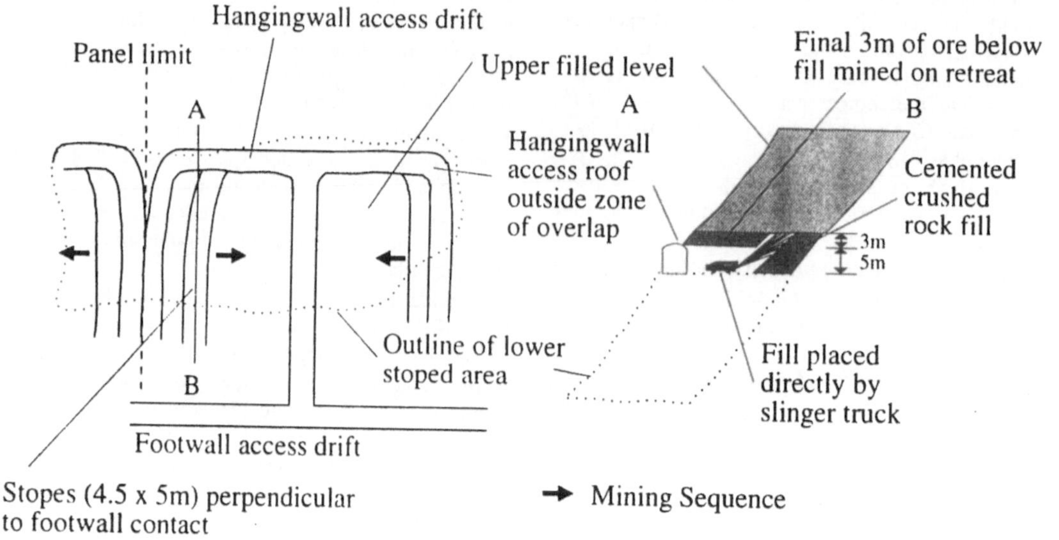

Figure 11. Method of mining the final lift of the sill pillars, Neves Corvo mine, Portugal (after Caupers et al 1998b).

advances have been made in the last 50 years in the development of understanding of, and in the techniques used for, rock support and reinforcement in underground mines. This has been part of the evolutionary development of rock mechanics and rock engineering over the same period. As Watson 1987) has noted, the application of geomechanics in the Australian mining industry has been part of the technological innovation that has sustained its growth and international competitiveness in difficult times.

But, as the author has noted previously, this apparently rosy picture does not tell the full story in the Australian underground metalliferous mining industry (Brown 1998). The continual need to reduce costs in order to remain competitive and, in some mining provinces at least, the need to improve safety underground, remain areas of concern. Concerns about costs and productivity on the one hand, and ground control and safety on the other, can appear to act in contrary directions, although this should not be the case (eg Schubert & Stewart 1999). Pressures to reduce costs and produce short-term profits can cause corners to be cut in the interests of financial expediency. Associated with this tendency there has been, in the Australian metalliferous mining industry, an apparent reduction in the continuing availability of specialist engineering skills on some minesites. The net result has been that the available mine design, geomechanics and support and reinforcement knowledge and technology is not always used to maximum advantage.

So, in the author's opinion, the greatest challenge that we face is not in making further incremental advances in knowledge, or even in solving the great unsolved problems, but in successfully applying existing knowledge and technology in practice. This is an issue of education (in its broader sense) and of training. It is reassuring to note that several papers to this conference discuss steps being taken at a number of minesites to overcome the generally negative tendencies just outlined (eg Chen & Collopy 1999, Harvey 1999, Li & Villaescusa 1999, Logan 1999, Rauert et al 1999, Sandy & Player 1999). This suggests that yet another (this time positive) change may be taking place in the Australian industry.

REFERENCES

Agricola, G. 1556. *De Re Metallica*. First edn. Trans. by H.C. Hoover & L.H. Hoover. New York: Dover, 1950.

Alexander, L.G. & A.D. Hosking 1971. Principles of rock bolting – formation of a support medium. *Proc. Symp. Rock Bolting*, Illawarra Branch, Australas. Inst. Min. Metall. Feb 1971: 1.1–1.21.

Andrews, K.E., A.R. McIntyre & R.H. Mattner 1964. Some aspects of high speed hard rock tunnelling in the Snowy Mountains. *Civ. Eng. Trans., Instn Engrs, Aust.* CE6(2): 51-69.

Bailey, D.C. & D.I. Hodson 1991. The Neves Corvo project - the successes and challenges. *Trans. Inst. Min. Metall., Sect. A - Min. Ind.* 100: A1-10.

Beaumont, E.K. 1903. Silver-lead ore mining and the various systems of stoping and timbering employed in Broken Hill, New South Wales. *Trans. Australas. Inst. Min. Engrs.* 9(1): 117-141.

Brady, B.H.G. & E.T. Brown 1985. *Rock mechanics for underground mining*, 1st ed. London: Allen & Unwin.

Brady, J.T., N.F. Owers & C.H. Annear 1969. Ore breaking and handling at the C.S.A. mine, Cobar. *Proc. Australas. Inst. Min. Metall.* 229: 7-15.

Brown, E.T. 1981. Putting the NATM into perspective. *Tunnels & Tunnelling*, 13 (10): 13-17.

Brown, E.T. 1998. Ground control for underground excavations – achievements and challenges. In N.I. Aziz & B. Indraratna (eds), *Proc. Int. Conf. on Geomechanics/Ground Control in Mining & Underground Construction*, Wollongong, 1: 13-31. Wollongong: Univ. Wollongong.

Bywater, S. & P.G. Fuller 1983. Cable support of lead open stope hanging walls at Mount Isa Mines Limited. In O. Stephansson (ed), *Proc. Int. Symp. Rock Bolting*, Abisko, Sweden, 539-555. Rotterdam: Balkema.

Caupers, D.J.D., L.F.P. Guerreiro, A.J.A. Salvador & N. Hepworth 1993. Design of stopes at Neves Corvo, Portugal. In L. Ribeiro e Sousa & N.F. Grossman (eds), *Proc. Int. Symp. Safety & Environmental Issues in Rock Engineering*, Lisbon, 499-507. Rotterdam: Balkema.

Caupers, D., M. Byrne, M. Lindqvist & P. Newman 1998a. Flow loop testing of different paste mixtures at Neves Corvo Mine, Portugal. In M. Bloss (Ed.), *Minefill 98, Proc. 6th Int. Symp. Mining with Backfill*, Brisbane, 87-97. Melbourne: Australas. Inst. Min. Metall.

Caupers, D., A. Salvador & J.P. Rodrigues 1998b. A exploração do último nível dos pilares doe soleira em Neves Corvo. *Proc. IV Simpósio Int. de Sulfuretos Polimetálicos da Faixa Peretosa Ibérica*, Lisbon, Jan 1998.

Cawdle, R.F. 1957. Some notes on recent changes in ore extraction practices in The Zinc Corporation and New Broken Hill Consolidated Mines. *Proc. Australas. Inst. Min. Metall.* 181: 45-57.

Clifford, R.L. 1974. Long rockbolt support at New Broken Hill Consolidated Limited. *Proc. Australas. Inst. Min. Metall.* 251: 21-26.

Conway, C.C. 1948. Roof support with suspension rods. *Min. Congr. J.* 34(6): 43.

Cummins, A.B. & I.A. Given (eds) 1973. *SME mining engineering handbook.* New York: SME of AIME.

Daemen, J.J.K. 1977. Problems in tunnel support mechanics. *Underground Space* 1: 163-172.

Davies, E. 1967. Mining practice at Mount Isa Mines Limited, Australia. *Trans. Instn Min. Metall., Sect. A - Min. Ind.* 76: A14-40.

Dickhout, M. H. 1973. The role and behaviour of fill in mining. *Proc. Jubilee Symp. Mine Filling*, Mount Isa, 1-11. Melbourne: Australas. Inst. Min. Metall.

Douglas, T.H. & L.J. Arthur 1983. *A guide to the use of rock reinforcement in underground excavations.* CIRIA Rep. 101. London: CIRIA.

Fuller, P.G. 1981. Pre-reinforcement of cut and fill stopes. In O. Stephansson & M.J. Jones (eds), *Applications of Rock Mechanics to Cut and Fill Mining*, 55-63. London: Instn Min. Metall.

Fuller, P.G. & R.H.T. Cox 1978. Rock reinforcement design based on control of joint displacement - a new concept. *Proc. 3rd Aus. Tunn. Conf.*, Sydney, 28-35. Canberra: Instn Engrs, Aust.

Gardner, F.J. 1971. History of rock bolting. *Proc. Symp. Rock Bolting*, Illawarra Branch, Australas. Inst. Min. Metall., 2.1-2.11.

Gürtunca, R.G. 1997. Rock engineering strategies and challenges for deep level mining. In H. Gurgenci & M. Hood (eds), *Proc. 4th Int. Symp. Mine Mechanisation & Automation*, Brisbane 2: A8.1 - A8.8. Brisbane: CMTE.

Hoek, E. & E.T. Brown 1980. *Underground excavations in rock.* London: Instn Min. Metall.

Hoek, E., P.K. Kaiser & W.F. Bawden 1995. *Support of Underground Excavations in Hard Rock.* Rotterdam: Balkema.

Horsemen, R.G. 1954. Roof bolting at John Darling Colliery, Belmont, New South Wales. *Proc. Australas. Inst. Min. Metall.* 172: 163-189.

Hutchins, W.R., S. Bywater, A. Thompson & C.R. Windsor 1990. A versatile grouted cable dowel reinforcing system for rock. *Proc. Australas. Inst. Min. Metall.* 295: 25-29.

Hyett, A.J., W.F. Bawden, R. Powers & P. Rocque 1993. The nutcase bolt. In W.F. Bawden & J.F. Archibald (eds), *Innovative Mine Design for the 21st Century, Proc. Int. Congr. Mine Design*, Kingston, Canada, 409-419. Rotterdam: Balkema.

Jaeger, C. 1972. *Rock mechanics and engineering.* Cambridge: Cambridge Univ. Press.

Jager, A.J. 1992. Two new support units for the control of rockburst damage. *Proc. Int. Symp. Rock Support Undergr. Min. Constrn.*, Sudbury, 621-631.

Jager, A.J., P.S. Piper & N.C. Gay 1987. Rock mechanics aspects of backfill in deep South African gold mines. In G. Herget & S. Vongpaisal (eds), *Proc. 6th Int. Congr. Rock Mech.*, Montreal 2: 991-998. Rotterdam: Balkema.

Landriault, D. & W. Lidkea 1993. Paste fill and high density slurry backfill. In W.F. Bawden & J.F. Archibald (eds), *Innovative Mine Design for the 21st Century, Proc. Int. Congr. Mine Design*, Kingston, Canada, 111-118. Rotterdam: Balkema.

Lang, T.A. 1961. Theory and practice of rockbolting. *Trans. Soc. Min. Engrs, AIME* 220: 333-348.

Lourence, P.B., P. Ball & M.W. Zeptner 1996. Evolution of shotcrete-based ground support designs at Kambalda Nickel Operations. *Symp. Shotcrete - Techniques, Procedures and Mining Applications*, Kalgoorlie. Pre-print, 10 pp.

McHugh, S. 1989. *The Snowy: the people behind the power.* Melbourne: Heinemann.

McKensey, S.B. 1953. Roof bolting in bord and pillar workings - Erlington and Hebburn No. 2 collieries. In *Coal in Australia, Proc. 5th Empire Min. Met. Conf.* 6: 155-186. Melbourne: Australas. Inst. Min. Metall.

Marshall, D. 1963. Hangingwall control at Willroy. *Can. Min. Metall. Bull.* 56: 327-331.

Muir Wood, A.M. 1993. Development of tunnel support philosophy. In J.A. Hudson (ed.), *Comprehensive Rock Engineering* 4: 349-368. Oxford: Pergamon.

Obert, L. & W.I. Duvall 1967. *Rock mechanics and the design of structures in rock.* New York: John Wiley & sons.

O'Grady, P., P. Fuller & P. Dight 1994. Cable bolting in Australian coal mines - current practice and design considerations. *Min. Engr.* 154 (9): 63-69.

Pacher, F. 1964. Deformationsmessungen im Versuchstollen als Mittel zur Erforschung des Gebirgsverhaltens und zur Bemessung des Ausbaues. *Felsmech. u. Ing. Geol.*, Suppl IV, 149-161.

Peele, R. (ed.) 1941. *Mining Engineers Handbook*, 3rd edn. New York: John Wiley & Sons.

Pender, E.B., A.D. Hosking & R.H. Mattner 1963. Grouted rock bolts for permanent support of major underground works. *J. Instn Engrs, Aust.* 35(7-8): 129-145.

Pinkerton, I.L., K.E. Andrews, A.N.G. Bray & A.C.H. Frost 1961. The design, construction and commissioning of Tumut 1 Power Station. *J. Instn Engrs, Aust.* 33(7-8): 235-252.

Pinkerton, I.L. & E.J. Gibson 1964. Tumut 2 underground power plant. *J. Power Div. Am. Soc. Civ. Engrs* 90(PO1): 33-58.

Rabcewicz, L. 1957. Modellversuche mit Ankerung in kohäsionslosem Material. *Die Bautechnik* 34(5): 171-173.

Rabcewicz, L. 1964. The New Austrian Tunnelling Method. *Water Power* 16: 453-457, 511-515.

Swindells, C.F. & T. Szwedzicki 1992. Mining geomechanics in Western Australia. In T. Szwedzicki, G.R. Baird & T.N. Little (eds), *Proc. Western Australian Conf. Min. Geomech.*, Kalgoorlie, 39-47. Kalgoorlie: Western Australian School of Mines.

Széchy, K. 1973. *The art of tunnelling*, 2nd edn. Budapest: Akadémiai Kiadó.

Terzaghi, K. 1946. Rock defects and loads on tunnel supports. In R.J. Proctor & T.L. White (eds), *Rock Tunnelling with Steel Supports*, 15-99. Youngstown, Ohio: The Commercial Shearing & Stamping Co.

Thompson, A.G., S.M. Matthews, C.R. Windsor, S. Bywater & V.H. Tillmann 1987. Innovations in rock reinforcement technology in the Australian mining industry. In G. Herget & S. Vongpaisal (eds), *Proc. 6th Int. Congr. Rock Mech.*, Montreal 2: 1275-1278. Rotterdam, Balkema.

Thorn, L.J. & D.S. Muller 1964. Prestressed roof support in underground engine chambers in Free State Geduld Mines Ltd. *Papers & Discns, Assn Mine Mngrs S. Afr.*, 411-428.

Villaescusa, E. 1996. Excavation design for bench stoping at Mount Isa mine, Queensland, Australia. *Trans. Instn Min. Metall., Sect. A - Min. Ind.*,105: A1-10.

Watson, B. 1987. The impact of research and development on productivity in the Australian minerals industry. In *Technology and Exports: The Role of the Technological Sciences in Australia's Export Performance*, 97-108. Melbourne: Aust. Acad. Technol. Sciences.

Williams, C.D. 1977. Mining technology at Mount Isa. *From Stump-Jump Plough to Interscan*, 48-56. Canberra: Aust. Acad. Sci.

Windsor, C.R. 1997. Rock reinforcement systems. *Int. J. Rock Mech. Min. Sci.* 34: 919-951.

Windsor, C.R. & A.G. Thompson 1993. Rock reinforcement - technology, testing, design and evaluation. In J.A. Hudson (ed.), *Comprehensive Rock Engineering*, 4: 451-484. Oxford: Pergamon.

Yates, C. & J. Holly 1956. Improvements in stoping practice at the Great Boulder Proprietary Gold Mines Limited. *Proc. Australas. Inst. Min. Metall.* 177: 33-48.

Rock Support and Reinforcement Practice in Mining, Villaescusa, Windsor & Thompson (eds)
© 1999 Balkema, Rotterdam, ISBN 90 5809 045 0

Support for very weak rock associated with faults and shear zones

Evert Hoek
North Vancouver, B.C., Canada

ABSTRACT: Controlling the stability of excavations in very weak rock associated with faults and shear zones requires the innovative use of combinations of support systems such as rockbolts, shotcrete, forepoles and, in some cases, yielding steel sets. Drainage and the sequence of excavation play critical roles in reducing the potential for failure. These approaches are illustrated by means of practical examples.

INTRODUCTION

Faults and shear zones present special challenges in tunnelling because they can lead to sudden and uncontrolled collapses unless appropriate action is taken as soon as they are encountered. The very weak and highly deformable nature of the materials and, in some cases, the presence of large volumes of high pressure water trapped behind the impermeable fault materials results in squeezing or flowing ground conditions. In order to control this behaviour, support must not only have sufficient capacity but it must be installed in a sequence that does not allow uncontrolled deformation of the tunnel.

Early detection of the presence of faults or shear zones is a very important component in the overall process of support design. When the presence of faults is suspected in the rock mass through which the tunnel is being excavated, a probe hole ahead of the advancing face should be made mandatory. This hole can be percussion drilled and the rate of penetration, colour and volume of return water and the character of the chippings should be monitored during the drilling. This will indicate significant differences in the rock mass character ahead of the face and, where very weak rock is indicated, a diamond-drilled probe hole can be used to explore the rock in more detail. In civil engineering tunnel driving these probe holes are typically drilled during weekend maintenance shifts and their length should be approximately the total advance during the week plus one tunnel diameter. This ensures that the rock has been explored for at least one tunnel diameter ahead of the advancing face.

When the presence and the approximate extent of a fault or shear zone has been confirmed, steps have to be taken to design a support system and a sequence of excavation and support installation to deal with the anticipated conditions. It is essential that all the required support elements should be available close at hand before the fault itself is exposed to any significant extent.

Depending upon the nature and extent of the fault and whether water is present, a variety of support systems and excavation sequences can be used. Before discussing these alternatives and considering some practical examples, it is necessary to consider some fundamental issues related to tunnelling through weak rock.

ROCK MASS STRENGTH / IN SITU STRESS

In the context of this discussion, a rock mass is considered to be weak when its in situ uniaxial compressive strength is less than about one third of the in situ stress acting upon the rock mass through which the tunnel is being excavated. This can be demonstrated by means of the plot of tunnel convergence versus the ratio of rock mass strength to in situ stress given in Figure 1. This plot shows a sudden increase in convergence for a strength/stress ratio of less than about one third. The plot was generated from a closed-form analysis of the development of rock mass failure surrounding an unsupported circular tunnel subjected to equal stresses in all directions. The analysis used follows that de-

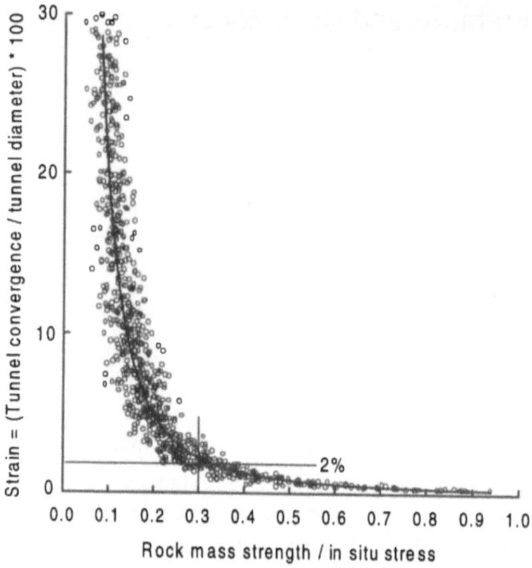

Figure 1: Plot of tunnel convergence against the ratio of rock mass strength to in situ stress.

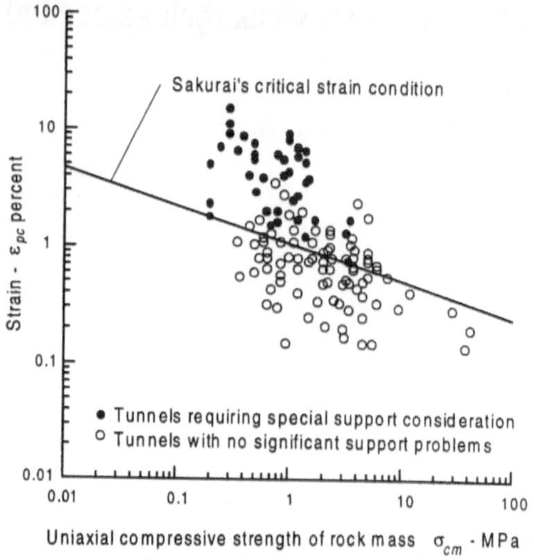

Figure 2: Percentage strain for different rock mass strengths. The points plotted are for the Second Freeway, the Pinglin and the New Tienlun Headrace tunnels in Taiwan.

scribed by Duncan-Fama (1993) and by Hoek, Kaiser and Bawden (1995).

A Monte Carlo simulation was used to carry out this analysis for 2000 iterations for uniform distributions of the rock mass properties, tunnel radius and in situ stress level. The rock mass properties were varied from fair to extremely poor, corresponding to the properties of weak sandstones and mudstones down to material that can almost be classed as soil. The in situ stresses were varied from 2 to 20 MPa, corresponding to depths below surface from 75 to 750 m, and the tunnel diameters were varied from 4 to 16 metres.

CRITICAL STRAIN

Sakurai (1983) has suggested that the stability of tunnels can be assessed on the basis of the strain in the rock mass surrounding the tunnel. The strain is defined by the ratio of tunnel convergence to tunnel diameter. A critical strain of approximately 2% represents the boundary between 'stable' tunnels that require minimal support and 'unstable' tunnels that require special consideration in terms of support design.

The application of this concept to practical tunnel problems is illustrated in Figure 2 that shows the percentage strain observed during the construction of three tunnels in Taiwan[1]. It can be seen that those tunnels categorised as requiring special support consideration fall above a line that is well defined by Sakurai's critical strain concept.

Note that all of the tunnels included in Figure 2 were constructed successfully, including those that suffered strains of approximately 10%. In some of these cases the tunnels had to be re-mined since the profiles were no longer adequate to accommodate the service structures for which they were designed. Sakurai's critical strain of 2% has been plotted in Figure 1. It can be seen that this corresponds well with the earlier conclusion that tunnels, excavated under conditions where the rock mass compressive strength is less that about one third of the in situ stress level, will suffer serious stability problems unless adequately supported.

ROCK MASS STRENGTH ESTIMATES

As a first approximation, the in situ stress can be assumed to equal the product of the depth below surface and the unit weight of the rock mass. The uniaxial compressive strength of the rock mass can be estimated from the Geological Strength Index (GSI) proposed by Hoek and Brown (1997) and extended by Hoek, Marinos and Benissi (1998). This descriptive index, which depends upon the structural characteristics and the surface conditions of discontinuities in the rock mass, is defined in Table 1.

[1] Information in this plot was supplied by Dr J.C. Chern of Sinotech Engineering Consultants Inc., Taipei.

Table 1: Table for estimating *GSI* (Hoek and Brown 1997, Hoek, Marinos and Benissi 1998)

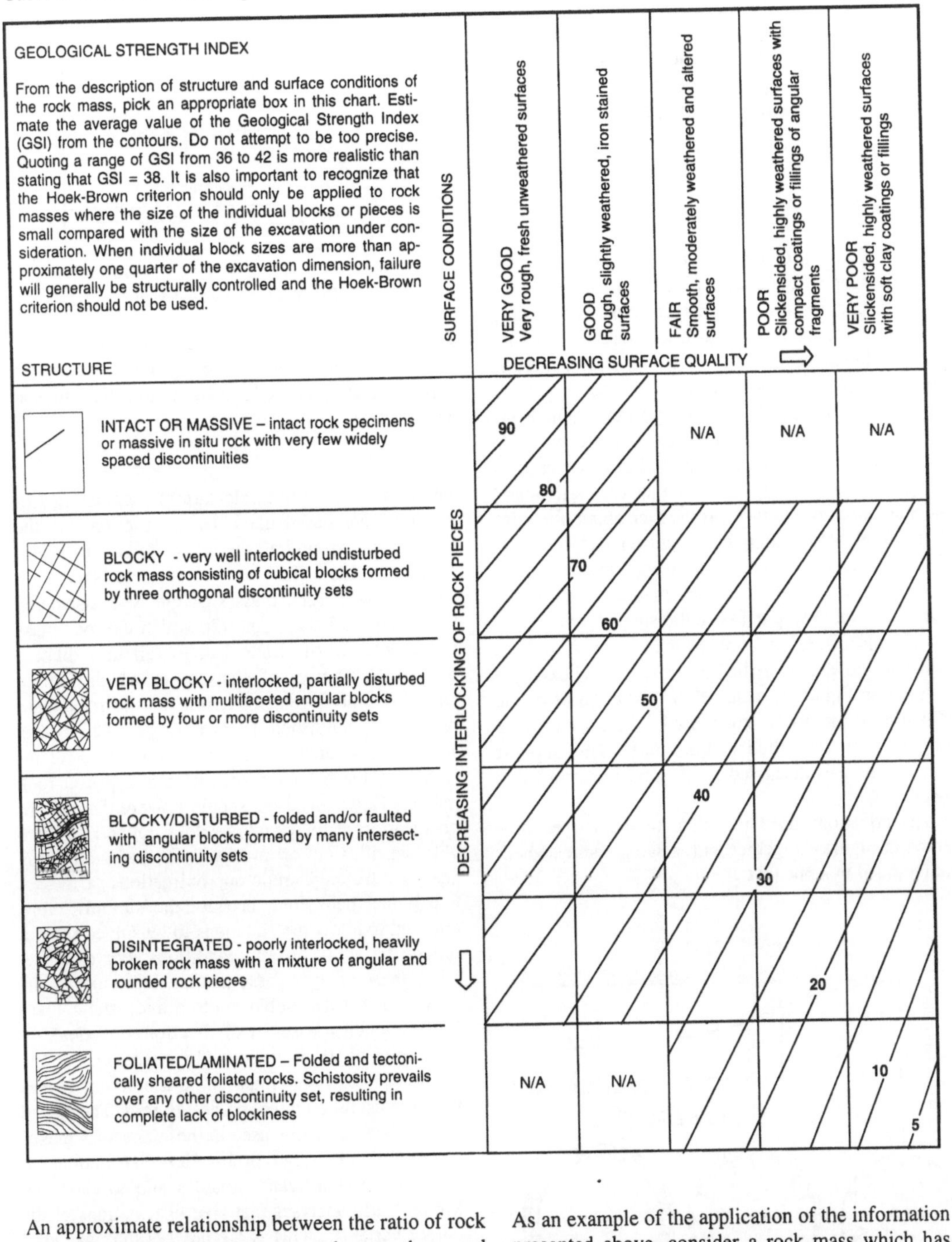

An approximate relationship between the ratio of rock mass strength σ_{cm} to laboratory rock strength σ_{ci} and the value of GSI is given by:

$$\sigma_{cm} = 0.019\sigma_{ci}\, e^{0.05GSI} \qquad (1)$$

As an example of the application of the information presented above, consider a rock mass which has been assigned a GSI = 23. The laboratory compressive strength of the rock mass is 5 MPa and hence, from equation 1, the rock mass compressive strength is estimated at 0.3 MPa. A tunnel being excavated at a depth of 150 m below surface in a rock mass with a unit weight of 0.025 MN/m^3 will be subjected to an in situ stress of approximately 3.8 MPa. Hence, the ratio of rock mass strength to in situ stress is approximately 0.08 and, from Figure 1, this is clearly in the category where serious stability problems can occur unless appropriate support is installed.

CHARACTERISTIC LINE CALCULATION

Fenner (1938) introduced the concept of calculating the convergence associated with the formation of a 'plastic' zone, or zone of damaged rock, surrounding an advancing tunnel. The basic elements of this concept are illustrated in Figure 3 which shows that the size of the plastic zone depends upon the equivalent support pressure p_i. In an unsupported tunnel the value of p_i reduces from the in situ stress value p_o to zero with distance from the face. The rock starts reacting to the oncoming tunnel about one-half tunnel diameter ahead of the face. At the face about one third of the total deformation has occurred and the final total deformation and complete formation of the plastic zone occurs about 1.5 tunnel diameters behind the face. The relationship between the support pressure p_i and the inward deformation δ of the tunnel walls is known as the characteristic line and a typical example is shown in Figure 4. Included in this are plots of the thickness of the plastic zone and of the reaction of a support system installed in the tunnel. This support reaction curve is discussed in the next section of this paper.

The equations used to calculate the curves presented in Figure 4 together with a sample spreadsheet are included in Appendix 1.

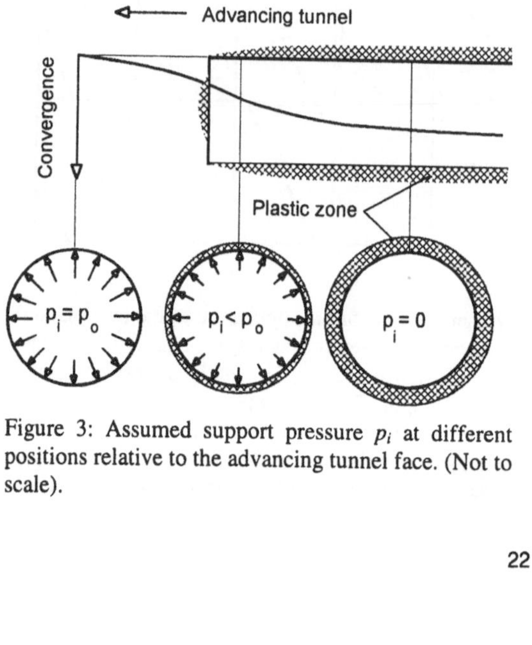

Figure 3: Assumed support pressure p_i at different positions relative to the advancing tunnel face. (Not to scale).

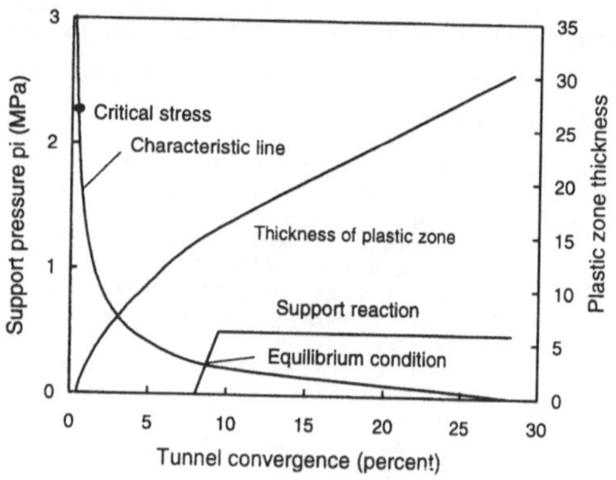

Figure 4: Example of characteristic line analysis including a plot of plastic zone thickness and of a support reaction curve for a 10 m diameter tunnel at 150 m depth.

The characteristic line calculations included in Appendix 1 are based upon the assumption that the rock mass surrounding the tunnel fails with no increase in volume. This is an appropriate assumption for very weak rock masses of the type associated with faults and shear zones in which the rock mass is likely to crush rather than to fail in a dilatant manner. A number of analyses in which dilation and also time-dependent failure of the rock mass have been published (Brown et al 1985, Panet 1993). These analyses can give a very useful insight into the behaviour of different types of rock masses. However, users should beware that the sophistication of the equations does not seduce them with an illusion of precision. In fact, all these analyses are only crude approximations of the actual tunnel behaviour, because the simplifications required to allow the equations to be solved seldom reflect the actual in situ conditions. For example, all of these analyses assume that the tunnel is circular and that it is subjected to a hydrostatic stress field in which stresses in all directions are equal.

In spite of its limitations, the characteristic line method has one very important attribute that makes it a very useful engineering design tool. The closed form of the equations used in these analyses makes it possible to assign probability distributions to each of the significant variables and to carry out Monte Carlo analyses that give distributions of the output variables. This is the process that was used to generate the points plotted in Figure 1. This type of probability analysis has also been used by Grasso et al (1997) to obtain an equivalent 'factor of safety' for tunnel support designs.

SUPPORT CHARACTERISTICS

When support is installed in a tunnel where plastic failure has occurred in the surrounding rock mass, the support pressure p_i provided by the support depends upon the stiffness of the support, its maximum load bearing capacity and the distance from the face when it was installed. The support acts very much like a system of springs and the support pressure increases with increasing deformation until the capacity of the system is exceeded. Hoek and Brown (1980) and Brady and Brown (1990) have published equations for calculating the stiffness and capacity of different support systems and these have been used to estimate the support characteristics given in Figure 5. Note that these characteristics are based upon the assumption that the support system is symmetrical around the tunnel. In other words, it is assumed that the steel sets and concrete or shotcrete linings are completely circular and that the rockbolts are installed in the roof, sidewalls and floor of the tunnel. These assumptions do not reflect typical support installations in the field and so, as in the case of the characteristic line calculations, they should be used to explore behaviour patterns rather than to calculate support characteristics to three decimal places.

As an example of the application of the information contained in Figure 5, consider the case of the tunnel behaviour illustrated in Figure 4. The total strain of the tunnel without support is approximately 28%, which means that the 10 m diameter tunnel will converge 2.8 m. From experience, I would suggest that the rock mass surrounding the tunnel would not be capable of sustaining this amount of convergence and that, without immediate support, this tunnel would collapse. Immediate support in this case means the installation of a suitable support system immediately behind the advancing face. As stated earlier, approximately one third of the final convergence has already occurred at the tunnel face. Consequently the earliest that conventional support can be installed is at a convergence of about 9%. It is probable that the face itself would also require support in the form of grouted fibreglass dowels or an umbrella of forepoles.

The plot in Figure 4 shows that the thickness of the plastic zone at a convergence of 5% is about 12 m. As a general rule, rockbolts or cables should have 1 to 2 m of anchorage in undisturbed rock outside the plastic zone. This means that rockbolts or cables of about 14 m would be required to provide support in this case and installation of these systems in a 10 m diameter tunnel is not a very efficient operation. In view of the uncertainty associated with the reliability of the anchorage in this poor quality rock mass, I suggest that rockbolts or cables are not an appropriate support system for this case and that steel sets or lattice girders should be considered.

From the information given in Figure 5, the support capacity of full-circle 200 x 200 mm wide flange ribs, 150 x 200 mm I section ribs, 124 x 108 mm TH section ribs and three-bar lattice girders at a spacing of 1 m in a 10 m diameter tunnel have support capacities of approximately 0.5 MPa. The maximum elastic strain that can be sustained by these girders is about 1% and, since they are installed at a convergence of 9%, this gives a total convergence of 10%. This support behaviour is plotted in Figure 4 as the support reaction curve.

Equilibrium conditions are achieved when the convergence of the tunnel and the support system are equal and, as shown in Figure 4, these conditions are defined by the intersection of the characteristic line and the support reaction curve.

The 'factor of safety' of the support system can be defined as the ratio of the maximum support capacity to the support pressure required for equilibrium. In this case, this factor of safety is approximately 2. While this may seem to be excessive, it must be remembered that it has been assumed that the full-circle steel support systems function perfectly. This may be too optimistic an assumption. As will be shown in the practical examples discussed later, there are times during the excavation of the tunnel and the installation of support when the support elements are subjected to unfavourable loading conditions and when they can be overstressed. Hence, some reserve capacity is appropriate.

Note that all of the support pressures in Figure 5 have been plotted for steel set or rockbolt spaced at 1 m and that, in order to determine the support pressures for other spacings, the equations given for each support type should be used.

When support types are combined, the total available support pressure can be estimated by summing the maximum allowable pressures for each system. However, in making this assumption it has to be realised that these support systems do not necessarily act at the same time and that it may be necessary to check the compatibility of the systems in terms of deformation. For example if lattice girders embedded in shotcrete are installed immediately behind the tunnel face, they will accept load immediately while the shotcrete will accept an increasing amount of load as it hardens (compare curves 24, 25 and 26 in Figure 5). Depending on the rate of advance of the tunnel, it is necessary to check that the capacity of the lattice girders is not exceeded before the shotcrete has hardened to the extent that it can carry its full share of the load.

Support type	Flange width - mm	Section depth - mm	Weight – kg/m	Curve number	Maximum support pressure p_{imax} (MPa) and average maximum strain $s_{max.av}$ for a tunnel of diameter D (m) and a support spacing of s (m)
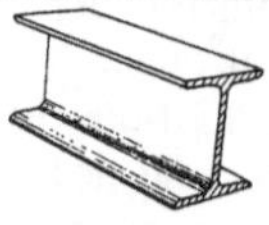	305	305	97	1	$p_{i\,max} = 19.9D^{-1.23}/s$
	203	203	67	2	$p_{i\,max} = 13.2D^{-1.3}/s$
	150	150	32	3	$p_{i\,max} = 7.0D^{-1.4}/s$
Wide flange rib					$s_{max.av} = 0.30\%$
	203	254	82	4	$p_{i\,max} = 17.6D^{-1.29}/s$
	152	203	52	5	$p_{i\,max} = 11.1D^{-1.33}/s$
I section rib					$s_{max.av} = 0.26\%$
	171	138	38	6	$p_{i\,max} = 15.5D^{-1.24}/s$
	124	108	21	7	$p_{i\,max} = 8.8D^{-1.27}/s$
TH section rib					$s_{max.av} = 0.55\%$
	220	190	19	8	$p_{i\,max} = 8.6D^{-1.03}/s$
3 bar lattice girder	140	130	18		$s_{max.av} = 1.35\%$
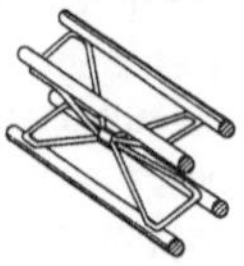	220	280	29	9	$p_{i\,max} = 18.3D^{-1.02}/s$
4 bar lattice girder	140	200	26		$s_{max.av} = 1.30\%$

Support type		Curve number	Maximum support pressure
	34 mm rockbolt	10	$p_{i\,max} = 0.354/s^2$
	25 mm rockbolt	11	$p_{i\,max} = 0.267/s^2$
	19 mm rockbolt	12	$p_{i\,max} = 0.184/s^2$
	17 mm rockbolt	13	$p_{i\,max} = 0.10/s^2$
Grouted rockbolts or cables spaced on a s x s metre grid. Maximum average strain is approximately 0.2%, excluding setting strain for faceplates and anchors and fibre-glass rods and cables.	SS39 Split set	14	$p_{i\,max} = 0.05/s^2$
	EXX Swellex	15	$p_{i\,max} = 0.11/s^2$
	20mm rebar	16	$p_{i\,max} = 0.17/s^2$
	22mm fibreglass	17	$p_{i\,max} = 0.26/s^2$
	Plain cable	18	$p_{i\,max} = 0.15/s^2$
	Birdcage cable	19	$p_{i\,max} = 0.30/s^2$

Support type	Thickness - mm	Age - days	UCS - MPa	Curve number	Maximum support pressure p_{imax} (MPa) for a tunnel of diameter D (metres)
	1m	28	35	20	$p_{i\,max} = 57.8D^{-0.92}$
	300	28	35	21	$p_{i\,max} = 19.1D^{-0.92}$
	150	28	35	22	$p_{i\,max} = 10.6D^{-0.97}$
Shotcrete or concrete lining. Maximum average strain ($s_{max.av}$) is approximately 0.1%.	100	28	35	23	$p_{i\,max} = 7.3D^{-0.98}$
	50	28	35	24	$p_{i\,max} = 3.8D^{-0.99}$
	50	3	11	25	$p_{i\,max} = 1.1D^{-0.97}$
	50	0.5	6	26	$p_{i\,max} = 0.6D^{-1.0}$

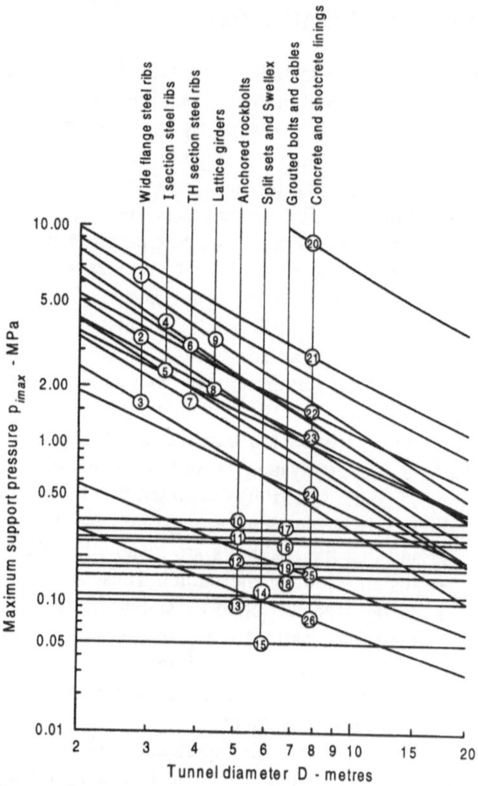

Figure 5: Approximate maximum capacities for different support systems installed in circular tunnels. Note that steel sets and rockbolts are all spaced at 1 m.

NUMERICAL ANALYSIS OF SUPPORT

The characteristic line calculation presented above is a very useful tool that is adequate for many practical support design problems. However, since it only gives an estimate of the final support capacity, it cannot be used to investigate the details of the excavation and support installation sequence required to deal with very difficult tunnelling problems such as those under consideration in this paper. In such cases, a numerical analysis will provide a more complete analysis.

Fortunately, there are several excellent programs available commercially that make it possible to carry out these analyses quickly and efficiently. Two of the best-known programs are FLAC[2], a very powerful finite difference program, and

[2] Available from the ITASCA Consulting Group Inc., Thresher Square East, 708 South Third Street, Suite 310, Minneapolis, Minnesota 55415, USA, Fax + 1 612 371 4717. Internet: http://www.itascacg.com.

PHASE2W[3] a simpler and more user-friendly finite element program. The program PHASE2W (Windows 95 version) was used to carry out a more detailed analysis of the problem discussed in the first part of this paper.

The tunnel to be analysed has a modified horseshoe shape of 10 m span and it is being excavated, as part of a drill-and-blast tunnel driving operation, through very weak rock at a depth of 150 m below surface. The properties of the rock mass, from the spreadsheet given in Appendix 1, are as follows:

Geological Strength Index GSI	23
Rock mass compressive strength σ_{cm}	0.28 MPa
Friction angle ϕ	22.15°
Cohesive strength *c*	0.1 MPa
Rock mass deformation modbulus E	473 MPa
Rock mass Poisson's ratio ν	0.3

It is assumed that the in situ stress field is hydrostatic, in other words the stresses are the same in all directions. This is a reasonable assumption for very weak rock such as that in a fault or shear zone, since this type of rock has already undergone failure and is incapable of sustaining significant stress differences. Hence, even if the far field stresses are asymmetrical, the stresses within the fault zone are likely to be approximately hydrostatic.

The first issue to be checked is the stability of the tunnel face. It has already been established, by means of the characteristic line calculation, that the 'plastic zone' surrounding the tunnel is very large and that substantial support is required to maintain its stability. What about the stability of the face and what support measures need to be taken to keep this stable?

Figure 6 shows the results of a three-dimensional analysis carried out using the axi-symmetric option in PHASE2W. This shows that the displacements of the face are of a similar order to those of the tunnel walls and this suggests that special measures will be required in order to prevent collapse of the face during excavation. These issues are discussed in detail in the following section.

TUNNEL FACE STABILITY

Water pressure and drainage

The first practical question to be considered is whether there is likely to be water pressure behind the face. Fault zones are generally less permeable than the surrounding rock mass and tend to act as dams. Consequently, there is a reasonable chance that a large volume of high-pressure water may be trapped behind the face. Driving the full face of a 10 m span tunnel into a fault zone with water trapped behind it is an invitation to disaster.

It is essential, when working in such ground, that a percussion-drilled probe hole be advanced well ahead of the face at all times. This will give warning of the presence of high pressure water and allow time for drainage measures to be set up before the fault material is exposed in the face. In general, drainage of the water is the most satisfactory solution and it may be necessary to install additional pumping capacity to deal with the water volume. In some cases, for example when mining or tunnelling under the sea or under a lake, drainage may not be the best option and grouting of the rock ahead of the face may have to be considered. The purpose of this grouting is to create a zone of impermeable rock in which the water pressure and flow can be controlled during tunnel driving.

Face support

Once the issue of water has been taken care of, the next question is how to prevent collapse of the face. Depending upon the severity of the problem, several options are available and these are reviewed below.

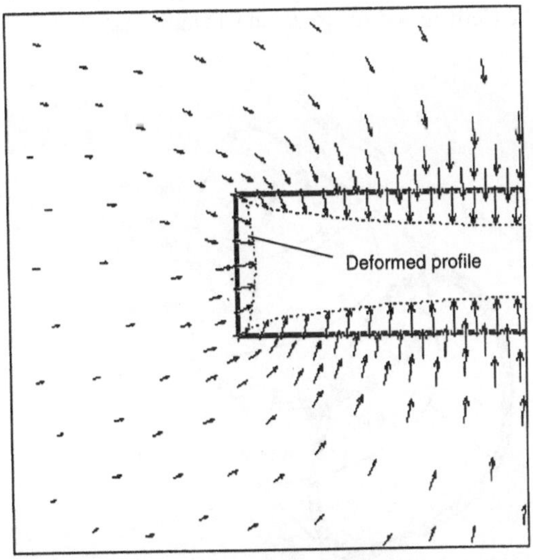

Figure 6: Vertical section through a three-dimensional finite element model of the failure and deformation of the rock mass surrounding the face of an advancing circular tunnel.

[3] Available from Rocscience Inc., 31 Balsam Avenue, Toronto, Ontario, Canada M4E 3B5, Fax + 1 416 698 0908. Internet: http://www.rocscience.com.

Grouted fibreglass dowels

The simplest option for face support is to reinforce the face with grouted fibreglass dowels which provide excellent support but are easy to mine through as the face is excavated. Experience suggests that these work best when the rock mass has a low clay mineral content and where friction rather than cohesion control its shear strength. This is because efficient operation of the dowels requires a high strength bond between the grout annulus and the rock mass and this is difficult to achieve in clay-rich cohesive materials. In the fault zone under consideration here, the cohesive strength is low (0.1 MPa) and the friction angle reasonably high (22°). Hence there is a good chance that fibreglass dowels would work well. However, given the magnitude of the 'plastic' zone in this case, it would be unwise to rely on such dowels as the only method of face support and I recommend that they be used to supplement other support systems rather than to replace them.

Partial face excavation

In tunnelling through weak ground it is generally accepted that the stability of the face depends upon the area exposed. Consequently, one commonly used technique for maintaining stability is partial face excavation in which the tunnel is driven in stages such that the area of each face is small enough to control. One of the many alternative ways of doing this is illustrated in Figure 7 where a technique favoured by the German and Austrian tunnel engineers is used.

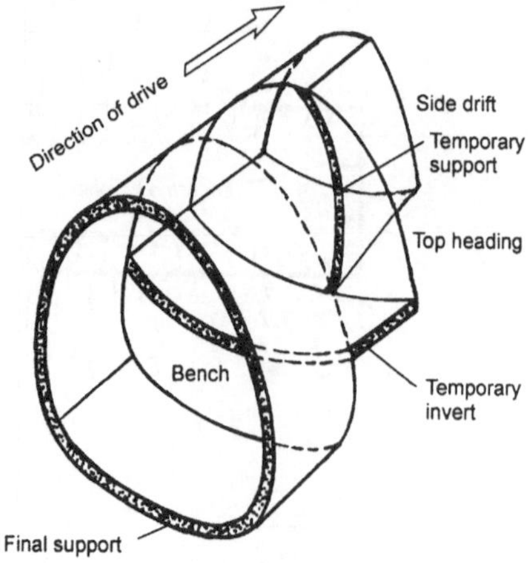

Figure 7: Partial face excavation method using a side drift followed by top heading and bench.

In this method, a side drift is excavated first and this is supported by installing the final support, such as steel sets embedded in shotcrete, against the outside wall and temporary support on the vertical wall and on the floor. Where possible, this temporary support should consist of fibre-reinforced shotcrete since this is easy to excavate when the tunnel is enlarged. However, in very heavy squeezing conditions, heavy weld mesh or steel ribs may have to be embedded in the shotcrete to provide sufficient support capacity. In many cases it is advantageous to drive the side drift all the way through the fault or shear zone, since it can be used to establish efficient drainage and ventilation arrangements before the main drive is attempted.

The excavation of the side drift is followed by opening the top heading to full span and this involves destruction of the temporary wall and the extension of the temporary invert to full span. This process should be carried out in steps of a few metres since one side of the top heading is effectively unsupported during the excavation process.

Excavation of the bench to form the full tunnel profile should also be carried out in steps of a few metres. The removal of the temporary invert leaves the top heading support 'suspended' until the lower legs of the steel sets can be installed and the final invert placed. In heavy squeezing conditions, a vertical bench parallel to the face is a good method of excavating the full profile since it allows systematic installation of the set and closure of the invert to be carried out with the benefit of support from the bench.

Placing steel sets and the application of shotcrete in a large tunnel requires heavy equipment to operate close to the face. This means that placing the final invert can present significant practical problems. Many tunnellers will attempt to advance the tunnel as far as possible and to leave the invert to be completed as an off-line activity tens of metres behind the face. In squeezing ground such as that under consideration here, this is always a serious mistake since it will allow floor heave and severe inward deformation of the installed roof and sidewall support. Failure to close the invert in time is probably the most common cause for the failure of support systems in squeezing ground.

The complete process of side drift, top heading and bench excavation has been simulated by numerical analysis and the results are presented in Figure 8. In spite of the approximations required in carrying out this analysis, the results provide a very useful guide to the adequacy of the excavation sequence and support systems.

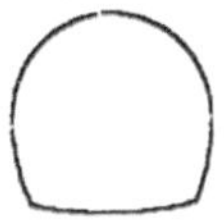

a. Excavation boundaries defined but no rock removed, model allowed to consolidate.

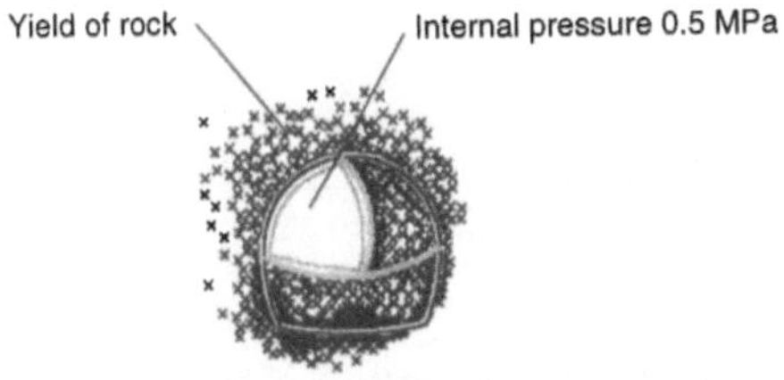

b. Excavation of side drift and application of an internal pressure of 0.5 MPa to simulate support provided by the advancing face.

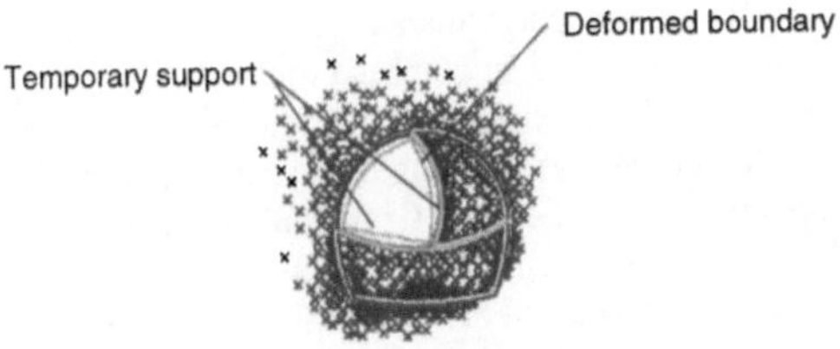

c. Installation of support in the form of steel sets embedded in shotcrete against the tunnel wall and temporary shotcrete invert and vertical partition. Removal of internal pressure.

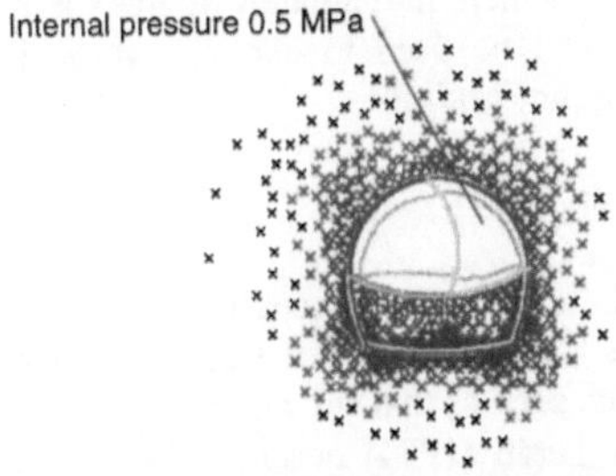

d. Excavation of top heading and application of internal pressure of 0.5 MPa to newly excavated right hand side to simulate support from face.

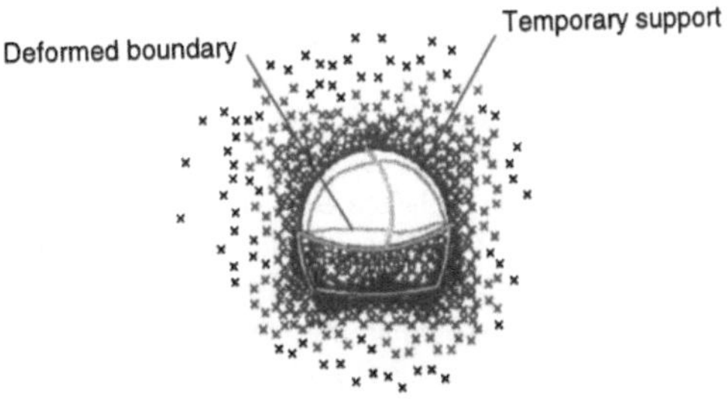

e. Installation of support in the form of steel sets embedded in shotcrete on tunnel walls and temporary shotcrete invert. Removal of internal pressure.

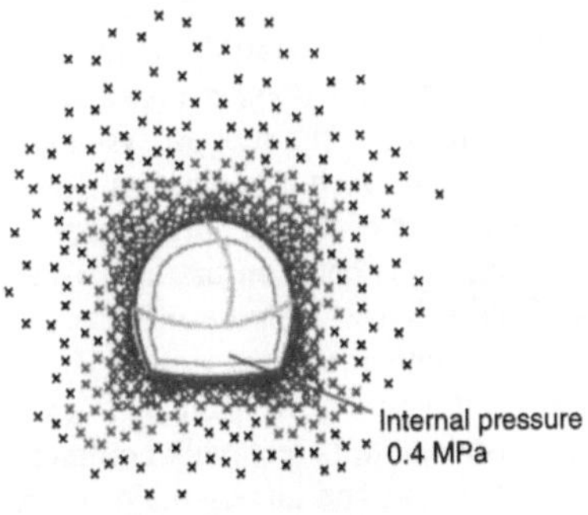

f. Excavation of lower bench and application of internal pressure of 0.4 MPa to simulate support provided by face.

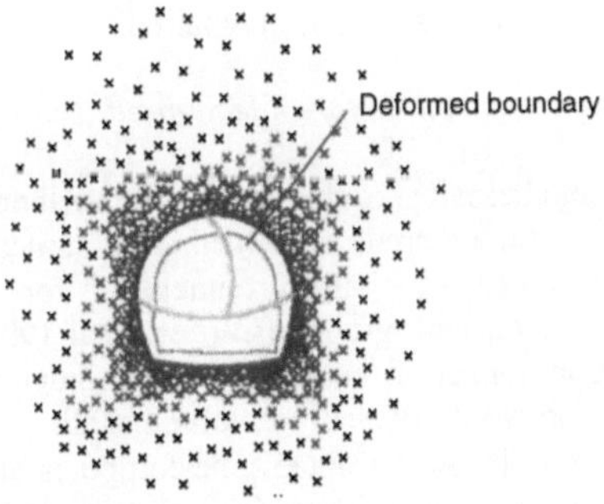

g. Completion of sidewall support and casting of concrete invert. Removal of internal pressure to allow full load on the completed support.

Figure 8 : Finite element analysis simulating the excavation sequence for mining a 10 m span tunnel through a fault at a depth of 150 m below surface. Final displacements for the roof are 250 mm, sidewalls 275 mm and floor heave is 190 mm.

Comparison of the results of the characteristic line calculation (Figure 4) and the finite element analysis (Figure 8) shows that the stepwise excavation and support installation sequence provides a significant improvement on the full-face excavation process depicted in Figure 4. Convergence is limited to about 5% and the maximum depth of the plastic zone in the roof is about 15 m. Note that this finite element model has been gravity loaded and the self-weight of the broken rock results in more failure in the rock above the roof than in that below the floor.

In the finite element analysis it was assumed that the final support consists of steel sets, typically 200 x 200 mm wide flange sections, embedded in shotcrete. While it is difficult to incorporate the properties of this composite system into currently available numerical models, it is possible to make a reasonable approximation based upon a consideration of when different elements of the support system are activated.

On the basis of such assumptions it was found that some minor yield of the support occurred, particularly at the connections between the upper and lower portions of the sidewall support and at the junctions of the sidewall support and the invert. This yield, which is associated with the sequence of excavation and support installation, is relatively common in this type of support system and has no major practical significance. It normally shows up as minor spalling in the shotcrete and this can easily be repaired by chipping out the damaged material and applying fresh shotcrete. Unless there is on-going time-dependent deformation, it is unlikely that this damage will recur once the damage has been repaired.

Advancing under a forepole umbrella

As an alternative to the partial face method described above, the umbrella arch method is sometimes used, particularly by Italian tunnellers, for advancing through difficult ground (Carrieri et al 1991). In a 10 m span tunnel of the type being considered here, the method would typically involve installing 12 m long 75 mm diameter grouted pipe forepoles at a spacing of 300 to 600 mm. These forepoles would be installed every 8 m to provide a minimum of 4 m of overlap between successive umbrellas. A sketch of a typical forepole umbrella is given in Figure 9.

A first step in this method usually involves drilling holes, up to 30 m ahead of the face, for drainage. This is followed by the drilling of the 12 m long holes and installation of the pipe forepoles to form the umbrella arch. In some cases, depending upon the nature of the rock mass being tunnelled through, jet-grouted columns are used rather than the grouted pipe forepoles.

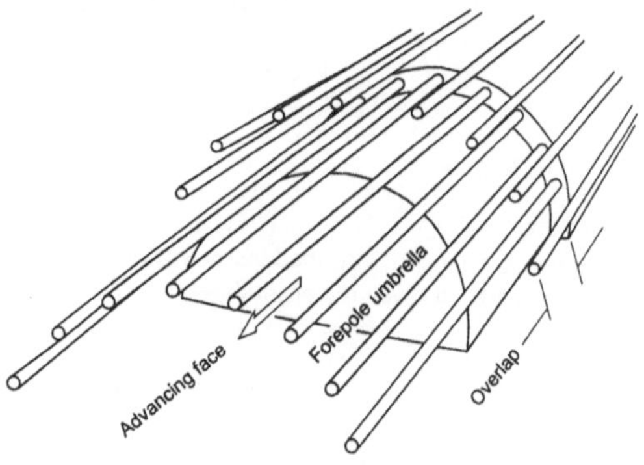

Figure 9: Sketch of tunnelling under the protection of a forepole umbrella.

The tunnel is then advanced 8 m before the cycle is repeated to create another protective umbrella.

This method is frequently used in combination with other support systems such as steel sets embedded in shotcrete, face stabilisation by grouted fibreglass dowels and the use of a temporary invert to control floor heave.

Analysis of the performance of this system, particularly when used in combination with other support systems, is extremely difficult. Simplified finite element analyses, using axi-symmetric models, have been described by Grasso et al (1993) and similar studies, using FLAC3D have been described by Itasca (1997). However, these analyses are certainly outside the type of studies that could be considered part of routine tunnel engineering design and a great deal of reliance has to be placed on judgement and experience. It is unlikely that this situation will change any time soon.

In spite of the lack of design guidelines and analytical tools, the umbrella arch is a very powerful tunnelling tool which, in the hands of an experienced operator, can be used to excavate through very difficult ground conditions.

Yielding steel sets

A method that has been used successfully in many tunnels involves the installation of steel sets fitted with sliding joints such as that shown in Figure 10.

Sánchez and Terán (1994) describe the use of yielding elements in steel ribs for the support of the Yacambú-Quibor tunnel in Venezuela – regarded by many as one of the most difficult tunnels in the world. This 5.5 m diameter water supply tunnel through the Andes is being excavated though weak rock masses, including graphitic phyllites, at a maximum depth below surface of 1200 m.

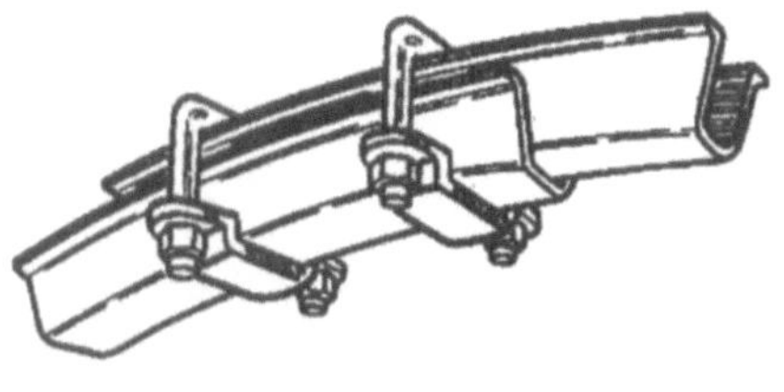

Figure 10: Assembly of a sliding joint in a Toussaint-Heintzmann or Top Hat section steel rib.

In the weakest rock sections, the support consists of WF6x20 steel sets, at 1 m spacing, with two sliding joints. These joints are set to lock when an additional tunnel closure of about 300 mm has been achieved. The sets are installed immediately behind the tunnel face and they are embedded in shotcrete, except for a 1 m wide 'window' that is left for each of the sliding joints. Once the joints have moved and locked, usually between 5 and 10 m behind the face, the 'windows' are closed to complete the shotcrete lining. This support system has proved to be very effective and measurements of tunnel convergence, carried out over several years, have shown that the tunnel is completely stable.

The installation of these sets immediately behind the face provides security for the men working at the face, in spite of the fact that the sets apply relatively little active support pressure to the rock mass during the sliding stage. The settings of the amount of sliding allowed in the joints is judged on the basis of the amount of tunnel convergence that can be tolerated before the full support reaction is activated (see Figure 4). Correct setting of the joints will achieve equilibrium between the tunnel convergence and the support reaction at much lower support pressures than for rigid steel sets. Consequently, provided that large tunnel convergence is acceptable, much lighter section support can be used than would otherwise be required.

Multiple tunnels

In one hydroelectric project in India, tunnelling through a large fault at depth proved to be extremely difficult and several collapses of the tunnel occurred as attempts were made to drive the tunnel at full size. Eventually a decision was made to split the tunnel into three smaller tunnels that would provide the same overall cross-section area for transmission of the water. Driving these three smaller tunnels, while still difficult was ultimately successful because, as shown in the case of the partial face excavation, dealing with smaller cross-sections has many practical advantages when tunnelling in difficult ground.

CONCLUSION

Tunnelling through very weak rock associated with faults and shear zones is a difficult problem, particularly when carried out under high in situ stress conditions. Conventional tunnel support such as rockbolts and shotcrete are seldom adequate to deal with heavy squeezing conditions that can occur when the rock mass surrounding the tunnel fails to a depth of several tunnel diameters.

Several alternative methods for maintaining stable face and tunnel stability have been explored in this paper. In some cases, limited theoretical analyses of the support systems are possible while, in other cases, reliance has to be placed on judgement and experience.

Tunnel driving costs, under the conditions described, are typically about three time average tunnel driving costs and advance rates seldom exceed about 1 m per day. Attempting to save money and time by adopting short-cuts or inadequate solutions invariably lead to even costlier failures.

Faults and shear zones exist in almost every rock mass and so it is inevitable that a tunnel engineer will be faced with the situations described in this paper at least once in his or her career. It is as well to attempt to learn from the experience of others rather than to attempt to sort out the problems when faced with a tunnel collapse. Reading about these problems is not a substitute for visiting a tunnel in which mining through a fault is in progress. Tunnel engineers should take every available opportunity to visit such tunnels.

REFERENCES

Brady, B.H.G. and Brown, E.T. (1985). *Rock mechanics for underground mining*. Allen and Unwin, London.

Brown E.T., Bray J.W., Ladanyi B. and Hoek E. 1983. Characteristic line calculations for rock tunnels. *J. Geotech. Engng Div., Am. Soc. Civ. Engrs* 109, 15-39.

Carrieri, G., Grasso, P., Mahtab, A. and Pelizza, S. (1991). Ten years of experience in the use of umbrella-arch for tunnelling. *Proc. SIG Conf. On Soil and Rock Improvement, Milan* 1, 99-111.

Duncan Fama, M.E. (1993). Numerical modelling of yield zones in weak rocks. In *Comprehensive rock engineering*, (ed. J.A. Hudson) 2, 49-75. Pergamon, Oxford.

Fenner, R. (1938) Undersuchungen zur erkenntnis des gebirgsdruckes. *Gluckauf.*

Grasso, P., Mahtab, M.A., Rabajoli, G. and Pelizza, S. (1993). Consideration for design of shallow tunnels. *Proc. Infrastructures souterraines de transport, Toulon,* 19-28.

Hoek, E., and Brown, E.T. (1980). *Underground excavations in rock.* Instn Min. Metall., London.

Hoek, E. and Brown, E.T. (1997). Practical estimates or rock mass strength. *Int. J. Rock Mech. & Min. Sci. & Geomech. Abstracts.* 34(8), 1165-1186.

Hoek, E., Kaiser, P.K. and Bawden. W.F. (1995). *Support of underground excavations in hard rock.* Balkema, Rotterdam.

Hoek, E., Marinos, P. and Benissi, M. (1998) Applicability of the Geological Strength Index (GSI) classification for very weak and sheared rock masses. The case of the Athens Schist Formation. *Bull. Engg. Geol. Env.* 57(2), 151-160.

Itasca (1997). Excavation and support for a shallow tunnel. *Company brochure on FLAC3D applications.*

Panet, M. (1997). Understanding deformations in tunnels. In *Comprehensive rock engineering*, (ed. J.A. Hudson) 1, 663-690. Pergamon, Oxford.

Sakurai, S. (1983). Displacement measurements associated with the design of underground openings. *Proc. Int. symp. field measurements in geomechanics, Zurich* 2, 1163-1178.

Sánchez Fernández, J.L. and Terán Benítez, C.E. (1994). Túnel de Trasvase Yacambú-Quibor. "Avance Actual de los Trabajos de Excavación Mediante la Utilización de Soportes Flexibles Aplicados a Rocas con Grandes Deformaciones". *Proc. IV Congreso Sudamericano de Mecanica de Rocas, Santiago* 1, 489-497.

APPENDIX 1: ANALYSIS OF PLASTIC FAILURE

It is assumed that the onset of plastic failure, for different values of the effective confining stress σ_3', is defined by the Mohr-Coulomb criterion and expressed as:

$$\sigma_1' = \sigma_{cm} + k\sigma_3' \qquad (A.1)$$

The uniaxial compressive strength of the rock mass σ_{cm} is defined by:

$$\sigma_{cm} = \frac{2c' \cos\phi'}{(1-\sin\phi')} \qquad (A.2)$$

and the slope k of the σ_1' versus σ_3' line as:

$$k = \frac{(1+\sin\phi')}{(1-\sin\phi')} \qquad (A.3)$$

where σ_1' is the axial stress at which failure occurs
σ_3' is the confining stress
c' is the cohesive strength and
ϕ' is the angle of friction of the rock mass

In order to estimate the cohesive strength c' and the friction angle ϕ' for an actual rock mass, the Hoek-Brown criterion (Hoek and Brown 1997) can be utilised. Having estimated the parameters for failure criterion, values for c' and ϕ' can be calculated.

Analysis of tunnel behaviour

Assume that a circular tunnel of radius r_o is subjected to hydrostatic stresses p_o and a uniform internal support pressure p_i as illustrated in Figure A.1. Failure of the rock mass surrounding the tunnel occurs when the internal pressure provided by the tunnel lining is less than a critical support pressure p_{cr}, which is defined by:

$$p_{cr} = \frac{2p_o - \sigma_{cm}}{1+k} \qquad (A.4)$$

If the internal support pressure p_i is greater than the critical support pressure p_{cr}, no failure occurs, the behaviour of the rock mass surrounding the tunnel is elastic and the inward radial elastic displacement of the tunnel wall is given by:

$$u_{ie} = \frac{r_o(1+\nu)}{E_m}(p_o - p_i) \qquad (A.5)$$

where E_m is the Young's modulus or deformation modulus and ν is the Poisson's ratio.

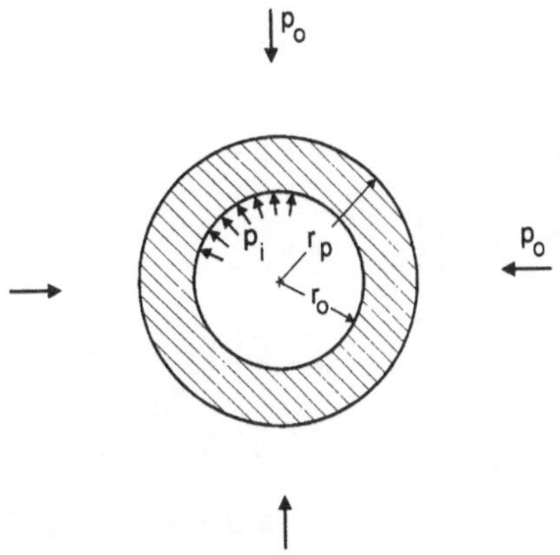

Figure A.1: Plastic zone surrounding a circular tunnel.

When the internal support pressure p_i is less than the critical support pressure p_{cr}, failure occurs and the radius r_p of the plastic zone around the tunnel is given by:

$$r_p = r_o\left[\frac{2(p_o(k-1)+\sigma_{cm})}{(1+k)((k-1)p_i+\sigma_{cm})}\right]^{\frac{1}{(k-1)}} \qquad (A.6)$$

For plastic failure, the total inward radial displacement of the walls of the tunnel is:

$$u_{ip} = \frac{r_o(1+\nu)}{E}\left[2(1-\nu)(p_o - p_{cr})\left(\frac{r_p}{r_o}\right)^2 - (1-2\nu)(p_o - p_i)\right] \qquad (A.7)$$

Appendix 1 (contd) - Characteristic line calculation for weak rock

Input:

Intact strength sigci =	5	MPa	Intact rock constant mi =	8		Geological Strength Index GSI =	23	
Poisson's ratio mu =	0.3		Tunnel diameter D =	10	m	In situ stress po =	3.8	MPa
Max. support pressure =	0.82	MPa	Initial convergence =	5	percent	Average maximum strain ($s_{max.av}$)=	1.35	percent

Output:

Rock mass constant mb =	0.51		Rock mass constant s =	0.0000		Rock mass constant a =	0.535	
Rock mass constant k =	2.21		Friction angle phi =	22.15	degrees	Rock mass cohesion coh =	0.10	MPa
Rock mass sigcm =	0.28	MPa	Rock mass modulus E =	472.6	MPa	Critical support pressure pcr =	2.25	MPa

Plot:

													Support reaction calculation		
Tunnel convergence ui (m) =	2.830	0.903	0.451	0.271	0.180	0.128	0.095	0.074	0.059	0.049	0.041	0.000			
Tunnel convergence (%) =	28.30	9.03	4.51	2.71	1.80	1.28	0.95	0.74	0.59	0.49	0.41	0.00	8	9.5	28.30
Support pressure pi (MPa) =	0.00	0.22	0.45	0.67	0.90	1.12	1.35	1.57	1.80	2.02	2.25	3.75			
Plastic zone radius rp (m) =	35.2	20.2	14.5	11.5	9.6	8.2	7.3	6.5	5.9	5.4	5.0	5.0			
Plastic zone thickness (m) =	30.2	15.2	9.5	6.5	4.6	3.2	2.3	1.5	0.9	0.4	0.0	0.0			
Avabailable support (MPa) =													0.00	0.5	0.5

Calculation :

										Sums
sig3	1E-10	0.18	0.36	0.5	0.71	0.89	1.07	1.25		5.00
sig1	0.00	0.77	1.21	1.59	1.95	2.28	2.60	2.91		13.31
sig3sig1	0.00	0.14	0.43	0.85	1.39	2.04	2.79	3.64		11
sig3sq	0.00	0.03	0.13	0.29	0.51	0.80	1.15	1.56		4

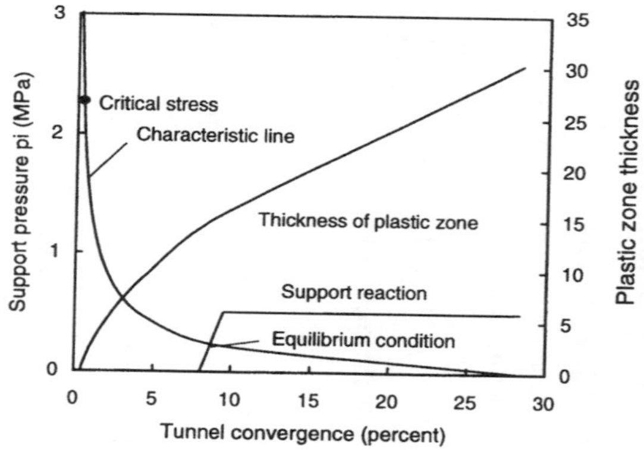

Cell formulae:

```
mb = mi*EXP((GSI-100)/28)
 s = IF(GSI>25,EXP((GSI-100)/9),0)
 a = IF(GSI>25,0.5,0.65-GSI/200)
sig3 = Start at 1E-10 (to avoid zero errors) and increment in 7 steps of  sigci/28 to 0.25*sigci
sig1 = sig3+sigci*(((mb*sig3)/sigci)+s)^a
 k = (sumsig3sig1 - (sumsig3*sumsig1)/8)/(sumsig3sq-(sumsig3^2)/8)
phi = ASIN((k-1)/(k+1))*180/PI()
coh = (sigcm*(1-SIN(phi*PI()/180)))/(2*COS(phi*PI()/180))
sigcm = sumsig1/8 - k*sumsig3/8
 E = IF(sigci>100,1000*10^((GSI-10)/40),SQRT(sigci/100)*1000*10^((GSI-10)/40))
pcr = (2*po-sigcm)/(k+1)
 rp = IF(pi<pcr,0.5*D*(2*(po*(k-1)+sigcm)/((1+k)*((k-1)*pi+sigcm)))^(1/(k-1)),D/2)
 ui = IF(rp>D/2,D*((1+mu)/E)*(2*(1-mu)*(po-pcr)*((rp/(0.5*D))^2)-(1-2*mu)*(po-pi)),D*(1+mu)*(po-pi)/E)
```

Support reaction calculation :

1. Pressure provided by support is zero at initial convergence which is the deformation at which the support is installed.
2. Maximum support pressure is achieved when the tunnel convergence equals the initial convergence plus the average maximum strain of the support system.
3. It is assumed that the support fails plastically and that the maximum support pressure remains constant after the support has reached its maximum capacity.

1 Design, analysis, testing and monitoring applications

Rock Support and Reinforcement Practice in Mining, Villaescusa, Windsor & Thompson (eds)
© 1999 Balkema, Rotterdam, ISBN 90 5809 045 0

Keynote lecture: Systematic design of reinforcement and support schemes for excavations in jointed rock

C.R.Windsor
Rock Technology, Perth, W.A., Australia

ABSTRACT: A simulation based procedure is presented for determining the rock mass demand of excavations in blocky rock. The procedure has been developed for use in the design of trial reinforcement and support schemes. It comprises a sequence of deterministic and probabilistic analyses of block shape, size and stability which results in a series of relative frequency distributions for the demand characteristics of the removable unstable blocks of rock that might possibly form at the excavation surface. Geometric constraints on block shape and size, imposed by the rock structure and the excavation, are introduced into the simulation process in an attempt to reduce the number of realisations required and increase simulation efficiency. Guidelines are given on the use of the relative frequency distributions in the selection and dimensioning of trial rock improvement schemes.

1 INTRODUCTION

The design of reinforcement and support for rock excavations requires consideration of many interrelated issues. If the process is simplified and thought of purely in mechanical terms then it could be considered to comprise six basic steps:

1. Formulation of a rock mass model.
2. Assessment of rock mass demand.
3. Dimensioning of trial reinforcement and/or support schemes.
4. Analysis of candidate reinforcement and/or support schemes.
5. Selection of an appropriate reinforcement and/or support scheme.
6. Performance assessment of the selected reinforcement and/or support scheme.

The design of reinforcement and/or support for an excavation in structured rock is particularly difficult due to a number of problems associated with completing steps 1 and 2. The mutual intersection of discontinuities in structured rock divides the rock into fully and partially formed blocks of rock. If an excavation cuts through this assembly of blocks, new sets of blocks are formed at the excavation surface. Some of these 'exposed' or 'surface' blocks will have a shape that will allow them to fall, slide or rotate into the excavation should the block driving forces exceed the block stabilising forces. In order to understand how such a rock mass may best be reinforced or supported, the assembly of blocks must be investigated. The outcome from this investigation would be to predict the exact shape, size, stability and spatial position of each block that forms around the excavation. These block characteristics define the rock mass demand and provide the information need to dimension trial reinforcement and or support schemes.

The ability to properly define the block characteristics is dependent on the quality and quantity of data that describes the rock mass surrounding the proposed excavation and the predictive capabilities of the current block analysis techniques. If the exact position of each discontinuity is known in advance, together with the parameters that describe its geometry and strength characteristics, then it is relatively easy to conduct an assessment and make a reasonable prediction. Unfortunately, these data are not available until after the excavation has been formed. It is also believed that, with the quality and quantity of structural data normally available, the current block analysis techniques are unable to make an accurate prediction of the shape, size and stability of the blocks.

However, it is suggested that an assessment of the block assembly can be made which results in a prediction of:

1. The block shapes that might possibly form.
2. The range of size of each possible block shape.
3. The stability of each size of each block shape that might possibly form.
4. The relative probability of each block size of each possible block shape forming.
5. The relative probability of each block size of each possible block shape forming and being unstable.

This discussion sets out a procedure that attempts to conduct such an assessment and make these predictions. The term relative probability is used because the aim here is to characterise the rock mass by comparing the characteristics of all the blocks that might possibly form. Thus the procedure involves identifying the possible block shapes, the possible range of size for each block shape and, the stability of all shapes over their individual size range and the relative probabilities of these occurrences.

2 BLOCK ANALYSIS AND THE DESIGN OF REINFORCEMENT AND SUPPORT

2.1 *Specific and Ubiquitous Block Analyses*

In any block analysis procedure, the types and numbers of different block shapes that are predicted to occur at the excavation surface depend on the assumptions made regarding the individual excavation faces, the discontinuities and the block shapes themselves. Of these, one of the most important aspects concerns the assumed nature of occurrence of the discontinuities in relation to the excavation surface and in relation to one other. The way this is treated gives rise to two fundamentally different analysis methods:

1. The Specific Approach. A given set of discontinuities and an excavation surface are arranged to occur at specific points in space. This unique arrangement means that the relative positions of all surfaces are fixed, resulting in one specific arrangement of the block assembly. This also dictates the size of each block that forms. The vector technique and computer program proposed by Warburton (1983) is the premier example of the 'specific approach'.
2. The Ubiquitous Approach. Given sets of discontinuities and an excavation surface are assumed to be able to occur everywhere and anywhere in space. This assumption means that all possible combinations of discontinuities and excavation faces are considered. This produces a list of all the possible block shapes that form around the excavation boundary. The stereographic projection technique (plus additional vector analyses) proposed by Goodman and Shi (1985) and the hemispherical projection technique proposed by Priest (1985) are the premier examples of the 'ubiquitous approach'.

The Warburton and the Goodman and Shi methods are equally important in that both cope with arbitrary polyhedra. However, the differences between 'specific' and 'ubiquitous' approaches have important consequences. In general, the specific approach will not, without an extreme number of analyses, provide the complete range of possible block shapes. However, it does provide the sizes of the specific blocks generated within each specific analysis. The ubiquitous approach can provide a complete range of possible block shapes but provides little information on the size of the possible block shapes that form. The recognition of these differences allows a third approach to be formulated that can be considered a hybrid of the specific and ubiquitous approaches.

2.2 *A Hybrid Specific-Ubiquitous Block Analysis*

A hybrid approach, which has been described previously by Windsor (1995), comprised four basic phases of block analysis:

1. Shape analysis.
2. Size analysis.
3. Stability analysis.
4. Shape-size-stability characterisation.

Vector analyses are used in all phases of this approach. This brings the mathematical efficiency needed to analyse many blocks very quickly but lacks the visual description, which aids comprehension, that is inherent in the Goodman and Shi and Priest graphical methods.

In the block shape analyses, the discontinuities and excavation surface are assumed to be ubiquitous. In the block size and stability analysis, each block shape may assume any size and occur anywhere on the excavation boundary within minimum and maximum geometric constraints set by the dimensions of the discontinuities and the excavation. The final phase is used to produce graphs that collectively characterise the block assembly in terms of the relative size and stability of the block shapes that are predicted to form at the excavation surface.

However, the approach proposed by Windsor (1995) made many unrealistic, simplifying assumptions. One of the most important was that it conveniently ignored the stochastic nature of the rock structure. This common characteristic dictates the shapes, sizes and stability of blocks. Consequently, the demands placed on a rock support or reinforcement scheme are also stochastic.

2.3 *Probabilistic Block Analyses*

The 'proper' design of reinforcement and support requires a probabilistic solution that takes into account variation in the discontinuity characteristics and provides for the probability of the occurrence of block shapes, sizes and their stability. Probabilistic block analyses have been investigated in the past by many workers along both specific and ubiquitous lines (e.g. Shapiro and Delport, 1991 and Kuszmaul, 1993). Some of the more recent investigations (in particular the approaches identified and being explored and developed by Kuszmaul (1994), Stone (1994) and Mauldon (1995)), show great promise and suggest that a formal solution to this important practical problem may one day be possible. In the past, the author has unsuccessfully attempted to obtain a partial solution to this problem using two simulation techniques:

1. Two-dimensional, specific rock structure simulation. A trace map of discontinuities was simulated from distributions of discontinuities at nucleation points generated in the face of an excavation. The block face areas for the different shapes were assessed and plotted as frequency histograms of block sizes.
2. Three-dimensional, specific rock structure simulation. The three-dimensional rock structure was simulated by distributions of discontinuities at nucleation points generated within a volume of rock. This volume was then cut by an excavation to reveal a two-dimensional trace map. Again, the resulting block face areas were assessed and plotted as frequency histograms of block sizes.

In both cases, the distribution of block shapes and sizes were found by analysing the blocks that form in the two-dimensional 'face map'. An example involving tetrahedral blocks is given in Figure 1 which shows a typical face map produced by the mutual intersection of three particular discontinuity sets and the block faces produced. Figure 2 shows the resulting frequency histogram of block face area. Both procedures involve considerable manipulation of geometric data in cumbersome mathematical operations; firstly for simulation and then for assessment. Furthermore, each realisation is specific in nature. Thus many face maps, or a very large face map area, have to be generated to obtain a reasonably precise estimate for the frequency distribution of block sizes.

Both the two- and three-dimensional procedures were found to be inefficient and too time consuming to be acceptable for use in standard practice. However, two important discoveries were made. Considerable computational time and effort was spent in:

1. Simulating multiple, specific realisations of a rock structure in what was an extremely large volume of rock mass relative to the volume characteristically associated with the blocks that formed.
2. Analysing specific circumstances where it was impossible for a block to actually form.

These discoveries suggested that, with some modifications, the hybrid approach described earlier might be useful in providing at least a partial solution to the stochastic problem of defining block shapes and block sizes.

2.4 *A Hybrid Possibilistic-Probabilistic Block Analysis*

The hybrid block analysis has been modified in an attempt to account for the variation in the rock structure. The block shape, size and shape-size-stability analysis phases described earlier are retained as 'deterministic engines'. In other words, they are core computer program modules that conduct a set of fundamental calculations for shape, size and stability. The engines may be driven once with a unique set of input parameters to provide a unique deterministic answer or a multiplicity of times with variation of the input parameters to produce a stochastic result. The way in which the proposed simulation is conducted distinguishes it from other probabilistic analyses. The main difference is the introduction of what will be termed a, for want of a better word, 'possibilistic' analysis. A possibilistic analysis is conducted to determined the extreme conditions within which the probabilistic analysis is then conducted.

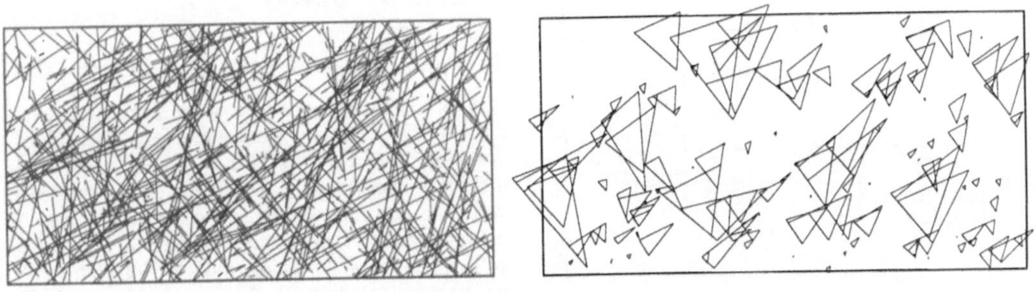

Figure 1. A realisation showing (a) the 2D trace map and (b) the resulting tetrahedral block faces.

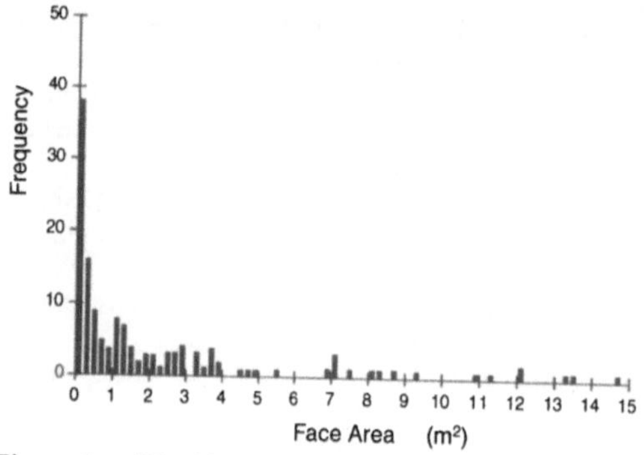

Figure 2. The block face area frequency histogram for the realisation given in Figure 1.

The block shape and block size engines are used to conduct three simulations:

1. Deterministic Simulation. This is conducted by driving the deterministic engines with mean values of the discontinuity characteristics to provide a measure of the mean of the outcome.
2. Possibilistic Simulation. This is conducted by driving the deterministic engines with extreme values of the discontinuity characteristics to provide a measure of the possibility limits or extrema of the outcome.
3. Probabilistic Simulation. This is conducted by driving the deterministic engines with distributions of values for the discontinuity characteristics to provide a measure of the relative probability of outcomes between the extrema.

The results from the mean and the possibilistic analyses are used set up the bounds within which the probability simulation is conducted, as shown schematically in Figure 3.

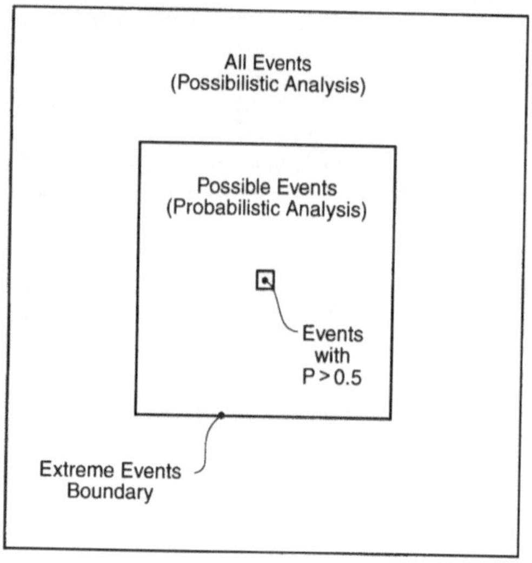

Figure 3. The probability analysis bounds.

Consequently, considerably smaller rock volumes are involved in each realisation and considerably less computational time and effort are invested in dealing with discontinuities that cannot be involved in block formation. In the balance of this discussion, the deterministic and possibilistic analyses of block shape, size and stability will be reviewed followed by a brief description of the probabilistic block shape-size-stability simulation.

3 BLOCK SHAPE ANALYSIS

3.1 Block Shape Analysis Program

A computer program called PBLOCK has been written to conduct the polyhedral block shape analysis. PBLOCK is used here as the 'deterministic engine' in both the possibilistic and probabilistic analyses.

3.2 Deterministic Block Shape Analysis

The input data required by PBLOCK comprises the orientations of the excavation faces and the discontinuities and whether the rock mass is situated above or below the excavation face. The results comprise a sequence of tables which list the input data, classify the block orders and kinematic modes and give details on each block analysed. For example, Figure 4 gives the tables of results from a block shape analysis for the flat roof of an underground excavation constructed in a rock mass intersected by 5 sets of discontinuities. The last type of table in this figure is the most important in that it lists details on each block shape. The numerical codes for kinematic modes used in Figure 4 are:

1. Inherently stable.
2. Inherently unstable, may fall vertically.
3. Candidate for sliding on a single plane.
4. Candidate for sliding on two planes.
5. Candidate for sliding on a multiplicity of planes.

The details on each block shape are automatically stored in data files that are used by other programs in the simulation sequence.

3.3 Possibilistic Block Shape Analysis

PBLOCK produces a set of block shapes for given discontinuity sets and excavation face orientations. Therefore, PBLOCK is able to be executed using different discontinuity orientations to produce variations in block shapes. The normals to sets of discontinuities often cluster together in accordance with a certain distribution model. A particularly tractable distribution model is the Fisher (1953) model which assumes that the normals cluster around

POLYHEDRAL BLOCK ANALYSIS	PROJECT: UNDERGROUND	DATE: 08-20-1996
EXCAVATION FACE AT 0 / 0	ROCK MASS SITUATED ABOVE EXCAVATION FACE	
PLANES	PLANE ORIENTATIONS	NORMAL ORIENTATIONS
1	47/135	43/315
2	39/ 10	51/190
3	43/ 87	47/267
4	61/170	29/350
5	28/225	62/ 45

BLOCK TYPES	INFINITE	FINITE	NON-REMOVAB.	REMOVABLE
TETRAHEDRAL BLOCKS	70	82	72	10
PENTAHEDRAL BLOCKS	55	81	66	15
HEXAHEDRAL BLOCKS	16	16	10	6
HEPTAHEDRAL BLOCKS	0	0	0	0
OCTAHEDRAL BLOCKS	0	0	0	0
ENNEAHEDRAL BLOCKS	0	0	0	0
DECAHEDRAL BLOCKS	0	0	0	0
TOTAL BLOCKS	141	179	148	31

REMOVABLE BLOCKS	BLOCK TRANSLATION MODE					
POLYHEDRAL ORDER	TYPE 0	TYPE 1	TYPE 2	TYPE 3	TYPE 4	TOTAL
TETRAHEDRAL BLOCKS	0	3	5	2	0	10
PENTAHEDRAL BLOCKS	0	3	8	4	0	15
HEXAHEDRAL BLOCKS	0	1	3	2	0	6
HEPTAHEDRAL BLOCKS	0	0	0	0	0	0
OCTAHEDRAL BLOCKS	0	0	0	0	0	0
ENNEAHEDRAL BLOCKS	0	0	0	0	0	0
DECAHEDRAL BLOCKS	0	0	0	0	0	0
TOTAL BLOCKS	0	7	16	8	0	31

HEXAHEDRAL BLOCK ANALYSIS	6 REMOVABLE BLOCKS				
BLOCK NO.	BLOCK FACE PLANES (DISCONTINUITY NO.)	BLOCK FACE CODE (A) ABOVE (B) BELOW	BLOCK TYPE	SLIDE PLANES	SLIDE DIRECT'N
26	1 2 3 4 5	B B A B B	2	3	43/ 87
27	1 2 4 3 5	A B B A B	3	3 1	42/102
28	1 2 5 3 4	B B B B B	1	-	90/ 0
29	1 3 4 2 5	A B B B B	3	1 4	45/114
30	1 3 5 2 4	A B B B A	2	1	47/135
31	1 4 5 2 3	B A B B B	2	4	61/170

Figure 4. Output tables from a block shape analysis.

a mean vector with the distribution defined by a single concentration parameter K_f. A clustering analysis of field orientation data can be used to obtain the mean normal vector to the set and the Fisher concentration parameter. From these results, a set of pseudo-random vectors, constrained by an associated K_f, may be generated around a mean discontinuity set orientation to simulate the discontinuity set. Polyhedra bounded by pseudo-random planes from particular discontinuity sets can then be analysed by PBLOCK.

Although this is useful in the probabilistic analysis, it is the bounding conditions, or the extremes of the vector clusters representing each set, that are of particular interest. The objective is to determine the relationships between the dispersion of the vector clusters and the associated block shape extrema.

Fortunately, Fisher, (1953) also proposed a probability equation that linked a vector chosen at random from the distribution with the acute angle it makes with the mean vector. This enables cones of different semi-apical angles to be constructed around the mean vector to represent the probability bounds within the distribution. Using Fisher's equation and the general equation for a cone, it is possible to derive vector equations for the cone surface at a particular probability level. The extrema of normals, planes, intersections and thus block shape may be determined at any given probability level (P) by vector analysis of the cone surfaces using compact vector equations. However, there is a considerable advantage in discussing these concepts in graphical terms. For example, the texts by Goodman and Shi (1985) and Priest (1985) both achieve a superior clarity of expression by the use of stereographic and hemispherical projection. Furthermore, many practitioners prefer and will probably continue to think in graphical terms.

The vectors on the probability cone surface form a probability zone that contains the normals to a set of planes with a probability of being oriented within a given angle to the mean normal to the set. On a stereographic projection, the cone surface plots as a circular locus and the corresponding envelope containing the great circles to that family of planes plots as an eccentric, circular annulus. This eccentric annulus represents a cluster of planes with a given probability relative to the mean plane. The intersection of two concentric annuli form a pair of intersection zones which plot as a pair of spherical quadrilaterals (in the lower half-space and upper half-space). The intersection zones represent the probability envelopes of the downward and upward directed intersections for the two families of intersecting planes.

In the case of a tetrahedral block, there are three eccentric circular annuli representing the block faces and there are three pairs of spherical quadrilateral zones representing the three block edges. Collectively, the intersections of the three eccentric, circular annuli produce a spherical triangular 'annulus' representing the block shape bounded at its corners by three of the spherical quadrilaterals. In summary, there are three types of zones that define the block shape envelope:

1. An eccentric, circular annulus which represents the orientation extrema for a block face at a particular probability level.
2. A spherical quadrilateral which represents the orientation extrema of a given block edge at a particular probability level.
3. A spherical triangular 'annulus' which represents the block shape extrema at a particular probability level.

The extrema that define the block shape envelope are best illustrated by considering the extreme block shapes for the underground excavation problem defined in Figure.4. The results from this analysis are given in graphical form as lower hemispherical projections in Figures 5, 6, 7 and 10 and as stereographic projections in Figures 8 and 9.

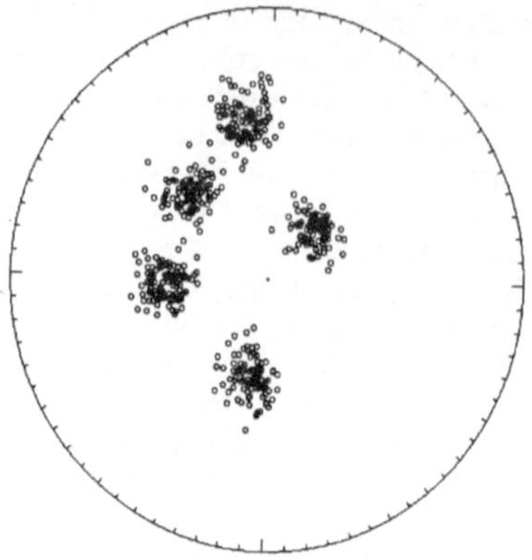

Figure 5. The simulation of 100 normals to each set obeying a Fisher distribution with $K_f = 100$.

Figure 6. The arcs of great circles of the planes corresponding to the 500 simulated normals.

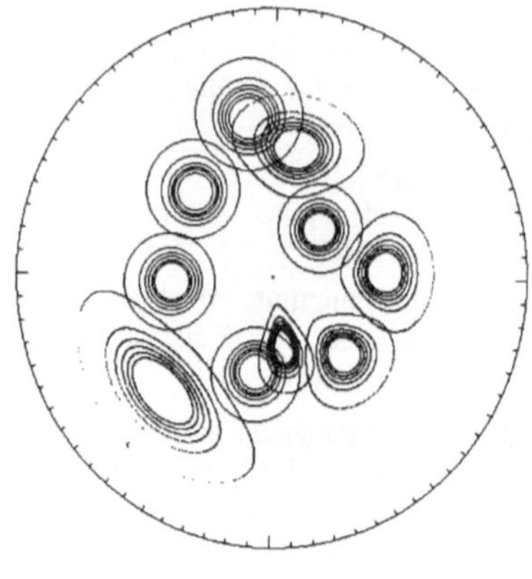

Figure 7. The probability cones and probability trumpets representing the normals and the vectors of maximum dip, respectively.

Figure 8. The 5 eccentric circular annuli containing the planes normal to the P = 0.95 probability cones.

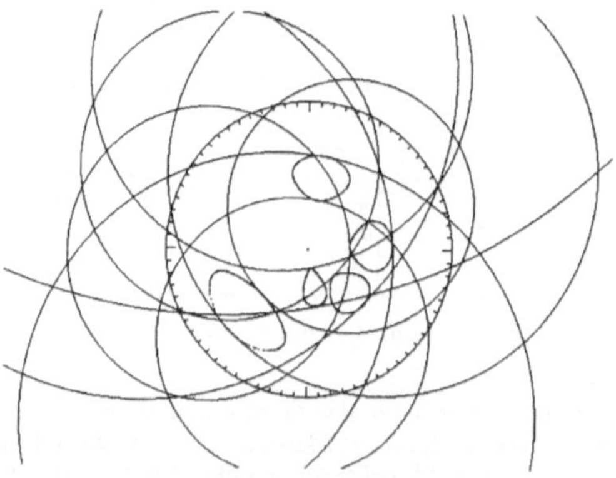

Figure 9. The extreme boundaries to the 5 circular annuli and the 5 maximum dip trumpets at P = 0.95.

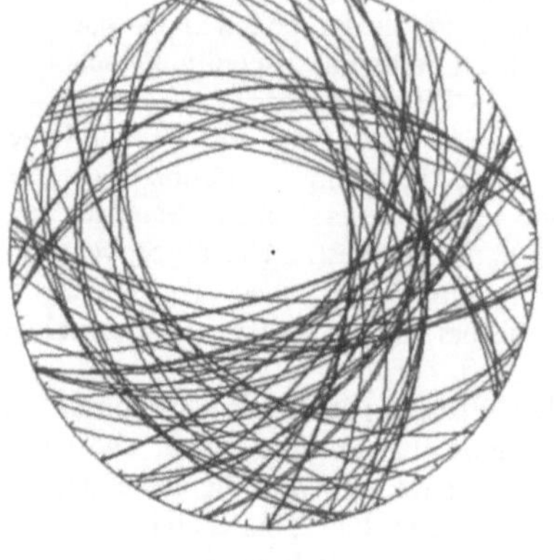

Figure 10. The circular arcs of great circles representing the planes involved in forming the ten extreme tetrahedral block shapes at P = 0.95.

The 5 sets of discontinuities are defined by a set number j, a dip and dip direction of the mean set normal, a Fisher constant and the number of discontinuities simulated or (j, α_{nj}, β_{nj}, K_{fj}, N_j):

(1, 43, 315, 100, 100)

(2, 51, 190, 100, 100)

(3, 47, 267, 100, 100)

(4, 29, 350, 100, 100)

(5, 62, 045, 100, 100)

Figure 5 gives the 100 normals simulated for each discontinuity set. Figure 6 shows the great circle arcs representing the planes to the simulated normals. Figure 7 shows circular loci representing probability cones and 'tear drop' shaped loci. The latter represent the probability zones containing the maximum dip vectors of the planes corresponding to the normal vectors in the cone surface. Probability levels from 0.95 to 0.999 in 0.05 intervals are shown. Figure 8 shows the 5 eccentric, circular annuli bounding all planes corresponding with the surface of each cone. These bounds are drawn at the 0.95 probability level. Figure 9 shows the extreme bounds from Figure 8 together with the 0.95 probability zones of maximum dip vectors. Figure 9 contains the 10 spherical triangular 'annuli' representing the 10 tetrahedral block shapes. The planes that are involved in forming the 10 extreme block shapes are shown in Figure 10.

In summary, the extremes in block shape are defined by a combination of extreme dispersions for each face. The possible extreme shapes may be determined at any given probability level based on the clustering and distribution analysis of field orientation data. Although the variation in block shape can be assessed graphically with the procedure given above, it is more efficiently conducted using computational vector methods. The results of the possibilistic block shape simulation are stored as vectors for use, firstly, in the block size possibilistic analysis and then in the block shape-size-stability probabilistic analysis.

4 BLOCK SIZE ANALYSIS

4.1 A Block Size Analysis Program

A computer program called SIZE is the deterministic engine used to conduct a size analysis on the list of removable block shapes. When SIZE is initialised, it is automatically loaded with a file containing the block shape information obtained from PBLOCK. In deterministic mode, the mean face orientations for each block shape are used. In possibilistic mode, the extrema in face orientations for each block shape are used.

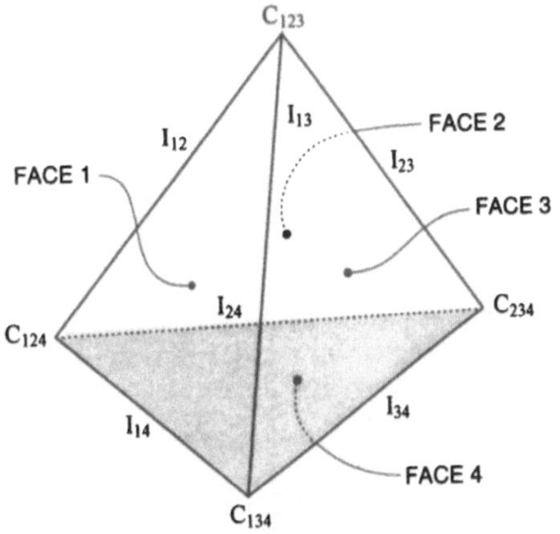

Figure 11. A typical tetrahedral block shape.

4.2 Block Size Limits

The range of block sizes that can possibly form for each shape will be determined by introducing a 'unit block size' and scaling this between 'limiting block sizes' imposed by dimensional data for the excavation and the discontinuities. The different block sizes will be explained using a typical tetrahedral block shape. The block is shown in Figure 11 annotated to define the four block faces (Face 1 to Face 4), the six block edges (I_{12} to I_{34}) and the four block vertices (C_{123} to C_{234}). The block face exposed at the excavation face (Face 4) is shown shaded. Tetrahedral blocks are relatively simple to visualise and define because they have one vertex completely contained within the rock mass (the apex C_{123}). The position of the apex, relative to the other vertices, is fully defined by the orientation of the planes involved in forming the shape. Thus, the size of a tetrahedral block shape may be defined by any characteristic dimension. This allows unit block sizes to be defined with the apex height, face area or volume set to unity. The unit block size may be scaled from very small to four limiting sizes:

1. The equilibrium limited block size.
2. The excavation limited block size.
3. The spacing limited block size.
4. The trace length limited block size.

These limiting block sizes define the possible extremes in block sizes that will be used later to bound the region within which the probabilistic analysis will be conducted.

4.3 The Unit Block Size

The unit block size or the 'unit block' is a block of given shape that has been dimensioned such that one of the dimensional characteristics of the shape (ie. altitude, face area or volume) is set to unity.

Once the size characteristic of the unit block has been decided, vector methods may be used to quickly obtain all other block characteristics. Given the orientations of the three discontinuities forming the block and the block code, the vectors of the six block edges (I_{12} to I_{34}), the coordinates of the four block vertices (C_{123} to C_{234}), the area of the four faces A_1 to A_4, the volume V and the altitude or apex height H may be found using vector calculus. Once all the dimensional characteristics have been defined, numerous scaling factors (K) allow any block of the same shape to be dimensioned from the scale factors for that block.

4.4 The Equilibrium Limited Block Size

The equilibrium limited block occurs at a natural equilibrium. That is, the natural disturbing forces acting on the block are just equilibrated by the natural stabilising forces. The equilibrium limited block is not valid for all block shapes (e.g. free falling blocks). The equilibrium equation is given by:

$$\gamma V_i \sin\psi_i - \Sigma_{j=1}^{j=n} c_j A_j + \gamma V_i \cos\psi_i \tan\phi_j = 0$$

where: n = 3 for a tetrahedral block.
- c_j and ϕ_j = shear strength parameters.
- γ = unit weight of rock
- V_i = block volume = $f(n_i, n_j, n_k, n_l$ and K).
- A_j = area of the translation surface.
  = $f\{K, n_l$ and one of $(n_i, n_j$ or $n_k)$, or
  = $f\{K, n_l$ and a doublet from $(n_i, n_j$ and $n_k)\}$.
- ψ_i = dip of the translation vector.
  = $f\{$one of $(n_i, n_j$or $n_k))$, or
  = $f\{$a doublet from $(n_i, n_j$ or $n_k)\}$.
- n_i, n_j, n_k = unit normals to the discontinuities.
- n_l = a unit normal to the excavation face.
- K = characteristic scaling parameter.

from which a value of V_i can be calculated to just satisfy the equilibrium equation using the scaling parameter K.

4.5 The Excavation Face Limited Block Size

The convexity, shape and dimensions of the excavation surface all affect the size of a block. There are four principal variations:

1. A finite or infinite, planar surface.
2. A convex or non-convex, edge or corner, formed by n planar surfaces .
3. A finite or closed curvi-planar surface.
4. A convex or non-convex, edge or corner, formed by n planar and curvi-planar surfaces.

The excavation limited block size is determined by the dimensions of the particular excavation face and the procedure for determining the block size is essentially the same in each case. For example, in a finite, rectangular excavation face the block size could be limited by either of the two fixed spans. For more complicated excavation surfaces, the face limited block may be controlled by any one, some or all of the dimensions of the excavation faces involved in forming the block. Figure 12 shows the cases of a single face, edge and corner block formed from the same discontinuity triple.

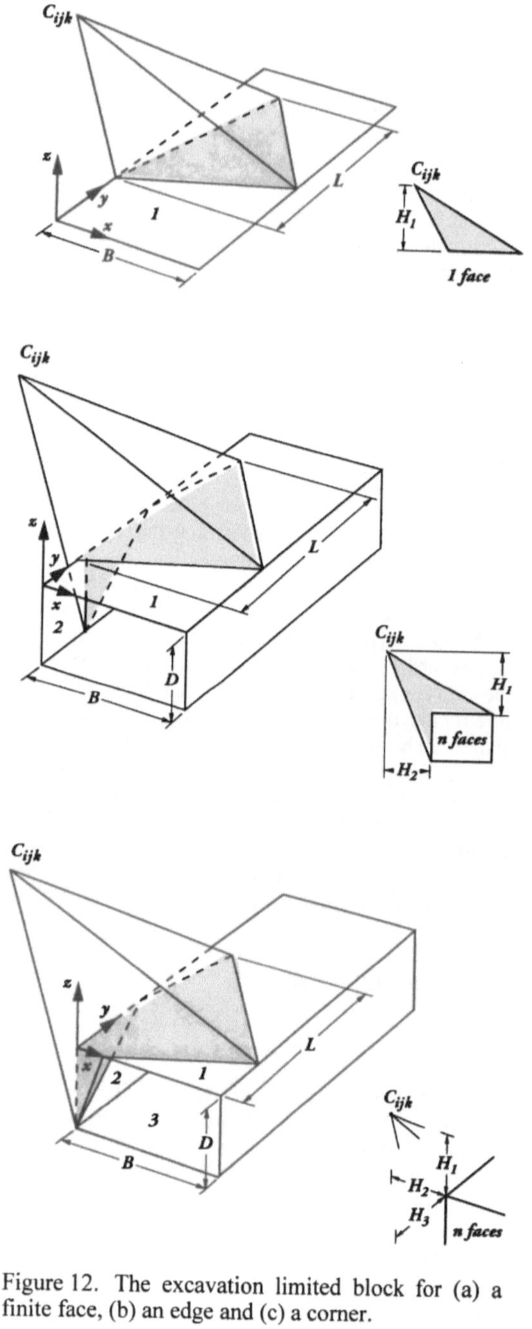

Figure 12. The excavation limited block for (a) a finite face, (b) an edge and (c) a corner.

4.6 *The Spacing Limited Block Size*

The spacing limited block size is the largest block size that can form without it being intersected by additional discontinuities that may result in other blocks being formed within that block. An estimate of this volume is determined by considering the spacing values of the discontinuity sets. For each discontinuity set, one discontinuity is placed to intersect the apex and form the block face associated with that set. The block is then scaled such that the vertex opposite the first discontinuity lies in the plane of a second discontinuity from the same set. This second discontinuity is placed at a perpendicular distance from the first equal to the set spacing. If the spacing chosen is the minimum likely spacing, it is unlikely that this block volume will be penetrated by additional discontinuities from this set. Each face is considered in turn to yield three candidate block sizes. The spacing limited block is the smallest of these three candidates.

4.7 *The Trace Length Limited Block Size*

Discontinuities of finite persistence can limit the size of a block that forms to less than the excavation limited block. The trace length limited block size is determined by an extreme arrangement of the discontinuities involved in forming the block. Basically, no line in a block face may be greater than some function of the maximum persistence specified for the discontinuity set associated with that face. The introduction of discontinuity persistence in an analysis demands that discontinuity shape also be considered. Here it will be assumed that the discontinuities are circular and that a maximum possible trace length can be estimated for each set. This results in discontinuities of either infinite or finite radii defined by the maximum tracelength attributed to each of the associated sets. For example, if 100 tracelengths are measured then the maximum length recorded is a more meaningful measure of maximum persistence than the assumption of continuity.

Consider Figure 13a which shows the intersection of three, circular-shaped discontinuities in plan view, looking from within the rock mass towards an excavation face. The discontinuities are arranged to intersect on their extreme edges at the point (C_{ijk}), (Note that this arrangement is possible, but extremely unlikely to occur). There are three lines of mutual intersection of the planes, radiating from C_{ijk} to the other three bounding intersections between each pair of planes at C_{ij}, C_{jk} and C_{ki}. These lines are vectors with magnitudes given by the distance from the point of common intersection C_{ijk} to C_{ij}, C_{jk} and C_{ki} respectively and orientations and senses given by the three unit vectors $\hat{I}_{ij}$, $\hat{I}_{jk}$ and $\hat{I}_{ki}$ respectively, as shown in Figure 13b.

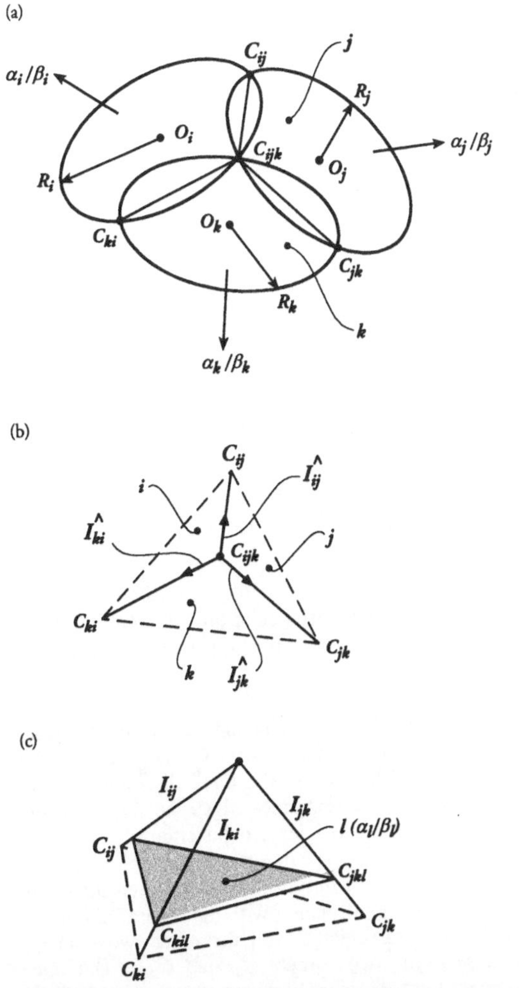

Figure 13. The intersection of three circular discontinuities in plan (a and b) and in isometric (c).

The three vectors form a vector triple and an open tetrahedral shape - 'open' because the fourth plane (excavation face) is not present. Now, if the position of C_{ijk} is such that all three vectors do intersect the excavation face (labelled l) in Figure 13c at the points C_{ijl}, C_{jkl} and C_{kil} respectively, then the open tetrahedral is completed and becomes a fully formed, tetrahedron block with apex C_{ijk}. The shape and size of the block is fully defined by the orientations of the four planes (i, j, k and l) and by the block edge vectors I_{ij}, I_{jk} and I_{ki}.

4.7.1 *Block Face Area A_i Constrained by the Maximum Tracelength of Planes i, j and k*

Now, consider Figure 14 which shows a discontinuity (circular plane i) isolated from Figure 13a.

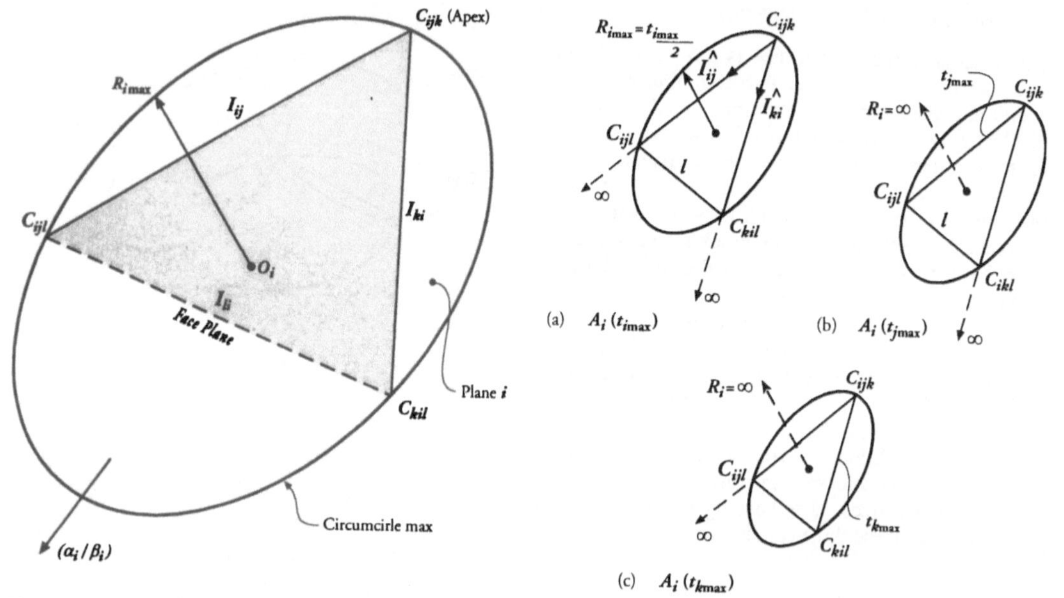

Figure 14. A plane triangular area in a circular discontinuity representing a potential block face and the three candidate variations (a),(b) and (c), one of which will limit the maximum area A_i that can form in plane i.

The position of point C_{ijk}, the three discontinuities and the excavation surface are all assumed to be ubiquitous. This allows the three discontinuities to intersect at their extreme edges at C_{ijk} and the excavation plane to intersect the circular planes i, j and k anywhere. The intersection of plane l with plane i produces a line in plane i of given direction. The intersection of plane l with the other two lines of intersection (associated with the other two planes j and k) forms two corners C_{ijl} and C_{ikl}. The three corners form the triangular face of a potential block. Similar events occur on planes j and k. The objective now is to determine the maximum plane triangular areas A_i, A_j and A_k that can form within discontinuities i, j and k; one of these will control the tracelength limited block. As an example consider area A_i associated with block face i. There are three candidate solutions, all illustrated in Figure 14:

1. A_i is constrained by the maximum tracelength of circular plane i (Fig. 14a).
2. A_i is constrained by the maximum tracelength of circular plane j (Fig. 14b).
3. A_i is constrained by the maximum tracelength of circular plane k (Fig. 14c).

To obtain the first candidate solution the possible controlling influence of the lengths of the lines of intersection with the other two planes are ignored. They are allowed to extend infinitely in the direction of the vectors I_{ij} and I_{kl}. Thus, if plane l can occur anywhere, then it may intersect the other two intersection lines at C_{ijl} and C_{kil} to form an infinitely large triangular area within plane i. However, the area that can form is constrained by the maximum diameter (maximum trace length) considered likely and attributed to joint set i. Given the directions of the three lines of intersection (I_{ij}, I_{jk}, I_{li}) and the maximum radius of plane i (R_{imax} = maximum semi-tracelength of Set i), then the maximum plane triangular area that may form can be determined. The other two candidate solutions for A_i are found by setting R_{imax} to infinity and A_i is then restricted by way of the maximum trace length of, firstly, plane j and then plane k as shown in Figures 14b and 14c. In summary, the maximum area A_i may be constrained by either R_{imax}, R_{jmax} or R_{kmax} and the minimum of these three candidate solutions defines the maximum tracelength limited, plane triangular area A_i associated with block face i.

A block with maximum plane triangular area A_i in face i does not corrupt the maximum tracelength criterion for any of the three discontinuity sets. The maximum plane triangular areas A_j and A_k associated with block faces j and k, respectively, are determined using the same approach.

4.7.2 *The Maximum Block Volume Limited by Areas Ai , Aj and Ak*

Three candidate block volumes can be determined from the maximum plane triangular areas of faces i, j and k and a scaling parameter K. The minimum block size from the three solutions defines the maximum tracelength limited block size. This is usually found to be controlled by the persistence of one of the sets. The maximum tracelength limited block size was defined in Figure 13 by setting the discontinuity diameters to the maximum tracelength and having them intersect at their extremities. As indicated previously, this arrangement is extremely unlikely but not impossible. In fact, if the characteristics of the discontinuities are independent, their trace lengths may vary independently and the point of common intersection C_{ijk} may occur anywhere within the plane and boundary of each. This has significant implications for the magnitudes of the vectors representing the lines of intersection between the planes and the ability of the vector triple and the excavation surface to form a valid, closed tetrahedral shape.

This general case, together with variations in orientation is the norm not an extreme and forms the basis of the probabilistic simulation. However, its worth remembering that the reason for the possibility analysis is to determine the extreme event within which all other events can occur. Thus, if maximum trace lengths can be attributed to the discontinuities, the maximum tracelength limited block does provide a bound imposed by structural geology to the probabilistic analysis.

4.7.3 *The Maximum Block Size*

The 'maximum block size' that may occur at the excavation face is given by either the excavation face limited block size or the trace limited block size, whichever is the smaller. The 'maximum block size' that may occur at the excavation face for each block shape is extremely important in that it defines a region, termed here the 'block existence zone', within which the probabllistic analysis is conducted.

4.7.4 *The Maximum Individual Block Size*

If the spacing limited block size is smaller than the 'maximum block size', then it represents the 'maximum individual block size' that can form. Under these circumstances, blocks larger that the spacing limited block size may be penetrated by additional discontinuities to produce multiple, smaller blocks within the larger blocks.

5 BLOCK SHAPE-SIZE-STABILITY ANALYSIS

In Sections 3 and 4, respectively, the mean and extreme shapes and the mean and extreme sizes for each block were determined. In this section the task is to describe the deterministic block stability analysis engine and the relationships between block shape, block size and block stability.

5.1 *A Block Stability Analysis Program*

With the range of block sizes defined, the stability of each possible block shape over the range of possible sizes is assessed by conducting a deterministic equilibrium analysis on the unit block size for that shape. A program module called ANALYSIS has been written to conduct this assessment. It takes into account the block shape, the rock density and the cohesive and frictional strength for any supporting surfaces. ANALYSIS is automatically loaded with standard data files from the other deterministic engines, PBLOCK and SIZE. The program prepares a large array of scaling parameters, based on the unit block, that are used to rapidly obtain block characteristics for any block size between the block size extrema.

5.2 *Scale Stability Diagrams*

There are various functional relationships between the shape, size and stability characteristics associated with each block (e.g. the relationship between block volume and out-of-balance force). The four limiting sizes discussed in Section 4 will truncate and divide these functional relationships into possibility limits:

- Existence Limits. The relations are truncated by minima and maxima of block sizes.
- Stability Limits. The relations are, where applicable, divided into stable and unstable ranges by the equilibrium sized blocks.
- Compound Limits. The relations are, where applicable, divided into sectors by the spacing limiting block sizes.

If the functional relations from the shape, size and stability analyses for all the blocks are combined into single diagrams they effectively characterise the behaviour of the overall rock mass in a way that is compatible with the original input data (i.e. a representative model). One possibility is a scale-stability diagram that summarises the functional relationship between one of the characteristics that define size (e.g. apex height, face area, volume or mass) and one of the characteristics that define stability (e.g. the out-of-balance force or the stability index).

For example, a plot of the stability index versus block face area in the excavation face for all the tetrahedral blocks for the underground excavation example is given in Figure 15. A horizontal line may be plotted at a stability index equal to unity to represent limiting equilibrium. If the planes of sliding are cohesionless, the relations are flat lines at a constant stability index; when the block is free-falling the stability index is zero. If the discontinuities possess some cohesion, the relations reduce with increasing scale and asymptote to a constant stability index. Some start with a stability index above unity and decay to below unity with increasing block size.

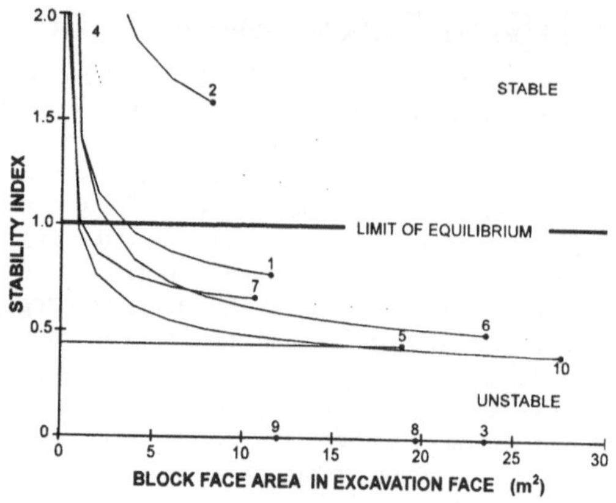

Figure 15. A scale-stability diagram of block face area versus stability index.

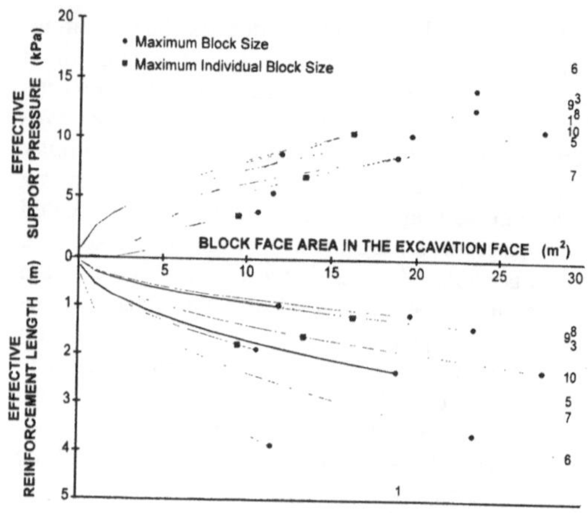

Figure 16. Block face area versus effective support pressure and effective reinforcement length.

Each relation is truncated by a maximum block size. The area of concern for design is limited to the relations below the limit of equilibrium and to the left of the maximum block size limit.

Another characteristic diagram could be drawn that combines ESP (Effective Support Pressure) and ERL (Effective Reinforcement Length) with block size. ESP is the ratio of the out-of-balance force to the block excavation face area. It is a vector quantity that acts parallel (but in the opposite sense) to the block translation vector and gives an indication of the effective face pressure needed to maintain block equilibrium. ESL is the apex height of the block plus some allowance for reinforcement anchoring. It is a scalar quantity in the direction of the normal vector to the excavation face and gives an indication of the minimum length of reinforcement required to anchor in the stable zone beyond the block. In contrast Support Pressure (SP) and Reinforcement Length (RL) are both functions of the chosen orientation of the reinforcement.

A characteristic diagram of ESP and ERL versus Block Face Area is shown in Figure 16 for all the removable and unstable tetrahedral blocks predicted to form in the excavation roof. The relations are drawn with a full line and terminated with a spot to represent the maximum size block. The relations for Blocks 3, 7 and 10 are also marked with additional spots which indicate the maximum individual block size limit.

Scale-stability diagrams offer the designer the opportunity to select a reinforcement and/or support scheme that will account for the complete range of possible block shapes, sizes and stabilities. Unfortunately, the current diagrams do not define the likelihood of a block shape forming and the likely size of that block; they simply define the possibility limits. Consequently, a rock improvement scheme might be formulated to account for the complete range of instability and then be found to be grossly conservative (e.g. if mostly smaller or mostly larger blocks actually formed). To formulate the most appropriate design, the designer also needs to know the relative probabilities of every block shape forming over their possible size ranges.

6 PROBABILISTIC BLOCK SHAPE - SIZE - STABILITY ANALYSIS

A closed-form solution for the probabilistic assessment of block shape, block size and block stability requires the solution of a number of severe intractable complexities associated with the stochastic nature of rock structures. The author's unsuccessful attempts to achieve an efficient solution using the two- and three-dimensional simulation techniques described in Section 2 have forced the development of an alternative approach using a combination of the possibilistic analysis to define the extreme conditions of block shapes and block sizes together with a simulation-based probability assessment that is conducted within those extremes. This assessment requires that the orientation, persistence, shape, spacing and the strength and deformation characteristics of each discontinuity set be simulated in terms of probability distribution functions. It also requires that the position of each discontinuity relative to the other discontinuities and the excavation surface be simulated. A sufficient number of realisations must be conducted to properly account for the possible variations in discontinuity characteristics and discontinuity positions.

6.1 A Summary of the Simulation Procedure

Unfortunately, the simulation procedure is necessarily complex and a complete, detailed description is beyond the scope of this discussion. A brief summary of the procedure is provided in the form of a sequence of steps. However, some of the more important features of the simulation are singled out for additional description in Section 6.2. In summary, the simulation procedure comprises a sequence of 16 main steps:

1. The orientation and geometry of the excavation surface are defined according to a global coordinate system. In some circumstances, it is convenient and more efficient to subdivide the excavation periphery into sections comprising single planar or curvi-planar surfaces, edges formed from planar and/or curvi-planar surfaces and corners formed from planar and/or curvi-planar surfaces.
2. A standard, deterministic block shape analysis is conducted. This analysis uses the mean discontinuity orientations to determine the list of removable blocks. Steps 3 to 14 are conducted for each shape in the removable block list.
3. A possibilistic block shape analysis is conducted. This analysis uses the extreme dispersion boundary of the orientation distribution for each discontinuity set. The analysis is used to determine the shape extrema that bound all possible shapes associated with that block.
4. A possibilistic block size analysis is conducted. This analysis uses the block shape extrema, the excavation dimensions and the trace length maxima to determine the extreme dimensions of the block. This results in a 'block existence zone' defined in the global coordinate system that bounds the shape and size for each combination of discontinuities involved in forming a removable block shape. There are two candidates for the block existence zone; the excavation limited block and the maximum trace limited block. The block existence zone is defined by the smaller of the two or, by either, if they are identical.
5. A point of nucleation (x, y, z) is placed at random within the 'block existence zone' according to the global coordinate system. This represents the intersection of at least three discontinuities of different orientations and coincides with the vertex (C_{ijk}) of a potential block.
6. A planar discontinuity from each relevant set is generated to intersect the point of nucleation. The orientation of each discontinuity is generated according to a given distribution function for the set. The shape and areal extent of the discontinuities are generated to comply with a given mathematical shape (e.g. circular) and with a dimension (e.g. diameter) selected according to a given distribution function for the persistence associated with each set. The coordinates of the centre of each discontinuity relative to the nucleation point are generated, according to a Poisson process, at a position within the plane of the discontinuity and constrained only by the shape and the areal extent of the discontinuity.
7. The three lines of intersection of the three planes are defined as vector triples radiating from the nucleation point and extending to the boundaries of the discontinuities. There are two vector triples (i.e. three lines of intersection, each represented by two vectors radiating in opposite directions) which define two open tetrahedra juxtaposed at the point of nucleation.
8. The vector triple associated with the 'removable open tetrahedral' is determined from the two candidates using the removable block code or the normals to the discontinuities. Note, that only one of the vector triples is congruent with three vectors representing the vertex of a removable tetrahedral block.
9. The equations describing the vector triple and the excavation surface are used to determine if a fully formed tetrahedral block occurs. The vector triple forms a tetrahedral block if all three vectors (representing potential block edges) radiating from the nucleation point (representing a potential apex) extend sufficiently to intersect the excavation surface.
10. The stability of the valid tetrahedral block is assessed using a force equilibrium analysis. This analysis is conducted using the specific shape and dimensional characteristics of the block together with data, simulated to obey the distribution functions, associated with the strength and deformation characteristics of the discontinuities.
11. The details describing the shape, dimensional characteristics and, if applicable, the out-of-balance force, the mode and mechanism of instability are stored for later analysis.
12. Relative frequency histograms and cumulative frequency diagrams are prepared for each characteristic describing the block (e.g. apex height, face area, volume, out-of-balance force).
13. Steps 5 to 12 are repeated until a sufficiently large number of realisations have been analysed. The simulation sequence is terminated on the basis of 'smoothness' and precision in the resulting relative frequency diagrams. After a certain number of simulations these distribution curves become 'smooth' and do not change appreciably. Both smoothness and precision default levels can be set automatically.
14. Steps 3 to 13 are repeated for each block shape appearing in the list of removable blocks.

15. The relative frequency histograms and the cumulative frequency curves are corrected for the relative probability of a nucleation point occurring in the block existence zone. This correction takes into account the relative probability of a vector triple occurring within the block existence zone.
16. The corrected relative frequency histograms, and the cumulative frequency diagrams for each of the removable blocks are redrawn on one diagram together with the total frequency distributions for the sum of all blocks. These diagrams are prepared for each block characteristic (e.g. altitude, face area, volume, out-of-balance force). This facilitates comparison between individual blocks and for the complete rock mass.

This 16 step sequence includes a number of features that deserve more than a cursory explanation and are discussed in more detail in Section 6.2.

6.2 *Features of the Simulation Procedure*

6.2.1 *Simulation of the Nucleation Point C_{ijk}*

One of the features that distinguish this approach from others concerns the nucleation point and its simulation. In the two- and three-dimensional simulation approaches discussed in Section 2 the discontinuities were generated at nucleation points around their respective centres. This produced a random arrangement of discontinuity intersections throughout the whole rock mass volume being investigated. A small number of these intersections involved three discontinuities and, of these, a considerably smaller number involved discontinuities that could actually intersect the excavation surface to form blocks. There are two important differences in this new approach:

1. The nucleation point represents the intersection of three discontinuities and not the discontinuity centres. This greatly increases the efficiency of the simulation because only planes involved in forming a discontinuity triple are simulated. Note that discontinuities that are not involved in a triple cannot be involved in block formation.
2. The nucleation point is generated to occur exclusively within the block existence zone defined by the extrema in shape and size for that block shape. This also greatly increases the efficiency of the simulation because only triples that could form a block are simulated. Note that the probability of a block forming from a triple situated outside the block existence zone is assumed to be zero.

Both of these features greatly reduce the number of simulations required. However, a correction factor must be applied to account for the probability of a particular triple forming relative to the formation of other triples. This correction factor is not used to proportion the number of nucleation points but is applied to the distributions that result from the simulation. The correction is one of the final steps and will discussed in Section 6.2.4.

6.2.2 *Simulation of a Discontinuity Centre Relative to C_{ijk}*

The centre of each discontinuity involved in the vector triple is simulated relative to the nucleation point. Before progressing to the general case it is convenient to firstly recall the maximum tracelength limited block described in Section 4.7. In that simpler, but somewhat unlikely case, the assumption of ubiquity was made such that the three intersecting discontinuities were allowed to intersect at their extreme edges and the excavation surface could intersect the three discontinuities at any position. This enabled the determination of the largest sized block that could form for a particular shape, constrained only by the maximum tracelengths of the three discontinuities. Now, in the more general approach, ubiety is assumed for the point C_{ijk} bounded by the block existence zone and ubiquity is assumed for the position of the centre of each discontinuity in relation to C_{ijk}. In other words, C_{ijk} must occur within the block existence zone but the discontinuity centre could occur inside or outside of that zone (e.g. deep within the rock mass or even within the excavation space).

The centre of each discontinuity relative to the nucleation point is generated to occur at a position within the plane of the discontinuity according to a Poisson process, constrained only by the shape and the areal extent of the discontinuity. A local, right-handed coordinate system is used in the plane of the discontinuity with the origin at the centre of the discontinuity and the local *z*-axis corresponding to a normal to the plane as shown in Figure 17.

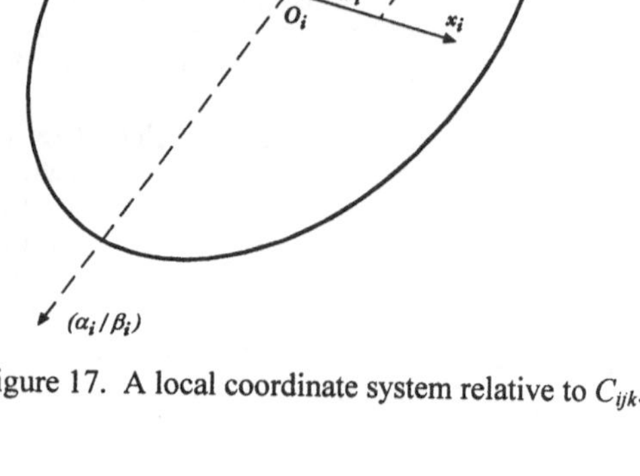

Figure 17. A local coordinate system relative to C_{ijk}.

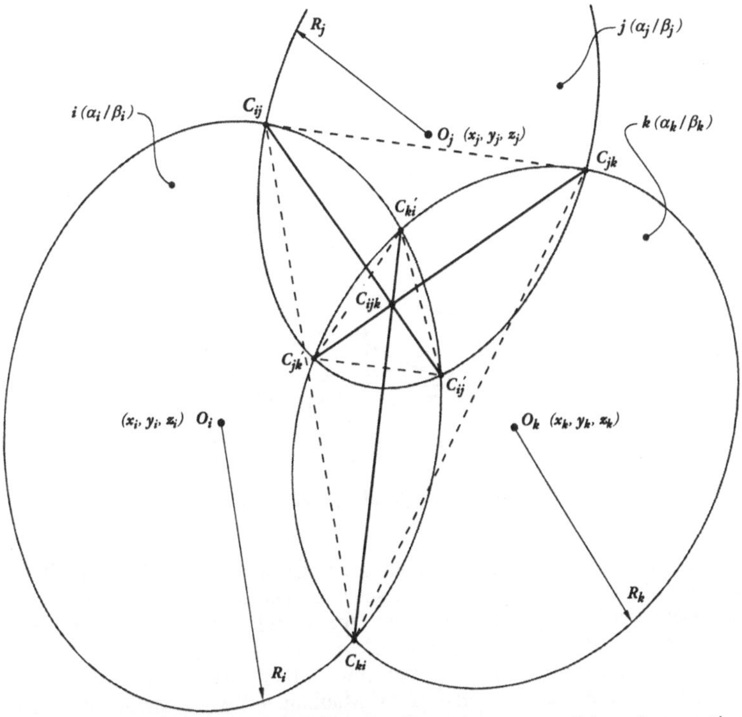

Figure 18. A simplified plan, viewed from above the apex, of three intersecting, independent discontinuities.

The 'local' coordinates of C_{ijk}, x' and y' ($z' = 0$), are generated in the plane of the discontinuity as two independent random variables with the constraint that the sum of their squares must be less than or equal to the square of the semi-trace of the plane. The global coordinates of the centre of the circular plane are then determined in a transformation from local to global coordinates using the orientation of the plane and the local x' and y' coordinates of C_{ijk}.

6.2.3 *The Determination of Valid Fully Formed Blocks*

In the general case, the orientation, shape, size and position of the centre of each discontinuity relative to C_{ijk} is independent of the other discontinuities. Consider Figure 18 which shows (greatly simplified) the intersection of three discontinuities i, j and k in plan, looking from within the rock mass, through C_{ijk}, towards the excavation surface. The lines of intersection of each pair of planes are given by $C_{ij} - C_{ij}'$, $C_{ki} - C_{ki}'$ and $C_{jk} - C_{jk}'$. In general, two open tetrahedra are formed juxtaposed about C_{ijk}. One is defined by upward directed, radiating vectors $C_{ijk} - C_{ij}'$, $C_{ijk} - C_{jk}'$, $C_{ijk} - C_{ki}'$ the other is defined by downward directed, radiating vectors $C_{ijk} - C_{ij}$, $C_{ijk} - C_{jk}$ and $C_{ijk} - C_{ki}$. The orientation, shape, dimension and global coordinates of each discontinuity centre allow these six vectors to be determined using standard vector operations. Only the vector triple comprising the three vectors directed towards the excavation ($C_{ijk} - C_{ij}$, $C_{ijk} - C_{jk}$ and $C_{ijk} - C_{ki}$) may intersect its surface to form a tetrahedral block. The chance of a tetrahedral block occurring depends on the orientation, shape, size and the position of the centre of each discontinuity relative to C_{ijk} and the orientation, shape, size and the position of the excavation surface relative to C_{ijk}.

Figure 19a shows how the dimensions of each intersecting plane and relative position of C_{ijk} affect the size of the plane triangular area that might form in the plane of each discontinuity. Figure 19b shows how the intersection of the plane with the excavation face may limit the formation of a plane triangular area in the plane of each discontinuity. The geometric arrangements shown in these two figures lead intuitively to two findings. Firstly, they support the suggestion made earlier that the maximum block is unlikely to occur. Secondly, they suggest that it would be more likely for smaller rather than larger blocks to form. Basically, to form a block, the three discontinuities must intersect the excavation free surface. The circular discontinuity planes and the finite excavation surface can all be described mathematically and the resulting system of equations can be solved within the geometric restraints needed to satisfy block formation.

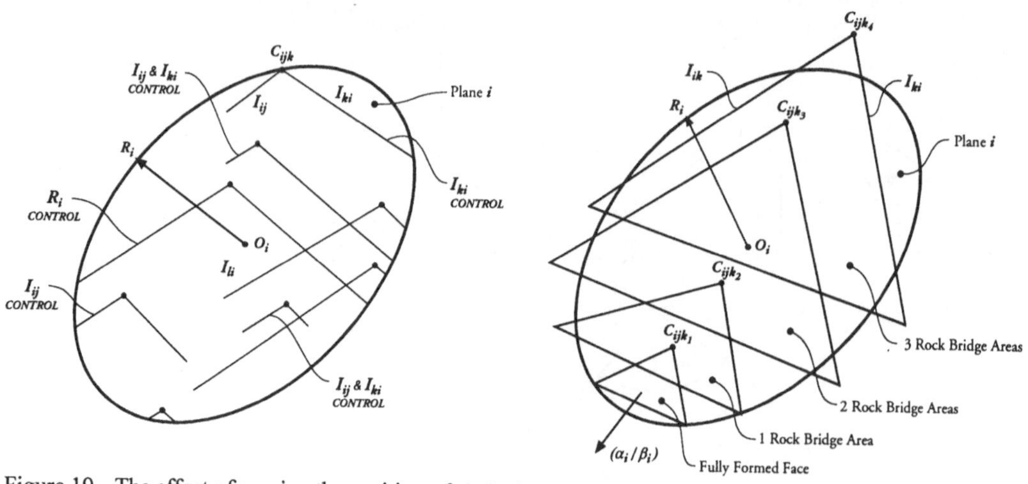

Figure 19. The effect of varying the position of a) the intersection point C_{ijk} and b) the excavation face.

If the excavation face is planar and infinite in all directions then the position of C_{ijk} need only be simulated in terms of a single coordinate which may be restricted to range from zero to the maximum altitude associated with the maximum size block for that block shape. If the excavation face is bounded by its intersection with other excavation faces, then the test for block formation not only includes the condition on the block altitude but the x, y, z coordinates of C_{ijl}, C_{jkl} and C_{kil} are also constrained by the orientation, dimensions and global position of these lines of intersection. If the excavation surface is considered to be a closed or partly-closed, polygonal region, then the vector triple radiating from C_{ijk} must intersect within the boundary of this polygonal region. If the excavation surface involves multiple edges and corners comprising the intersection of planar or curviplanar surfaces, then the vector triple must intersect the surface within the bounds set by the surfaces, as previously suggested in Figure 12.

6.2.4 *Corrections for the Relative Probability of Occurrence of the Nucleation Point*

The potential block vertex is simulated to occur at a unique point (x, y, z) relative to the position of the excavation surface in a volume V_{ijk} defined by the particular block shape and size extrema. If the nucleation points for all possible discontinuity triples are assumed to be distributed randomly throughout the rock mass, then a correction is required associated with the relative probability of the different triples occurring. Clearly, the relative chance of each discontinuity triple forming is affected by the relative persistence and spacing of the three discontinuity sets involved. For example, comparatively less triples will be produced if the triple includes an impersistent, widely-spaced set than if that set were persistent and closely-spaced. The correction for the relative probability of triples forming is taken from an elegant theory for the relative probability of intersections of three planes developed by Mauldon (1994). Mauldon's work, central to the implementation of the proposed procedure, produced two compact solutions; one for infinite planes and one for finite planes. These solutions are used to determine the correction factors. The corrections to proportion the number of triples generated in V_{ijk} are not applied early in the sequence because a certain number of simulations are required to achieve precision in the distributions. They are applied to the block characteristic distributions to allow the distributions for different blocks to be compared on a single diagram.

6.3 *A Probabilistic Block Simulation Program*

A computer program has been prepared that conducts steps 1 to 16 and provides the relative frequency graphs. The block simulation program is best illustrated by returning to the underground excavation example. The results from the block shape analysis were given in Figures 4 to 10 and the results from the shape-size-stability analysis were given in Figures 15 and 16. Some sample outputs from a simulation exercise conducted on this example are given in Figures 20 to 31. The simulation comprised 1 million realisations on each of the 10 possible block shapes. The input requirements for the simulation comprise:

- The excavation orientations and dimensions.
- The rock unit weight distribution and associated statistical parameters.
- The number of discontinuity sets and, for each:
    - The mean orientation, Fisher's constant and the chosen confidence level.

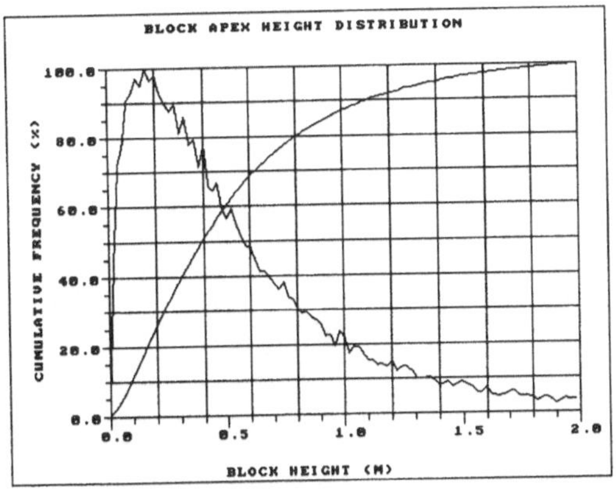

Figure 20. Apex height relative frequency and cumulative frequency distributions for block 1.

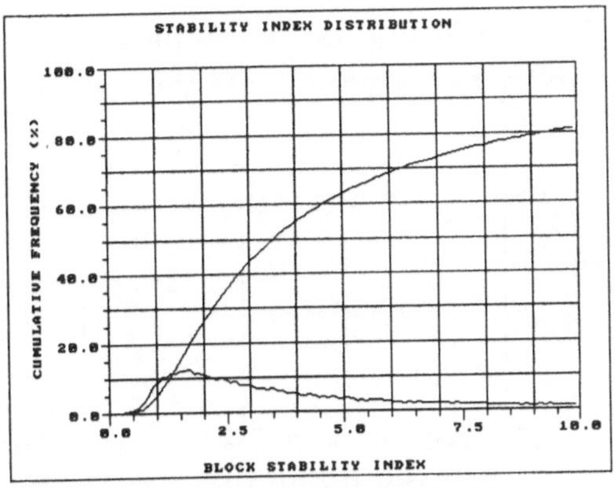

Figure 21. Stability index relative frequency and cumulative frequency distributions for block 1.

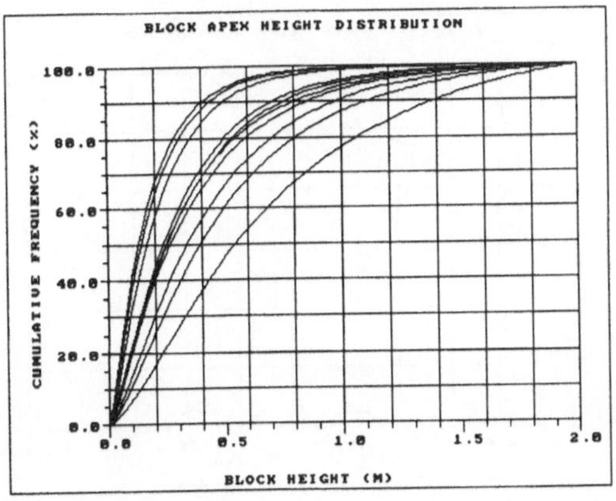

Figure 22. Cumulative frequency of apex height for all removable blocks.

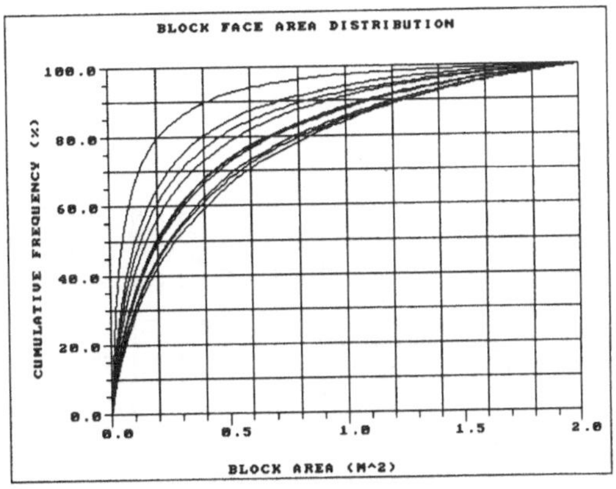

Figure 23. Cumulative frequency of face area for all removable blocks.

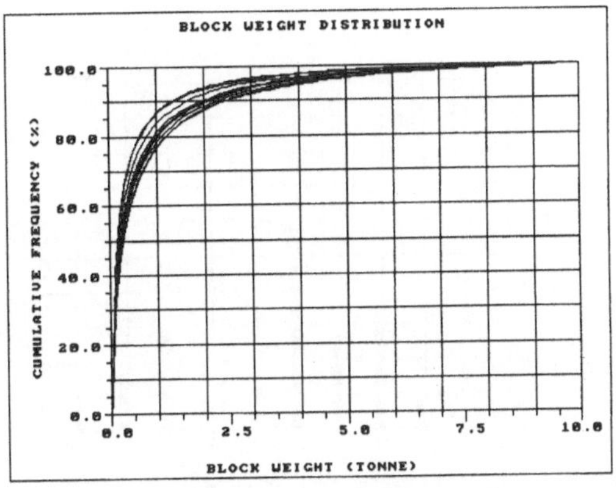

Figure 24. Cumulative frequency of weight for all removable blocks.

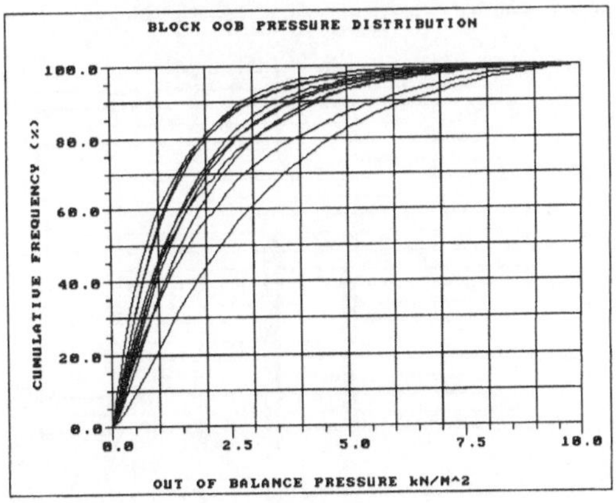

Figure 25. Cumulative frequency of out-of-balance pressure for all removable blocks.

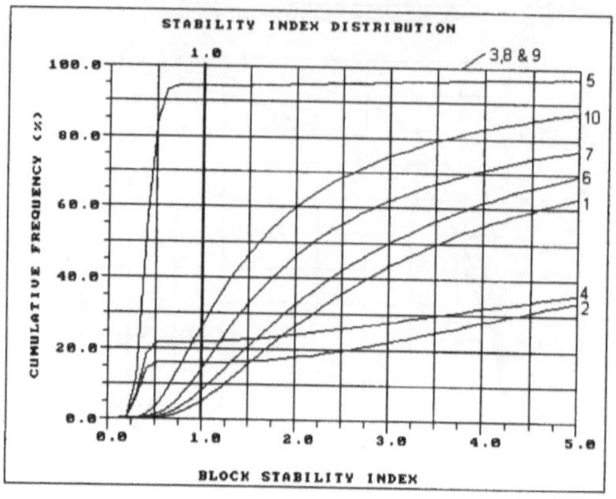

Figure 26. Stability index cumulative frequency graph for all removable blocks.

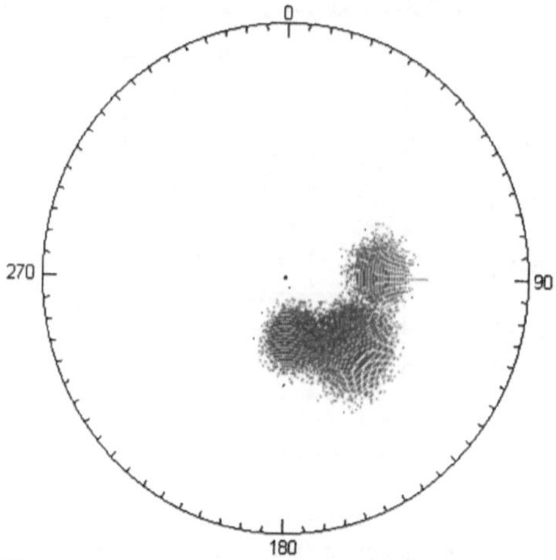

Figure 29. Unstable block translation vectors (Note that there is a high density of 90/000 vectors).

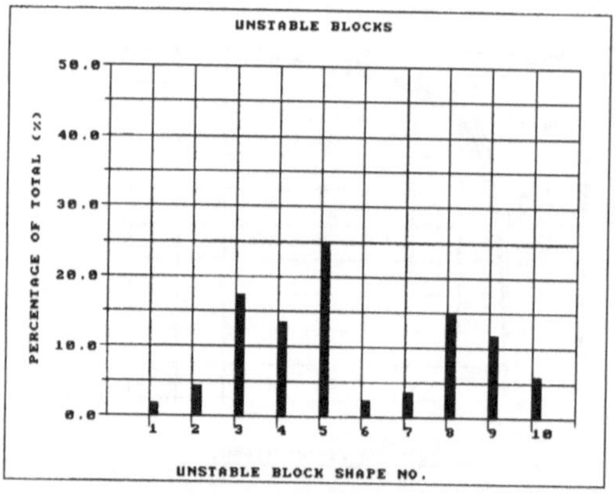

Figure 27. Percentage of the total number of unstable blocks associated with each block shape.

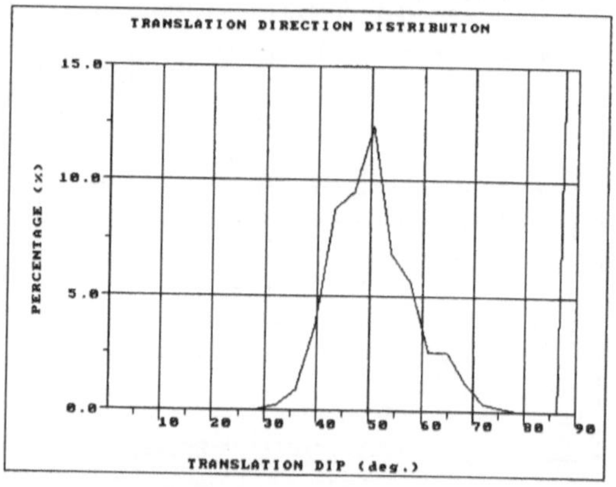

Figure 30. Distribution of translation dip (Note that there is a high percentage of dips at 90°).

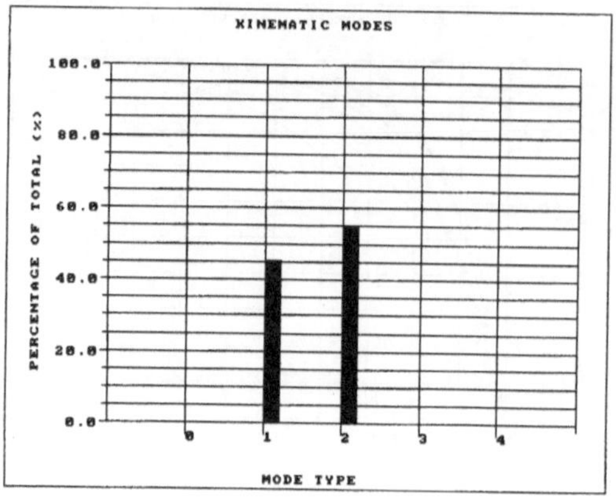

Figure 28. The kinematic modes for all unstable blocks.

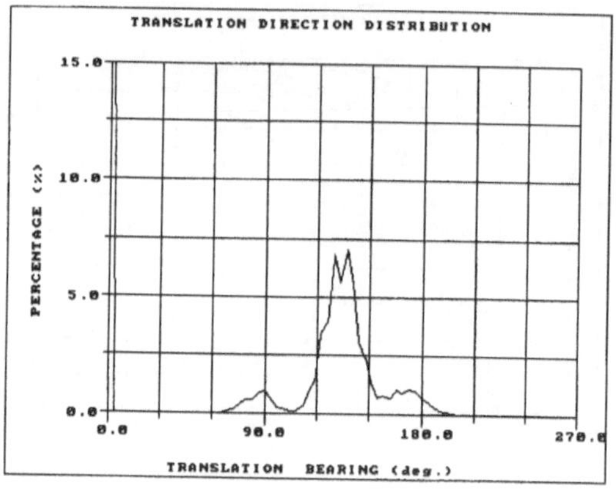

Figure 31. Distribution of translation azimuth (Note that there is a high percentage of azimuths at 0°).

- The trace length distribution and associated statistical parameters.
- The spacing distribution and associated statistical parameters.
- The joint friction distribution and associated statistical parameters.
- The joint cohesion distribution and associated statistical parameters.

These variables may be simulated using the uniform, exponential, normal and log-normal distributions.

For this example, the intersections were assumed to be distributed at random through V_{ijk} with equal spacing distributions enabling the relative probability corrections to be set to unity. The other characteristics of the five sets were all simulated according to the following distributions:

- Orientation - Fisher distribution ($K_f = 100$).
- Tracelength - Exponential distribution (mean = 5m).
- Rock Unit Weight - Normal (mean = 27 kN/m).
- Friction Angle - Normal (mean = 25° all sets).
- Cohesion - Normal (means = 0, 2.5, 5.0, 7.5 and 10.0 kPa for sets 1 to 5 respectively).

The 10,000,000 realisations produced a total of 258,197 removable blocks (i.e. 2.58%) of which, 91,982 were unstable blocks (i.e. 0.92%). Figures 20 and 21, respectively, show the relative frequency distribution and the cumulative frequency distributions of apex height and stability index for block number 1. The relatively frequency distribution in Figure 20 has been normalised to the maximum frequency to better show the shape of the distribution. Figures 22 to 26 show the cumulative frequency distributions for apex height, face area, weight, out-of-balance pressure and stability index for all removable blocks. Figure 26 clearly shows that some but not all of the removable blocks are unstable. Figures 27 to 31 concern the unstable blocks and should be viewed with the scale-stability diagrams (Figures 15 and 16) in mind. Figure 27 gives the proportion of unstable blocks for each block shape. Figure 28 shows that the kinematic modes for all unstable blocks comprise either free-falling (type 1) or single plane sliding (type 2). Clearly, the variations in block shapes causes variations in kinematic modes which changes the stability of the blocks. Figures 29 to 31, respectively, show the translation vectors for the unstable blocks as a hemispherical projection, the distribution of maximum dip and the distribution of the azimuth of maximum dip.

The 10 million realisations took a total of 847 minutes (14.1 hours) to complete on a standard PC. However, it was noted in this exercise (and in work on other problems) that the distributions seem to become relatively smooth and precise after about 250,000 realisations. This implies that for 'design work' on a problem involving 5 sets of discontinuities, a simulation of 1,250,000 realisations followed by storage and plotting of results could be comfortably completed within 1 working day.

7 SELECTING AND DIMENSIONING TRIAL ROCK IMPROVEMENT SCHEMES

A process for design of reinforcement and support for rock excavations was defined previously in the introduction. Sections 2 to 6 have discussed the second step in that process for the case of blocky rock masses; namely, assessment of the rock mass demand. The demand for blocky rock was determined in terms of the relative frequency distributions for the block characteristics of the different shaped unstable blocks over their anticipated range of occurrence. The objective was to enable the tasks involved in step three of the design process to be conducted; namely, the selection and dimensioning of trial reinforcement and/or support schemes.

7.1 Demand Characteristics

Consider Figure 32 which shows the relative frequency distributions and the cumulative frequency distributions for two particular blocks i and j and for both blocks. A distribution can be drawn for any given block demand characteristic Ω. These characteristics define particular aspects of demand which enable particular related aspects of a trial rock improvement scheme to be selected and dimensioned.

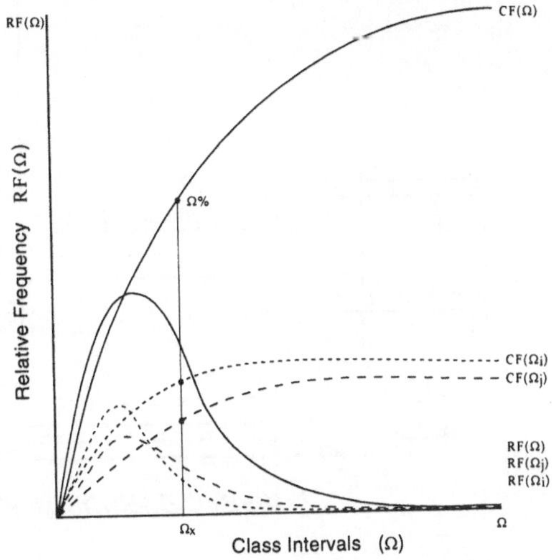

Figure 32. Relative and cumulative frequency graphs for two blocks and for the aggregate of both blocks.

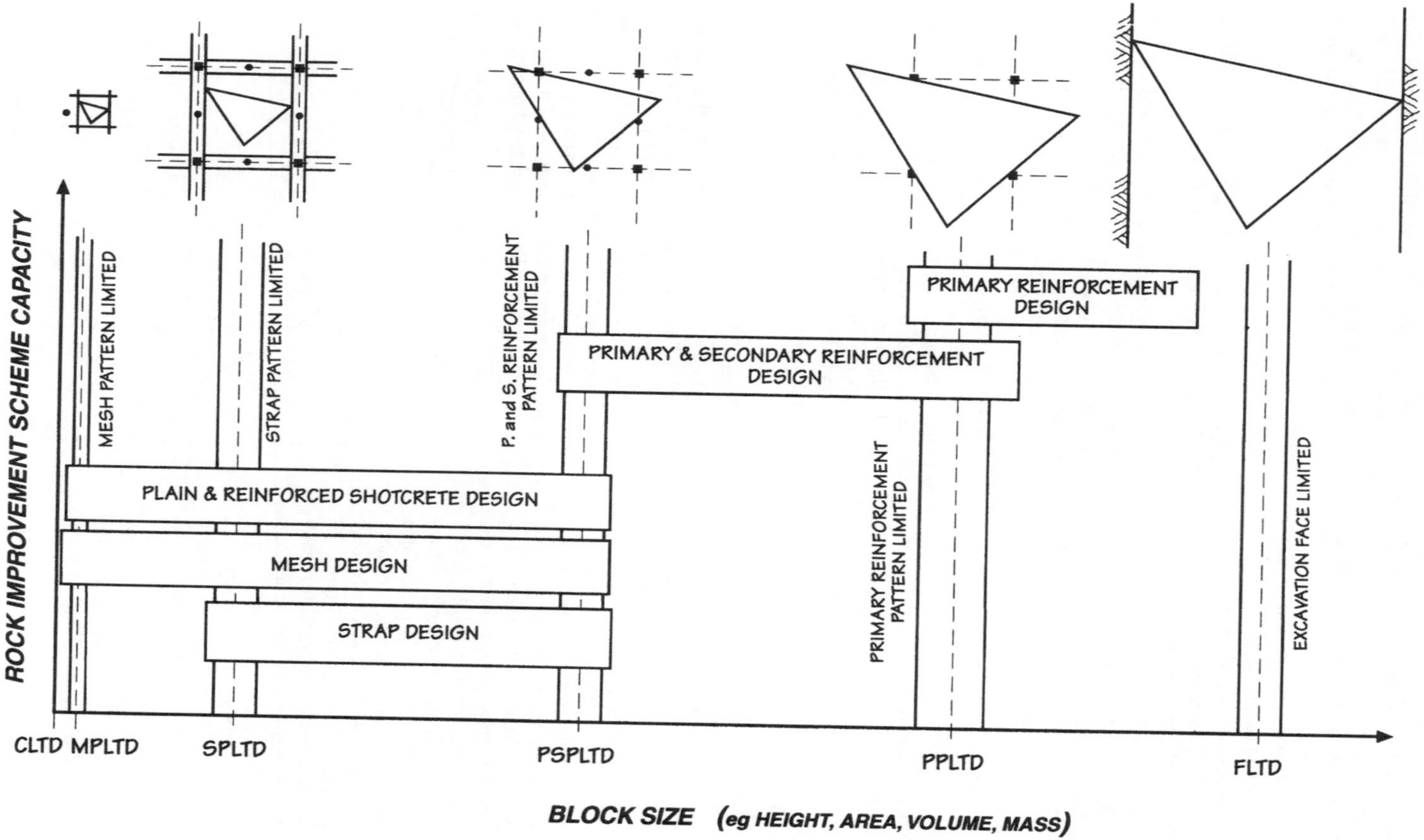

Figure 33. A schematic of the relationship between system capacity and certain limiting block sizes.

The characteristics found to be particularly useful in rock improvement design include:

- Altitude (H) - perpendicular distance from the excavation face to the apex.
- Face area (A_f) - block area in the excavation face.
- Out-of-balance force (OBF) - difference between driving and resisting forces
- Translation vector (α_d/β_d)-movement direction.
- Face perimeter (L_p) - perimeter of the face area.
- Shear surface area (A_s) - sum of shearing areas.
- Force demand $F_{\alpha\beta}$ - Force in a given direction required to maintain equilibrium.
- Pressure demand $P_{\alpha\beta} = F_{\alpha\beta} / A_f$.
- Internal shear demand (S_{in}) = OBF/ A_s
- External shear demand (S_{ex}) - OBF/ (L_p x 1mm)

These may be used for dimensioning of:

- Reinforcement systems and schemes (e.g. rock bolts, cable bolts and ground anchors) in terms of axial and shear capacity, length, orientation, density, pattern type and pattern dimensions.
- Support systems and schemes (e.g. shotcrete, polymer membranes, sets, liners) in terms of shear, tensile, flexural and bond capacity and cover area, cover dimensions and cover sequence.
- Restraint systems and schemes (e.g. mesh, straps, laces) in terms of shear, tensile, and flexural capacities, grid dimensions, grid orientation, cover area and cover sequence.

7.2 Use of the Demand Characteristics in Design

The use of the relative frequency distributions in rock improvement scheme design requires careful consideration of a number of issues:

1. It is extremely important to recognise that a 'trial' rock improvement scheme must be dimensioned before it can be assessed for reliability. The distributions given here are for use in dimensioning trial solutions before the design process enters the infinitely more complex reliability assessment step. The reliability assessment is discussed in Section 7.3.
2. The distributions provide a measure of the relative probability of the block characteristics occurring for each removable block. Consider the case of a rock mass containing one unstable removable block. The actual portion of the shape envelope that tends to produce say, blocks with large apex heights does not necessarily coincide with the portion of the envelope that tends to produce say, blocks with large pressure demand. Consequently, if the altitude distribution is used to dimension the length of reinforcement and the pressure demand distribution is used to select reinforcement capacity and array dimensions, then the resulting trial design is somewhat inconsistent but appropriate across the range of possible variations. The blocks with large altitude will be supplied with appropriate length reinforcement and the blocks with large pressure demand with be supplied with an appropriate reinforcement capacity at a suitable density. However, the design will be more appropriate for some blocks than others. This is one reason why the trial design must be analysed for reliability after it has been selected and dimensioned.
3. The relative frequency distributions of Ω tend to a negative exponential form or comprise a peak frequency at a low value of Ω and decay to very low frequencies at higher values of Ω. The type of curve depends on how the first class interval of the histogram is treated. However, the main feature is that all the distributions decay asymptotically from low Ω to high Ω. This means that the higher values of Ω are all poorly defined. The scale-stability diagrams define that demand Ω is continuous over a range in which the lower extreme (Ω_{min}) is either an infinitesimal block or the equilibrium block and the upper extreme (Ω_{max}) is the maximum block size. Unfortunately, the designer must choose an upper design limit (Ω_{max}) in the relatively-ill defined asymptotic region. It is more helpful to consider this problem in terms of cumulative frequency. An X% cumulative frequency may be chosen to account for X% of blocks. This means that the design must be dimensioned to account for all blocks in the range $\Omega_{min} < \Omega < \Omega_{X\%}$. However, this decision requires consideration of a number of additional issues concerning the project requirements (e.g. safety, serviceability, and economics). These are not discussed here and the author is unable to recommend a value of X% that will satisfy all projects. The author uses 95% level (i.e. $\Omega_{95\%}$) to achieve reliability for the larger blocks.
4. The distributions of Ω may require a rock improvement scheme design that comprises a combination of reinforcement systems, support systems and restraint systems. For example, Figure 33 is a schematic representation of the relationship between the size of a particular shaped block and the approximate force and size capacity of primary and secondary reinforcement (e.g. cable bolts and rock bolts), shotcrete, mesh and straps. The diagram also shows a number of limiting block sizes that are constrained by the capacity of the rock improvement scheme design. For example, the 'primary reinforcement pattern limited block' is the largest size block that can possibly move through the primary reinforcement pattern. These block limits are a subset of 'capacity limits' that are defined by practical issues associated with rock improvement schemes (e.g. restrictions on the dimensions and capacities of commercial systems, installation restrictions on pattern geometry, dimensions and reinforcement densities).

The capacity limits divide the block size range into sections which, together with the distribution of Ω in the range Ω_{min} to $\Omega_{X\%}$, indicate what type of rock improvement systems are required. Clearly, a number of alternative designs involving different types of rock improvement may be selected. However, the selection is again somewhat simplified by considering other issues such as project requirements in terms of serviceability, longevity, installation equipment and compatibility within the excavation and construction sequence.

Some procedures for dimensioning reinforcement and/or support capacity to satisfy demand have been given previously by Windsor (1996). However, the choice of dimensions will invariably result in a design that is more appropriate and more reliable for some block shapes than for others. This supports the suggestion that trial designs must be analysed for reliability.

5. The distributions for all removable, unstable block shapes may be drawn on one diagram together with an aggregated distribution for all the removable, unstable block shapes. Issues 2, 3 and 4 were discussed with respect to one unstable block but can be equally applied to the case of multiple unstable block shapes. Basically, the design must account for the demand for all the blocks in the range Ω_{min} to $\Omega_{X\%}$. However, some of the demand characteristics are affected by the translation vector (α_d/β_d). For example, force demand $F_{\alpha\beta}$ and pressure demand $P_{\alpha\beta}$ are a function of the direction in which a block moves α_d/β_d and the direction in which the capacity is supplied α_c/β_c.

Now reconsider Figures 29 to 31 which show the variation in (α_d/β_d) for each block and for all blocks. Force, pressure and shear capacities must be dimensioned and oriented to satisfy these vector demands in the range Ω_{min} to $\Omega_{X\%}$. Some of the dimensioning procedures needed to account for this are given by Windsor (1996). However, the choice of orientation will invariably result in a design that is more appropriate and more reliable for some block shapes than for others. Again, this requires that the trial designs must be analysed for reliability.

7.3 Additional Steps in the Design Process

Once a trial rock improvement scheme design has been selected and dimensioned, it must be assessed to ensure that it satisfies a certain reliability level of stability in terms of force equilibrium and displacement compatibility. In most but not all circumstances, the stability of each block shape is markedly affected by its position in relation to the rock improvement scheme. For example, consider Figure 34 which shows how the block position in a reinforcement array will affect the number of reinforcement penetrations through the block. Similarly, Figure 35 shows that the position of a block in relation to the points of fixture for a sheet of mesh will affect the loading of the mesh, and the position of a block in relation to a shotcreted surface will affect the geometry of a yield line crack pattern and thus the moment capacity of the shotcrete. The reliability assessment requires a complex force-displacement analysis in which the position of each block shape (over its complete size range) is allowed to vary in relation to the rock improvement scheme.

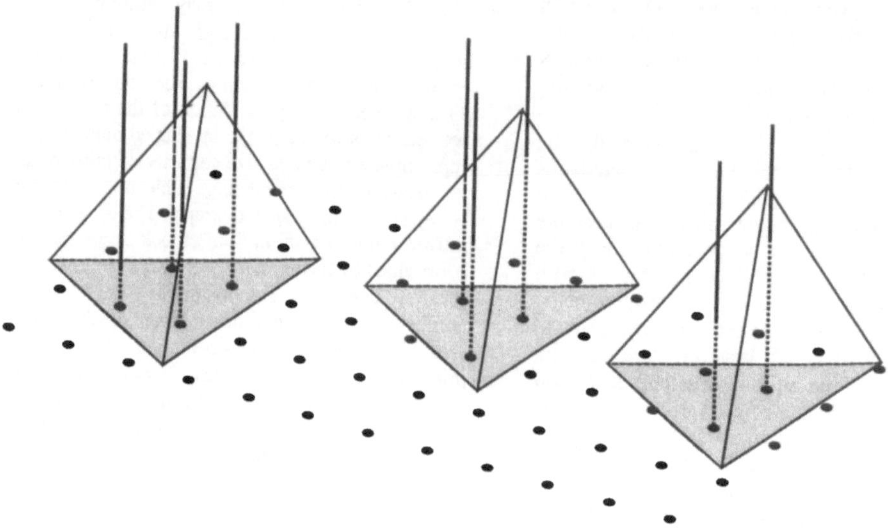

Figure 34. The severe consequences of varying the relative position of the block within a reinforcement array.

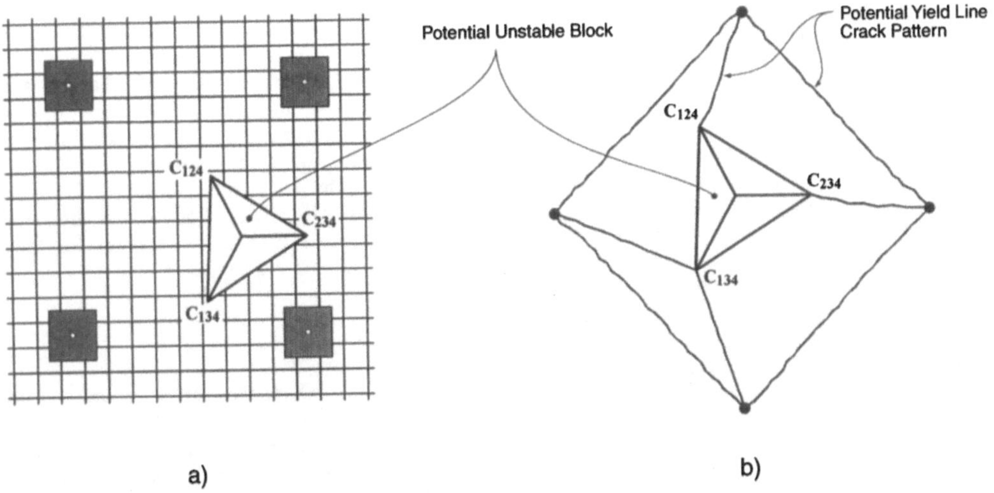

Figure 35. The effect of block shape and position relative to fixture points on a) the loading and performance of mesh and b) the yield line crack patterns and thus the capacity of a shotcrete lining.

Fortunately, the work conducted to obtain the demand characteristics can be used to simplify the reliability analysis. The demand distributions for the different block characteristics may be approximated by standard statistical distributions in which the random variable Ω is continuous over a finite interval. These distributions may then be used to simulate a distribution of block characteristics for each unstable block shape. The blocks may occur anywhere within a region of the rock improvement scheme constrained by the block existence zone associated with that block shape. This is facilitated by selecting a block vertex in the excavation face and positioning it within the polygonal region of the rock improvement scheme constrained by the surface expression of the block existence zone. The vertex position could be generated within this region according to a Poisson point process for the three vertex coordinates (x, y, z). However, here one must consider the objective of this exercise and the reporting of results.

The end aims are to define the reliability of the design in terms of block stability and rock improvement scheme performance. Recall that there could be many different shaped blocks involved and these give rise to many different forms of demand, all of which must be satisfied. Consequently a collective measure of the probability of failure is of limited value. Furthermore, the results must be in a format that will provide the designer with sufficient information to modify inappropriate designs in order to achieve both reliability and efficiency - the two are distinctly different concepts.

A reliable and efficient design is one in which:

1. Rock mass stability criteria are satisfied (e.g. maximum block displacement).
2. Rock improvement scheme performance criteria are satisfied (e.g. maximum and minimum force capacity utilisation)

The author suggests a grid simulation approach where the chosen block vertex is sequentially placed throughout a grid of specific positions in relation to the block existence zone and the rock improvement scheme. This enables three-dimensional surfaces to be drawn with the grid axes as abscissae and stability or performance characteristics as the ordinate (e.g. displacement of the block, the force, moment, displacement utilisation of the scheme). This provides the designer with clues on how to improve the design. Some additional notes on the stability assessment procedure have been given by Windsor (1996). However, a more complete description of the reliability assessment procedure will be described in a later publication.

8 CONCLUDING REMARKS

A procedure has been presented for determining the rock mass demand of excavations in blocky rock for use in the design of trial reinforcement and support schemes. It comprises a sequence of analyses which result in a series of relative frequency distributions for the demand characteristics of the removable unstable blocks of rock that might possibly form at the excavation surface. Some of the problems encountered in earlier unsuccessful attempts at defining block sizes led to an alternative approach in which certain 'limiting or possibility constraints' are introduced into the simulation process. These modifications produce marked improvements in both simplicity and efficiency. Guidelines were also given on the use of the relative frequency distributions in the selection and dimensioning of trial rock improvement schemes.

However, there are a number of problems with the procedure proposed here. Firstly, the author has simply followed others and prepared distributions that are reasonably useful in design. However, a new set of demand diagrams are really required that show demand as vectors rather than lumped scalars. The orientation of the vector demands may be shown in stereographic projections or in Cartesian space (i.e. vertical angle versus horizontal angle). The magnitudes of the demand are treated as densities at given orientations which are then contoured. This approach will be described in a latter publication.

Secondly, and more importantly, the distributions given provide a measure of the relative probability of the block characteristics occurring, not a measure of the absolute probability of them occurring. It is suggested that measures of absolute probability of occurrence associated with blocks of rock will remain elusive due to the data limited nature of rock structure problems. This suggestion is particularly relevant in the design of reinforcement and support which is undertaken at the pre-excavation stage when, unfortunately the availability of structural information is characteristically low.

Consequently, the very notion of probability and in particular, absolute probability in the area of rock reinforcement and support design must be considered in the context of this restriction. This leads directly to the idea that the reinforcement and support design process should be cyclical and based on tentative design followed by performance assessment with feedback and, where necessary, design modification. The performance assessment (i.e. step 6 listed in the introduction) also needs to include observations on the rock mass conditions. In jointed rock these will include the characteristics of the discontinuities and the blocks that actually form. This information is fed directly back into the design process at steps 1 and 2 in a dynamic process that should quickly determine if the assumptions explicitly and implicitly embodied in the demand assessment procedure are appropriate.

It is hoped that this work will assist in the search for consistent, formal procedures for use in the design of rock improvement schemes. However, it is recommended that the standard rock mechanics practice of design, implementation, observation and, if necessary, modification should continue to be adopted.

ACKNOWLEDGMENTS

The author is most grateful for the support and assistance provided by AMIRA and the many Australian and international companies that have enabled this work to be conducted. The author thanks his ex-colleagues; Mr Wayne Robertson and Ms Rosalie Caliva for completing many of the arduous tasks involved in the initial 2D and 3D simulation trials. The author would also like to express special thanks to his colleagues; Dr Alan Thompson and Mr Glynn Cadby and his Phd supervisor Professor Ernesto Villaescusa for their continual help, advice and support.

REFERENCES

Fisher, R. 1953. Dispersion on a sphere. Proc. R.Soc. London. A. 217. 295-305.

Goodman, R.E. & Gen-hua Shi 1985. Block theory and its application to rock engineering, Englewood Cliffs, New Jersey: Prentice-Hall.

Kuszmaul, J.S. 1993. A probabilistic method for estimating keyblock sizes in underground excavations. Ph.D. Dissertation, Dept. of Civil Eng. University of California, Berkeley, USA.

Kuszmaul, J.S. 1994. Application of the Unit-Cell method to determine the optimum tunnel orientation. Proc. 1st N. American Rock Mech Symp. Austin, TX, USA. Rotterdam: Balkema , 515 - 529.

Mauldon, M. 1994. Intersection probabilities of impersistent joints. Int. J. Rock. Mech. Min. Sc. & Geomech. Abstr., V31, 2, 107-115.

Mauldon, M. 1995. Keyblock probabilities and size distributions: A first model for impersistent 2-D Fractures. Int. J. Rock. Mech. Min. Sc. & Geomech. Abstr., V32, 6, 575-583.

Priest, S.D. 1985. Hemispherical projection methods in rock mechanics, London: George Allen & Unwin.

Shapiro, A & J.L. Delport 1991. Statistical analysis of jointed rock data. Int. J. Rock. Mech. Min. Sc. & Geomech. Abstr., V28, 5: 375-382.

Stone, C.A., 1994. A matrix approach to probabilistic key block analysis. Ph.D. Dissertation, Michigan Tech. University, USA.

Warburton, P.M. 1983. Applications of a new computer model for reconstructing blocky rock geometry - Analysing single block stability and identifying keystones. 5th ISRM Int. Cong. on Rock Mech., F225-F230, Melbourne: A.G.S.

Windsor, C.R. 1995. Block Stability in Jointed Rock Masses. Fractured and Jointed Rock Masses. Rotterdam: Balkema , 59-66.

Windsor, C.R. 1996. Rock reinforcement systems. Int. J. Rock. Mech. Min. Sc. & Geomech. Abstr., V34, 6: 919-951.

Rock Support and Reinforcement Practice in Mining, Villaescusa, Windsor & Thompson (eds)
© 1999 Balkema, Rotterdam, ISBN 90 5809 045 0

Formulating a site specific support system design methodology

A. Daehnke
CSIR Mining Technology, Johannesburg, South Africa

ABSTRACT: The current methodology to design stope support systems in South African gold and platinum mines is based upon the tributary area concept. The discontinuous nature of the hangingwall rock is not adequately addressed and mechanisms leading to rock mass failures between adjacent support units are poorly understood. The work reported here aims to formulate a basis for quantifying support mechanisms and gain insights into the influence of rock discontinuities on stable hangingwall spans.

Numerical models are used to qualitatively investigate the stress transfer from the support units to the discontinuous hangingwall. Stable hangingwall spans are quantified by considering two failure mechanisms, namely (i) beam buckling, and (ii) shear failure due to slip at the abutments.

The output of the proposed design methodology is appropriate support spacing, based on discontinuity spacing and orientation. The method is particularly suited to mines in intermediate and great depth, where typically the hangingwall is highly discontinuous due to face parallel mining induced fractures. The results of parametric studies show that in-situ compressive hangingwall stress, beam thickness and the friction angle play an important role in hangingwall stability.

1 INTRODUCTION

Rock mass instabilities account for the single largest cause contributing towards the toll of injuries and fatalities suffered by the workforce during gold and platinum mining operations in South Africa. Approximately half of the number of fatalities are associated with rockfalls, whilst the remainder are a consequence of the failure of dynamically loaded rock during seismic events and rockbursts.

Stope support systems, consisting of props and packs, are used to stabilise the rock mass surrounding the mining excavation and reduce the risk of rockfalls and rockbursts. In South African gold and platinum mines, the methodology currently used to design stope support is based upon the tributary area concept, and does not adequately address the fact that the rock being supported may be fractured and jointed. Due to the complexities associated with the behaviour of a discontinuous rock mass, the interaction of the support system with a fractured and jointed rock mass is poorly understood. Relatively little progress has been made in quantifying the effect of support spacing on the rock mass stability, and typically it is the responsibility of the rock engineer to estimate support spacing based upon past experience. In order to improve safety and continue mining at increasing depth, it is important to understand the mechanisms of the support – rock interaction, the zones of support influence and the role of rock mass discontinuities.

The objective is to formulate a basis for quantifying support mechanisms, and more specifically to gain insights into the influence of rock discontinuities such as joints, fractures and bedding planes on stable spans. Instabilities generally initiate at discontinuities, which are shown to be of prime importance when spacing support units. An attempt is made to evaluate the interaction of the support with a discontinuous rock mass, with a view of developing a support system design methodology, which maximises support spacing, whilst maintaining a stable hangingwall span between support units. Due to widely varying rock mass conditions and behaviour of the reefs extracted by the gold and platinum mines, the methodology is influenced by local geological conditions, and local rock and discontinuity types, as well as spacing and orientation thereof.

The ultimate output of this work is a design tool for the gold and platinum mining industry to assist the rock mechanics engineer to improve support design, with particular emphasis on optimised support spacing and support performance requirements.

2 CLASSIFICATION OF ROCK MASS DISCONTINUITIES

The behaviour and deformation of the rock mass surrounding stopes is influenced by mining induced and geological discontinuities. Hence, in order to gain insights into the support – rock interaction, an understanding of typical rock mass discontinuities is required.

Investigations into mining induced stress fractures in intermediate and deep level gold mine stopes have revealed that two main types of fractures are present in the hangingwall (Adams et al., 1981):

- *Shear Fractures*: These fractures are associated with highly stressed rock, and thus are found in intermediate and deep level mines. It is estimated that the fractures initiate 6 to 8 m ahead of the advancing stope face and separate the rock into blocks of relatively intact material. They are oriented approximately parallel to the stope face and are regularly spaced at intervals of 1 to 3 m. Shear fractures usually occur in conjugate pairs in the hangingwall and footwall, and typically reveal distinct signs of shear movement. Their dip in the hangingwall is generally towards the back area at angles of 60 to 70 degrees (Esterhuizen, 1998).
- *Extension Fractures*: These fractures initiate ahead of the stope face and are smaller than shear fractures. They form after shear fractures have propagated and are generally truncated by parting planes. Extension fractures normally do not exhibit relative movement parallel to the fracture surface and are typically oriented parallel to the stope face. They are commonly spaced at intervals of 10 cm with lower and upper limits of 5 to 50 cm, respectively. The strike length is typically 3 m, where lower and upper limits of 0.4 and 6 m have been observed (Esterhuizen, 1998). Extension fractures normally dip between 60 and 90 degrees, where the direction of dip (i.e. towards or away from the back area) can be influenced by the hanging and footwall rock types (Roberts, 1995)

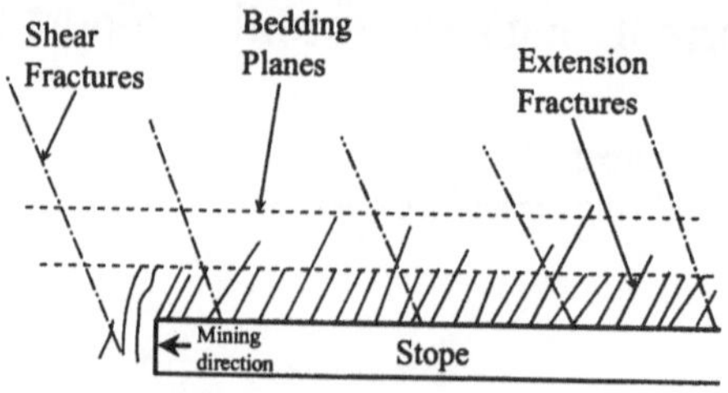

Figure 1: Simplified schematic showing the three most prevalent discontinuity types in intermediate and deep level mines.

Most gold reef extraction takes place in bedded quartzites. Reef parallel bedding planes typically represent weak interfaces between adjacent strata, and provide little cohesion and low frictional resistance. Bedding planes are generally spaced at 0.2 to 2.0 m intervals. The rock fallout height is commonly governed by the position of bedding planes.

The three most prevalent discontinuity types, extension fractures, shear fractures and bedding planes, are illustrated in Figure 1. Their influence on the rock mass behaviour and stability thereof is considered in this study, and an attempt is made to quantify their effect on support spacing and rock mass stability.

3 NUMERICAL SIMULATIONS OF THE SUPPORT INTERACTION WITH A DISCONTINUOUS ROCK MASS

The finite-discrete element program ELFEN (Rockfield, 1996) was used as a tool to investigate the interaction of support units with a discontinuous rock mass. Various ELFEN models were constructed, incorporating several discontinuity types. The aim of the numerical models was to obtain qualitative insights into the stress trajectories arising due to the load transfer from the support elements to the discontinuous hangingwall rock. The boundary conditions of the numerical models were a uniformly applied load along the top of the model and horizontal displacement confinement along the left hand and right hand sides of the model. The fracture surfaces were modelled with no cohesion and a friction angle of 40 degrees.

Figure 2 shows the principal stress distribution associated with two prop like support elements loaded by a hangingwall slab discretised into blocks by closely spaced vertical fractures. It is evident that the stresses are transmitted from the support units, though the long axis of the blocks defined by the vertical discontinuities, into the more competent (in this case non-fractured) medium above the bedding plane.

In the case of a hangingwall discretised by dipping extension fractures in conjunction with a discontinuity modelling a shear fracture, the stress trajectories follow the paths shown in Figure 3. As noted before, the stresses seem to be transmitted mainly parallel to the discontinuities, through the fractured layer, into the competent layer above the bedding plane.

In intermediate and deep mines, fracturing ahead of the stope face induces rock dilation, leading to compressive hangingwall stresses parallel to the skin of the excavation (Squelch, 1994). The compressive stresses can only be maintained if the hangingwall beam is continuous, i.e. if the hangingwall is confined at both ends. This is generally the case in intermediate and deep gold mines, and compressive hangingwall stresses contribute significantly towards the rock mass stability.

Figure 4 shows the stress distribution associated with a model loaded in the vertical and horizontal direction. The magnitude of the horizontal stress is 1 MPa, whilst the total load carried by each support unit is 200 kN; these are typical load magnitudes as measured underground. Although the stress field is complicated by the addition of the horizontal stresses, the basic stress field consists of the superposition of the vertical and horizontal stress components. Stress arching in the competent layer is evident (shown in Figure 4), leading to the conclusion that most of the stress transmitted by the support units follows a path parallel to the discontinuities, as observed in Figure 3.

To summarise, simplified numerical models representing a discontinuous hangingwall beam, discretised by extension fractures, a shear fracture and a bedding plane, have shown that the load carried by support units is generally transmitted in a direction parallel to the fracture orientations. Comparatively little stress is transmitted across the fractures, and the hangingwall rock between adjacent support units is essentially unstressed. When the hangingwall is clamped by compressive stresses acting parallel to excavation surface, the resulting stress field can be approximated by the linear superposition of horizontal and vertical

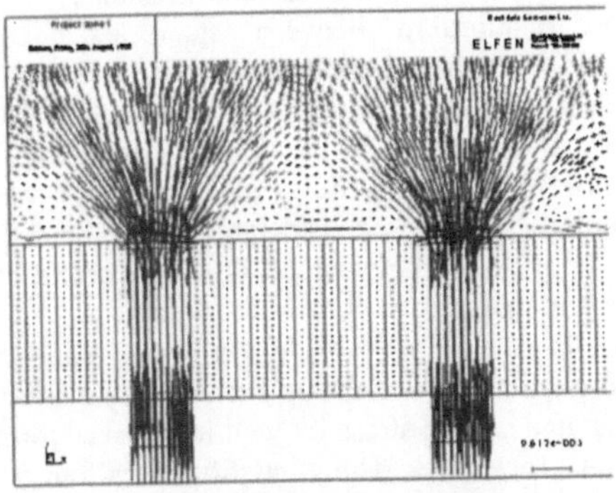

Figure 2: Stress distribution associated with a discontinuous hangingwall interacting with two support units.

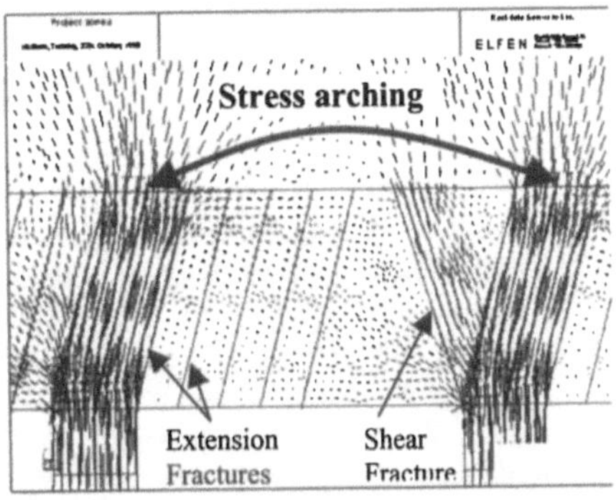

Figure 3: Stress transfer through a hangingwall beam discretised by discontinuities modelling extension and shear fractures.

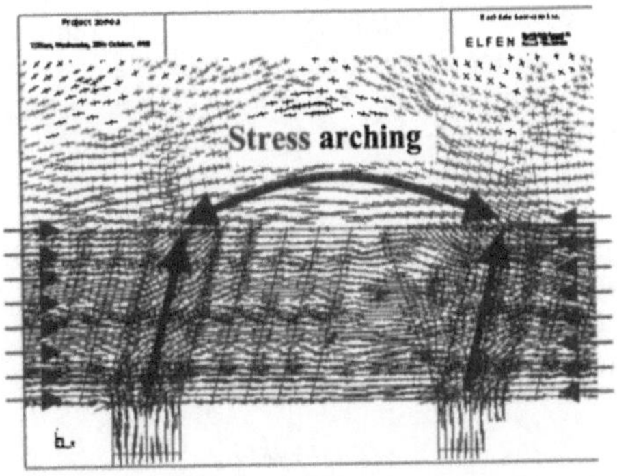

Figure 4: Vertical and horizontal loading of a discontinuous hangingwall beam.

loading, i.e. the majority of the stress induced by the loaded support units continues to be transmitted parallel to fracture surfaces.

Hence, in subsequent results given in the following sections, the compressive stresses induced by support units are assumed to be transmitted parallel to the fracture surfaces. Support units do not directly stress the hangingwall rock between adjacent support units, and hangingwall stability in this region is governed by compressive in-situ stresses and the buckling potential of the hangingwall beam.

4 QUANTIFYING STABLE HANGINGWALL SPANS BETWEEN SUPPORT UNITS

The qualitative insights gained from the numerical simulations are used to develop a simplified conceptual model describing the rock mass stability and quantifying stable spans between adjacent support units. The model is suitable to apply in a support design methodology in order to optimise support spacing, whilst maintaining high levels of safety.

Two failure mechanisms are considered, namely instabilities due to (i) beam buckling, and (ii) shear failure due to slip at the abutments.

4.1 *Hangingwall beam buckling*

The design procedure followed here is based on that developed by Evans (1941), and subsequently modified and extended by Beer and Meek (1982), Brady and Brown (1985), and Hutchinson and Diederichs (1996). The solution technique, which is based on the Voussoir beam, follows the intuitive idea that in a discontinuous hangingwall beam the central transverse crack determines the deformational behaviour (Figure 5). In the buckling mode the beam becomes unstable to form a 'snap-through' mechanism.

In analysing the stability of the Voussoir beam the following assumptions are made:

- As the beam deflects, a parabolic compression arch develops in the beam.
- Deflection of the beam occurs before slippage at the abutments. Stability against slippage (see Section 4.2) is determined after the compression arch develops.
- The abutments are stiff, i.e. they do not deform under the arching stress. Each abutment is subjected to the same distributed load as the ends of the beam, however the loaded area is small compared with the beam span. Therefore, elastic compression of the abutments will be small compared with the beam compression, and may be neglected (Brady and Brown, 1985).

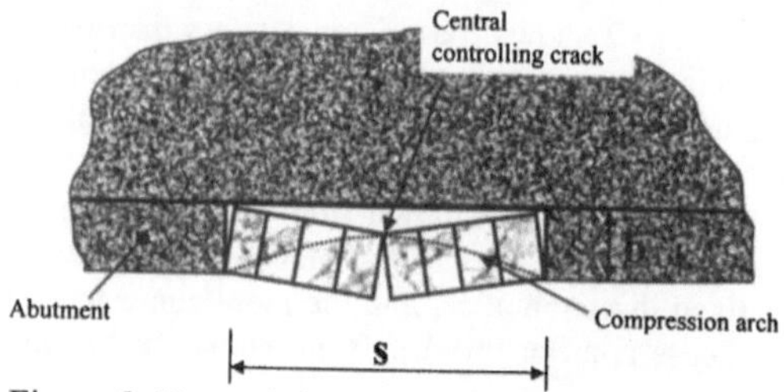

Figure 5: Voussoir beam geometry for hangingwall beam analysis.

The Voussoir beam problem is statically indeterminate, i.e. no explicit solution is available and an iterative process is followed to determine the beam equilibrium position. The solution procedure is given in texts such as Brady and Brown (1985), Hutchinson and Diederichs (1996) and is not repeated here.

Previously documented results of this solution have used an absolute snap-through limit, which is the limit of stable deflection according to the mathematical formulation. This limit is extremely sensitive to beam thickness, a difficult parameter to accurately and reliably estimate, and one which may change as deflection and layer separation occurs. Hutchinson and Diederichs (1996) recommend a design snap-through limit which is reached when the mid-span deflections reach 10 % of the beam thickness. Beyond this deflection, small differences in thickness have an unacceptably large influence on stability, and the beam is considered unstable.

The relationship between span and beam thickness is highly dependent on the in-situ rock mass stiffness parallel to the excavation surface. The in-situ rock mass stiffness is predominantly governed by the stiffness of the rock mass discontinuities, and is considerably lower than the stiffness of solid rock, which is characterised by Young's modulus.

Bandis et al. (1983) made use of experimental data to establish a relationship between normal joint stiffness and normal stress for well interlocked joints in various rock types. The joint stiffness is found to increase with increasing normal stress. For rock mass discontinuities in a typical gold or platinum hangingwall, where the compressive hangingwall stresses are generally less than 5 MPa, a discontinuity stiffness of 40 MPa/mm (Bandis et al., 1983) is assumed for the purposes of this study. It is recognised, however, that further in-situ discontinuity stiffness measurements are required to obtain more accurate and representative stiffness data.

The joint stiffness is incorporated in the buckling analysis procedure. The effective rock modulus is taken as the product of the normal joint stiffness and the arch shortening during beam deflection. The analysis is modelled with a minimum of three discontinuities, which act as springs in series, and each discontinuity is compressed equally.

Taking Hutchinson and Diederichs's (1996) design snap-through limit into account (max. 10 % beam deflection at mid-span), the span versus thickness relations shown in Figure 6 give the stability envelopes for hangingwall beams with 1, 3, 5 and 10 joints per meter of hangingwall length.

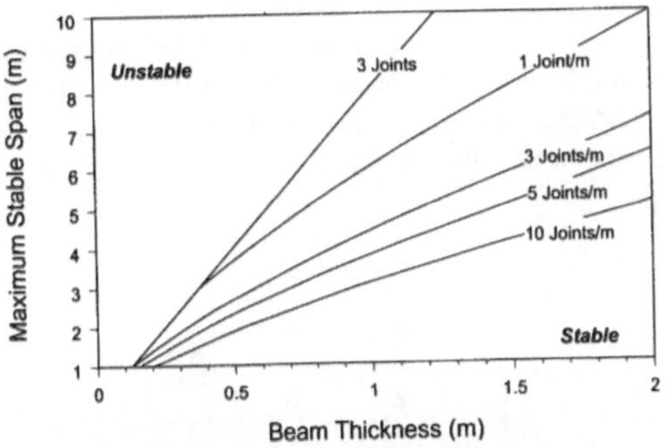

Figure 6: Buckling stability envelopes of a discontinuous hangingwall beam.

4.2 *Shear failure by slip at the abutments*

The second failure mechanism considered in this study is shear failure by slip at the abutments. Figure 7 shows a simplified schematic indicating the main parameters governing the stability of hangingwall keyblocks prone to shear failure, where W is the block weight, b is the beam thickness, s is the span between adjacent support units, α and β are angles of extension and shear fractures, and σ_x is the magnitude of compressive hangingwall stress. The hangingwall stresses are generated by two mechanisms, namely:

- In intermediate and deep level mines, the rock dilation associated with fracturing immediately ahead of the stope face generates compressive stresses parallel to the excavation surface.
- The block rotation associated with the 'snap-though' failure mechanisms described in Section 4.1 generates compressive hangingwall stresses.

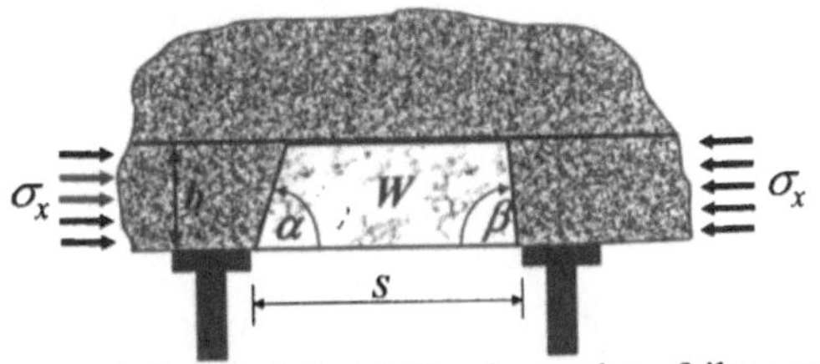

Figure 7: Keyblock instability due to shear failure at the abutments.

For the keyblock to be stable, the lateral thrust at the abutments due to in-situ compressive hangingwall stresses must mobilise a frictional resistance sufficient to provide the abutment shear force. The frictional resistance for either side of the keyblock can be calculated using the following expressions:

$$V_I = \sigma_x\, b\left(\frac{\mu \sin\alpha - \cos\alpha}{\sin\alpha + \mu\cos\alpha}\right),$$

$$V_{II} = \sigma_x\, b\left(\frac{\mu \sin\beta - \cos\beta}{\sin\beta + \mu\cos\beta}\right). \tag{1}$$

The coefficient of friction, μ, is an important parameter governing the resistance to shear. Typically, discontinuity types encountered underground have closely matched surfaces, this being particularly the case with mining-induced fractures. The associated apparent friction angle can be relatively high, and friction angles of 30 to 50 degrees are considered realistic.

The keyblock is considered unstable when $W > V_I + V_{II}$. If the frictional resistance exceeds the weight of the hangingwall block ($V_I + V_{II} > W$), stability is ensured only if:

- $V_I > \frac{1}{2} W$ and $V_{II} > \frac{1}{2} W$ (Figure 8 a and b), or
- $V_I > \frac{1}{2} W$ *or* $V_{II} > \frac{1}{2} W$ and the angle of the opposing discontinuity prevents any rotation (Figure 8 c and d). Thus, if the frictional resistance on either discontinuity is less than half the keyblock weight, the keyblock is stable only on condition that kinematically no block rotation is possible.

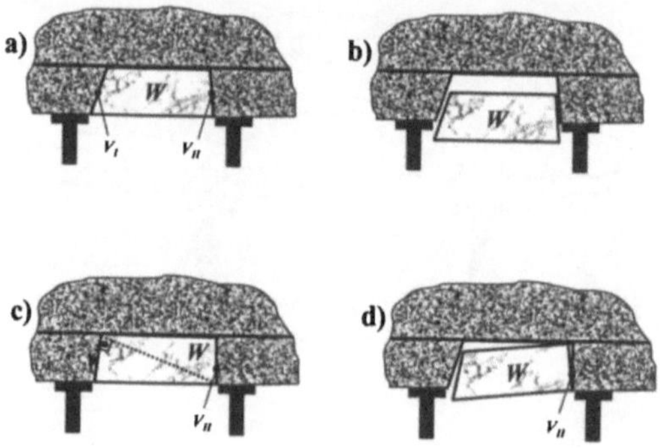

Figure 8: Schematic diagrams showing possible failure modes due to shear at discontinuity interfaces.

5 SUPPORT DESIGN METHODOLOGY

A methodology is proposed to design for maximum stable hangingwall spans between adjacent support units. The design procedure consists of two phases, namely:

1. Tributary area theory is applied to determine the load requirements of the support system.

2. Maximum support spacing and stable spans between adjacent support units are determined by calculating the limiting equilibrium due to the two predominant failure mechanisms outlined in Sections 4.1 and 4.2, i.e. (i) beam buckling, or (ii) shear failure due to slip at the abutments.

The following assumptions are made with regard to the design methodology:

- The rock mass fall-out thickness is determined by the position of prominent bedding planes and/or back analyses of previous rockfall accidents. Work by Roberts (1995) has shown that the fall-out thickness is reef dependent, and varies between 1.0, 1.2 and 1.4 m for the Carbon Leader, Vaal and Ventersdorp Contact Reef, respectively.
- The hangingwall rock mass is discretised by face parallel extension and shear fractures. At this stage no attempt has been made to include discontinuities of geological origin; unlike mining induced fractures, these are generally not face parallel and need to be considered in future work.
- The loading requirements due to tributary area load distributions are calculated based on a two dimensional plan view of the stope and associated support system (Figure 9 a).
- To calculate maximum stable spans and prevent hangingwall failure due to buckling or shear failure, the cross-section (plane strain is assumed) of the hangingwall rock is taken along the strike direction, i.e. face perpendicular (Figure 9 b).
- Deflection of the beam occurs before shear failure at the discontinuity interfaces. Stability against slippage is determined after the compression arch has developed in buckling mode.
- Pre-existing (before any beam deflection) compressive hangingwall stresses do not affect the behaviour of the Voussoir beam, i.e. the abutments are fixed and hence any compressive hangingwall stresses do not contribute towards the beam stability. This conservative assumption is based on the fact that the beam hinges at the abutments in the immediate vicinity of the hangingwall skin, and in many cases the local hangingwall stresses will be low.
- Horizontal stresses induced by beam rotation in the buckling mode are added to the pre-existing stresses when considering failure due to shear slip at the abutments.

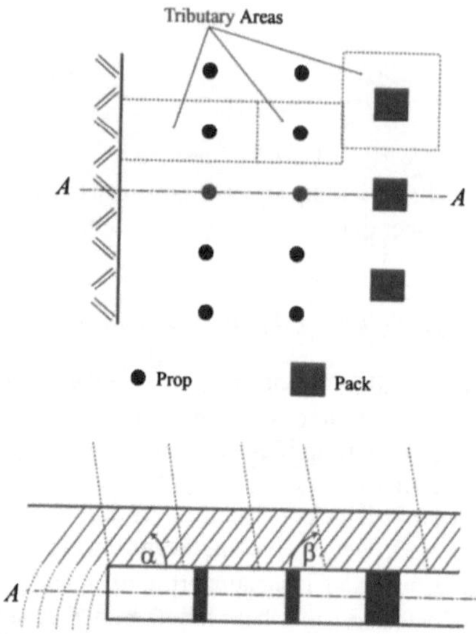

Figure 9 a): Plan view applied for tributary area load requirements (top), and b) face perpendicular cross-section used to determine maximum stable hangingwall spans in strike direction (bottom).

The solution procedure can readily be coded in the form of a computer program to facilitate the rapid and convenient evaluation of various support systems and associated spacing of support units.

6 INFLUENCE OF VARIOUS PARAMETERS GOVERNING STABLE HANGINGWALL SPANS

The parameters defining the maximum stable hangingwall span are the discontinuity angles (α and β), bedding height (b), friction coefficient (μ), and hangingwall clamping stresses (σ_x). A parametric study is conducted to illustrate the stable span dependency on the above-mentioned parameters.

The maximum span was determined using the proposed rockfall design methodology for an intermediate depth mine with 5 fractures per meter. Failure due to buckling, as well as shear at the discontinuity surfaces, is taken into account. Compressive stresses at the abutments due to deformation in the buckling mode are added to the in-situ hangingwall stresses due to fracturing ahead of the stope face.

The limiting equilibrium of the keyblock is governed by one of two failure mechanisms: (i) shear failure due to slip at the abutments and/or block rotation, and (ii) buckling failure. Figure 10 gives stability envelopes for the hangingwall at limiting equilibrium for $\sigma_x = 1.0$ MPa, $b = 2.0$ m and $\mu = \tan 40°$. Three prominent stability zones can be identified:

1. **Zone I:** At these discontinuity angle combinations, no shear resistance is active, i.e. all keyblocks, irrespective of span size, are unstable. According to Equations (1), the transition from Zone I to Zone II (when shear resistance becomes active) occurs when $V_I = -V_{II}$. For the friction angle applied here ($\mu = \tan 40°$), it can be shown that shear resistance becomes active when $\alpha + \beta > 100°$. Generally, shear resistance is active when the sum of the two discontinuity angles ($\alpha + \beta$) exceeds a constant angle, the magnitude of which is linearly related to the friction angle. Figure 11 gives the minimum angle of $\alpha + \beta$ versus friction angle.

2. **Zone II:** Here the failure mechanism is predominantly shear failure at one of the discontinuity interfaces, leading to instability due to block rotation. Note that for the parameters displayed in Figure 10, the maximum stable span in Zone II is approximately 3 m, which is typical of stable spans as observed underground.

3. **Zone III:** In Zone III the shear resistance increases rapidly. The angles of both discontinuities are greater than 50° and the keyblock is effectively clamped by in-situ compressive hangingwall stresses. The maximum stable span is governed by the buckling resistance, and for the parameters shown in Figure 10, the maximum stable span is 6 m.

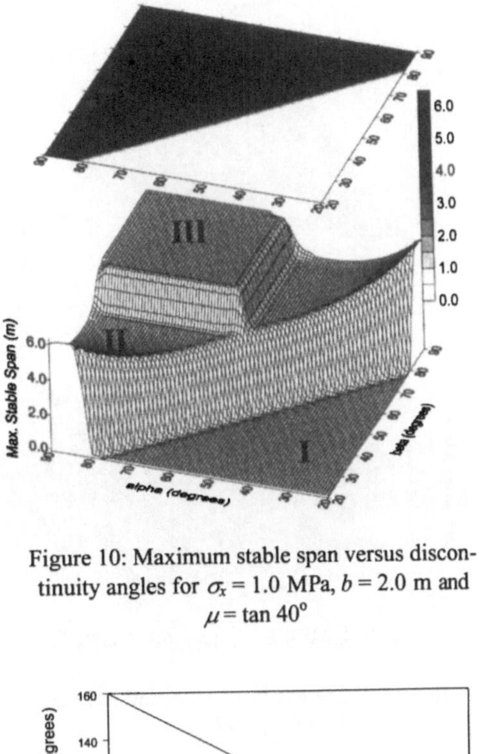

Figure 10: Maximum stable span versus discontinuity angles for $\sigma_x = 1.0$ MPa, $b = 2.0$ m and $\mu = \tan 40°$

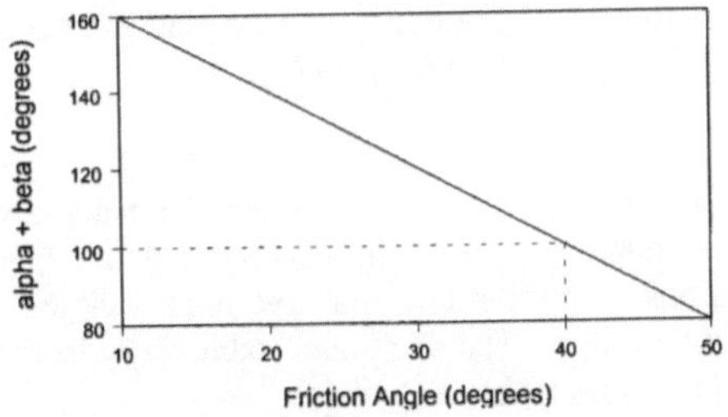

Figure 11: Values of $\alpha + \beta$ versus friction angle to generate positive shear resistance.

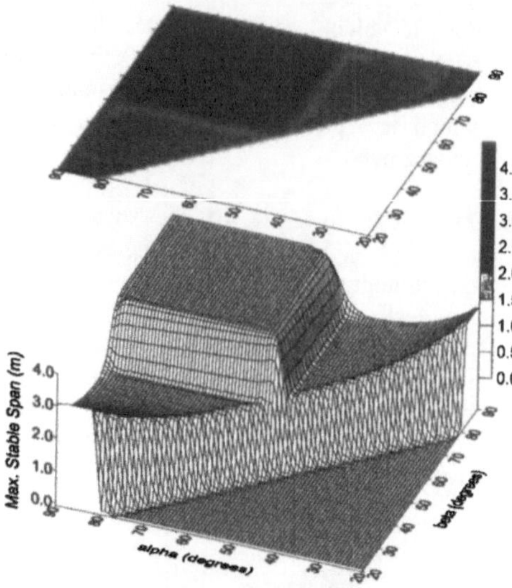

$\sigma_x = 1.0$ MPa, $b = 1.0$ m and $\mu = \tan 40^o$

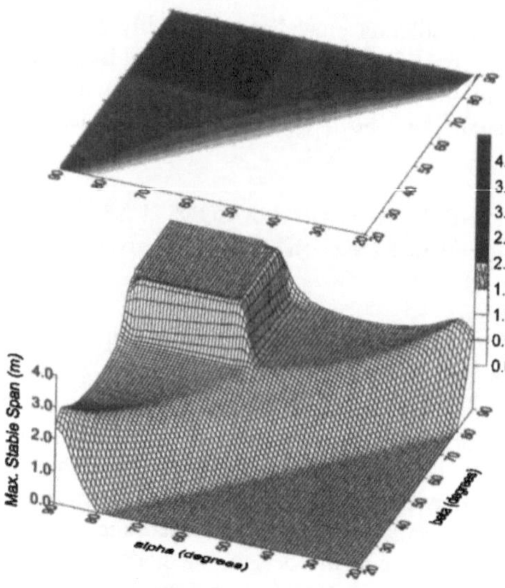

$\sigma_x = 0.1$ MPa, $b = 1.0$ m and $\mu = \tan 40^o$

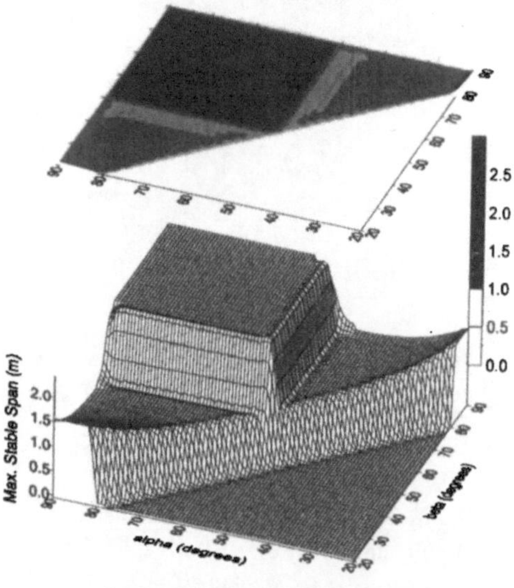

$\sigma_x = 1.0$ MPa, $b = 0.5$ m and $\mu = \tan 40^o$

Figure 12: Effect of reduced hangingwall beam thickness (b).

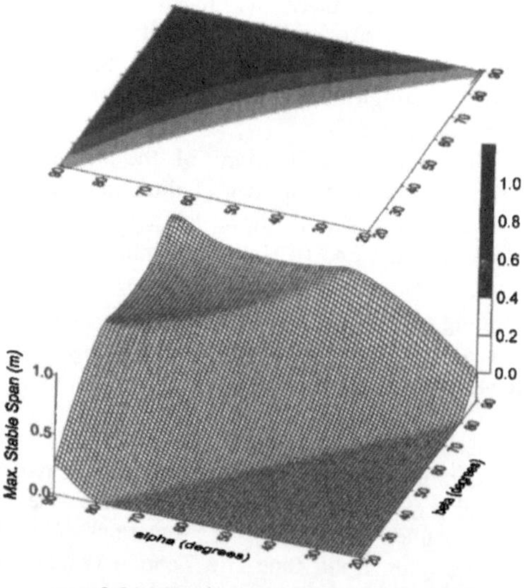

$\sigma_x = 0.01$ MPa, $b = 1.0$ m and $\mu = \tan 40^o$

Figure 13: Effect of reduced hangingwall compressive stresses (σ_x).

Figure 12 shows the influence of hangingwall beam thickness (b) on the maximum stable span at limiting equilibrium for various angles of discontinuities. The maximum stable span decreases with decreasing beam thickness.

The effect of compressive hangingwall stresses on the maximum stable span is given in Figure 13. It is evident that compressive stresses increase the stability, and larger spans between adjacent support units are possible. In shallow mines less fracturing occurs ahead of the stope face. Hence, less rock dilation occurs, and the compressive hangingwall stresses are generally low in magnitude. The analysis presented here assumes 5 fractures per meter of hangingwall. In shallow mines the rock is generally less densely fractured, and hence the stable spans can be increased. The methodology proposed here is particularly suited for comparatively densely fractured hangingwall rock, as typically encountered in intermediate and deep level mines. In shallow mines the rock engineer needs to assess the spacing of discontinuities and base the maximum stable span on typical block sizes.

7 CONCLUSIONS

Preliminary concepts of a support design methodology suitable for rockfall conditions is proposed. The design procedure consists of two stages: (i) a tributary area analysis determines the general support resistance requirements for the support system as a whole, and (ii) a stability analysis considering failure due to beam buckling and shear at the beam abutments gives suitable spacing of individual support units.

The methodology is particularly suited to mines in intermediate and great depth, where typically the hangingwall is highly discontinuous due to face parallel mining induced fractures. Although the method can give valuable insights into support requirements in shallow mines, additional input in terms of past experience and engineering judgement is required from the rock engineer to determine typical block sizes and the support thereof.

The support design method gives insights into spacing and associated stable hangingwall spans in strike direction only. Due to the face parallel mining induced fracture orientation in intermediate and deep level mines, the hangingwall rock is generally less prone to failure between two support units in dip direction, compared to failure between units in strike direction. Hence, from a conservative design point of view, the support spacing in the dip direction is recommended to be similar to the spacing in strike direction.

It is recommended to conduct additional work to quantify the effect of arbitrarily oriented discontinuities of geological origin on support spacing in strike and dip directions. Further work could re-address the influence of the modified hangingwall stress distribution due to loading by the stope face, support units and backfill. Particularly in the case of shallow dipping fractures, stresses transmitted across discontinuities could stabilise the hangingwall, leading to wider permissible spans.

Parametric evaluations of the proposed support design methodology show that, for most discontinuities observed underground, i.e. dipping between 50° and 90°, the design procedure provides realistic stable spans, which agree with current mining practice and post rockfall/rockburst underground observations and back analyses. However, further back analyses are required to calibrate and verify the design method.

8 ACKNOWLEDGEMENTS

The author would like to acknowledge the South African Department of Mineral and Energy Affairs for permission to publish, and M.D.G. Salamon, A.J. Jager, M.K.C. Roberts and J.A.L. Napier for their technical advice and input.

9 REFERENCES

Adams, G.R., Jager, A.J. & Roering, C. 1981. Investigations of the rock fracture around deep-level gold mine stopes. 22nd U.S. Symposium on Rock Mechanics, Massachusetts Inst. of Technology, June 28-July 2.

Bandis, S.C., Lumsden, A.C. & Barton, N.R. 1983. Fundamentals of rock joint deformation. Int. J. Rock Mech. Min. Sci. & Geomech. Abstr. 20(6), pp. 249 – 268.

Beer, G. & Meek, J.L. 1982. Design curves for roofs and hangingwalls in bedded rock based on voussoir beam and plate solutions. Trans Instn Min. Metall. 91, pp. A18 – 22.

Brady, B.H.G. & Brown, E.T. 1985. Rock Mechanics for Underground Mining. George Allen & Unwin, pp. 214 – 218.

Daehnke, A. 1997. Stress wave and fracture propagation in rock. Ph.D. Thesis, Vienna University of Technology.

Esterhuizen, E. 1998. Improved support design through a keyblock approach. Interim Report for SIMRAC Project GAP 330, Stope Face Support Systems, prepared by CSIR Mining Technology and University of Pretoria, pp. 1 – 83.

Evans, W.H. 1941. The strength of undermined strata. Trans Instn Min. Metall. 50, pp. 475 – 532.

Hutchinson, D.J. and Diederichs, M.S. 1996. Cablebolting in Underground Mines. BiTech Publishers, pp. 265 – 269.

Roberts, M.K.C. 1995. Stope and gully support. Final Report for SIMRAC Project GAP 032, CSIR Mining Technology, pp. 1 – 45.

Rockfield 1996. ELFEN User Manual, Version 2.6, University College of Swansea.

Squelch, A.P. 1994. The determination of the influence of backfill on rockfalls in South African tabular mines. M.Sc. Dissertation, University of Witwatersrand, Johannesburg.

Rock Support and Reinforcement Practice in Mining, Villaescusa, Windsor & Thompson (eds)
© 1999 Balkema, Rotterdam, ISBN 90 5809 045 0

Corrosivity classification of the underground environment

C.Li
Luleå University of Technology, Sweden

K.Lindblad
Vattenfall Hydropower AB, Sweden

ABSTRACT: Two corrosivity classification systems are proposed for the underground environment with relation to the corrosion of rock bolts. The corrosivity of the underground environment is judged on the basis of a number of corrosion-related parameters. The first classification system is for wet rock conditions. The corrosion-related parameters used in this system are the pH value of the water, the dissolved oxygen and the resistivity, as well as the ambient temperature, rock mass quality and precipitation of solid substances on the rock bolts. The second classification system is for dry rock conditions. The corrosion-related parameters used are the deposition rate of sulphur-and-nitrogen oxides and chloride and the relative humidity of the atmosphere, as well as the ambient temperature.

1 INTRODUCTION

Steel elements, such as steel rock bolts, are commonly used for reinforcement of underground structures in rock. The consumption of rock bolts amounts to tens of thousands of tonnes each year. In a humid or water-bearing underground environment, rock bolts corrode and gradually lose their load-bearing capacity. The corrosion of rock bolts depends on both the properties of the steel material and the environment in which the bolts are installed. In this study, our interest is focused on the latter, i.e. the corrosivity classification of the underground environment.

Corrosion occurs in just about any environment. However, the corrosion rate depends on the type of steel being used. Electrochemical corrosion is one of the most common types of corrosion for carbon steels in the underground environment. For a metal that corrodes electrochemically, there are many small corrosion cells on the metal surface. A corrosion cell is composed of two electrodes (the anode and the cathode), an electrolyte and an electric circuit. For a rock bolt installed in rock, the water in the rock mass acts as the electrolyte, and the bolt itself as the electric circuit. The concentrations of hydrogen ions (represented by the pH value) and of the oxygen dissolved in the water are two key parameters for the electrochemical corrosion of steels. In the process of corrosion, the oxidised species in the water solution have to be transported to the cathode. The resistivity of the water solution, therefore, is also a parameter that affects the process of corrosion. The pH value, the concentration of dissolved oxygen and the resistivity are the three basic parameters governing the corrosion process. Apart from these three parameters, there are a few others, such as temperature, the rock mass quality and precipitation, which have indirect effects on corrosion.

In most cases, rock bolts are installed in wet soil or rock masses, but sometimes they are installed in dry rock masses. In a dry rock condition corrosion can still occur on bolts if the atmosphere is polluted by gases from blasting and exhaust emissions of diesel engines. The governing parameters for corrosion in dry rock conditions are different from those in wet conditions. The most important parameters for corrosion in dry rock conditions are the content of sulphur and nitrogen oxides in the air and the relative humidity (Barton, 1976; Shreir, 1976; Mansfeld, 1987).

2 SYSTEM I - CLASSIFICATION IN WET ROCK CONDITIONS

2.1 *Corrosion-related parameters*

(1) The pH value

In water having a pH value below 5, i.e. in an acidic solution, carbon steels will corrode at a considerable rate, even in the absence of oxygen. At higher pH values the corrosion rate of steel is, however, determined by other factors. The relationship between the corrosion rate of carbon steel and the pH value is demonstrated in Figure 1.

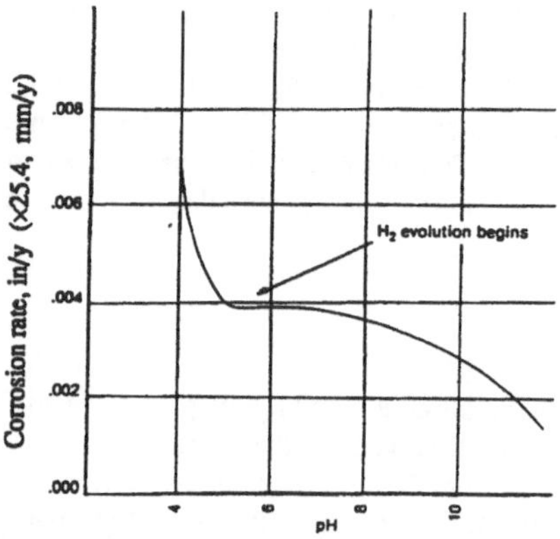

Figure 1. Effect of the pH value on the corrosion of steel (NACE, 1984).

(2) Oxygen concentration

At ordinary temperatures in neutral or near-neutral water, dissolved oxygen is necessary for appreciable corrosion of iron. In air-saturated water, the initial corrosion rate may reach a value of about 0.46 mm/year. This rate declines over a period of days as the iron oxide (rust) film is formed and acts as a barrier to further oxygen diffusion. The steady-state corrosion rate may be 0.046-0.12 mm/year. As shown in Figure 2, the corrosion rate of iron is proportional to the oxygen concentration.

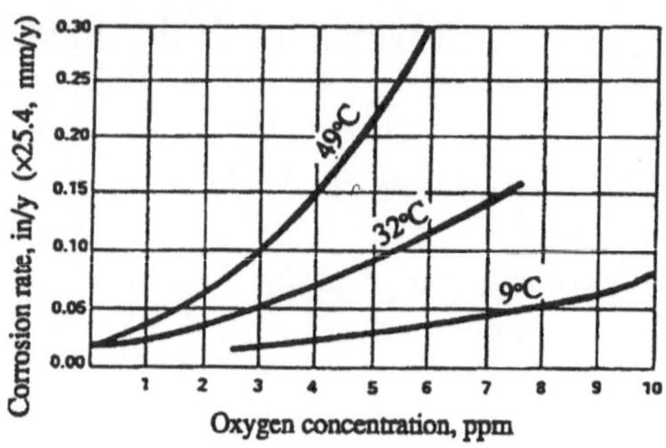

Figure 2. Effect of the oxygen concentration on the corrosion of steel in tap water (NACE, 1984).

(3) Resistivity

The resistivity of soil moisture is determined by the concentration of different ions and their mobility. The conductivity (i.e. the reciprocal of resistivity) is approximately linearly related to the ion concentration.

Resistivity influences the current in the corrosion cell. The risk of localised corrosion is great if the soil resistivity is below 1000 Ω-cm, but little if it is greater than 5000 Ω-cm.

Figure 3 illustrates how the resistivity of soil, as well as the corrosivity of soil, varies with the water content. It is seen in the figure that the soil resistivity may be used to express the corrosivity of soil only in unsaturated soils. The importance of the water content for the oxygen transport explains why the corrosion rate is considerably higher above than below the groundwater level.

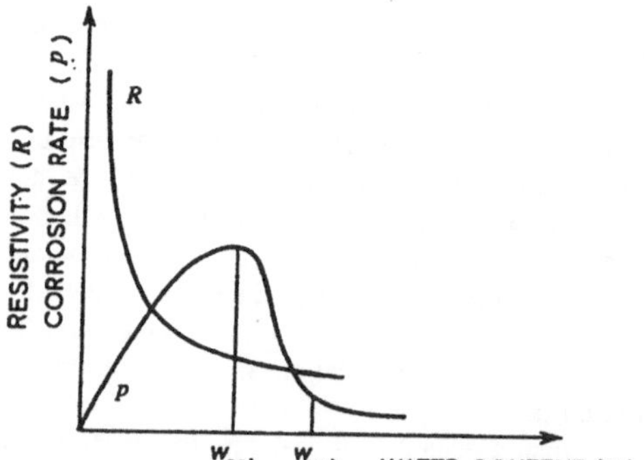

Figure 3. Resistivity, *R*, and corrosion rate, *p*, versus the water content in soil. W_{opt} = optimal water content, W_{sat} = saturated water content (Wranglen, 1985).

(4) Temperature

It has been observed that a 10°C rise in temperature doubles the corrosion rate of carbon steel (NACE, 1984). The increase in the corrosion rate caused by temperature reaches its maximum at about 60-80°C, see Figure 4.

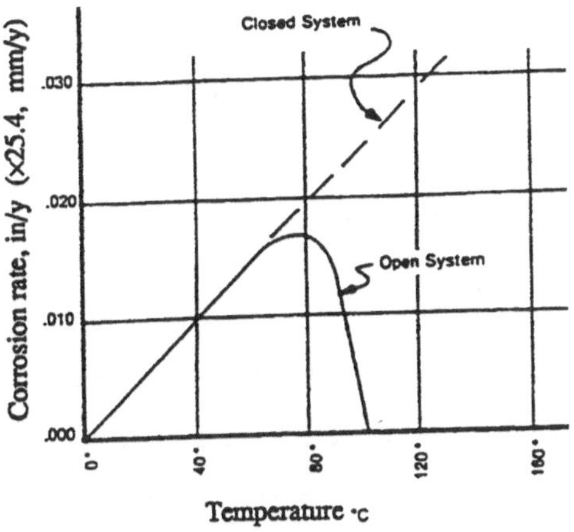

Figure 4. Effect of temperature on corrosion of steel (NACE, 1984).

(5) Rock Mass Quality

The rock mass quality does not influence the corrosion of steels directly, but indirectly via open joints and fractures in the rock mass. The open joints and fractures provide channels for flowing water. The rock mass quality should be taken into account in the corrosivity classification.

(6) Precipitation on rock bolts

The risk of corrosion is also dependent on the possibility of precipitation. Precipitation acts as a protective layer on the metal surface. At pH values higherthan the so-called saturation pH, the water can deposit calcium carbonate. Conversely, at lower pH values the water does not possess this characteristic and will dissolve any existing calcium carbonate precipitate.

2.2 *Corrosivity classification*

Six parameters are used in the corrosivity classification in wet rock conditions. They are the pH value, the concentration of dissolved oxygen, the resistivity of the water solution, the ambient temperature, the rock mass quality and the precipitation of solid substances like calcium carbonate on the rock bolts. The first three are the basic parameters for the classification. The other three are called the influencing parameters. The rule for the classification is that a class number is given to the environment in question based on the total rating, calculated using the ratings of the basic parameters and the two influencing parameters, the temperature and the rock mass quality. The class number is then adjusted for precipitation.

The three basic parameters are rated in accordance with their influence on corrosion. The total rating, W_{wet}, is obtained through the following calculation:

$$W_{wet} = (N_{pH} + N_{O2} + N_R)\, K_t\, K_R \qquad (1)$$

where

W_{wet} = total rating for corrosivity in wet rock conditions,
N_{pH} = rating for the pH value,
N_{O2} = rating for the dissolved oxygen,
N_R = rating for the resistivity,
K_t = factor for the ambient temperature,
K_R = factor for the rock mass quality.

The ratings of the basic parameters are listed in Table 1. The total rating, W_{wet}, and the corresponding class numbers are summarised in Table 2. The class number needs to be adjusted according to the rule described in Table 2 when precipitation occurs on the rock bolts.

The upper limit for the temperature T in calculating the temperature factor K_t, see Table 1, is set at 60°C, since corrosion would decrease when the temperature is above 60-80°C, see Figure 4. The reference temperature in the equation of K_t is 20°C. This means that K_t is equal to 1 at T = 20°C. The factor K_t is increased or decreased by a multiplication factor of 2 with every 10°C rise or drop in temperature.

Bieniawski's Rock Mass Rating (RMR) is used here to classify the rock mass quality. Not all parameters in the RMR system are equally important for the corrosivity of the environment. Thus, besides RMR, we list only three parameters of the rock mass which are thought to be most closely related to the corrosivity of the environment in Table 1. The three parameters are RQD, spacing of discontinuities and water inflow. When the factor for rock mass quality is considered in the corrosivity classification, either RMR or any single one of the above three parameters can be used to determine the rock quality factor K_R. In Table 1, the rock quality factor K_R = 1 corresponds to a rock mass with very good quality (RMR = I), implying that RMR = I is the reference rock quality for the rating of corrosivity. In a rock mass with very poor quality (RMR = V), there is a great potential risk that the corrosivity of the environment will be increased due to the water flow through discontinuities in the rock mass. The rock quality

Table 1. Rating system for the parameters for wet underground rock conditions.

	Parameter	*Ranges of the Value*				
		Basic parameters				
1	pH value:	>11	11-9	9-7	7-5	<5
	Rating, N_{pH}:	*0*	*1*	*2*	*3*	*4*
2	Oxygen (ppm):	<1	1-2	2-4	4-7	>7
	Rating, N_{O2}:	*1*	*2*	*3*	*4*	*5*
3	Resistivity (Ω-cm):	>10 000	10 000-2 000	2 000-1 000	1 000-500	<500
	Rating, N_R:	*0*	*0.25*	*0.5*	*1*	*2*
		Influencing parameters				
4	Temperature Factor, K_t:	$K_t = 2^{\left(\frac{T-20}{10}\right)}$,		(T= 0-60°C)		
5	RMR:	I	II	III	IV	V
	Rock mass quality:	Very good	Good	Fair	Poor	Very poor
	RQD	100-95	95-75	75-50	50-25	<25
	Spacing of rock joints	>2 m	2-0.6	0.6-0.2	0.2-0.06	<0.06
	Water inflow/10 m tunnel length, (*l*/min):	<1	1-10	10-25	25-125	>125
	Factor, K_R:	*1*	*1.25*	*1.5*	*1.75*	2

The Redox potential is considered an appropriate parameter to replace the parameters of the pH value and oxygen. It has the following descriptive relation with the corrosivity of soils:

Redox potential (mV):	>400	400-200	200-100	<100
Description:	No corrosion	Little	Moderate	Severe

Table 2. System I: total rating (W_{wet}) and the corresponding class number of corrosivity.

Total Rating, W_{wet}:	2	2-4	4-7	7-10	>10
Class Number:	I	II	III	IV	V
	* Adjustment of Class Number for precipitation: Class Number drops one when precipitation occurs.				
In descriptive terms:	No or very little corrosion	Little corrosion	Moderate corrosion	Severe corrosion	Very severe corrosion
Reference corrosion rate of carbon steel, (mm/y):	<0.05	0.05-0.10	0.10-0.15	0.15-0.30	>0.30

Table 3. Mean values of the water and rock property parameters measured in the field.

No.	pH	Redox Potential (mV)	Conductivity (S/cm)	Water flow (*l*/min/10m)	Joint spacing (m)	Temperature (°C)
GT	9.6	106	27	0.8		6
KM1	7.9	127	116	5.3	0.2 - 0.6	10
KM3	8.1	146	134	40	0.06 - 0.2	10

GT = Glödberg Tunnel, KM1 = Kiruna Mine, at 795 m level, KM3 = Kiruna Mine, section B at 1080 m level.

factor K_R is designated as 2 in this case, meaning that the corrosivity of the environment is doubled due to the poor quality of the rock mass. For the other classes of RMR, the factor K_R takes a value between 1 and 2.

2.3 *Examples*

The rock masses in a road tunnel at Glödberg and at two mining levels in the Kiruna Mine, Sweden, were investigated to determine their corrosivity. The properties of the water and the rock masses measured in the field are summarised in Table 3.

Using Table 1 and equation (1), we obtain the rating of individual parameters, the total rating and the corrosivity class numbers as follows:

	GT	KM1	KM3
Oxygen rating, N_{O2}:	3	3	3
pH rating, N_{pH}:	1	2	2
Resistivity rating, N_R:	0.25	0.25	0.25
Rock quality factor, K_R:	1	1.25	1.75
Temperature factor, K_t:	0.38	0.5	0.5
Total Rating, W_{wet} =	**1.62**	**3.28**	**4.59**
Class Number: (no precipitation)	**I**	**II**	**III**
Class Number: (with precipitation)	**I**	**I**	**II**

The oxygen concentration was not measured in the field. It is assumed in the above evaluation that the concentration of oxygen is between 2 and 4 ppm, resulting in N_{O2}= 3 according to Table 1.

Without precipitation, the Kiruna Mine should be ranked as class II at 795 m level, but as class III in section B at the 1080 m level. With precipitation the class number is reduced to I and II, respectively.

3 SYSTEM II - CLASSIFICATION IN DRY ROCK CONDITIONS

3.1 *Corrosion-related parameters*

(1) Sulphur and nitrogen oxides and chloride

The corrosion rate of carbon steel in atmospheric conditions depends on the interaction of a number of parameters, but the concentration of sulphur and nitrogen oxides and chloride in the air is the most important one. As shown in Figure 5, the corrosion rate increases with the sulphur content.

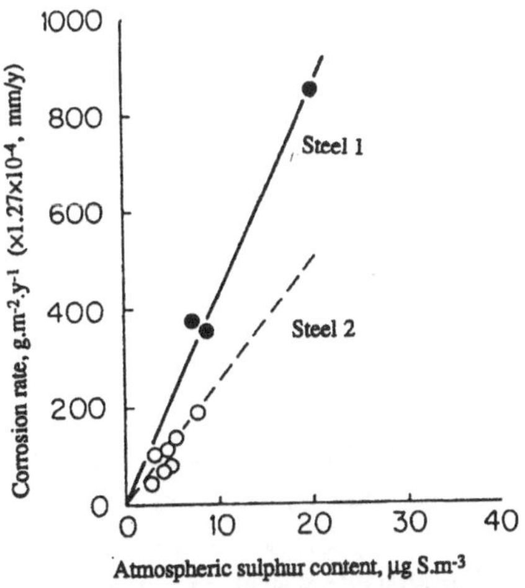

Figure 5. The relationship between the sulphur content in the atmosphere and the corrosion rate of steel (Barton, 1976).

(2) Relative humidity

The relative humidity (r.h.) is defined as the percentage ratio of the water vapour pressure in the atmosphere compared with that which would saturate the atmosphere at the same temperature. It is known that there is a critical humidity in relation to atmospheric corrosion. Below the critical humidity, the corrosion rate is generally negligible, but above it the corrosion increases noticeably with increasing relative humidity, see Figure 6. The critical humidity depends on the metal as well as any surface pollution. It has been observed that the outdoor atmospheric corrosion of practical importance takes place only at a relative humidity higher than 80%.

In the case of iron and steel it appears that there may even be a tertiary critical humidity (Shreir, 1976). Thus, at about 60% r.h. rusting commences at a very slow rate (primary value); at 75-80% r.h. there is a sharp increase in the corrosion rate, probably attributable to capillary condensation of moisture within the rust. At 90% r.h. there is a further increase in the rusting rate corresponding to the vapour pressure of the saturated ferrous sulphate solution, ferrous sulphate being identifiable in rust as crystalline agglomerates.

(3) Temperature

The temperature also influences atmospheric corro-

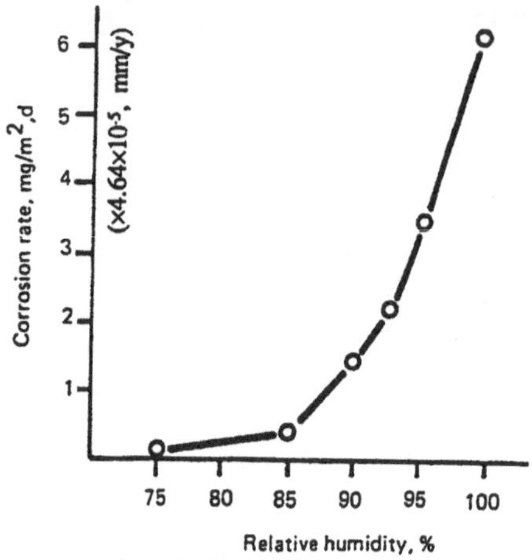

Figure 6. Corrosion rate of steel, preexposed for a month in an urban atmosphere, when exposed in the laboratory at different relative humidities (Mansfeld, 1987).

sion. On the one hand a rise in temperature will increase the corrosion rate, but, on the other hand, the duration of the wetness will be decreased, thus counteracting corrosion. The combination of the two reactions results in the atmospheric corrosion rate reaching its maximum at a certain temperature. Below the freezing point of water the corrosion rate

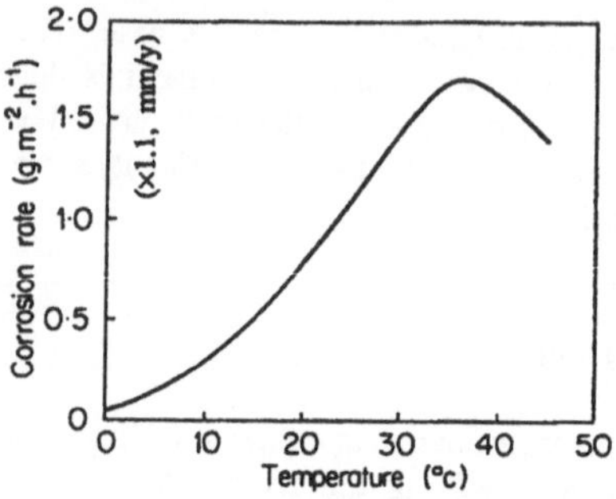

Figure 7. The temperature dependence of the rate of atmospheric corrosion of iron (Barton, 1976).

is generally negligible. Figure 7 shows an example of the influence of the air temperature on atmospheric corrosion.

3.2 *Corrosivity classification*

The three parameters used in the classification system in dry rock conditions are the deposition rate of sulphur and nitrogen oxides and chloride (SO_2 + NO_X + Cl^-), the relative humidity of the atmosphere, and the ambient temperature. The first two are the basic parameters for the corrosivity classification, and the third one, the temperature, is

Table 4. Rating system for the parameters for dry underground rock conditions.

	Parameters	*Ranges of the Value*			
1	$SO_2 + NO_x + NaCl$:	rural atm.	urban atm.	industrial and marine atm.	
	Concentration, (g/m³):	<100	100-200	200-350	>350
	Deposition rate, (mg/m²/day):	<20	20-100	100-200	>200
	Rating , N_{Ox}:	*1*	2	3	*4*
2	Relative humidity (%):	<60	60-75	75-100	dew
	Rating, N_{rh}:	*0*	*1*	2	*3*
3	Temperature Factor, K_t:	$K_t = 2^{\left(\frac{T-20}{10}\right)}$,	(T= 0 - 40°C)		

Table 5. System II: total rating (W_{dry}) and the corresponding class number of corrosivity.

Total Rating, W_{dry}:	0-6	6-10	>10
Class number:	I	II	III
In descriptive terms:	No or very little corrosion	Little corrosion	Moderate to severe corrosion
Reference corrosion rate of carbon steel, (mm/y):	<0.05	0.05-0.10	>0.10

an influencing parameter. The total rating, W_{dry}, is obtained by multiplication of the ratings of the two basic parameters and the temperature factor K_t:

$$W_{dry} = N_{Ox} N_{rh} K_t \quad (2)$$

where

W_{dry} = total rating for corrosion in dry rock conditions,

N_{Ox} = rating for sulphur and nitrogen oxides and chloride,

N_{rh} = rating for relative humidity,

K_t = factor for ambient temperature.

The ratings of the individual parameters are listed in Table 4. The total rating, calculated by equation (2), and the corresponding class numbers are summarised in Table 5.

The temperature factor K_t is calculated in the same way as in classification system I for wet rock conditions, but its valid range is from 0 to 40°C, see Table 4.

4 CONCLUDING REMARKS

Two corrosivity classification systems are proposed for the underground environment with relation to the corrosion of rock bolts. System I is for wet rock conditions, while system II is for dry rock conditions. The selection of the corrosion-related parameters for corrosivity classification was based on the knowledge of the corrosion of carbon steel reported in the literature. The systems must be verified, and even modified, through a number of case studies. The verification work will be conducted soon through examining the corrosion of rock bolts seated in rock masses for a number of years.

Acknowledgements –The authors greatly appreciate the financial support provided for the research work by Atlas Copco Rock Drills AB, as well as the company's permission to publish the results. Thanks are extended to Dr. Sten Davidsson, Mr. Daniel Krzysztalowicz and Prof. Bengt Stillborg for their constructive comments and discussions during the process of the work.

REFERENCES

Barton, K. (1976). *Protection Against Atmospheric Corrosion.* John Wiley & Sons, London.

Mansfeld, F. (1987). *Corrosion Mechanisms.* Marcel Dekker, Inc., New York.

NACE (1984). *Corrosion Basics.* National Association of Corrosion Engineers, Texas.

Shreir, L.L. (1976). *Corrosion Vol.1: Metal / Environment Reactions.* Newnes-Butterworths, London.

Wranglen, G. (1985). *An Introduction to Corrosion and Protection of Metals.* Chapman and Hall, London.

Rock Support and Reinforcement Practice in Mining, Villaescusa, Windsor & Thompson (eds)
© 1999 Balkema, Rotterdam, ISBN 90 5809 045 0

A study of corrosion in underground reinforcement at Mount Isa Mines, Australia

J. Robinson & D. B. Tyler
Mount Isa Mines Limited, Qld, Australia

ABSTRACT: Since 1963 the Mining Research rock mechanics group at Mount Isa Mines has collected and analysed a large database of rock properties, ground behaviour, fill performance and blast vibrations. This data has been used to develop predictive mine design tools. In 1995 a program of mine water sampling, pull testing and corrosion observation was undertaken. The aim of this study was to identify regions where corrosion may impact on mining activities and to determine suitable forms of corrosion resistant ground support.

1 INTRODUCTION

The Mount Isa Mining field lies in North West Queensland approximately 900km due West of Townsville and produces 8mT of copper and silver-lead-zinc ore per annum. At present, four mines are in operation on the mining lease with the fifth (George Fisher) under development.

Different ground support systems have been developed for use across the lease with both temporary and permanent reinforcement systems to suit the local requirements. Permanent ground reinforcement is currently achieved by the use of full column, cement-grouted dowels and cablebolts (if required).

In the presence of aggressive water, black steel will corrode. In addition, the strength and life expectancy of the cement grout can also be compromised. Corrosion of a reinforcement system reduces its overall effectiveness and creates a number of operational difficulties:

- safety hazards to employees
- damage to equipment
- impact on development and production economics, and
- scheduling problems incurred by rehabilitation works.

At Isa Lead and Enterprise mines, drainage water is re-circulated and re-used by the backfill plant and underground machinery. The re-use of mine water can result in concentration of contaminants in the system. These contaminants can reduce the strength of cement grouts, retard cement grout curing and enhance the corrosion rate of rock reinforcement.

Several corrosion studies have been performed at Mount Isa between 1995 and 1998 (Murie 1995, 1996a & b, Krebb 1997 a, b, c, d & e, Steven & Armstrong 1998). The major objectives were to locate high corrosion risk regions by water sampling and to identify corrosion resistant reinforcement systems suitable to MIM ground conditions.

2 CORROSION OF REINFORCEMENT SYSTEMS

Water encountered in underground mining comes from many different sources. Some of these include natural ground water, development and production drilling, drainage of wet fill and damaged water pipes. Mining of sulphides also by nature can contribute to corrosive ground water.

Measuring corrosion and locating high risk areas throughout one of the world's largest underground mining complexes is not a simple task. The original focus of the studies was to determine the major causes of corrosion and how they influence the rate of ground support deterioration. It was found that the degree of corrosion is dependent on a number of factors which include:

- presence of water and its chemical make-up
- rock mass quality
- type and properties of the ground reinforcement system
- age and history of the excavation/ground reinforcement, and
- type of fill material.

The corrosion of ground reinforcement is an important issue for underground workings. Corrosion failure may occur due to several conditions:

2.1 *Chemical / temperature corrosion mechanisms*

- Corrosion of mild steel in neutral water is due to the presence of oxygen. The corrosion rate is highly dependent on the rate of dissolved oxygen diffusion to the surface of the mild steel. Mild steel can also corrode in acidic conditions (pH < 5) without the presence of dissolved oxygen.
- The corrosion rate of mild steel rock bolts will not change in the pH range 5 to 9.5. Note however, that at pH levels < 7, the rate of zinc corrosion increases and at pH levels < 5 the use of galvanised rockbolts is not recommended.
- An increase in temperature of water will increase the corrosion rate. At 47°C the corrosion rate of mild steel and galvanised rockbolts is approximately double that compared to water at room temperature (22°C). Galvanised rockbolts can not be used at temperatures >55°C as zinc becomes cathodic to mild steel.
- The highly aggressive nature of mine waters comes in the main form of dissolved ionic species. Minik (1987) observed salinity and sulphite ions as being the most aggressive. The presence of chloride ions is generally regarded as the main source of pitting in steel. Combine this with oxidising salts (Fe and Cu) and high temperatures, the resulting water becomes highly aggressive (corrosion rate ~ 1mm yr^{-1} m^{-2}).. Dissolved salts effect the corrosion rate by:
    1. increasing the conductivity of the solution allowing anodic and cathodic sites to operate at a distance, and
    2. precipitating a less protective layer
- Due to the high temperatures underground at Mount Isa there is a low concentration of dissolved oxygen (4.5 ppm) in mine water sampled. Reduction reactions of Cu and Fe cause iron on the surface of rockbolts to react ($Fe \Rightarrow Fe^{2+}$) with little or no oxygen present.

2.2 *Physical corrosion mechanisms*

In addition to chemical/temperature constrains, several physical mechanisms will increase the rate of corrosion:

- Stress Corrosion Cracking (SCC) may increase the rate of corrosion induced rock reinforcement failure. SCC occurs when metal under stress is in contact with an ionic medium and suffers brittle cracking. This is due to stresses within the metal increasing the thermodynamic energy of the atomic bonds, creating high energy areas, which become anodic sites for preferential accelerated corrosion attack.
- Poor grouting of rockbolts can result in preferential corrosion. Water can percolate into gaps in the grout column and set up localised concentration cells that can rapidly pit mild steel. Pitting cells are autocatalytic and can corrode mild steel at the same rate as acid.
- Grout is susceptible to cracking if it is exposed to sulphate ions, which can penetrate its porous concrete structure. Sulphate ions can form crystals in the pores of the grout column, which when dry can increase their size by upto 227%. The resulting stresses can crack the grout column allowing preferential corrosion paths.

2.3 *Mineralogical effects*

Precipitated minerals associated with rockbolt corrosion and mine water were analysed to determine the nature of the minerals. The aim of this was to:

1. estimate the corrosiveness of mine water, and
2. determine if corrosion has taken place

The presence of corrosive metal salts were identified underground by observing the colour of the minerals precipitated:

- White coloured salts - Halite (NaCl) and Gypsum ($CaSO_4 H_2O$). These two salts produce solutions with similar corrosion rates of around 0.1mm yr^{-1}.
- White coloured salts with blue and green staining. This is due to blue copper (II) salts and green copper (I) salts. The different colours indicate oxidation and reduction occurring. The presence of these salts greatly increases the corrosion rate to >1mm yr^{-1}.
- Brown / orange colouration is attributed to the presence of iron in the precipitated salts. They are explained by the presence of hydrated iron oxides and iron hydroxides which are typical corrosion products (i.e. rust).

2.4 *Examination of corroded rockbolt samples*

Corroded rockbolt plates and shanks have been studied from across the lease. In general old bolts from X41 and Isa Lead Mine show complete oxidation of plates and shanks. Newer bolts from Enterprise Mine tend to display localised corrosion (pitting). The orange areas observed on bolts and plates are iron hydroxides and oxides (corrosion products), from anodic sites of the corrosion reaction ($Fe \Leftrightarrow Fe^{2+} + 2e^-$). Purple areas are the cathodic sites. Generally at cathodic sites none of the mild steel is rusted and there is a noticeable build up of salt scale. At Enterprise Mine it is thought that the rock and cablebolt plates may be acting as cathodes with respect to the bolt / cable (anode) depending on the extent of water contact.

2.5 *Interpretation*

Sulphide minerals such as pyrite and chalcopyrite react readily with water in hot temperatures, oxidising and producing a more acidic solution. The oxidation process occurs when water has high concentrations of dissolved oxygen. If the pH of the water decreases to below 5, corrosion of black steel can still happen even if the dissolved oxygen content is minimal. In areas of low pH, preferential corrosion can occur even if galvanised steel is used. If the pH is less than five, galvanised ground support should be avoided.

3 WATER SAMPLING PROGRAM

In total 175 underground water samples have been collected at the four mines since 1995. Murie (1996 b) collected 23 samples, Krebb (1997e) collected 39 samples, Steven and Armstrong (1998) collected 113 samples. All of these samples have been collected from an number of locations in each mine. Although many samples were collected from each of the four mines, the database does not yet adequately cover the whole operation. The samples were analysed for key corrosive indicators and then classified according to potential corrosiveness. Care was taken not to sample drain holes, as contamination from oil and diesel was likely. Results of pH, copper, iron, chloride ions and sulphate ions were obtained.

During collection, a pH-mV-Temp meter was used to determine the acidity, conductivity and temperature of the sample. On the days they were collected, water samples were sent to the water laboratory for testing. A database was established including all results and the sample location. A simple method was devised by Steven and Armstrong (1998) to group the samples according to their level of corrosiveness based on the guidelines outlined below in Table 1. Level 1 being the most corrosive.

Table 1 Guidelines for corrosion Level.

		Level
1	pH < 5	1
	pH > 10	5
2	Fe or Cu > 50 ppm	1
3	Cl/SO_4 < 4000ppm	5
	4000 < Cl/SO_4 > 5500	4
	5500 < Cl/SO_4 > 10000	3
	10000 < Cl/SO_4 > 25000	2
	Cl/SO_4 > 25000	1
4	Fe or Cu > 1ppm then subtract one from the Level number	

A representative sample from each of the five corrosive levels was sent to be analysed for its mild steel corrosion rate using a PC3 Gamry Potentiostat. The corrosion rate of mild steel in a 3.5% saline solution is 0.164mm yr^{-1} m^{-2} (i.e. corrosion of a 20mm diameter mild steel dowel in 699 days) One sample from each corrosion level was tested using a PC3 Gamry Potentiostat to determine their relative corrosion rates. Results from these samples indicated that Level 1 and Level 5 will corrode a 20mm diameter mild steel dowel in 244 and 377 days respectively.

At Isa all stopes are filled to provide regional support. Wet filling of stopes introduces a significant amount of water into the mine. Grice, 1986 measured drainage at the base of Isa lead Mine open stopes. He found that 76% of the total fill water reported to the base of the stope. Typical wetting zones around filled open stopes are minimal (between 5m and 10m) in the upper levels of the stope. At the drawpoints, however, the wetting zone often extends upto 20m from the stope. A total of 10 samples were collected around stopes which were either currently being filled or had been filled. The pH and ion content of fill water are highly dependent on the type of fill being run. Generally the pH of cemented fill mass water lies anywhere between 7 and 12, but will not have a pH lower than 7 (due to retardation of cement in acidic conditions). Chemical analysis indicated all samples were in the Level 4 to 5 range, indicating a low corrosive environment. However, although waters found in wetting zones around fill masses have relatively low corrosion levels, the volume of water produced can lead to significant corrosion problems.

3.1 *Interpretation*

Enterprise Mine is relatively dry, and only 10 ground water samples were collected. All but one of the 10 samples were classified as Level 1. This was predominantly due to the high Cu and Cl ion concentrations in the water.

Dissolved salts increase the conductivity of ground water and increase the corrosion of ground reinforcement by reacting with the metal itself. Chloride and sulphate ions are the most aggressive ions. The combination of high Cl^- ions and high water temperatures, as exhibited in Enterprise Mine samples creates a very agressive solution. The main source of the chloride is from the basement Greenstones, a chloritised basalt. Sodium and calcium ions, to a lesser extent, also contribute to the corrosion process by pitting of metal surfaces in low dissolved oxygen samples. The other two ions present which indicated and influence high corrosion rates respectively are those of iron and copper. The main ion concentration indicating high corrosive levels are chlorides, sulphates, iron and copper.

Compared to Enterprise Mine, the ground water at X41 Mine is less aggressive. Most samples collected were classified at corrosion Levels 4 or 5. Water collected from fill bulkheads at X41 was classified as corrosion Level 3. It is thought that the aggressive nature of fill water is highly influenced by the material it permeates through. Leaving sulphide material at the base of the stope, or using the stope as a dump prior to filling will introduce additional ion species.

At Isa Lead Mine variable corrosion levels were recorded for all waters samples collected throught the mine. Near surface samples, (above 7 level) were recorded at corrosion levels 2 to 3, due to the high level of dissolved ions.

At Hilton Mine the majority of the sampled water were at Levels 4 and 5. The most agressive waters sampled were located at the plat (on 14 level), where corrosion Levels 2 and 3 were recorded (due to high levels of disolved salts). In addition corrosion levels of 2 and 3 were recorded in the footwall drives and down dip from the old wet filled stopes.

The main objective of the water sampling was to produce level plans showing high corrosion rate areas to aid in rehabilitation and design of reinforcement systems in these areas. Although many samples were taken over each mine, coverage did not give satisfactory results. Additional sampling will be initiated in December 1998 to increase coverage.

4 CORROSION RESISTANT GROUND REINFORCEMENT

Generally, all steel rock reinforcing systems will rust. The most important points to consider are the rate of corrosion and how it may be controlled to provide the design life required (Baxter 1997).

Conclusions drawn by Sundholm and Forsen (1995) were that cement grout acts as a protective layer for rebar if the grout column is intact. The protective properties were reduced if there were defects within the grout column, i.e. cracks and / or poor encapsulation etc.

4.1 *Types of system investigated*

Investigations to identify corrosion resistant rockbolts, materials, and reinforcement systems for use in wet aggressive areas of the mine were undertaken. Initial costing studies at Mount Isa on different systems (O'Hare 1994) suggested that hot-dip galvanised cables should be used as a resistant alternative to standard black steel cablebolts. The cost increase for a 6m twin strand system was calculated at 101%.

The use of hot dip galvanised reinforcement was dropped following the ground water sampling program due to the aggressive nature of the water in the Deep Copper mine.

Additional laboratory strength testing work was carried out by Steven and Armstrong (1998) on the following different systems:

- Flowline® 302 (plastic) coated rockbolts
- Flowline® 302 coated single strand Garford bulb cablebolts, and
- CT bolts.

The use of Ferritic stainless steel (3CR12) dowels and / or Duplex stainless steel (SAF 2205) dowels were not considered due to their high cost (approximately $80/unit). Three types of protected ground support were tested through a series of standard laboratory pull tests. The bolts included the Ingersol Rand, CT Bolt the ANI Arnall deformed rebar and single strand Garford bulbed cable bolt. The CT bolt is post grouted, which is done through a special spherical washer with the grout flowing between the deformed bar and the plastic sheath to the toe of the hole. Grout is then returned down between the sheath and the drill hole, escaping through the face plate, ensuring complete encapsulation. The standard deformed bar and single strand Garford bulb cable bolt were coated with a chemically stable epoxy resin by Flowline Pty Ltd. The epoxy coating was applied directly to the steel. There were two types of epoxy coatings applied to the deformed bar. One was much smoother than the other, hereafter referred to as smooth plastic coated.

4.2 *Surface Testing - pull loads*

All three rock bolts were subjected to a series of embedment length pull tests in the laboratory. The bolts were grouted in steel pipes with a 64mm internal diameter and 4mm wall thickness. Pipes were plugged to prevent water bleed and the bolts were centralised during grout curing. All samples were pulled at 892N/s in a Avery Universal testing machine. The tests were conducted over 0.5 metre and 1.0 metre embedment lengths. The epoxy coated cable bolts were installed so that one bulb was positioned 100mm from the toe of the hole for the 0.5 metre embedment test, while two bulbs were located at 100mm and 750mm from the toe respectively for the 1.0 metre embedment tests. The pull tests were conducted with a water to cement ratio of 0.35 (ie 7 litres of water for each 20kg of Portland 10 cement). For the cablebolt tests a weaker 0.40 water cement ratio was used. Grout was mixed in a Minepro Cam Grout Pump using the toe-to-collar method of grouting. Grout curing times were 6, 8, 10, 12 and 24 hours.

After the pull tests were conducted, each sample was sectioned through the centre of the pipe to observe the bond between the element and the grout.

4.3 *Results*

Table 2 indicated the pull loads achieved by the Flowline® coated rockbolt.

Table 2 Pull load results in tonnes for Flowline® treated rockbolts.

Curing time, hrs	0.5m*	1.0m*	1.0m#
8	9.4 (7.7)	12.7	6.2
12	10.5	12.7	11.9
24	11.3	11.8	12.4

* Flowline treated rockbolt. # black steel rockbolt. Value in brackets is for smooth plastic coated Flowline® treated bolt.

When the samples were cut in half, two observations were made. Firstly, the smooth plastic coating appeared to have an insufficient bond between the bolt and the grout. This was evident by the striations in the grout where the bolt had slipped, but with the plastic coating remaining intact (see Figure 1).

Figure 1 Striation observed in grout during bolt slipping.

Secondly, following failure, the coarse plastic coating exhibitted brittle cracking radially along the bolt. Although acceptable loads were obtained with the coarse coating, the radial cracks could result in rapid, localised corrosion if water came into contact with the bolt.

Table 3 below indicates the results for the Flowline treated Garford bulbed cablebolts. During application of the epoxy on the cable bolts, the amount of void at the bulb was markedly reduced. It was felt that full grout penetration into the bulbs would be difficult and a 0.4 WC ratio grout was used for these tests.

Table 3 Pull load results in tonnes for Flowline® treated Garford cablebolts.

Curing time, hrs	0.5m	1.0m
8	6.8	16.7
10	7.2	14.9
12	8.0	14.9
24	6.4	15.8

Measured pull loads varied considerably for each embedment length tested. After sectioning of samples it was evident that the bulbs had been pulled straight and the epoxy coating cracked off (see Figure 2). This was due to minimal grout penetration of the bulb.

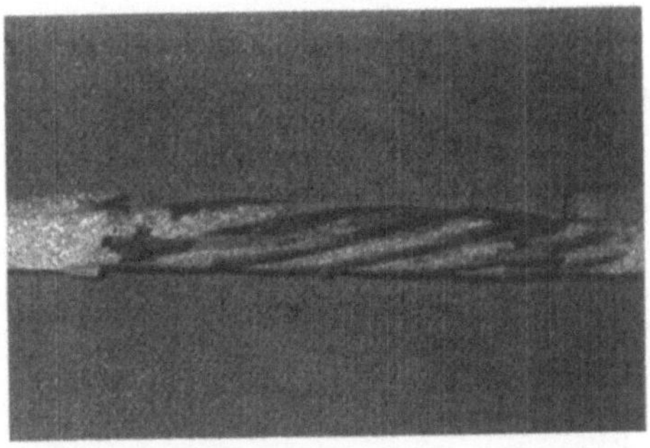

Figure 2 Crushing and stripping of a single strand Flowline® treated Garford bulb cablebolt.

The final series of tests were on CT bolts. The results of this work is presented below in Table 4.

Table 4 Pull load results for for CT bolts.

Curing time, hrs	0.5m*	1.0m*	1.0m#
8	10	15.9	6.2
12	13.3	22.7	11.9
24	18.6	23.5	12.4

* CT bolt. # black steel rockbolt

After sectioning of the samples a good bond between both the grout and bolt, and the grout and polyetherane coating was observed. The polyetherane sheath was much thicker than the Flowline® plastic coating and it appeared less brittle. No cracking was observed in the CT bolts polyetherane sheath following sectioning. It is considered that this sheath would act as a good inhibitor to reduce the risk of aggressive water reaching the rockbolt shank.

5 CONCLUSIONS

1. In pillar recovery operations, the corrosion of rock reinforcement is a major problem. Attempts have been made to measure and locate regions of aggressive mine waters at MIM. Corrosion rates have been determined along with their local mechanisms.
2. The high corrosion rates determined for Enterprise Mine samples are related to both high ambient temperatures and Cu and Cl ion concentrations.
3. At X41 Mine the greatest source of aggressive mine water is from cemented fill masses (or unconsolidated material left in the stope).

4. At both Isa Lead and Hilton Mines high corrosion rates were associated with old wet filled stopes.
5. Corrosion underground at the Isa lease is predominantly influenced by
   - chloride and sulphate ions
   - dissolved oxygen
   - temperature and pH
6. Testing of grout additives along with different protective coatings on rock and cablebolts gave mixed results.
7. The use of galvanised reinforcement elements is not practicable in Enterprise Mine, due to high water temperatures in combination with low pH's and high Cl ion levels.
8. The use of Ferritic or Duplex Stainless steel is not economically viable (approximately $80 per element).
9. Flowline® coated rockbolts appear to achieve equal of better loads than black steel during pull tests (see Table 2). However, radial cracking observed in the plastic coating may result in locallised corrosion of the bolt.
10. Flowline® coated garford bulbed cablebolts failed to achieve the required pull loads.
11. The best economically available corrosion resistant ground reinforcement element tested during the project was the CT bolt. At present no deep reinforcement element (ie bulbed cable bolts) has shown significant corrosion resistance.

ACKNOWLEDGMENTS

The authors would like to thank the management at Mount Isa Mines for permission to prepare this paper and their colleagues for their assistance in its preparation.

REFERENCES

Murie, SA 1995. pH and Ion species analysis for Isa and Hilton. MIM Ltd internal memorandum (25/4/95).

Murie, SA 1996a. Ground support - environmental monitoring. MIM Ltd internal memorandum (14/3/96).

Murie, SA 1996b. Environmental data for ground support testwork. MIM internal memorandum (29/4/96).

Krebbs, D 1997a, b, c, d & e. U/G corrosion project reports. MIM Ltd internal reports.

Steven, A & Armstrong, B 1998. MIM Ltd internal report - Corrosion Project.

Minik, GA et al 1987. Corrosion in the mineral industry. ASM International.

PC3 Potentiostat / Galvanostat / ZRA by Gamry Instruments Inc, 734 Lousi Drive, Warminster, PA 18974. USA.

Grice, AG 1986. MIM Technical Report, number Min Res 64. N673 stope fill project - fill and drainage report.

O'Hare, S 1994. MIM internal filenote. Project progress report - corrosion of cablebolts in laboratory pull tests II.

Flowline 302. Acid and corrosion resistant lining. Distributed by Flowline (Pty) Ltd, PO Box 68545 Bryanston 202. South Africa.

Baxter D A 1997. Rockbolt corrosion under scrutiny. Tunnels & Tunnelling International, July 1997.

Sundhold, S and Forsen, O 1995. Corrosion tests on cement grouted rockbolts. Bergmekeukdagen, Stockholm, March 1995.

Rock Support and Reinforcement Practice in Mining, Villaescusa, Windsor & Thompson (eds)
© 1999 Balkema, Rotterdam, ISBN 90 5809 045 0

An Australian perspective to grouting for cablebolt reinforcement

E. Villaescusa
Western Australian School of Mines, Kalgoorlie, W.A., Australia

ABSTRACT: This paper describes recent developments in grouting practices for cablebolt reinforcement in underground hard rock mining in Australia. Advantages and disadvantages for conventional and toe to collar grouting techniques are discussed. Equipment, grout mix design, training and quality control measures implemented at a number of mine sites are reviewed. The experimental developments have significantly improved the quality and efficiency of the grouting operations.

1 INTRODUCTION

Grouting is the procedure by which a drillhole in the boundary of an excavation is filled with a cement paste in order to anchor a reinforcement element securely inside a rockmass. This allows load transfer from a potentially unstable section of a rockmass (at an excavation boundary) to a stable section deep within the rockmass via the reinforcing element, as described in the load transfer concept (Windsor and Thompson, 1992). The strength of the grout is critical in order to minimize the length of embedment needed to mobilize the ultimate steel tendon capacity of a particular cablebolt system. In general, failure by slippage at the steel-grout interface may be experienced when weak grouts are used. Alternatively, rupture of the tendons can occur when using thick, strong grouts in conjunction with stiff reinforcement systems.

Hutchinson and Diederichs (1996) undertook a comprehensive review of grouting practices as part of their cablebolting manual. The conclusions reached were based on visits to mines, laboratory test data, theoretical models and feedback from mine site questionaires. This paper provides an Australian perspective to grouting for cablebolt reinforcement. The results are based on practical experience and testing at a number of Australian mine sites. The methodologies and practices described in this paper are somewhat different to those reported by Hutchinson and Diederichs (1996). The equipment, reinforcement systems and working practices used in Australia have produced unique results and conclusions.

2 CONVENTIONAL GROUTING

Traditionally, collar to toe grouting methods that require a breather tube have been used in Australia. These methods were initially developed in conjunction with piston-based grout pumps. The water/cement (w/c) ratio for such grouts can range from 0.40 to 0.55. This method requires a breather tube, generally 13 mm (inside diameter) to be attached to the cablebolt element before it is installed into a hole. A permanent collar packing is placed at the hole collar for the wet grout to be kept inside the hole. A short grouting hose of approximately 1.0m in length is also placed permanently in the collar of the hole. The grout is pumped through the grouting tube, and when it reaches the upper end of the breather tube it begins to flow back through this tube. When grout emerges from

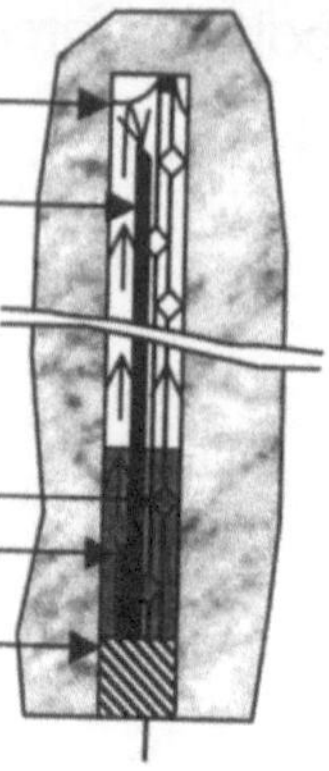

Figure 1. Conventional collar to toe grouting method (not to scale).

the end of the breather tube it indicates that full encapsulation of the steel tendon has been achieved (See Figure 1).

One disadvantage of this method is all the additional work and material costs required to successfully grout the cables. Weak grouts are necessarily utilized with this method. The dimensions of the breather tube require that grouts having a high w/c ratio are used. In addition, leaks from vertical upholes are common, especially in heavily broken rockmasses.

A typical piston-based pumping system usually requires that the mixing and pumping are carried out within the same container. This is called a *one-stage* grouting system. Mixing of the grout is achieved using paddles that rotate around a vertical axis, similar to the rotating mechanism on a typical washing machine (See Figure 2). This may create a settlement of the cement particles into the area where the pumping is taking place, i.e. the bottom of the mixing tank, potentially affecting pumpability if the grout is too thick. Having a single container can lead to changes in water/cement ratios during the grouting operations.

Traditionally, no accurate devices to measure the amount of water being poured into the mixing tank have been fitted to most conventional one-stage grouting systems. Consequently, following an initial mix design in which the water cement ratio was probably correct, additional water is added (while still pumping and grouting) before a corresponding amount of cement is added to the mix. This problem can be avoided if each mix is pumped separately. However this process can be too slow, and may affect productivity. Figure 3 shows a 28-day Uniaxial Compressive Strength result for grout mixes carried out using a one-stage grouting process. The results indicate a significant variation in the compressive strength as a result of varying w/c ratios during conventional grouting operations.

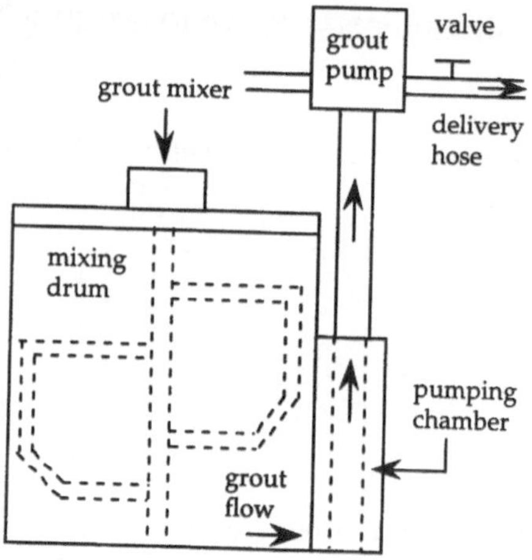

Figure 2. A schematic of a conventional one-stage grouting system.

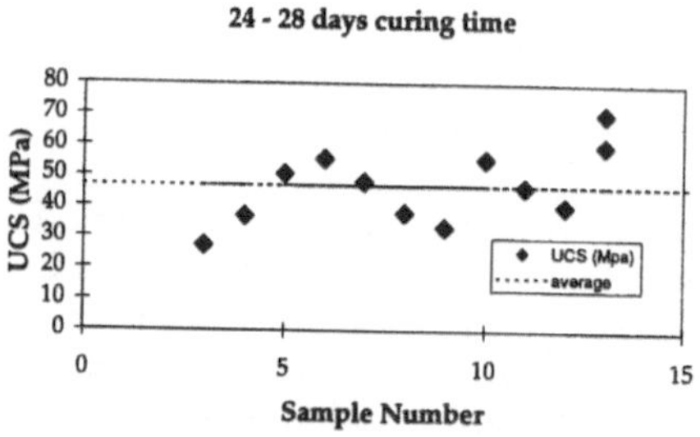

Figure 3. Grout strength variability.

3 ADVANCED GROUTING

Advanced grouting (toe to collar) practices have been successfully developed and implemented at a number of underground operations in Australia since the early 1990's. The technique consists of inserting a cablebolt *without a breather tube* into the hole to be grouted, (the need for the collar plug is also eliminated), followed by a subsequent grouting of the cable by means of a *self-retracting* grouting hose. The grout pushes the grouting hose out of the hole as the grouting process is undertaken. The optimal grouting rate is such that a self-retracting hose should be in minimal contact with an advancing grout paste inside the hole. Underground testing and experience suggest that at anyone time, only 30 cm of hose should be in contact with the grout, in order to minimize voids at the collar end when the hose is pulled out of the hole (See Figure 4). In order to achieve this, the typical water/cement ratios required usually range from 0.30 to 0.32.

This type of grouting is achieved routinely at a number of Australian mines. The method provides many advantages, including a much faster initial cable placing and pre-grouting preparation time (also, saving on materials), faster rate of grouting, and a significant increase in grout strength. The increased strength of the grout may effectively decrease the required embedment length to achieve the nominal steel failure capacity. Strong and thicker grouts do not leak into voids and crevices encountered along the drillhole axis, thereby minimizing wastage of cement.

This technique has been successfully implemented in Australia mainly due to the use of modern grout pumps that allow for a high degree of quality control on the mixing, pumping and water/cement ratios used. A disadvantage of this method is the potential for poor cablebolt encapsulation that may result if the grouting operator retrieves the hose while grouting is being done. Trial grouting using transparent pipes suggests that a minimal pull of the grouting hose may leave some grout gaps along the hole axis. The dimensions of the hole, the cable geometry, the length and diameter of the grout hose, the consistency of the grout and the type of grout pump being used are likely to influence the self-retracting behavior of a grout hose. Sometimes an initial (minimal) pull will start the self-retracting behavior, and the hose should not be pulled until it reaches the final 30-cm at the collar end.

4 SURFACE TESTING PROGRAM

A number of experiments can be conducted at a mine site to select the most appropriate equipment and determine the best grout mix design. A series of controlled experiments using steel pipes are recommended. Several hanging pipes can be grouted to a height of approximately 15 to 20m from the floor (in order to simulate the longest possible cable needed underground), using 0.40, 0.35 and 0.30 water/cement ratio grouts.

The initial test usually consists of pumping

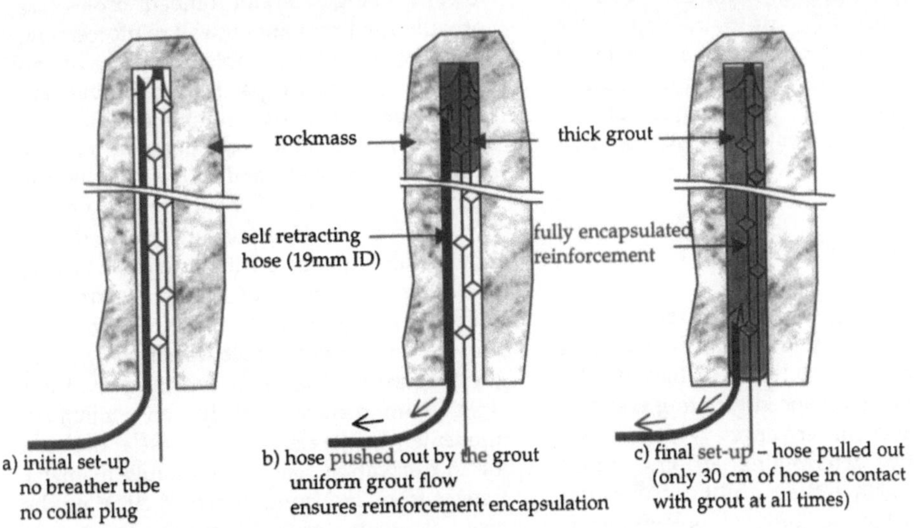

Figure 4. Advanced grouting (toe to collar) method (not to scale).

the grout through a 25mm inside diameter grouting hose that can be raised to the upper portion of the hanging pipes. This simple exercise allows a comparison between piston-based pumps and monopumps. These toe to collar grouting tests can be undertaken using a method that requires the grouting hose to be left permanently inside the hanging pipes. The pump tests are simply stopped when grout reports at the collar of the pipes. Encapsulation of the cables along the pipe axis can be determined by slicing the pipe every 5 cms along its entire length (Stjern, 1995).

Another test can be devised to determine under what conditions the grouting process can be carried out using a self-retracting hose. In order to understand this process, grouting experiments using transparent pipes can be implemented. Grout consistency, pumping rates and plain and modified geometry cables can then be analyzed. Encapsulation and flow of grout into bulbed cables is of particular interest and importance. The use of transparent pipes allows full inspection during grouting and helps to determine if any problems are likely to be experienced. Results from transparent pipes have suggested that the optimum bulb size should range from 29-31mm in diameter in order to ensure the flow of thick grout into the modified cable geometry.

In general, the use of transparent pipes for trial grouting has proven successful, with the additional benefit of training the underground operators who are able to develop a 'feeling' for the grouting process. Figure 5 shows a photograph of an experimental grout trial using transparent pipes.

Figure 5. Experimental set-up used to grout transparent pipes.

5 UNDERGROUND TESTING

In-situ pull testing is usually required to determine the degree of encapsulation and load transfer capabilities achieved by conventional and advanced grouting systems. The advantages of advanced grouting are quantified by determining drilling rates, cablebolt installation time, grout consumption and strength for the different hole-cablebolt configurations available. A description of the different cablebolt systems currently available in Australia is beyond the scope of this paper. However, single strand bulbed cables are normally used for hangingwall reinforcement, while twin bulbed cables that have an ultimate combined capacity of 50 tonnes are used for permanent back reinforcement.

Each individual cable has a nominal capacity of 20 tonnes at yield and an ultimate tensile strength of 25 tonnes. For back reinforcement, two cables per hole are usually installed on a 2x2m drilling pattern, with each cable having a modified geometry of 2 bulbs/metre. The bulbs are staggered with respect to each cable, and the final installed bulb density is 4/metre (See Figure 6). This bulb density provides a stiff reinforcement likely to minimize movement of the reinforced blocks at the back of an excavation. The optimal bulb diameter ranges from 29-31mm, thereby facilitating the use of thick cement grouts (having a

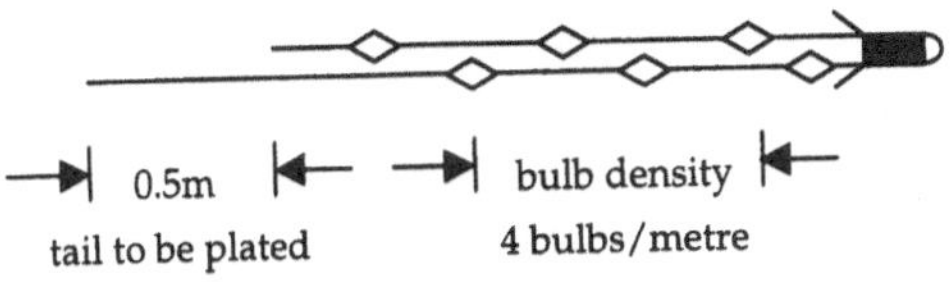

Figure 6. A typical two-strand cable used for back reinforcement in Australia.

Table 1. Relative average drilling time per hole (4.5m hole length).

Drillhole Diameter (mm)	Drilling Time	Rod Handling Time	Total Time
73	1.00	1.00	1.00
64	0.90	0.96	0.91
57	0.76	1.70	0.90

water/cement ratio ranging from 0.3 to 0.32) that can effectively penetrate the bulbs. Underground observations from exposed sections of cables indicate that air pockets are never a problem with two strand bulbed cables that are grouted with thick grouts.

In Australia, the typical hole size where the cables can be properly installed and *post grouted* ranges from 57 to 73mm, depending upon the cable type, i.e. single or double strand. The hole size chosen influences the drilling rates, the ease of cablebolt installation, the ease of grouting, and the grout consumption. In-situ time trials can be used to quantify the difference in drilling times for various drill bit sizes. The typical bits used are 57mm, 64mm and 73mm for typical post-grouting cablebolting operations. Table 1 shows a summary of relative average drilling times for different hole diameters, using the 73mm diameter as a reference (Prentice, 1994). The results suggest a substantial decrease in drilling time with a decrease on drillhole diameter. However, manual handling of R32 type rods (57mm) was considerable higher than the mechanized T 38 type rods (64 and 73mm).

The initial cablebolt installation time is significantly reduced during advanced grouting operations, as virtually no preparation of the cables is required before

Table 2. Relative grout consumption and grouting rate.

Drillhole Diameter (mm)	Grouting Time	Grout Consumption
73	1.00	1.00
64	0.73	0.76
57	0.60	0.59

being inserted into the holes. Table 2 shows the results of grouting time and total grout consumption for a typical range of holes using a typical monopump grouting pump machine. A substantial reduction in both grouting time and grout consumption was determined for a reduced cablebolt hole diameter.

In addition, in-situ pull tests can be undertaken on specially debonded cables that are covered with a plastic sleeve, except for the area to be mobilized by the pull test (Villaescusa, 1999). The effective embedment length is set to one metre at the toe of the cable. Pull tests can then be carried out on a number of hole diameters to determine whether a reduced hole size would adversely affect cablebolt load strength for the same w/c grout ratios. The results given by Prentice (1994) indicate that the pull-out strength for 4.5m long single strand cables (1 bulb per metre) was not influenced by the cablebolt hole diameter. In general, the testing program determined that 57 and 64mm holes are recommended for single and double strand cables (having a modified geometry and 30mm diameter bulbs), respectively.

6 EQUIPMENT ISSUES

The ability to install modified strand cablebolts (i.e. bulbed) in conjunction with very thick grouts depends upon the type of grout pump being used, the additives, the size of the grouting hose, the distance of grouting and the method of grouting. Experience from surface and underground testing at a number of underground operations in Australia suggests that a two-stage grout pumping machine such as the GP

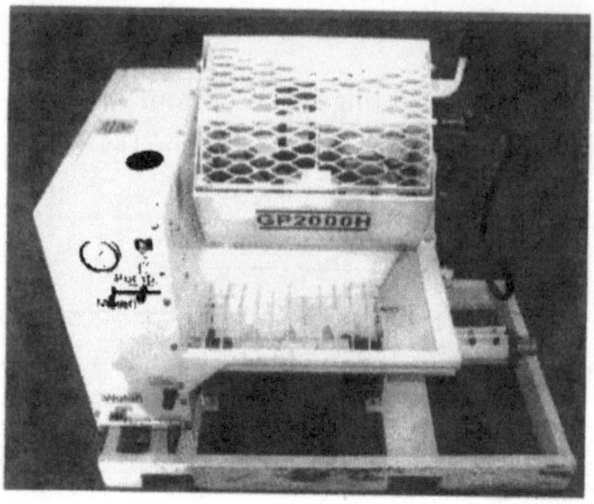

Figure 7. The GP 2000 grout pumping machine (courtesy of Master Builders Technology, 1998).

2000 is recommended for efficient advanced grouting (See Figure 7). This machine can routinely deliver thick grouts having a 0.30 w/c ratio, with the cement paste penetrating the cablebolt bulbs without any problem.

Two-stage means that the machine has *separate* mixing and pumping containers, which can be operated simultaneously or independently, thereby significantly increasing productivity. The ability to mix independently of pumping allows for a constant water cement ratio throughout a grouting operation. This enables a high degree of quality control, as an accurate water meter allows a controlled water addition to a precision of one tenth of a litre. Cement, water and additives are mixed in the paddle mixer and then discharged into the lower hopper, where a variable speed drive coupled to a rotor-stator pump (monopump) discharges the grout at the desired rate. Significant increases in productivity have been achieved using the two-stage pumping technique at a number of Australian operations.

A number of experiments have shown that a two-stage, monopump-based grouting machine is more powerful than most air-driven standard piston-based pumps. Piston-based pumps had limited success mixing and pumping thick grouts to any significant lengths, heights or required pumping rates during several experimental stages. The machine shown in Figure 7 has very few moving parts, which means low yearly maintenance costs, even with significant daily usage. All the working parts of the machine are within easy access, and a maximum of 15 minutes (by a single operator) are required for equipment clean up at the end of the shift.

The machine can be placed permanently on a mobile platform and taken to the grouting areas. A mobile type of transportation set-up is recommended, in order to minimize the length of grouting hose at all times. The maximum length of hose recommended is approximately 15 metres (for cablebolt grouting); however for rockbolt grouting a short 5-m hose is optimal. It is recommended that the grouting platform is moved close to the grouting position rather than adding an additional segment to the grouting hose. The platform should provide enough space for a palette of cement to be placed near the machine, without creating any tripping hazards for the operator.

A two-operator set-up is recommended with one operator mixing and another working with the grouting hose. At mine sites where compressed air is not readily available, the GP 2000 machine can be connected to the hydraulics of the transporting vehicle, and thus facilitate mobility and operational capability. This type of set-up has been successfully implemented at the McArthur River Mine.

7 GROUT MIX AND ADDITIVES

Several aspects require consideration during the selection of the most appropriate mix design to suit a particular operation. For example, the ability of a machine to mix the required water/cement ratio in a reasonable amount of time must be considered. For example, in order to achieve a required consistency, a mixing time of approximately 5 minutes is needed with the GP 2000 after all the constituents are added to the grout mix. Paddles that rotate around a horizontal axis facilitate better mixing than paddles that rotate around a vertical axis.

The flow rate or pumpability of the grout paste is also critical. A flow rate of 9 to 11 litres per minute is achieved routinely with an GP 2000 grout pumping machine using a 25mm (inside diameter) grout hose and a 0.3 w/c paste.

The volume of the grout that can be efficiently mixed is also important. At best, most conventional piston-based pump-mixer systems can only efficiently mix 4 cement bags (20 kg each). On the other hand a GP2000 machine can mix 8 bags of cement with no problems, even at very low water/cement ratios. A total of 8 bags can be mixed, while another 8 bags are being pumped giving a total of 320 kg of cement grout being active at any one time. The productivity gains with such an efficient activity are very significant, especially when a considerable area is to be grouted.

The use of the additive Methocel is recommended for efficient grouting of cablebolts. Methocel prevents segregation of water and cement, while reducing grout shrinkage during curing. Preventing water-cement segregation at the toe end of the hole is very important in order to achieve the required anchorage according to the load transfer concept (Windsor and Thompson, 1992). Laboratory tests where Methocel is not used show greater slippage in the upper portion of a split-pipe test (Villaescusa et al, 1992), suggesting that water-cement segregation is an important mechanism and it must be minimized. Methocel also improves grout pumpability, allowing thick grout mixes to be used. The recommended dosage for Methocel is 0.2% per weight of cement (i.e., 40 grams of Methocel per each 20 kg cement bag).

Recently, further improvements to grout mix design have been achieved with the introduction of fine sand to the grout mix. A cost reduction of approximately 30% can be envisaged due to the introduction of fine sand to the grout mix, depending upon sand availability. A typical mix design used routinely at Mount Isa Mines for a number of years consists of 3 cement bags(20 kg each), 120 grams of Methocel and 4 bags of fine sand(20 kg each). Water in the range of 24 to 26 litres was added. The exact amount of water required always depends upon the moisture in the sand. A typical range of size distributions for such fine sand is presented in Table 3.

Table 3. A typical size distribution of fine sand for grouting.

Cumulative % Passing	Size (μm)
100 - 100	4760
99 - 100	2380
97 - 99	1190
83 - 89	595
47 - 63	297
11 - 27	149
5 - 5	74

Table 4. Relative grout strength (7 days)

Grouting Method	UCS (MPa)
Piston-based (0.55 w/c)	0.67
Piston-based (0.45w/c)	1.00
Monopump (0.30 w/c)	1.40
Monopump (4 sand:3 cement)	1.56
Monopump (3 sand:3 cement)	1.80

Table 4 shows the relative Uniaxial Compressive Strength values for the grout tested at 7 days for typical piston-based and GP 2000 pumps. The results indicate that a significant increase on strength can be achieved by adding fine sand to the grout mixtures.

8 CONCLUSIONS

Grout quality is likely to be affected by the grouting equipment and the level of training provided to the underground workers. The experimental developments that have taken place in Australia over the last five years or so have significantly improved the quality and efficiency of the grouting operations at a number of mine sites. Toe to collar grouting using a self retracting hose in conjunction with very thick grouts and modified geometry cables is routinely undertaken.

9 ACKNOWLEDGEMENTS

The author acknowledges the financial support from the Australian Centre for Geomechanics and Curtin University of Technology for his position as Professor of Mining Geomechanics at WASM. The Australian Centre for Geomechanics has received funding for this position from the Government of Western Australia, Centres of Excellence Program.

10 REFERENCES

Hutchinson, D.J and M. Diederichs, (1996). Cablebolting in underground mines. BiTech Publishers, 406 p.

MBT (1998), GP 2000 Operating Manual, 20p.

Prentice, A. (1994). Cable bolting at Mount Isa Mine Limited. Undergraduate Thesis, the University of Auckland, 36p.

Stjern, G., (1995). Practical performance of rock bolts, PhD Thesis, The University of Trondheim, 178p.

Villaescusa, E., M. Sandy and S. Bywater, (1992). Ground support investigations and practices at Mount Isa. Proc. Int. Symp. on Rock Support. Sudbury, Kaiser and McCreath (Eds), 185-193, Balkema, Rotterdam.

Villaescusa, E. (1999). The reinforcement process in underground mining. Proc. Int. Symp. Rock Support and Reinforcement Practice in Mining, Kalgoorlie.

Windsor, C.R. and A. Thompson, (1992). Reinforcement design for jointed rock masses, Procc. 33rd US Symposium on Rock Mechanics, Santa Fe, pp521-530, Balkema, Rotterdam.

Rock Support and Reinforcement Practice in Mining, Villaescusa, Windsor & Thompson (eds)
© 1999 Balkema, Rotterdam, ISBN 90 5809 045 0

The performance of grouted split tube rock bolt systems

Alan G.Thompson
Rock Technology, Perth, W.A., Australia

David J.Finn
WMC Resources Limited, St.Ives Gold, Kambalda, W.A., Australia

ABSTRACT: Split tube friction rock stabilisers are widely used for rock reinforcement. The load transfer between the split tube and the borehole is critically dependent on the borehole diameter being smaller than the diameter of the split tube. One method of improving the load transfer is to fill the borehole and split tube with cement grout. The factors influencing both the ungrouted and grouted performance of split tube rock bolts have been investigated theoretically and in an extensive testing program. The testing results are in accordance with the theoretical considerations. Suggestions are made with regard to the implementation of grouted split tube rock bolt systems.

1 INTRODUCTION

The advantages of friction rock stabiliser reinforcement systems are related mainly to their ease of installation and their ability to sustain large rock mass displacements due to slip at the interface between the device and the borehole wall. However, consistent performance is difficult to achieve due to the sensitivity of the system to variations in borehole diameter caused by different bit diameters and configurations, drilling equipment and rock types. In some cases, the force at which slip occurs has proven not to be sufficient to prevent blocks of rock from detaching from the surrounding rock mass.

It has been found operationally that the choice of bit diameter for a particular rock type is a compromise between reducing the borehole diameter to increase the frictional load transfer and the ability to drive the split tube bolt to full depth. Another method of increasing the load transfer is to fill the borehole and split tube with thick cement grout at some time after initial installation. The effects of grouting have been shown to increase the load transfer by up to 200 % (Fuller and Dugan, 1992).

The mechanisms of basic frictional load transfer and increased load transfer due to grouting have been investigated theoretically and experimentally. The testing program was undertaken to quantify the different load transfer mechanisms associated with grouted split tube reinforcement systems, including those in which a length of flexible strand is also placed in the borehole to extend beyond the length of the split tube bolt. The results are presented and critically evaluated in terms of the implementation of this rock reinforcement system in the mining environment.

2 REVIEW OF MECHANICS OF ROCK REINFORCEMENT SYSTEM ACTION

All reinforcement comprises a system of four principal components:

0. The rock.
1. The element.
2. The internal fixture.
3. The external fixture.

The four principal components are shown schematically in Figure 1. Each component of the system is involved in two load transfer interactions.

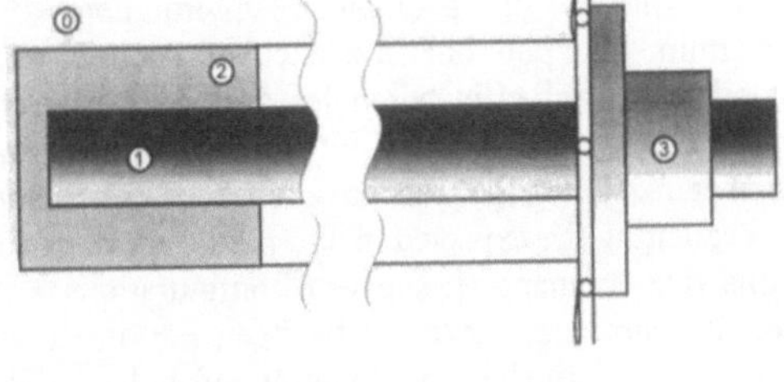

Figure 1. Schematic representation of a reinforcement system.

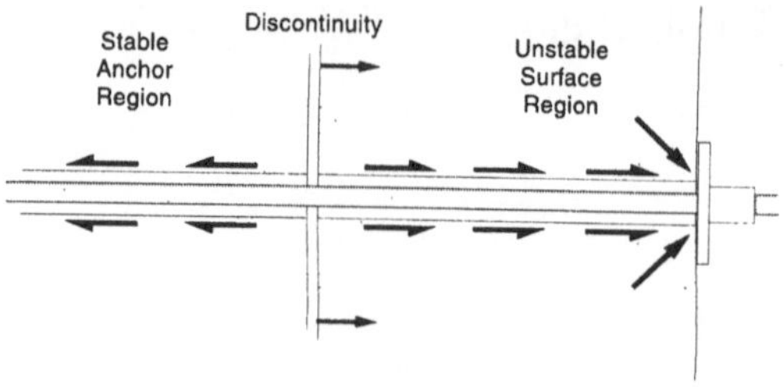

Figure 2. Schematic diagram showing the load transfer concept for a rock reinforcement system.

The concept of a reinforcement system is extremely important because the overall behaviour of the reinforcement system is dictated by the aggregated result of the component interactions of the system. Consequently, the concept of a system of principal components holds the key to understanding the mechanical behaviour of reinforcement.

The idea of a reinforcement system leads directly to the load transfer concept. That is, the transfer of load from unstable rock through the reinforcement system to stable rock as shown in Figure 2. The load transfer concept for reinforcement systems has been discussed by Windsor and Thompson (1993).

The reinforcement system and the load transfer concepts have been used to define three fundamental types of reinforcement system:

1. Continuously Mechanically Coupled (CMC) Systems.
2. Continuously Frictionally Coupled (CFC) Systems.
3. Discretely Mechanically or Frictionally Coupled (DMFC) Systems.

This classification results from considering the method of load transfer between the reinforcement element and the rock.

3 PREVIOUS SPLIT TUBE ROCK BOLT ANALYSIS AND TESTING

A large number of tests have been performed in a wide variety of geological environments to determine the pull out resistance of friction rock stabilisers installed in boreholes drilled in different rock types with known bit configurations and diameters. Often the tests have not been conclusive in evaluating the expected in situ performance of the bolts due to inappropriate test configurations. In recent years, tests have mostly been performed on standard, relatively long, lengths of bolts. The results from these tests often do not result in slip of the bolt. The rate of frictional load transfer between the rock and the bolt cannot be determined and the maximum forces recorded in these tests often do not highlight potentially low load transfer.

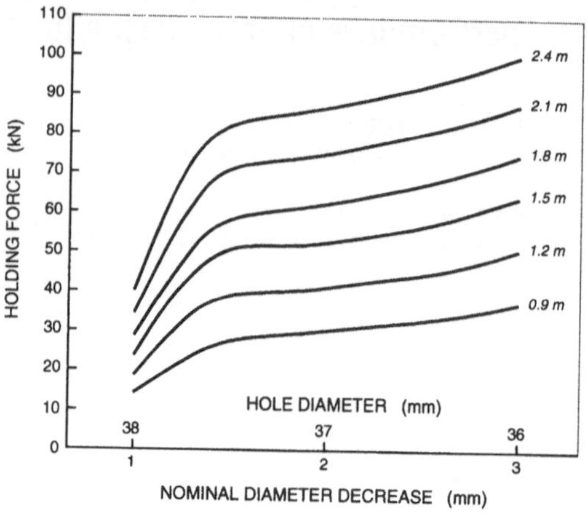

Figure 3. Relationships between pull out capacity and borehole diameters for SS39 Split Sets (reproduced from Ingersoll Rand catalogue).

The continuing need for large numbers of tests more that 20 years after the introduction of the Ingersoll Rand Split Set is in some regards questionable and reflects to a certain degree on the lack of a good understanding of the load transfer mechanism and the factors which influence it.

Ingersoll Rand provided pull test results (Figure 3) which demonstrated that there was a proportional increase in load transfer as the borehole diameter was reduced. However, at a certain borehole size (~1.5 mm less than the bolt diameter) the rate of increase in load transfer changed to a lesser rate.

Davis (1979) developed a theoretical solution and an associated computer program (Haas et al., 1979) to quantify the behaviour of Split Sets for different bolt materials, borehole diameters and friction coefficients for the interface between the bolt and the rock. The work of Davis showed that the reduction in the rate of increase of load transfer was due to yielding of the steel. Unfortunately, this work was published in a specialist journal with a small reader base within the mining industry and the charts were not produced in forms that were readily useable by mining personnel. Also, the solution required the borehole diameter as input. No instruments were made available commercially to enable the borehole diameters to be checked on a regular basis at mine sites. The combination of these facts has resulted in the continued use of the pull out test as the sole means of determining the required range of bit diameters to be used in different rock types and for quality control.

Fuller and Dugan (1992) reported on the performance of standard friction stabilisers, friction stabilisers in which the interior was filled with cement grout and friction stabilisers which included a cement grouted length of strand. The friction stabilisers were either the Ingersoll Rand Split Set (IR SS46) or the Hills Hardi Friction Rock Stabiliser (HH FRS). The result of filling the Split Set with cement grout was an increase in load transfer from about 3 tonnes to 15 tonnes for a 360 mm test length. The improvement for the HH FRS was not as demonstrable (< 2.5 tonnes to about 5 tonnes). This is to be expected as the HH FRS comprises a tube with a groove running the length of the element and, unless the grout is pumped to fill the groove, there is not direct contact between the grout and the rock. The addition of the strand produced a small improvement for both reinforcement systems. It is assumed that in these tests the Split Set was pulled and not the strand (it is not stated in the paper). The authors did not make any suggestions as to the cause of the increased pull out load.

Villaescusa and Wright (1997) also conducted tests on ungrouted and grouted Split Sets (IR SS46) at a number of different Australian mines. In one series of tests, pull out tests were performed at 2, 3, 4, 5.5 and 7 hours after grouting. The results showed that in excess of 10 tonnes/metre of load transfer could be achieved at 5.5 hours after grouting and, after 7 hours, the load transfer increased to more than 15 tonnes/metre. At another mine site, the load transfer 3 days after grouting exceeded 13 tonnes/metre. The authors attributed the increased load transfer to the '...partial dowelling and interlocking effect achieved by the high strength grout being in contact (bonded) with the rock all the way along the split axis of the bolt'. The grout was stated to have a water/cement ratio of 0.3. Further, they stated that the frictional resistance along the entire bolt axis is also increased by the high compressive strength grout increasing the resistance of inward deformation in the bolt required to allow bolt slippage.

The writers agree with both these proposed mechanisms of increased load transfer. It will be shown in Section 5 that the first mechanism is most likely to be the cause of the increased load transfer. The second effect does cause some increase in load transfer as evidenced by the results reported by Fuller and Dugan for the HH FRS in which there may not have been direct load transfer between the grout and the rock. This second effect will be dependent on both the straightness of the borehole and the uniformity of its diameter along its length.

4 GROUTED SPLIT TUBE AND STRAND REINFORCEMENT SYSTEM

The load at which slip occurs in a friction rock stabiliser that is forced into an undersized borehole is directly related to the borehole diameter and the length of contact between the device and the borehole wall. Other factors which influence the slip load are the surface conditions and properties of the materials either side of the contact interface.

In some circumstances, the resistance and the length of a bolt in the anchor region may be insufficient to stabilise the rock surrounding an excavation. The deficiency in force capacity can be increased by simply grouting the split tube bolt as described in the previous section. Both the length and force capacity can be increased by drilling the borehole longer than the split tube bolt and installing a length of steel strand within the split tube and to the full depth of the borehole as shown in Figure 4.

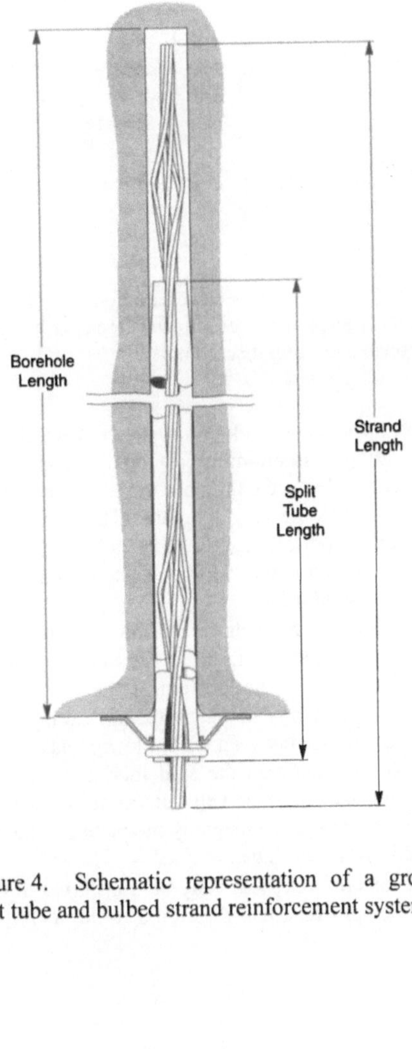

Figure 4. Schematic representation of a grouted split tube and bulbed strand reinforcement system.

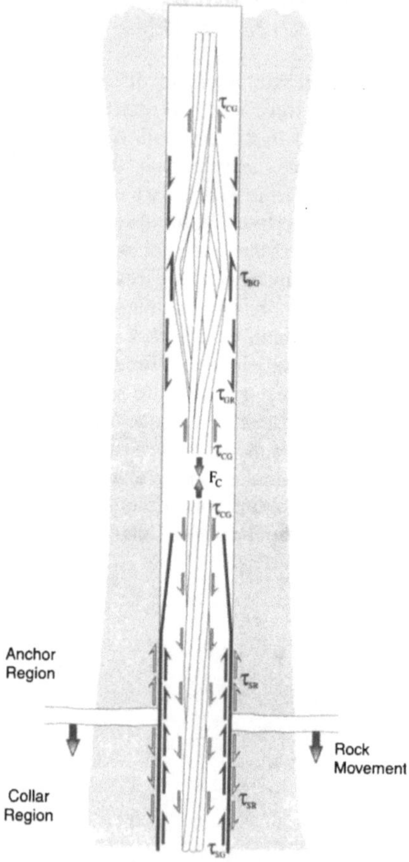

Figure 5. Schematic representation of the load transfer mechanisms when the rock bolt system intersects a dilating discontinuity.

Figure 5 shows the rock bolt system either side of an intersecting discontinuity in the rock. The interactions between the split bolt, strand, cement grout and rock are very complex. The arrows given in Figure 5 indicate the probable directions of load transfer between the various system components near the discontinuity.

In the collar zone, the mechanics of this system involves load transfer from the rock to the split tube along its length. The location of the discontinuity relative to the collar and the rates of load transfer at the outer surface between the split tube and rock, the inner surface between the split tube and grout and the grout and the rock will determine whether any load is required to be taken by the plate and the ring welded to the split tube.

In the anchor zone, load may be transferred through both the split tube and the strand to the rock. Any force in the strand will be transferred through the grout to the rock in the anchor zone. Within the split tube, the strand force will be transferred to the grout annulus and then to the inside surface of the split tube and to the rock in the area of the slot.

The actual distribution of forces and the capacity of the system will depend on the absolute and relative stiffnesses (i.e. the force per unit displacement response) of the component materials and the various load transfer mechanisms between these components. The formal analysis of the system is beyond the scope of this paper.

5 SPLIT TUBE REINFORCEMENT SYSTEM PERFORMANCE

Some theoretical considerations are now presented regarding the performance of the individual components of the system and the overall performance of the complete system in response to dilation of a discontinuity intersecting the complete system.

In the following sections, shear stress or shear strength (τ) is the mean stress acting to resist relative movement at an interface between any two components of the reinforcement system. Unit load transfer (t) is assumed to have units of force per unit length and is obtained by multiplying the shear stress at an interface by the perimeter over which it is assumed to act. Load transfer (T) has units of force and is given by multiplying the unit load transfer by the length over which this is assumed to act.

5.1 Slot width

The slot width is an important geometric parameter affecting the behaviour of the grouted split tube rock bolt system. The slot width (w) after installation for a split tube compressed into an under sized borehole is given approximately by:

$$w = w_0 - \pi(D_S - D_H) \quad (1)$$

where

w_0 = undeformed slot width

D_S = external diameter of undeformed split tube.

D_H = borehole diameter.

Note that the slot is theoretically closed when the difference between the split tube and the borehole diameters is (w_0/π). For an Ingersoll Rand SS46, $w_0 \approx 20$ mm and the approximate borehole diameter which will cause closure of the slot is ~40 mm.

5.2 *Unit load transfer between split tube and rock*

The reduction in the cross-section of the split tube causes radial forces to be developed at the interface between the tube and the borehole wall. Davis (1975) detailed how the split tube contact with the borehole wall is limited to the two edges of the slot and part of the surface diametrically opposite to the slot. However, the average unit load transfer at the interface between the element and the rock (t_{SR}) can be assumed to be given by:

$$t_{SR} = \tau_{SR}\left(\pi D_H - w\right) \tag{2}$$

where:
τ_{SR} = the friction between the split tube and rock.
w = slot width after installation.

It is worthwhile noting at this point the emphasis given to the frictional resistance to pull out after installation. However, it is also important to consider the resistance to driving during installation. The resistance increases as the length of tube within the borehole increases. Additional resistance to driving is caused by the initial constriction of the tube close to the collar. in some cases, the total resistance may be sufficient to cause buckling or prevent the tube being driven to full depth. In both cases, this results in a reduction of effective anchor length and prevents the bolt providing effective restraint to a plate at the collar.

5.3 *Unit load transfer at the split*

The maximum unit load transfer (t_{GR}) at the interface between the rock and the grout is given by:

$$t_{GR} = \tau_{GR}\, w \tag{3}$$

where:
τ_{GR} = the shear strength of the interface between the grout and the rock.

5.4 *Load transfer between strand and grout*

The approximate unit load transfer between plain strand and grout (t_{CG}) is given by:

$$t_{CG} = \tau_{CG}\, \pi D_C \tag{4}$$

where:
τ_{CG} = the shear stress at the interface between the strand and the grout.
D_C = diameter of the strand.

If the grouted length of strand includes a bulb, then the total load transfer (T_C) is markedly increased and is given by:

$$T_C = T_{BG} + \tau_{CG}\left(L - L_B\right) \tag{5}$$

where:
T_{BG} = load transfer at the bulb
L = total load transfer length
L_B = length of bulb

5.5 *Unit load transfer from grout to split tube*

Forces are developed in the cement grout annulus due to load transfer from the rock and the strand. These forces will tend to cause relative movement between the grout annulus and the split tube. The unit load transfer between the grout and the inside of the split tube (t_{SG}) is given by.

$$t_{SG} = \tau_{SG}\left(\pi D_{SI} - w\right) \tag{6}$$

where:
D_{SI} = internal diameter of the Split Set.
τ_{SG} = shear strength at the internal Split Set/grout interface.

5.6 *Unit load transfer from grout to rock*

The unit load transfer between the grout and rock in the length beyond the split tube (t_{CR}) is given by:

$$t_{CR} = \tau_{GR}\, \pi D_H \tag{7}$$

where the symbols have been defined previously.

5.7 *Equilibrium requirements*

Equilibrium of the rock bolt system shown in Figure 5 requires that:

$$F = F_S + F_C \tag{8}$$

where:
F = demand or the force applied to the rock bolt system below the discontinuity.
F_S = force transferred to the split tube above the discontinuity.
F_C = force in the strand at the end of the Split Set.

Using equation 8 and knowing the geometry of the reinforcement system and the depth to the discontinuity, it is possible to determine the minimum requirements for the respective unit load transfers.

Consider the following example. A potentially unstable, 2 metre thick extensive slab of rock is reinforced with 2.4 metre long, split tube, friction rock stabilisers (capacity 180 kN) installed vertically in 44 mm diameter boreholes on a 1.5 metre square pattern. The unit weight of rock is 30 kN/m^3.

The force demand is calculated to be 67.5 kN which is less than the axial capacity of the split tube. However, the questions to be answered are:

- Will ungrouted split tube bolts be satisfactory?
- If not, what minimum additional unit load transfer is required for equilibrium?
- Can this be achieved by grouting alone or will over-drilling of the borehole length and installation of a length of strand be required?

Allowing for a domed plate at the collar and the taper of the split tube, the effective load transfer length in the anchor region is approximately 300 mm. The minimum required frictional unit load transfer between the split tube and rock is therefore estimated to be 225 kN/m. Based on previous testing, this load transfer is unlikely to be achieved.

Assuming now that the frictional unit load transfer is 40 kN/m. The required additional load transfer between the cement grout and rock through the slot in the anchor region is 185 kN/m. Using equation 1, the slot width is estimated to be 13.7 mm (w_0 = 20 mm) and the minimum required shear strength between the grout and rock is calculated from equation 3 to be 13.5 MPa. This may or may not be achieved depending on the strengths of the grout and rock and the roughness of the interface between them.

If the shear strength at the cement grout/rock interface is assumed to be 8 MPa, then a strand will be required to pass beyond the end of the split tube. The force required to be carried by the strand is approximately 22.6 kN (assuming the frictional load transfer is 12 kN and the grout/rock load transfer is 32.9 kN). This is clearly very much less than the minimum tensile strength (250 kN) of 15.2 mm superstrand. If the strand is assumed to pass 300 mm beyond the end of the split tube, the minimum unit load transfers between the strand and grout and the grout and rock are ~75 kN/m. This leads to the required minimum shear strengths at the strand/grout interface of ~1.6 MPa and at the grout/rock interface of ~0.5 MPa. Previous testing has indicated that these minimum shear strengths are easily achieved.

The required minimum load transfer inside the split tube is calculated using equation 6 to be very low with approximately 2.4 metres available inside the split tube to transfer the strand force of 22.6 kN.

It is clear from the equations and example presented in this section that testing is required to determine the unit load transfers for the various interfaces. The strengths of the various interfaces may then be calculated from the equations relating the unit load transfer to the shear stresses which may then be used to perform similar calculations for design.

6 SPLIT TUBE REINFORCEMENT SYSTEM TESTING

The split tube testing program involved a number of important aspects which were critical in attempting to understand the behaviour of the reinforcement system. It was deemed important to obtain relationships between bit size and borehole diameter for the different rock types encountered at the mine site. It was also important to define the strengths of the various interfaces in the system.

The testing program involved:

- Measurement of bit size.
- Profiling of borehole diameter with depth in the borehole.
- Installation of a 600 mm long split bolt with loading ring.
- Pull out test on ungrouted bolt.
- Grouting of the bolt.
- Pull out test on grouted bolt.
- Axial pull test on 600 mm length of bulbed strand from within the split bolt and from within rock.

The load transfer lengths were selected on the basis of previous testing to be sufficiently short to expect slip to result during testing or, in the case of the strand tests, to be representative of the likely configuration to be implemented in practice.

6.1 Bit size measurement

Bit size is measured by an electronic vernier with a resolution of 0.01 mm. For a button type bit, the diameter is the average of the measurements taken across several diameters associated with the number and layout of the buttons.

6.2 Borehole profiling

In view of the critical dependence of split tube bolt load transfer on borehole diameter, it is necessary to measure the actual borehole diameter. A prototype electronic gauge has been developed to measure the diameter of boreholes at 100 mm intervals along their length. The instrument, shown in Figure 6, measures the borehole size across 2 orthogonal diameters. The electronic readout indicates the average diameter to a resolution of 0.01 mm.

It will be shown that the borehole diameter varies with the rock type and there are major variations along the borehole as it passes through different hardness zones in the rock. These variations of borehole diameters are thought to be the reason for the variable load transfers measured in pull out tests conducted for apparently similar conditions.

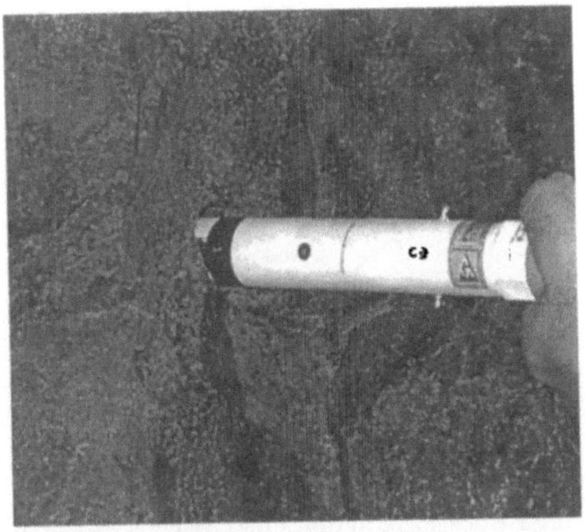

Figure 6. Prototype electronic borehole diameter measuring equipment.

6.3 *Split bolt pull test*

As indicated previously, it is necessary to cause gross slip to provide a reliable indication of load transfer. Displacement measurement is required. Reduction of load with time is not a reliable indicator of slip. Factors related to the hydraulic and other equipment may result in gradual load loss not related to the performance of the reinforcement system.

6.4 *Strand pull tests*

Two types of strand pull tests were performed. The first type, shown in Figure 7, pulled a length of strand relative to both the split tube bolt and the rock. This test was designed to measure the load transfer that could be expected in the collar region of the borehole and to assess the need for a plate on the strand. The other type was a conventional strand pull test.

Figure 7. Test set up to pull strand from inside a split tube bolt.

7 TESTING PROGRAM RESULTS

The testing program involved Ingersoll Rand galvanised SS46 (46 mm nominal diameter) bolts and various lengths of 'bulbed' 15.2 mm diameter, steel superstrand.

The Ordinary Portland Cement grout used in the tests was mixed at water/cement ratios which varied between 0.310 and 0.325. The compressive strength of the grout was not measured. However, previous work (Thompson and Windsor, 1998) would suggest that the unconfined compressive strength of cement grout at 28 days would be about 60 MPa with a range of about ±10 %. The strengths at 1 day and 7 days are estimated to have been about 40 % and 80 % of the 28 days strength, respectively.

7.1 *Bit sizes*

Two different sizes of 6 button bits were used to drill the boreholes. Table 1 details the designation numbers for the bits used to drill the various length boreholes in the four rock types. A smaller diameter bit (designated Bit #1) was used to drill Holes #1 and #2 in each of the four rock types. The bit size started at 43.65 mm and was measured to be 43.37 mm at the completion of drilling.

Two, larger size bits were used to drill the remaining boreholes. Bit #2 started at 45.38 mm, reduced to 45.19 mm after drilling 7 holes in the Dyke and further reduced to 45.00 mm after drilling the boreholes in Gabbro and Porphyry. Bit #3 was measured to be 45.25 mm prior to the drilling of the boreholes in Ore.

Table 1. Summary of borehole depths and drill number used to drill the boreholes.

B'hole No.	B'hole Depth (m)	Dyke	Gabbro	P'phyry	Ore
1	1	Bit #1	Bit #1	Bit #1	Bit #1
2	1	Bit #1	Bit #1	Bit #1	Bit #1
3	1	Bit #2	Bit #2	Bit #2	Bit #3
4	1	Bit #2	Bit #2	Bit #2	Bit #3
5	1	Bit #2	Bit #2	Bit #2	Bit #3
6	1	Bit #2	Bit #2	Bit #2	Bit #3
7	1	Bit #2	Bit #2	Bit #2	Bit #3

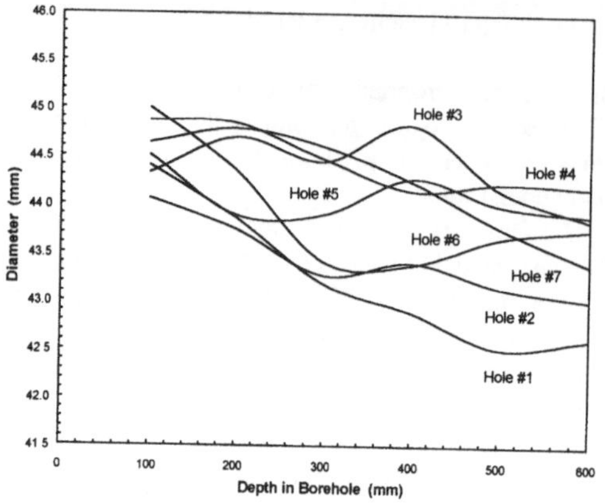

Figure 8. Diameter profiles for boreholes drilled in the Dyke.

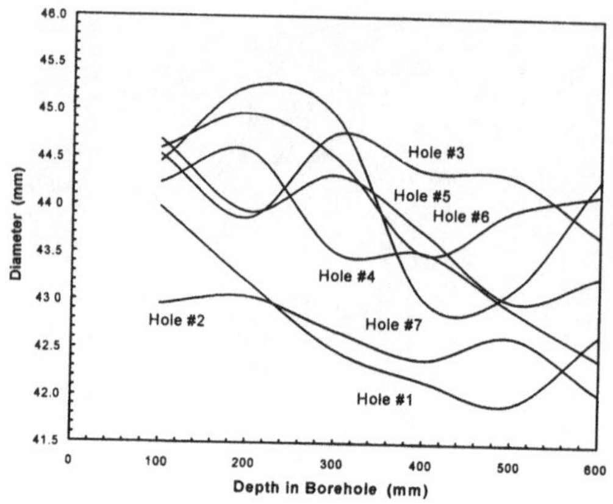

Figure 10. Diameter profiles for boreholes drilled in the Porphyry.

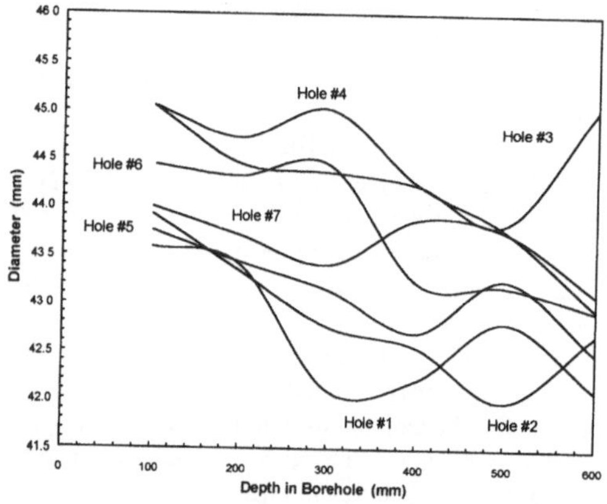

Figure 9. Diameter profiles for boreholes drilled in the Gabbro.

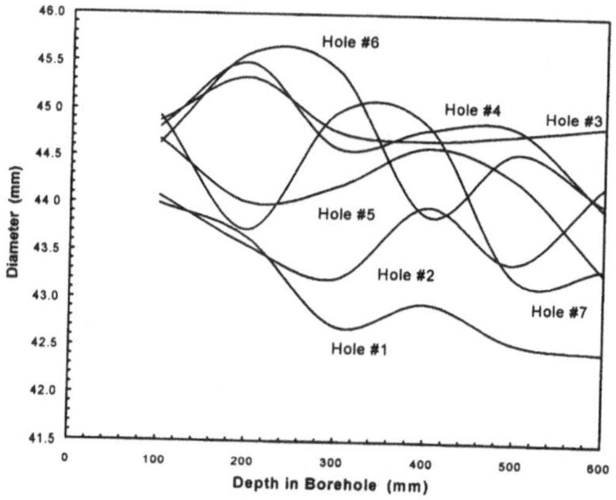

Figure 11. Diameter profiles for boreholes drilled in the Ore.

7.2 Borehole profiles

The diameter profiles, to a depth of 600 mm, for all the boreholes used for the Split Set testing are shown in Figures 8 to 11. As indicated previously, the diameters varied significantly both in diameter and with depth in the borehole.

Table 2 gives the mean borehole diameter calculated from the profile measured immediately after drilling each borehole and prior to the installation of the Split Sets. The borehole profiles indicated that the diameters were often less than the measured bit size. This is thought to be due to hole closure after drilling caused by stress redistribution and rock relaxation.

Table 2. Summary of mean borehole diameters.

Borehole Number	Dyke (mm)	Gabbro (mm)	Porphyry (mm)	Ore (mm)
1	43.25	42.70	42.74	43.05
2	43.44	42.87	42.63	43.74
3	44.38	44.49	44.25	44.88
4	44.46	44.29	43.53	44.75
5	44.08	43.13	43.82	44.18
6	43.93	43.76	44.27	44.69
7	44.24	43.64	44.18	44.18

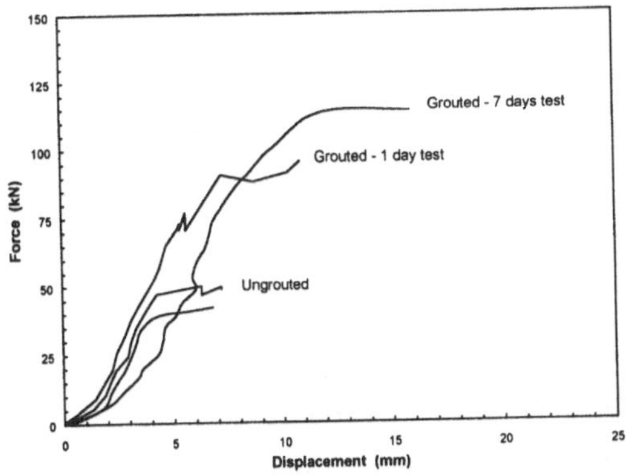

Figure 12. Typical pull test results for ungrouted and grouted Split Sets installed in the smaller boreholes drilled in the Dyke.

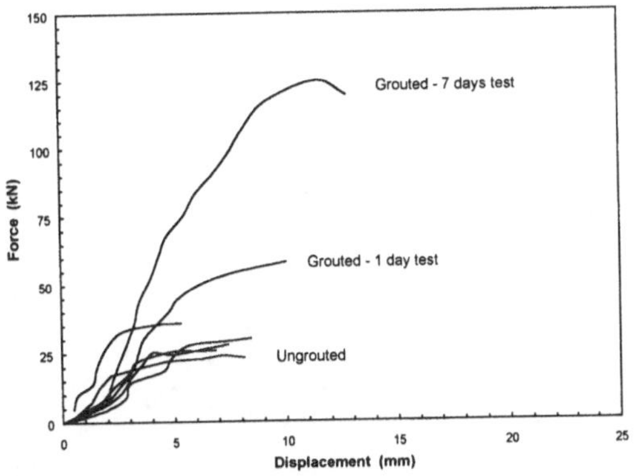

Figure 13. Typical pull test results for ungrouted and grouted Split Sets installed in the larger boreholes drilled in the Dyke.

7.3 Split Set pull test results

Force-displacement responses for the ungrouted and the grouted Split Sets in the Dyke are given in Figures 12 and 13, respectively. These responses are typical of those obtained in the other three types of rock.

Table 3 gives a summary of the pull out tests on the 600 mm long Split Sets before and after grouting. The maximum pull out loads given in Table 3 for the ungrouted Split Sets were converted to values of load transfer expressed in terms of kN/metre of embedment length and then plotted versus mean borehole diameter as shown in Figure 14. Note that, allowing for the taper and outstand at the collar, the effective load transfer length between the Split Set and rock was assumed to be about 450 mm.

The load transfers given in Figure 14 show some inconsistencies but a general trend for the load transfer to increase with reduction in borehole size and bit size. The load transfer also appears to vary with the rock type. The variations in load transfers are not unexpected when the borehole profiles are also taken into consideration. It is thought that in some cases the tube would not have been in continuous contact over the length within the borehole. This means that locally the radial contact forces between the tube and rock may be higher but the actual transfer length is smaller. The overall effect on the measured load transfer is indeterminate.

Table 3. Summary of peak loads recorded in pull tests on Split Sets before and after grouting.

Borehole Number	Dyke		Gabbro	
	Ungrouted (kN)	Grouted (kN)	Ungrouted (kN)	Grouted (kN)
1	50.0	95.6	34.8	52.3
2	42.1	114.4	44.3	>123.1
3	24.1	58.2	29.5	84.1
4	28.0	120.2	27.1	>147.2
5	26.0	-	32.9	-
6	35.8	-	31.1	-
7	30.2	-	28.3	-
Borehole Number	**Porphyry**		**Ore**	
	Ungrouted (kN)	Grouted (kN)	Ungrouted (kN)	Grouted (kN)
1	38.5	87.3	32.2	108.5
2	31.8	>143.6	29.4	>131.4
3	27.6	94.7	23.2	82.6
4	25.7	>115.2	31.5	>151.3
5	32.0	-	26.4	-
6	29.6	-	30.9	-
7	26.7	-	33.0	-

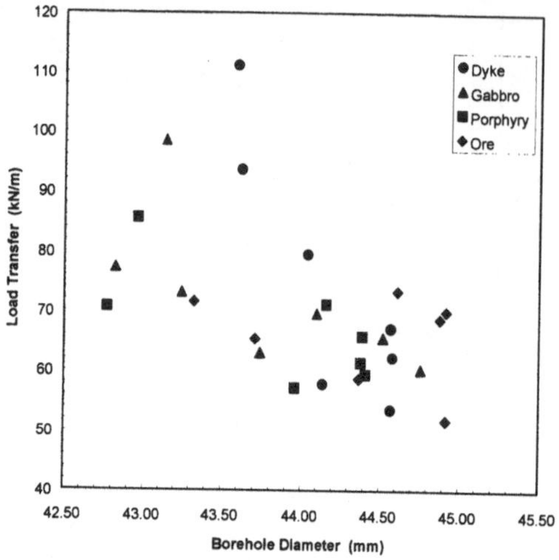

Figure 14. Variation of Split Set load transfer with borehole diameter.

7.4 *Strand pull test results*

As indicated previously, two types of strand pull test were performed. One type of test (shown in Figure 7), pulled a 600 mm length of strand relative to both the Split Set and the rock. A bulb was located 300 mm from the collar end of the Split Set. This test was designed to measure the load transfer that could be expected in the collar region of the borehole when loaded as shown in Figure 5.

Table 4 gives a summary of the maximum pull out loads recorded in the tests on bulb strand embedded inside a 600 mm long Split Set. The maximum load that could be applied to the end of the strand was limited by when the end of the Split Set started to deform excessively. If slip did not occur, the peak load given in Table 4 is prefixed with a '>'.

The other tests involved a conventional pull test of a 600 mm coupled length of strand. A bulb was located at the middle of the coupled length which started approximately 1.5 metres from the collar of the borehole. Table 5 gives a summary of the maximum load recorded in these pull tests.

Table 4. Summary of peak loads recorded in pull tests on bulbed strands embedded inside Split Sets.

Borehole Number	Dyke (kN)	Gabbro (kN)	Porphyry (kN)	Ore (kN)
5	90	100	>95	>120
6	80	>110	>150	>130
7	>85	>130	>150	>150

Table 5. Summary of peak loads recorded in pull tests on bulbed strands embedded in rock.

Borehole Number	Dyke (kN)	Gabbro (kN)	Porphyry (kN)	Ore (kN)
8	140	>110	160	>180
9	>200	>200	>200	>200
10	>200	>200	>200	>200

7.5 *Interpretation of results*

Table 6 gives an overall summary of the load transfer for Split Sets ungrouted and at 1 day and 7 days after grouting. The internal grouting of Split Sets significantly increases the load transfer from around 5 - 10 tonne/metre to over 25 tonne/metre. That is, the area of grout in contact with the rock and the increased resistance to radial deformation combine to produce an increase in load transfer of more than 150 kN per metre. This result is consistent with the theoretical considerations presented in Section 5. In particular, the increase in load transfer due to grouting appeared to be greater for the larger diameter boreholes in which the slot could be expected to be wider as indicated by equation 1.

If it is assumed that the full capacity of strand needs to be transferred within the length of a 2.4 metre long Split Set, equations 3 and 6 can be used to predict that the combined load transfer from the strand inside the Split Set has to exceed about 105 kN/m (250 kN/2.4 m) to suggest that the strand does not need to be plated. For the pull tests involving the bulb strand embedded inside a 600 mm long Split Set, the peak loads given in Table 4 all exceeded 80 kN (8 tonne) with the weakest grout (tests in the Dyke). With the slightly lower water/cement ratio grouts, the strength exceeded 100 kN at 1 day and 150 kN at 7 days. Therefore, the lowest load transfer is greater than ~130 kN/m (80kN/0.6m). From this, it is inferred that plating of the strand is unnecessary when used in conjunction with a Split Set and plate.

Table 6. Overall summary of the range of load transfers for Split Sets before and after grouting.

Bit Size	Ungrouted (kN/m)	Grouted (1 day) (kN/m)	Grouted (7 days) (kN/m)
Larger	50 – 70	130 – 215	>250
Smaller	70 – 100	115 – 240	>250

In the pull tests involving the 600 mm length of bulb strand grouted in rock, the peak loads given in Table 5 exceeded 110 kN (11 tonne) at 1 day with the weakest grout (in the Gabbro and the Dyke). For the slightly lower water/cement ratio grouts, the pull out strength at 1 day exceeded 150 kN. In all cases, the strength exceeded 200 kN at 7 days. For the tests at 7 days, application of load was stopped prior to causing rupture of any wires in the strand. In typical applications, it is unlikely that the forces imposed on the system would exceed 200 kN. However, in wider spans where strand longer than 3 metres would be required, the increased embedment length in the anchor region would ensure that the capacity of the strand (250 kN) could be achieved.

8 IMPLEMENTATION OF GROUTED SPLIT TUBE BOLTS

The outcomes from the testing program suggest that a flexible ground support strategy may be adopted using split tube bolts and mesh as the basic ground control scheme. That is:

- Ungrouted split bolts may be used in walls to restrain mesh.
- Grouted split bolts may have sufficiently improved load capacity to be used in backs formed in better quality rock.
- A bulbed strand grouted inside a split bolt inside a 3 metre deep borehole can be used to form a 20 tonne capacity bolt with an anchorage more than 2.4 metres beyond the excavation face.
- Longer boreholes can be used as required in excavations in which large spans are created. In longer boreholes, the reinforcement system capacity can be expected to be at least 20 tonne.

To implement the above ground control schemes, the following suggestions are made:

- A water/cement ratio of 0.3 should be used. This will result in thick grout that will remain inside the split tube bolt in vertical up holes and results in higher early strength.
- The grout should be mixed for at least 5 minutes prior to pumping to ensure adequate pumpability is achieved with the thick grout.
- The grout tube must be withdrawn at a rate that matches the pumping rate. Simple calculations and trials may need to be undertaken to ensure that an appropriate rate of tube withdrawal is used.

If strand is placed in the same borehole as the split tube, then:

- To ensure adequate load transfer, a bulb must be located in the grouted section of the borehole at the toe of the borehole.
- The tests indicated that the strand would not need to be plated if a plate was used with the split tube bolt.
- The tests and theoretical considerations indicate that a bulb may not be required within the split bolt. However, for additional security, it is suggested that a bulb be located near the collar of the borehole inside the split bolt.

9 CONCLUDING REMARKS

Testing and associated theoretical considerations have shown that the load transfer of split tube friction rock stabilisers may be significantly improved by pumping cement grout into the borehole to fill the centre of the tube. The improved load transfer is attributed mainly to the additional load transfer between the grout and the rock at the slot in the tube.

Where large rock mass displacements occur, the improved load transfer may be sufficient to prevent slip of the split tube relative to the borehole and the axial capacity of the split tube may be exceeded. That is, the ability of the split tube friction rock stabiliser to sustain large rock mass displacements may be lost.

Strand may be used to supplement the split tube. This will usually only be warranted on the basis that the depth of rock that is required to be secured exceeds the length of the split tube (e.g. in development intersections and wide span entry stopes). For typical development applications, the rock mass force demand is unlikely to warrant the additional force capacity provided by the strand as suggested by the 'length-capacity relationship' that has evolved for rock reinforcement systems (Windsor and Thompson, 1996).

ACKNOWLEDGEMENTS

The authors acknowledge WMC Resources Ltd – St Ives Gold for their financial support and kind permission to publish this information. The test program was conducted at Junction Mine, Kambalda and the writers wish to thank the miners for their involvement in establishing the reinforcement for testing.

REFERENCES

Davis, R.L. 1979. Split-set rock bolt analysis. Int. J. Rock Mech. Min. SC. & Geomech. Abstr., V16, 1-10, Pergamon: London

Fuller, P.G. & K.J. Dugan 1992. Factors affecting ground support performance in mining. Fifth Underground Operators' Conference, Ballarat, 139-144, AusIMM: Melbourne.

Haas, C.J., R.L. Davis, D. Keith & W.C. Patrick 1979. An investigation of the interaction of rock and types of rock bolts for selected loading conditions. BuMines OFR 29-79, US Bureau of Mines: Washington, DC.

Thompson, A.G. & C.R. Windsor 1998. Cement grouts in theory and practice. Proc. 3rd North Amer. Rock Mech Symp., Paper AUS-330-2, Cancun, Mexico, Pergamon: London.

Villaescusa, E. & J. Wright 1997. Permanent excavation support using cement grouted Split Set bolts. Proceedings of the International Symposium on Rock Support, Lillehammer, 660-670, Norwegian Society of Chartered Engineers: Norway.

Windsor, C.R. & A.G. Thompson 1993. Rock reinforcement-technology, testing, design and evaluation. Comprehensive Rock Engineering - Principles, Practice and Projects (Ed. J. A. Hudson), V4, Excavation, Support and Monitoring, 451-484, Pergamon: Oxford.

Windsor, C.R. & A.G. Thompson 1996. Terminology in rock reinforcement practice. Rock Mechanics-Tools and Techniques (Eds. M. Aubertin, F. Hassani and H. Mitri)), V1, 225-232, Balkema: Rotterdam.

Rock Support and Reinforcement Practice in Mining, Villaescusa, Windsor & Thompson (eds)
© 1999 Balkema, Rotterdam, ISBN 90 5809 045 0

Performance of the Swellex bolt in hard and soft rocks

C. Li
Luleå University of Technology, Sweden

U. Håkansson
Jacobson and Widmark AB, Sweden

ABSTRACT: In general Swellex bolts strengthen the rock mass through a combination of friction and mechanical interlock at the rock-bolt interface. The anchoring mechanism of the Swellex bolt is actually different in hard and soft rocks. In hard rock, the secondary contact stress, induced by the mechanical interlock of the asperities at the borehole wall, plays a major role in the anchoring, whereas the primary contact stress, created by the bolt expansion does not contribute much. However, in soft rocks, it is the friction, and thus the primary contact stress, which determines the anchoring capacity of the bolt.

1 INTRODUCTION

The Swellex is a type of rock bolt that strengthens the rock mass through a combination of friction and mechanical interlock at the rock-bolt interface. The use of the Swellex has grown rapidly in a world-wide scope during recent years, not only in hard rocks (high strength and high deformation modulus) but also in soft rocks (low strength and low deformation modulus). The Swellex is unique in providing an instant reinforcement action in soft rocks where time is a crucial factor. Underground openings in soft rocks demand instant support following the excavation. The Swellex can satisfy this demand and practical applications have demonstrated its potential in dealing with stability problems in weak rock engineering. To achieve an optimum reinforcement effect, the detailed knowledge of the performance of the bolt is needed.

The emphasis of this paper is put on the performance of the Swellex bolt in hard and soft rocks. The roles of the primary contact stress and the roughness of the borehole wall are examined in detail. The limits for the primary contact stress and the roughness angle of the wall are also discussed.

2 WORKING PRINCIPLES

Swellex rock bolts were introduced by Atlas Copco in the 1980's (Wijk and Skogberg, 1982). The Swellex bolt is made from a folded thin-wall steel tube. Bushings are pressed onto both ends of the bolt, which are then sealed by welding. The lower bushing has a small hole through which water is injected into the bolt at high pressure to expand the bolt. During the expansion process, the Swellex bolt compresses the rock surrounding the hole and adapts its shape to fit the irregularities of the borehole, see Figure 1. After installation the bolt is held in the hole by the contact stress between the bolt and the borehole due to the elastic recovery of the rock material, as well as the mechanical interlock due to the roughness of the borehole (Stillborg, 1994).

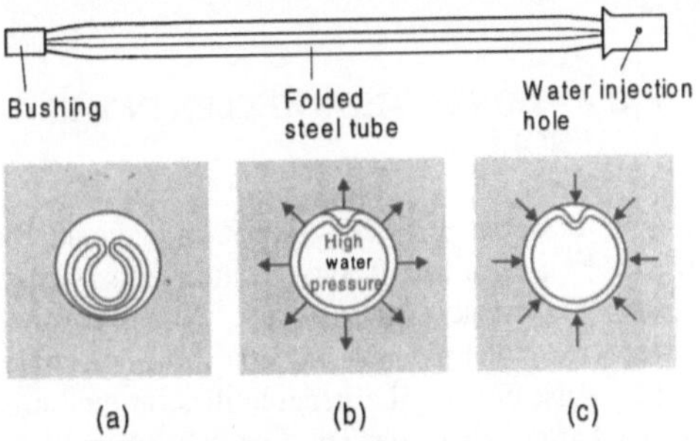

Figure 1. Sketches illustrating the Swelllex bolt and the interaction between the rock and the Swellex bolt: (a) The Swellex bolt is placed in the hole; (b) The bolt is expanded under high water pressure; (c) The water pressure is released and the surrounding rock contracts, thereby providing the Swellex locking effect. After Stillborg (1994).

The reinforcement effect of the Swellex bolt can be represented by its pull-out resistance, F_{pull}, expressed as the pull-out load per metre. The failure of the bond between the rock and the Swellex bolt is either in the form of sliding along the asperities on the borehole wall, and/or in the form of breakage of the asperities. The pull-out resistance can be generally expressed as

$$F_{pull} = \min(R_f, S) \tag{1}$$

where R_f refers to the frictional resistance at the rock-bolt interface, and S to the total strength of the sheared-off asperities. Expression (1) means that the pull-out resistance F_{pull} is the smaller one between R_f and S. The terms R_f and S can be expressed, respectively, as:

$$R_f = \pi d(q_1 + q_2)\tan(\phi + i) \tag{2}$$

and

$$S = \tau A \tag{3}$$

where
d = the diameter of the borehole,
q_1 = the primary contact stress at the rock-bolt interface, created by the installation,
q_2 = the secondary contact stress, induced by the mechanical interlock,
ϕ = the friction angle between the rock and the bolt,
i = the roughness (or dilation) angle of the borehole wall,
τ = the shear strength of the rock,
A = the total area of all sheared-off asperities.

3 EXPANSION OF THE SWELLEX IN THE BOREHOLE

In order to achieve an effective reinforcement the Swellex bolt must satisfy two requirements: firstly, a primary contact stress must be established between the borehole wall and the bolt; secondly, the bolt must fully fit the irregularities on the borehole wall after expansion. The first requirement is to enhance the frictional anchorage of the bolt, while the second is to achieve a mechanical interlock. The primary contact stress is built up due to the difference between the stiffness of the bolt and that of the borehole. By considering the elastic recovery of the borehole as well as that of the bolt tube, the expression for the contact stress on the borehole wall is obtained as (Håkansson and Li, 1997):

$$q_1 = \frac{K_s}{K_r + K_s}\left[P_i - \frac{K_r}{K_f}\left(P_{pm} - P_i\right)\right] \tag{4}$$

where

K_s = the radial stiffness of the bolt, $K_s = \dfrac{K_b K_f}{K_b + K_f}$,
K_f = the radial stiffness of the tube ring, $K_f = \dfrac{E_s}{1-\nu_s^2}\dfrac{t}{r_i}$,
K_b = the stiffness of the Swellex tongue (see Figure 2). It is a function of the expansion degree of the bolt.
K_r = the stiffness of the rock, $K_r = \dfrac{E_r}{1+\nu_r}$,
q_1 = the primary contact stress at the rock-bolt interface,
P_{pm}= the maximum pump pressure,
P_i = the borehole pressure, *i.e.* the pressure on the borehole wall during the bolt installation,
t = the thickness of the bolt tube,
r_i = the radius of the borehole,
E_s = the Young's modulus for steel,
E_r = the Young's modulus for rock,
ν_s = the Poisson's ratio for steel,
ν_r = the Poisson's ratio for rock.

Expression (4) means that the primary contact stress on the borehole wall is a function of the stiffness of the rock and that of the Swellex bolt. The stiffness of the Swellex depends on the length of the bolt tongue, a short tongue resulting in a high stiffness. Therefore, the primary contact stress q_1 is also a function of the degree of expansion. Figure 3

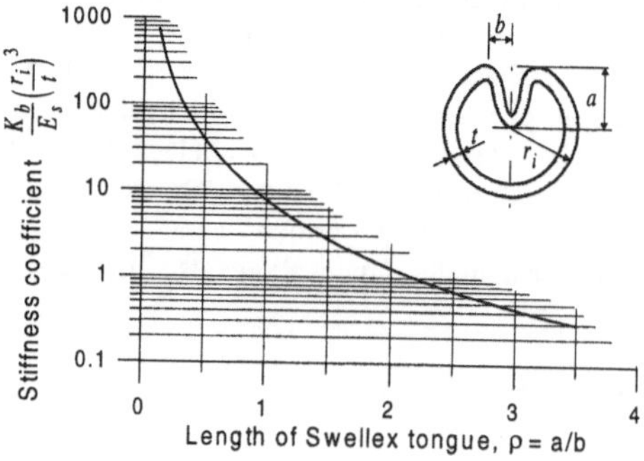

Figure 2. The relationship between the stiffness coefficient K_b and the length of the Swellex tongue.

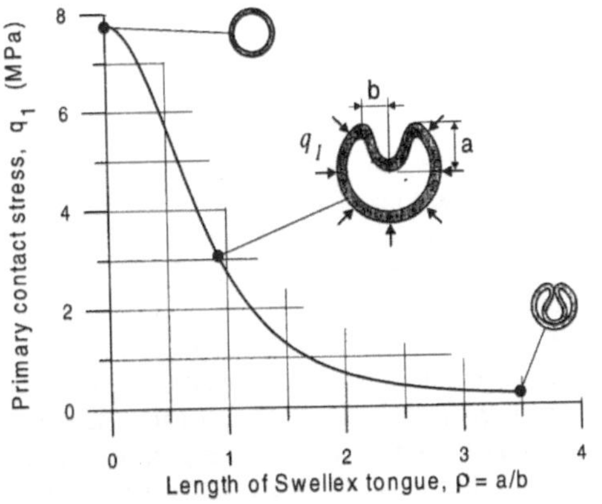

Figure 3. Primary contact stress versus the expansion of the Super Swellex bolt. Parameter values used for the calculation: E_s= 210 GPa, E_r = 10 GPa, P_{pm}= 30 MPa, P_i= 15 MPa.

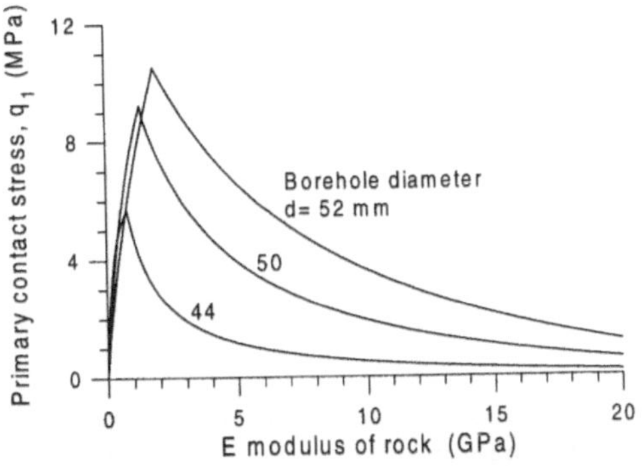

Figure 4. Primary contact stress versus the Young's modulus of rock in hard rocks. Parameter values used in the calculations for the Super Swellex: D=54 mm, t=3 mm, b = 10 mm, E_s = 210 GPa, ν_s = 0.3, ν_r = 0.2, P_{pm} = 30 MPa, P_i = 15 MPa, $\phi = 30^o$.

shows an example illustrating the relationship between the primary contact stress and the tongue length.

4 PERFORMANCE IN HARD AND SOFT ROCKS

The primary contact stress is dependent on the Young's modulus of the rock in hard rocks, but in soft rocks it is dependent on both the Young's modulus and the strength of the rock, since yielding may occur in a limited area around the borehole during installation. As an example, Figure 4 illustrates the theoretical solution of the primary contact stress when the Young's modulus of the rock varies. In the calculations, it is assumed that the Young's modulus is related to the uniaxial compressive strength (in soft rocks) by $E_r = 80\sigma_c^{1.4}$ (Aydan et al., 1995). The primary contact stress increases with the Young's modulus in soft rocks, while in hard rocks it decreases. It can be seen that it is in relatively soft rocks that high primary contact stresses are achieved.

As stated before, the anchorage of the Swellex rock bolts is achieved by a combination of friction and mechanical interlock between the borehole wall and the bolt. In soft rocks the asperities on the borehole wall are either crushed during the bolt installation or sheared off afterwards when relative movements occur between the borehole and the bolt. This means that the mechanical interlock makes only a small contribution to the anchoring of the bolt, but it is friction that plays the main role in this case. The friction is directly proportional to the primary contact stress on the borehole wall. Thus, the primary contact stress is of vital importance to enhance the anchorage of the Swellex bolt in soft rocks.

In hard rocks, however, the asperities on the borehole wall are sheared off only with great difficulty, and therefore the mechanical interlock plays a significant role in the anchorage. It can be seen in Figure 4 that the primary contact stress is low in hard rocks. A secondary contact stress has to be provided in order to enhance the anchorage of the bolt. This secondary contact stress is achieved when the bolt tends to slide over the asperities on the borehole wall in hard rocks. Suppose the roughness (or dilation) angle of the borehole wall is i, see Figure 5. The radial contraction, u, of the bolt tube is related to the axial movement of the bolt, x, as follows:

$$u = x \tan i \qquad (5)$$

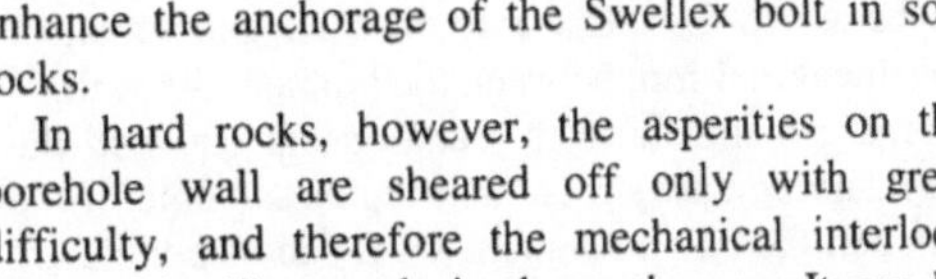

Figure 5. A schematic illustration showing the mechanical interlock between the borehole wall and the Swellex rock bolt.

On the other hand, it is known from elasticity theory that the radial contraction of the bolt will bring about a radial resistant stress, *i.e.* the secondary contact stress, q_2. The relationship between the radial displacement and the secondary contact stress is given by:

$$u = \frac{q_2 r_i}{K_s} \quad (6)$$

Combining the above two expressions, one obtains the expression for q_2 as:

$$q_2 = K_s \tan i \frac{x}{r_i} \quad (7)$$

Let us look at the contribution of the roughness of the borehole wall to the secondary contact stress through an example. Assume that a Super Swellex bolt is installed in a borehole with the radius $r_i = 24$ mm. The roughness angle of the borehole wall is assumed to be $i = 1^{\circ}$. The calculation using expression (7) shows that the secondary contact stress q_2 can be up to 0.9 MPa only with an axial relative movement of 1 mm between the bolt and the wall.

Figure 6 illustrates the relationship between the secondary contact stress, the roughness angle of the wall, and the borehole diameter. It can be concluded that in hard rocks it is not the primary contact stress but the secondary contact stress that dominates the anchorage capacity of the Swellex bolt. For this reason, the borehole wall must be rough enough to induce a secondary contact stress, as long as the bolt is subjected to an axial pull-out load.

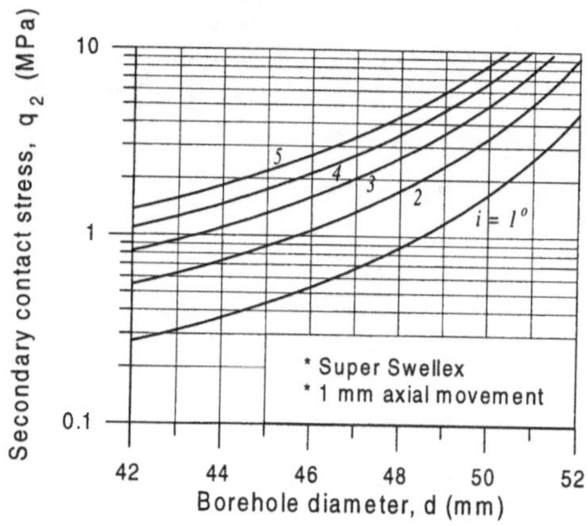

Figure 6. Secondary contact stress versus the diameter of the borehole at different roughness angles during a 1 mm axial movement of the Swellex bolt.

5 *IN-SITU* LOAD

The load exerted on the *in-situ* rock bolts is caused by rock deformation. The distribution of the shear stress and also the axial tensile load along the entire length of the rock bolt can be calculated based on the rock deformation (Li and Stillborg, 1997). The loading process is the same for all types of rock bolts, but the characteristic of the Swellex bolt is that the shear stress on the bolt can be maintained at the level of the ultimate shear strength when slip is triggered at the rock-bolt interface. At this stage the maximum anchorage capacity of the Swellex bolt is reached. As illustrated in Figure 7, the shear stress on the bolt is directed toward the tunnel wall on the portion close to the wall, while it is directed toward the opposite direction on the portion located further inside the rock. The maximum axial tensile load in the bolt occurs at the neutral point where the shear stress on the bolt is zero. It is the anchor length of the bolt that determines the maximum axial load. When the rock deformation is large enough, sliding may occur on the anchoring section of the bolt, *i.e.* along the anchor length of the bolt.

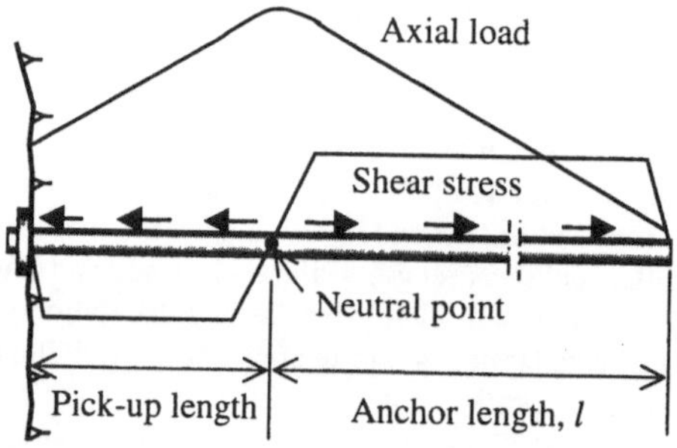

Figure 7. A sketch illustrating the shear stress and the axial tensile load along the Swellex bolt with a face plate, corresponding to the maximum anchorage capacity of the bolt.

6 CAPACITY

In contrast to traditional types of bolts, a limited amount of sliding movement on the borehole wall does not make the anchorage of the Swellex bolt fail, but on the contrary can activate the full load bearing capacity of the bolt. The evidence for this statement is displayed in the pull-out tests where the ultimate pull-out load of the Swellex bolt remains constant even after a long displacement. In this case the shear strength of the rock-bolt interface is mobilised along the entire length of the bolt. This feature of the Swellex bolt, which means that it can tolerate a long displacement without loss of its ultimate load bearing capacity, makes it unique in stabilising rock masses with large deformations. The ideal performance of the Swellex bolt should be that the bolt slides, instead of ruptures, in the case of large rock deformation. This performance demands that the anchor length of the bolt has a limit that depends on the bond strength, thus the total contact stress, at the rock-bolt interface. In the case where the asperities on the borehole wall are not sheared off, a general criterion exists for determining the anchor length of the bolt is:

Max. pull-out load on bolt < Tensile strength of bolt

i.e.

$$\pi d l (q_1 + q_2)\tan(\phi + i) < T \quad (8)$$

where the parameter l stands for the anchor length of the bolt, see Figure 7. In soft rocks, the roughness of the borehole wall as well as the secondary contact stress can be assumed to be zero, *i.e.* $i = 0$ and $q_2 = 0$. Thus, the anchor length is obtained from expression (8) as:

$$l < \frac{T}{\pi d q_1 \tan\phi} \quad (9)$$

In hard rocks it is obtained by assuming that $q_1 \approx 0$:

$$l < \frac{T}{\pi d q_2 \tan(\phi + i)}$$

or

$$l < \frac{T}{2\pi x K_s \tan(\phi + i)\tan i} \quad (10)$$

Let us look at the influence of the primary contact stress and the roughness angle on the anchor length through an example involving the Super Swellex. The minimum breaking load of the Super Swellex is $T = 200$ kN. Assume that

Friction angle: $\phi = 35^\circ$
Borehole diameter: $d = 48$ mm
Bolt tube stiffness: $K_s = 340$ MPa (calculated using the equation for K_s)

In the soft rock, the anchor length of the bolt, expressed by (9), is illustrated in Figure 8a versus the primary contact stress. For a primary contact stress of 1 MPa, an anchor length of about 2 m can guarantee to make use of the maximum anchoring capacity of the bolt. In the hard rock, the anchor length of bolt, expressed by (10), is illustrated in Figure 8b versus the roughness angle of the borehole wall. For a roughness angle of 5.5°, an anchor length of about 1.2 m is long enough to make use of the maximum anchoring capacity of the Super Swellex. The pneumatic drilling can easily make a very rough borehole wall in hard rocks. This was proved by a field pull-out test in granite. In the test, the tensile strength of the Super Swellex was reached at an anchor length of 0.75 m. It is seen in Figure 8b that the roughness angle of the borehole wall is about 8° corresponding to an anchor length of 0.75 m if the primary contact stress is neglected.

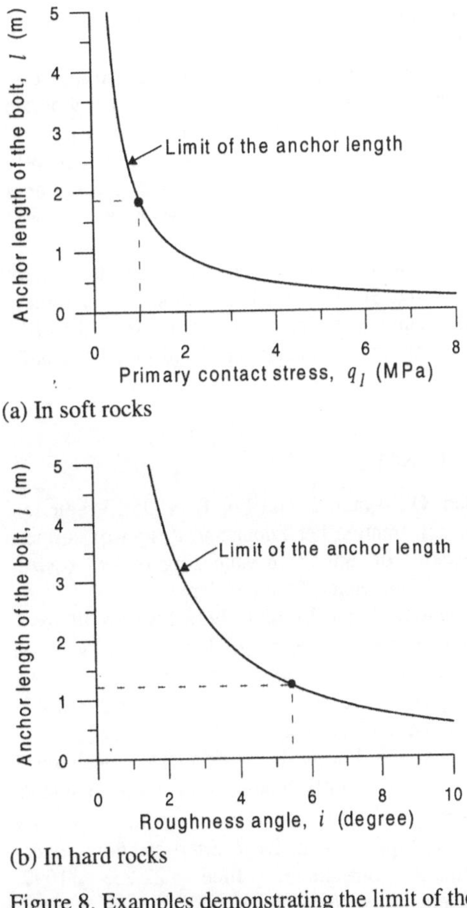

(b) In hard rocks

Figure 8. Examples demonstrating the limit of the anchor length of the Super Swellex bolt with relation to the primary contact stress in soft rocks, and to the roughness angle i on the borehole wall in hard rocks. Parameter values used for the calculation: $d = 48$ mm, $t = 3$ mm, $x = 1$ mm, $\phi = 35^\circ$, $T = 200$ kN.

The two diagrams in Figure 8 are for the purpose of demonstration. The best way to determine the anchor length of the bolt, the primary contact stress as well as the roughness of the borehole is to perform field pull-out tests.

7 CONCLUSIONS

(1) In soft rocks, it is the primary contact stress that plays a significant role in the anchorage of the bolt.
(2) In hard rocks, the borehole wall must be rough enough and the bolt tube must be tightly matched with the irregular surface of the bore-hole wall. It is the secondary contact stress induced due to the roughness of the borehole wall that mainly determines the anchorage of the bolt in hard rocks.
(3) The Swellex bolt can provide a strong rein-forcement to the rock as long as a moderate primary contact stress, such as between 0.5 and 1 MPa, is achieved in soft rocks, or a relatively rough borehole wall, such as with a roughness angle of about 6°, is made in hard rocks.

Acknowledgements –The authors would like to tank Prof. Bengt Stillborg for his comments and review of the manuscript. Thanks are also extended to Mr. Pether Jonsson for providing the result of the pull-out tests.

REFERENCES

Aydan, Ö., Akagi, T. and Ito, T. (1995). Prediction of deformation behaviour and design of support systems of tunnels in squeezing rocks. *Report*, Tokai University, Shimizu, Japan.

Håkansson, U. and Li, C. (1997). Swellex in weak and soft rock — design guidelines. *Int. Symp. on Rock Support - Applied Solution for Under-ground Structures*, June 22-25, 1997, Lillehammer, Norway, 555-568.

Li, C. and Stillborg, B. (1997). The mechanical behaviour of fully bonded rock bolts in tensile loading — A theoretical analysis. *Int. Symp. on Rock Support - Applied Solution for Under-ground Structures*, June 22-25, 1997, Lillehammer, Norway, 178-192.

Stillborg, B. (1994). *Professional Users Handbook for Rock Bolting* (2nd edition), Tans tech Publi-cations.

Wijk, G., and Skogberg, B. (1982). The SWELLEX® rock bolting system. *14th Canadian Rock Mechanics Symposium*, Canadian Institute of Mining and Metallurgy, 106-115.

Rock Support and Reinforcement Practice in Mining, Villaescusa, Windsor & Thompson (eds)
© 1999 Balkema, Rotterdam, ISBN 90 5809 045 0

Reinforcement of underground excavations using the CT Bolt

E. Villaescusa
Western Australian School of Mines, Kalgoorlie, W.A., Australia

J. Wright
Ingersoll Rand (Australia) Limited, Sydney, N.S.W., Australia

ABSTRACT: This paper presents laboratory and in-situ pull test results on a new bolt system called the CT Bolt. The laboratory results were designed to determine the limiting strength of the ungrouted bolt components. Tests were carried out on the anchor, bolt and plate/hemispherical nut assembly. Tests were also carried out to determine the grouted strength behaviour. Split-pipe tests were set-up to determine load transfer data and to check the likely corrosion protection provided by the PVC tube around the CT Bolt. In-situ pull testing was carried out in different geological environments to determine the anchor, bolt and rock interactions for ungrouted and grouted cases.

1 INTRODUCTION

The C-Tube bolt (CT bolt) is a relatively new bolt that has been designed to overcome all the deficiencies of a standard point-anchored reinforcement system. The CT bolt consists of a 20mm diameter steel bar (of varying lengths) installed with point anchor expansion shells in conjunction with face plates. The tension to the bolt is provided by tightening a nut-hemispherical washer and a plate against the rock on the exposed end of the bolt.

The initial bolt installation can be mechanized to provide immediate reinforcement. In addition, grouting after the initial bolt installation provides long-term reinforcement. The bolt is installed in a specially designed corrugated polyethylene tube, to facilitate grouting following bolt installation. The hemispherical washer is hollow, and the grout is injected through a 16mm hole in the washer. A grout mix having a 0.35 to 0.4 water/cement ratio is recommended to facilitate grout flow and to achieve complete full column grouting.

Long term corrosion of the bolt is minimized by the tube, which effectively covers the steel bar, protecting the bolt from aggressive elements that may penetrate cracks developed in the grout due to subsequent ground movement. Figure 1 is a schematic long and cross section through a typical CT bolt, showing the components of the system.

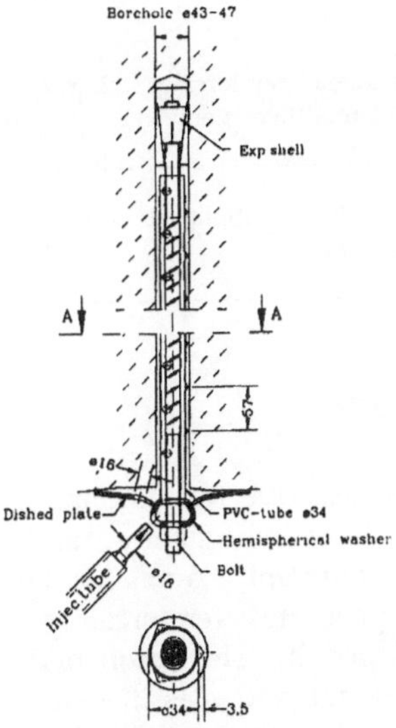

Figure 1. The components of the CT bolt.

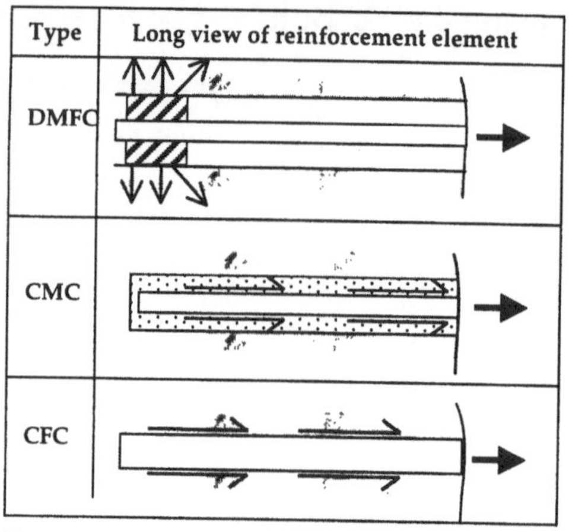

Figure 2. Classification of reinforcement action (After Thompson and Windsor, 1992).

Table 1. Classification of typical reinforcement devices, (After Thompson and Windsor, 1992).

Type	Description
DMFC	Mechanical anchors (ungrouted CT and HGB bolts, expansion shell, slot and wedge) Single cement/resin cartridge anchors (paddle bolt, deformed bar)
CMC	Full column cement/resin grouted bars (Grouted Ct bolt, deformed bar, thread bar) Cement grouted reinforcing cables (plain strand, modified geometry)
CFC	Friction stabilizers (Split-set Bolt, Friction Bolt, Swellex)

2 TERMINOLOGY

A classification to describe the different forms, functions, basic mechanics and behaviour of the different commercially available rock support and reinforcement systems was developed by Thompson and Windsor (1992). The method classifies the existing reinforcement systems by dividing them into three basic categories in order to explain the basic mechanisms of load transfer between the reinforcing elements and the rock mass.

A description and comparison of devices within a particular category or between separate categories is facilitated by the method. This is particularly important when comparing their performance or during the evaluation of new reinforcement devices such as the CT bolt. The categories are shown in Figure 2 and are described as discrete mechanical and friction coupled (DMFC), continuous mechanical coupled (CMC) and continuous friction coupled (CFC). Some of the typical reinforcing devices, including the CT bolt, are grouped according to this classification in Table 1.

2.1 *Discrete mechanical and friction coupled*

A discrete mechanical and frictional coupled (DMFC) device transfers load at two discrete points, namely the borehole collar and the anchor point, which is located at some depth in the borehole. The length of the element between the two discrete points (plate and anchor) is actually decoupled from the rock mass. The load transfer is then limited to a relatively short anchor length. Load transfer at the anchor point can be achieved by either mechanical (grouted anchor) or frictional means (expansion shell). An ungrouted CT bolt is a typical example of a discrete friction coupled reinforcement element.

The strength of an expansion shell may be limited by the strength of the rock at the borehole wall, and these devices are best suited to hard rock applications. Grouted anchors may be used in soft rock masses, where a high load transfer can be achieved over a short length, provided the grout does not migrate into pre-existing cracks in the rock mass.

2.2 *Continuous mechanical coupled*

A continuous mechanically coupled (CMC) reinforcing element relies on a fixing agent, usually a cement or resin based grout, which fills the annulus between the element and the borehole wall. The main function of the grout is to provide a mechanism for load transfer between the rock mass and the reinforcing element.

The reinforcing elements are usually manufactured with variable cross-sectional shapes in order to increase the element to grout bond strength. A mechanical key is effectively created by the geometrical interference between the element and the grout along the entire reinforcement length. The corrugations on the PVC tube that is used to grout and protect the CT bolt is a typical example. The element is defined as continuously coupled to the rock mass by way of interlock with the grouting agent (Thompson and Windsor, 1992).

A grouted CT bolt is a typical example of a continuous mechanical coupled reinforcing action. The CT bolt has an additional advantage due to the mechanical interlock likely to be developed between the end anchor and the grout column.

2.3 *Continuous friction coupled*

A continuous frictionally coupled (CFC) reinforcing element is installed in direct contact with the rock mass. The mechanism of load transfer is a function of the frictional forces developed between the reinforcing element and the borehole wall. The load transfer is limited by the radial pre stress set up during the initial element insertion. The bond strength is a function of the element diameter, the borehole diameter and any geometrical irregularities occurring at the borehole wall.

The radial stress can be related to a force along the length of the reinforcing element, and is achieved by deforming the cross sectional area of the element to suit the borehole. This can be achieved by either contracting an oversized element section into an undersized borehole (Split Set stabilizer) or by expanding an undersized element section into an oversized borehole (Swellex bolt). A modification of this reinforcing action can be achieved by cement grouting of the split-set bolts as described by Villaescusa and Wright (1997).

3 ONE PASS ROCKBOLT SYSTEMS

In general, most underground mining operations may require the installation of temporary and long-term reinforcement schemes. A reinforcement scheme is an arrangement of reinforcement systems (such as CMC, CFC or DMFC) in a variety of dimensional and spatial configurations (Windsor, 1997).

In-situ geologic conditions, induced stresses and blast damage are likely to control the stand-up time and overall behaviour of an excavation in rock. While it is sometimes possible to install long-term reinforcement in one step, in some cases, the installation of temporary reinforcement is required. In this context, the mechanised installation of both temporary and long-term reinforcement in one step (in conjunction with meshing capabilities) with minimum exposure of the operating personnel is highly advantageous.

This reinforcement scheme is called a one pass bolting system, with the temporary reinforcement provided by a point anchor component while the long-term reinforcement relies on cement grouting.

Mechanised installation is achieved by means of specialised mining equipment, capable of installing mesh and providing high initial tensions to the bolt. A mechanised system minimises exposure of the operating personnel to potentially unstable spans during bolt installation.

Grouting can be performed before the actual installation of the bolt, such as in a Slot and Wedge bolt, or post installation such as in the Hollow Groutable Bolts (HGB) or CT Bolts.

4 POINT ANCHORED REINFORCEMENT

The CT bolt is frictionally coupled to the rock mass by means of an expansion shell anchor. Because of this short internal coupling, the actual point anchor strength is limited by the strength of the rock around the borehole. Consequently, laboratory and in-situ testing is required to ensure that the initial reinforcement provided by the frictional device is adequate for temporary reinforcement purposes. Laboratory punch tests on the plate and nut/washer assembly are also required (ASTM: F432-95). The combined results can be used to determine the weakest member of the discrete friction coupled system. The global rating for temporary reinforcement is then set equal to the lowest of the individual strength values established for the plate, bar and anchor.

4.1 *Laboratory testing*

The expansion shells of point anchored bolts must be tested for their recommended hole diameter range as per ASTM Standard F432-95. This test consists of applying axial loading to the bolt with the expansion shell tightened against a thick wall tube. The bolt is examined at the yield and rupture loads. In both cases, the plug of the expansion shell must be able to be screwed out of the bolt by hand. Figure 3 shows the results of a number of successful ASTM F432-95 load deformation tests carried out at the Western Australian School of Mines (WASM) for point anchored CT bolts tested to yielding.

The capacity of a face-plate is critical to the performance of a point anchored system. Plates are used to secure the exposed end of the bolt, and hence are in direct contact with the rock mass. The plate, bar and anchor are an integral part of the reinforcement system

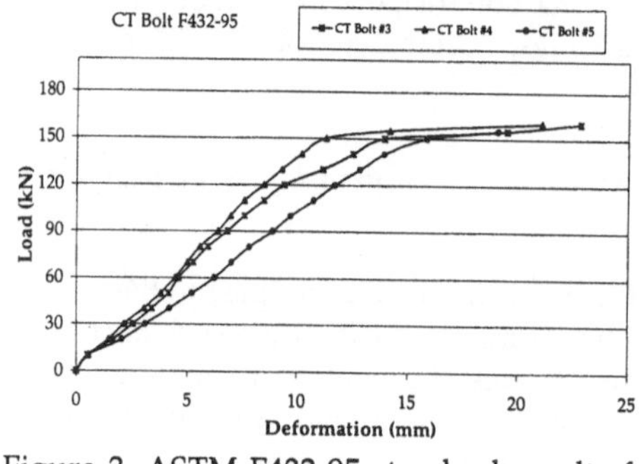

Figure 3. ASTM F432-95 standard results for CT bolt anchor tested up to bolt yield.

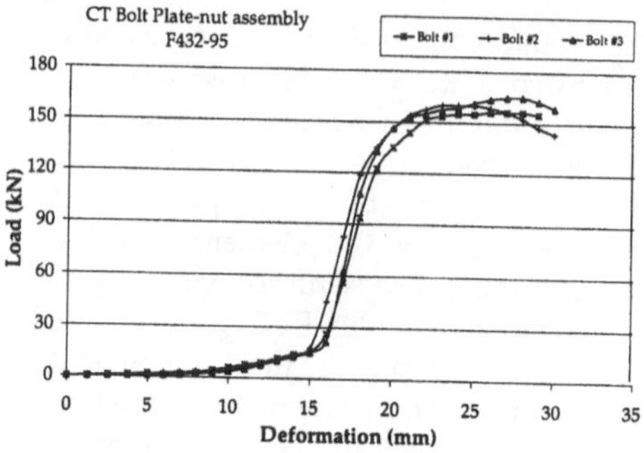

Figure 4. Punch test results on a typical CT bolt plate, nut-hemispherical washer assembly.

required to control deformation around an excavation surface. Consequently, the plates must be designed to exceed the yield strength of the steel bar. Failure of the plate is usually described as plate inversion and its caused by the nut/washer punching through the plate. The recommended procedure for testing a bearing plate is described in the ASTM F432-95 standard. Figure 4 shows the result of a number of laboratory punch tests carried out at WASM on a CT bolt plate assembly.

All point-anchored systems must be installed and designed in such a way that if failure ever occurs, it must always occur along the steel axis. This necessarily implies that the plate/nut and expansion shells are designed to be stronger than the bolt itself. The results of the laboratory tests carried out at WASM indicate that the CT bolt system is properly designed. Each of the system components can be rated at 15 tonnes, equal to the yield strength of the bar.

5 LONG-TERM REINFORCEMENT

Mechanical anchored bolts should not be considered effective long-term reinforcement unless the elements are grouted. Previous experience indicates that mechanical anchored bolts are not consistently reliable beyond three to six months. Consequently, the long-term reinforcement for a CT bolt system is provided by means of cement grouting. In other one pass bolt systems such as the Slot and Wedge bolt system, grouting is undertaken prior to the insertion of the bolt in the hole. The bolt is then inserted and tensioned before the grout sets. In the Hollow Groutable and CT bolt systems, cement grouting is undertaken after the bolts have been installed and tensioned.

Grouted systems can be susceptible to poor practices from operators, who sometimes fail to grout the bolts correctly. Corrosion may develop on non grouted bolts and may prevent grouting altogether, effectively eliminating their long term reinforcing capabilities. A potential disadvantage of a Slot and Wedge type system is that the required initial anchoring strength may never be developed before the grout sets. This is due to the inherent weakness in the slot and wedge geometries, where the full capacity of the steel bar may never be mobilized following the initial mechanical installation.

A problem with all one-pass bolt systems is that the bolts may be susceptible to blast damage when they are installed very close to a blasting face. The damage is localised at the hole collar, where the bolt may be hit by fly rock from the blast, or at the anchor end, where some slippage of the anchor system may occur due to blast vibrations. To minimise damage at the collar, the bolts should be installed with minimum protruding tail (ideally less than 10-cm) and to minimise anchor slippage, an initial tension of approximately 5-7 tonnes should be established.

5.1 *Load displacement characteristics*

Laboratory and in-situ pull tests can be used to determine the load deformation characteristics of cement grouted CT bolts. Laboratory tests provide a controlled environment where comparisons of capacity and load transfer for a given amount of deformation can be established. These tests can be specifically designed to determine and compare the tensile strength of a reinforcement scheme, and the load deformation characteristics prior to tensile failure or slippage at the steel grout interface.

The grout-steel bond strength and the load deformation characteristics prior to and during failure of the grout steel bond can also be determined. The concepts of load transfer and embedment length are critical to the understanding of any load deformation results (Windsor and Thompson, 1993). Pull tests carried out underground determine the in-situ performance of a particular reinforcement scheme, and also help to understand how the reinforcement matches the rockmass response to mining.

5.2 Laboratory results

Laboratory testing of grouted CT bolts was conducted at WASM by grouting a number of devices inside steel pipes (50-67mm diameter, 5mm thick). The pipes were split across their long axis to simulate a geological discontinuity, and a similar double embedment length of 1.0 metre was defined either side of the simulated discontinuity as shown in Figure 5. The bolts were inserted into the PVC tubes and installed within the split-pipe without their anchoring devices.
The grout was allowed to cure for 7 days prior to testing, and the specimens were pulled until failure of the steel bar or the grout/steel bond occured. The loads and the displacement at the pipe split were continuously monitored throughout the tests. Damage to the PVC tube for different combinations of load and displacement was also monitored during the tests.

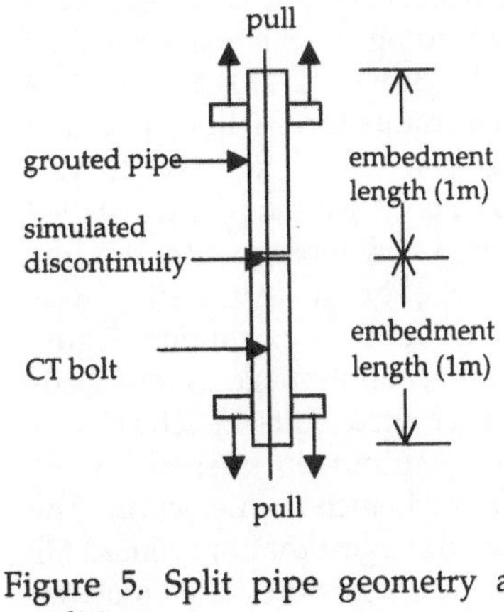

Figure 5. Split pipe geometry and loading conditions.

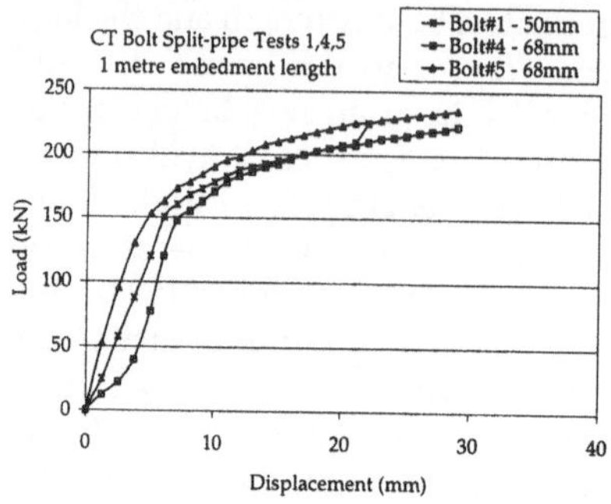

Figure 6. Load deformation results from split-pipe testing of CT bolts.

Figure 6 shows some typical results of the double embedment test carried out at WASM. The results suggest that the nominal steel capacity is reached at very short embedment lengths, despite the bolts being tested without the anchoring device. The corrosion protection capability provided by the plastic tube was evident during the tests. Significant crack opening on the split-pipe was observed without damage to the PVC tube. Split-pipes having 50 and 68mm diameter and grouted with 0.35 water/cement ratio grout were used.

6 IN-SITU PULL TESTING

A series of in-situ pull tests were carried out to determine the actual strength achieved by the anchor-rock interaction in a number of geological environments. All pull tests were carried out according to the recommended ISRM standards (Brown, 1978). Figures 7, 8 and 9 show the results for CT bolts tested at the Cannington Mine in North-West Queensland. The bolts were initially installed as point anchored and subsequently grouted using a 0.40 w/c grout. All testing was carried out 2 days following grouting. Figure 7 suggests that anchor slippage at low loads occurred for ungrouted bolts in schist. This confirms that ungrouted strength is a function of the rock-anchor interaction. The results in Figure 8 suggest that ungrouted CT bolts installed in pegmatite and footwall (F/W) zinc are likely to produce better results than those found in schist. Finally, Figure 9 shows the benefits of post grouting. An increased strength due to bolt-grout, anchor-grout and grout-rock interlock was found for each rock type including the schist. Similar

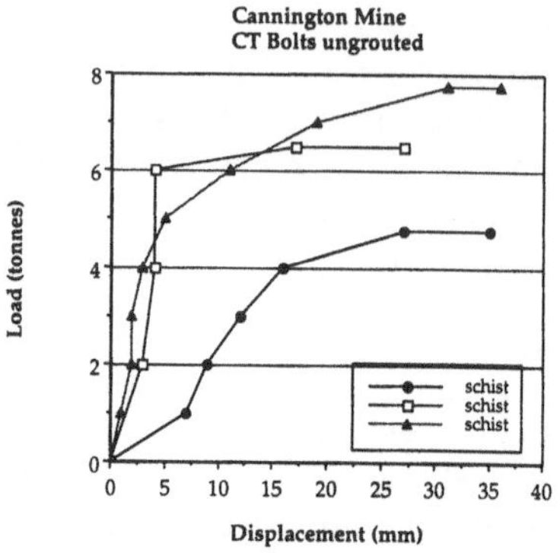

Figure 7. Ungrouted CT bolt strength on schist.

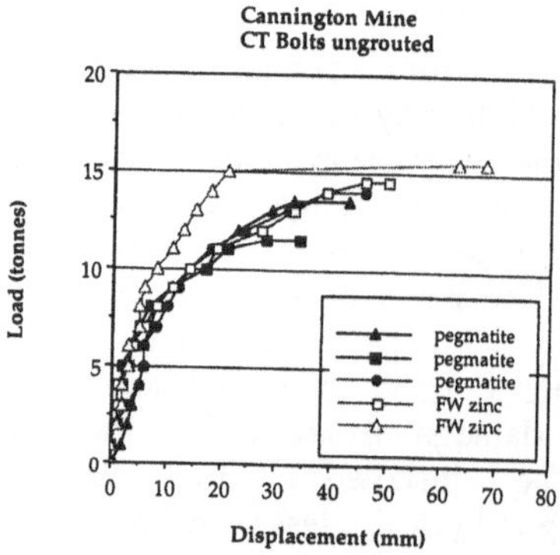

Figure 8. Ungrouted CT bolt strength on pegmatite and F/W zinc.

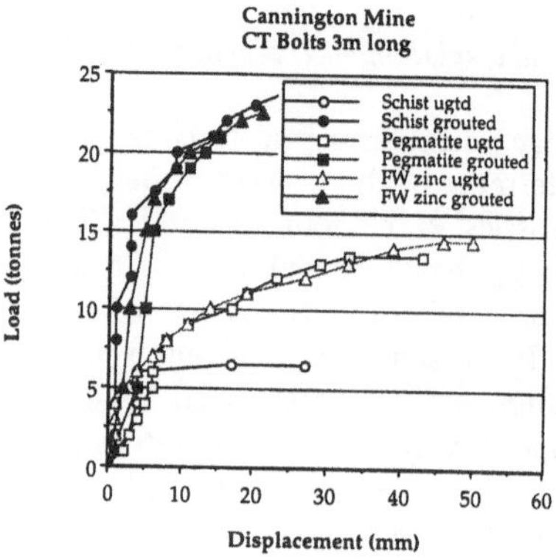

Figure 9. Grouted an ungrouted CT bolt strength.

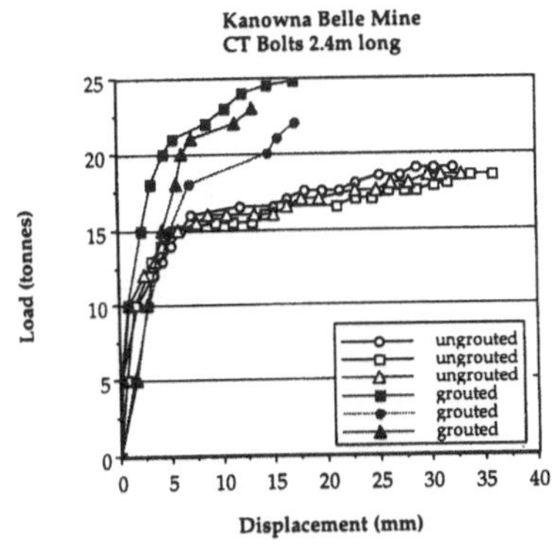

Figure 10. A comparison of grouted and ungrouted strength for CT bolts.

results were found from pull tests carried out at Kanowna Belle, in Western Australia. Figure 10 suggests an ungrouted strength greater than 15 tonnes per reinforcement element.

The grouted strengths on Figures 9 and 10 can not be directly related to bolt capacity per metre of embedment length. This is because the bolts were initially installed with anchors. Consequently, it can be assumed that the pull tests mobilised the anchor against the grout column along the bolt axis. As a result, a series of in-situ pull tests were carried out on CT bolts installed with one metre of embedment length and no anchors. The in-situ pull test results are presented in Figure 11 and suggests a grouted strength in excess of 15 tonnes per metre of embedment length. The pull tests were carried out in Kambalda 56 hours following grouting.

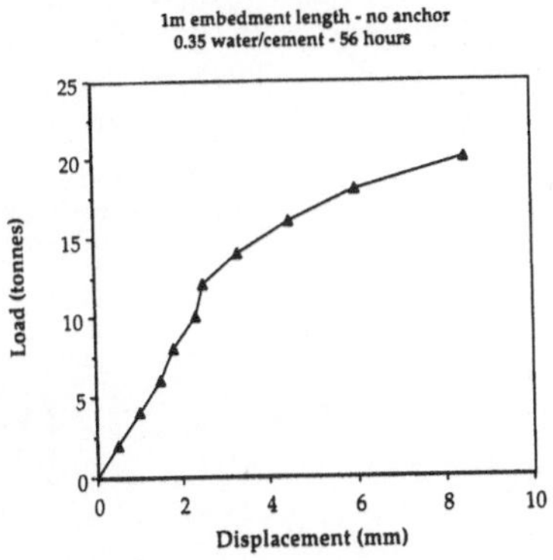

Figure 11. Typical grouted strength per metre of embedment length.

7 CONCLUSIONS

Laboratory testing has established a minimum strength of 15 tonnes for each of the components of an ungrouted CT bolt. However, field tests suggested that the actual ungrouted strength depends upon the rock type. Testing also indicated that grouting of the CT bolt significantly increases the available system capacity. Grouted strengths in excess of 15 tonnes per metre of embedment length were determined from laboratory and in-situ testing.

8 ACKNOWLEDGEMENTS

The first author acknowledges the financial support from the Australian Centre for Geomechanics and Curtin University of Technology for his position as Professor of Mining Geomechanics at WASM. The Australian Centre for Geomechanics has received funding for this position from the Government of Western Australia, Centres of Excellence Program.

9 REFERENCES

ASTM: F432 1995. Standard specification for roof and rock bolts and accessories.

Brown, E.T., (Ed) 1978. Rock characterization testing and monitoring. Commission on Testing Methods, International Society for Rock Mechanics, Pergamon Press, Oxford.

Thompson, A. & C.R. Windsor 1992. A classification system for reinforcement and its use in design. Procc. Western Australian Conference on Mining Geomechanics, (Swedzicki, Baird and Little, Eds), 115-125.

Villaescusa, E. & J. Wright 1997. Permanent excavation reinforcement using cement grouted split set bolts. The AusIMM Proceedings., No 1., 65-69.

Windsor, C.R. & A. Thompson 1993. Rock reinforcement – technology, testing, design and evaluation. Comprehensive Rock Engineering. J. Hudson (Ed). Vol 4, Chp 16, 451-484, Pergamon Press, Oxford.

Windsor, C.R. 1997. Rock reinforcement systems. Int. J. Rock Mech. Min. Sci. Vol 34, No. 6, 919-951.

2 Mesh and membrane support systems

Rock Support and Reinforcement Practice in Mining, Villaescusa, Windsor & Thompson (eds)
© 1999 Balkema, Rotterdam, ISBN 90 5809 045 0

Performance assessment of mesh for ground control applications

Alan G.Thompson, Christopher R.Windsor & Glynn W.Cadby
Rock Technology, Perth, W.A., Australia

ABSTRACT: Mesh is widely used to provide areal rock restraint in mining excavations. The mesh used currently is intended for concrete reinforcement and the optimal mesh requirements for mining applications are unknown. Previous mesh testing methods and the results obtained have been reviewed. A testing program was developed to more closely resemble in situ mesh restraint and loading conditions. The results from these test have been used to propose requirements for modelling the behaviour of mesh. The models for mesh are described and the results from a preliminary analysis, using a simple model, are presented and compared with the corresponding testing results.

1 INTRODUCTION

Ground control for mining excavations may consist of combinations of reinforcement and support. Reinforcement comprises rock bolts and cable bolts which are installed in boreholes. Support may comprise point, strip or areal restraint at the excavation surface. Plates and straps are used to provide point and strip restraint, respectively. A variety of methods may be used to provide areal support to the rock mass. The most common methods are sprayed coatings and mesh. The sprayed coatings may be either based on concrete (e.g. shotcrete or fibrecrete) or polymeric based materials (e.g. Wojno, 1998). Mesh used in mining is generally manufactured from steel wires that may either be woven together to form a flexible, continuous sheet (which is usually supplied in rolls) or welded to form a relatively stiff, flat sheet.

The most commonly used mesh is of the welded variety. It is used because the relatively stiff sheets are more amenable to mechanised placement; either by drilling jumbo or, more recently, using purpose built machines developed by various equipment manufacturers. This mesh is formally called welded wire reinforcing fabric and consists of longitudinal wires with cross-wires welded to them. The mesh has been specifically developed for its intended use as reinforcement in concrete. Its manufacture and mechanical properties are specified by AS 1304-1991 (Standards Australia, 1991). The standard assumes that the mesh will be encased in concrete. The important mechanical properties are the tensile strength of the wires and the shear strength of the welds. The current code AS 1304-91 and a new draft code DR 98119 (Standards Australia, 1998) both effectively specify the minimum weld strength as approximately 50% of the nominal yield strength of the wire. For mining applications, the mesh is supported by plates restrained by reinforcement and loaded by slabs or blocks of rock. In addition to axial loading, the rock loading causes bending and shear near the plates and distortion of the regular grid of wires. Other complex responses of the mesh are caused by irregularly shaped pieces of rock attempting to be forced between the grid of wires.

The requirements of welded mesh for mining applications are being investigated as part of an Australian Mineral Industries Research Association Limited (AMIRA) sponsored research and development project. The investigation was prompted by a number of observations relating to the typical usage of mesh in mining. For example, it was found that bolt spacings were being governed by the dimensions of the mesh. This often results in bolts being more closely spaced than would otherwise have been required. Bolt spacings without mesh are typically of the order of 1.2 m. The use of standard 2.4 m wide sheets with an overlap at the edges may result in a decreased bolt spacing of about 1.1 metre. That is, approximately 9% more bolts may be required per lineal metre of development.

Many questions may be posed with regard to the required configuration and grade of mesh for particular rock conditions and bolting patterns. Some of these are:

- What is the maximum volume of potentially unstable rock located between the reinforcement?
- What wire size (diameter) is appropriate for a given reinforcement pattern?
- What is the acceptable maximum displacement of the mesh?

The behaviour of mesh is being investigated with a view to defining the optimum wire sizes and their configurations. The research involves:

- mechanical tests of mesh sheets with different bolting and loading patterns
- theoretical investigations and associated computer program development.

The theoretical investigations are required since it is not feasible to mechanically test for the large number of possible combinations of bolting and loading conditions with various mesh configurations. It is estimated that previous testing programs described in the next section have involved in excess of 150 tests. While these have provided useful information regarding the behaviour of existing mesh types and configurations, the data do not allow for the evaluation of the effects of modifying some of the parameters with a view to developing meshes that are possibly more suitable for use in underground excavations mined in rock.

2 REVIEW OF PREVIOUS TESTING

An early study of different types of mesh was conducted by Ortlepp (1983). The test program included 'diamond' meshes (stainless and galvanised steel), square woven stainless steel, stranded wire weave, hexagonal twisted, weld mesh and polyethylene rope net. The mesh was supported on a 1.1 m square steel frame and restrained on all sides. Rock loading was simulated by the application of an articulated load over a central area of 0.42 m by 0.52 m. The mesh behaviour was characterised by measuring the load-displacement response. It was stated that failure invariably occurred at a mesh intersections or 'cross-overs'. At these points, different stress concentrations and material strength impairment occurred depending on whether the cross-over was linked, woven, welded or knotted. The least impairment to the wire strength occurred for the inter-linked diamond mesh.

Tannant (1995) has reported the results for tests on welded-wire meshes. In these tests, the mesh was restrained by four plates placed in square or 'diamond' patterns and bolted to a steel frame. The rock loading was simulated by pulling a 300 mm square plate with rounded corners upwards 'through' the mesh. Wire with gauges of #9, #6 and #4 (nominal diameters of 3.77 mm, 4.88 mm and 5.72 mm, respectively) were used for the welded-wire mesh. The tension in the bolts holding down the plates was varied between 10 kN and 30 kN. The stiffness of the response to loading was found to vary mainly due to the orientation of the rock bolt pattern relative to the wires of the welded mesh and the number of wires transferring load between the loading plate and the plates holding down the mesh. The initial stiffness was reported to be more or less independent of the wire diameter for the diamond pattern. However, the stiffness of response for the square pattern was greatest for the largest wire diameter. The response stiffness was found to decrease due to slip of the mesh relative to the plates holding down the mesh at lower bolt tensions. As expected, the peak load recorded in the tests was greatest at the largest wire diameter.

A more recent South African study (reported by Ortlepp and Stacey, 1997 and Stacey, Ortlepp and Kirsten, 1998) involved dynamic testing of various types of welded and chain-link (diamond) wire mesh, either alone or in combination with wire rope lacing. The mesh was supported by four plates and rock bolts suspended from steel beams and was also restrained at the four edges to simulate continuity of the mesh beyond the edge of the tested area. The mesh supported an arrangement of concrete blocks on to which various masses were dropped from different heights to simulate loading from rock bursts of different energy intensities. Weld mesh was found to have a relatively low capacity to absorb energy. In almost all tests wires broke and in some cases welds also broke. The diamond meshes tested were found to have a 50% greater capacity than weld mesh to absorb energy. However, a tendency was observed for the diamond mesh to unravel once a wire had failed and this allowed the concrete blocks to spill through the mesh. It was also found that standard plates had a detrimental effect on the mesh performance. This was attributed to the sharp edges causing stress concentrations in the wires.

In summary, the previous mesh testing has identified a number of important features associated with behaviour of weld mesh subjected to both static and dynamic loading. However, some of the observations are related to the particular mesh configuration, the arrangements of bolts and the size and location of the loading area. A testing program was designed to provide data to supplement previous testing with the main objective of developing a model for welded-wire mesh.

3 MESH TESTING PROGRAM

The initial mesh testing program involved measuring the performance of a specific weld mesh commonly used in the Australian mining industry. The mesh consists of nominally 5.6 mm diameter wires. The mesh is fabricated by electric arc welding of cross-wires to longitudinal wires. Both sets of wires are spaced at 100 mm centres. The mesh is supplied in sheets 2.4 m wide and in lengths from ~3 m to ~6 m.

3.1 *Equipment and Test Procedure*

Figure 1 shows the arrangement for applying load to the mesh. A square loading frame, fabricated from steel angle section, was used to simulate a 'rigid' slab or block of rock. A sheet of mesh, trimmed to a convenient size (approximately 100 mm beyond the edge of the plates), was placed over the loading frame and held between 10 mm thick, 200 mm square steel plates secured with M20 anchor bolts screwed into inserts in the concrete slab floor. A torque wrench was used to apply 200 Nm of torque to the bolts to provide a consistent pre-tension estimated to be ~45 kN. The loading frame was attached by four bolts to a cross-head which was supported, through a universal coupling, by a threaded bar. The entire assembly of loading frame, cross-head and threaded bar was raised by rotating a capstan on the threaded bar.

The capstan rested on a thrust bearing to reduce rotational friction. An electronic load cell placed between the thrust bearing and cross beam was used to measure the force applied to the mesh. The displacement was measured by counting the number of revolutions of the capstan (one revolution of the capstan resulted in 10 mm of vertical displacement). An assessment of the probable mesh behaviour was used to estimate the rate of applying displacement to the steel frame. In general, displacement was applied in increments of 5 mm or 10 mm (i.e. 0.5 or 1 revolution of the capstan, respectively).

The slip of the mesh relative to the plates was estimated by marking the wires at the start of the test and recording any changes during the test.

Note that no restraint was provided at the edges of the mesh sheet. In actual practice, the restraint beyond the line of the bolts is largely unknown and depends on the location of loading relative to the mesh and the bolts. For example, an additional line of bolts may be located all round a loading to the middle of a sheet so that the 'edges' of the mesh would be restrained by the mesh outside the line of bolts. On the other hand, the loading may occur near the edge of the sheet, in which case the edge is unrestrained laterally but may overlap another sheet and the three other edges may have varying degrees of both horizontal and vertical restraint. For these reasons, it was deemed preferable to have the edges of the mesh free for the purposes of having the simplest configurations to facilitate comparison between the test results and the computational models.

Figure 1. Photograph of mesh testing facility.

Figure 2. Photograph showing arrangement of mesh and bolts and load frame raised by bolts passing from near the corners to the cross-head.

Figure 3. Photograph showing arrangement of mesh and bolts and load frame raised by bolts passing from the middle of the edges to the cross-head.

3.2 Mesh, bolt and loading configurations

The test program involved varying:

- the size and orientation of the area of mesh loading
- the bolt spacings
- the orientation of the mesh relative to the bolts.

A system of uniquely identifying the tests was developed. Each test was identified by 13 alpha-numeric characters (MmBbxxyyPpsss) in which:

Mm = mesh orientation – m = s(quare)/o(blique)
Bb = bolt orientation – b = s(quare)/o(blique)
xx = bolt spacing in dm (1 dm = 0.1 m)
yy = bolt spacing in dm
Pp = plate orientation – p = s(quare)/o(blique)
sss = square plate side length in cm

The initial test program geometric arrangements are shown in Table 1. The details for each test are given in Table 2.

In practice, mesh bolting patterns have usually been designated as being either 'square' or 'diamond' with the implication that bolts are equally spaced. A more preferable terminology to account for unequal bolt

Table 1 Geometric arrangements corresponding to the test designations given in Table 2.

$M_SB_S1010P_O075$	$M_SB_S1515P_O075$	$M_SB_S2020P_O075$	$M_OB_O1515P_O075$
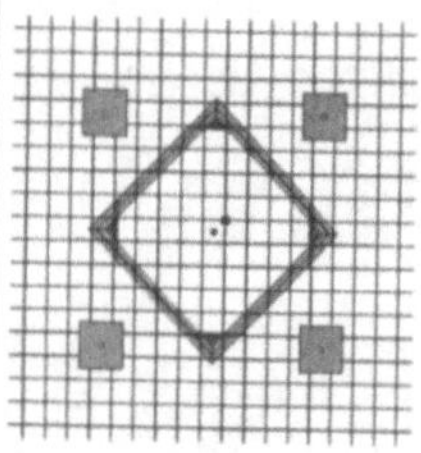	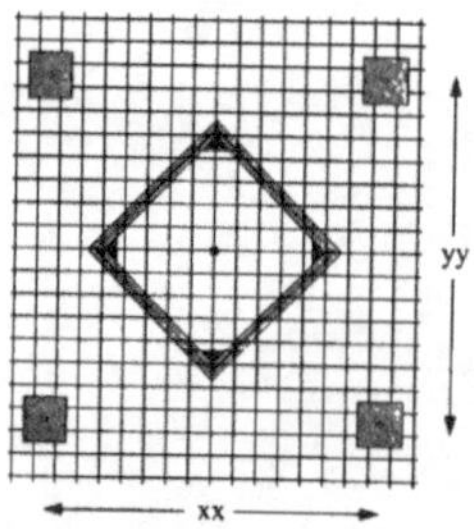	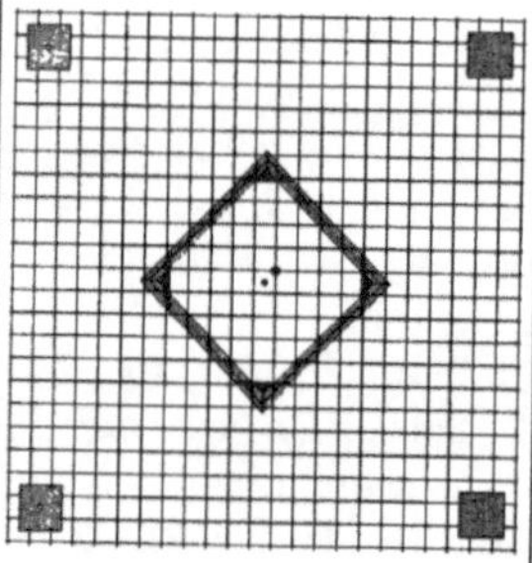	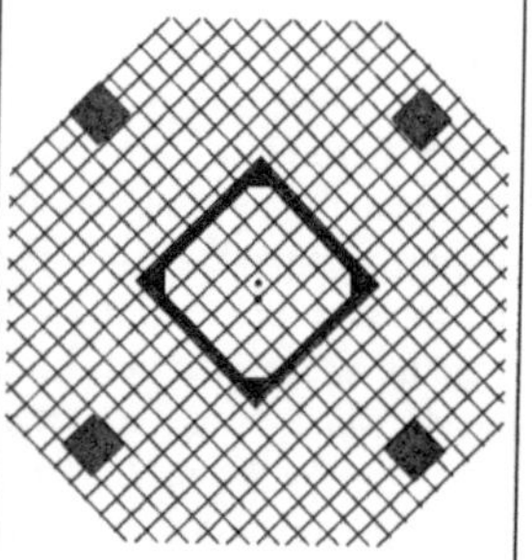
$M_SB_S1515P_S075$	$M_SB_S2020P_S075$	$M_OB_O1515P_S075$	$M_OB_O1520P_S075$
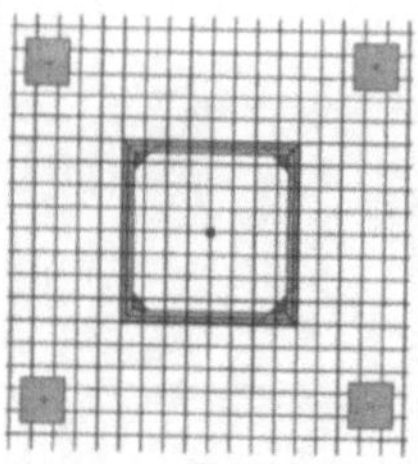	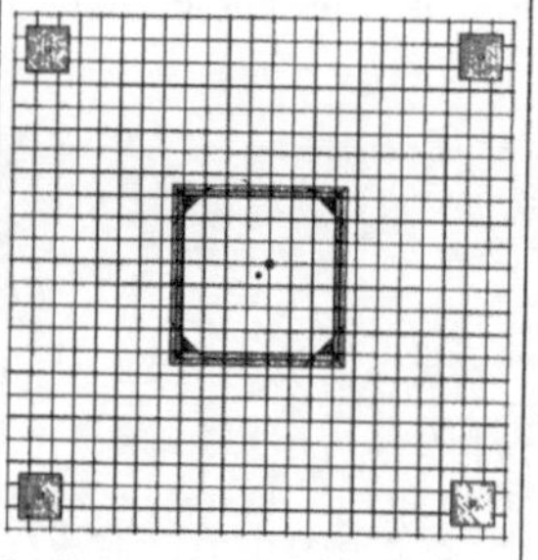	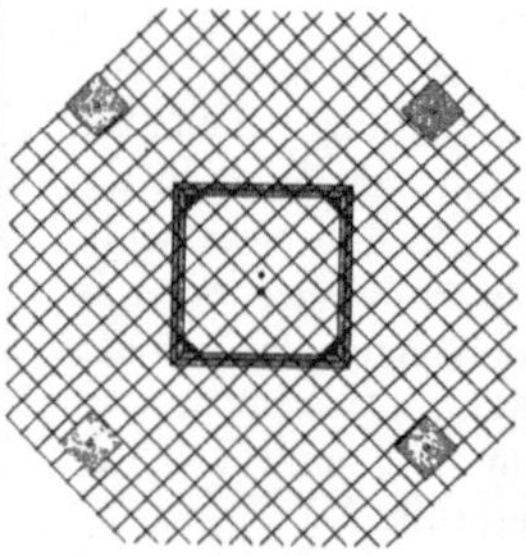	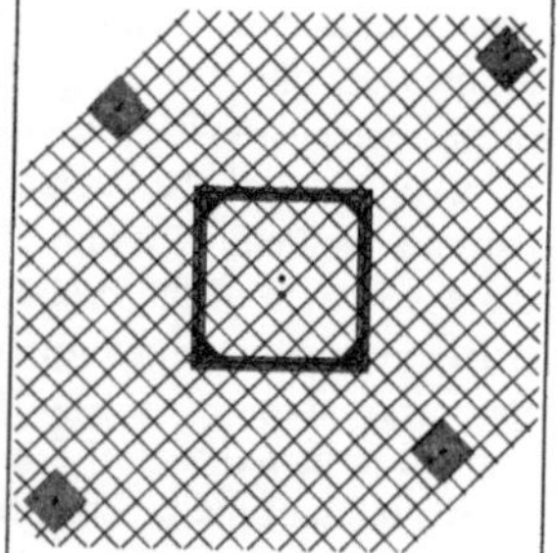
$M_SB_S1515P_O105$	$M_SB_S2020P_O105$	$M_SB_S2020P_S105$	$M_OB_O1515P_O105$
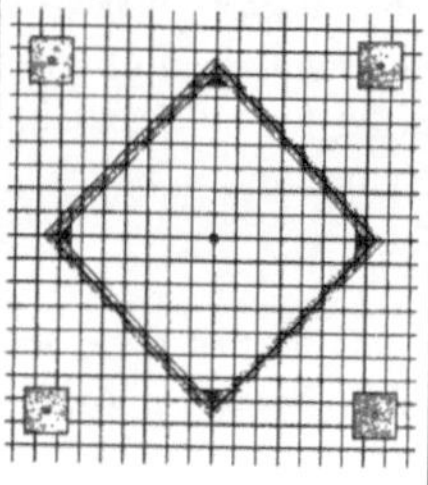	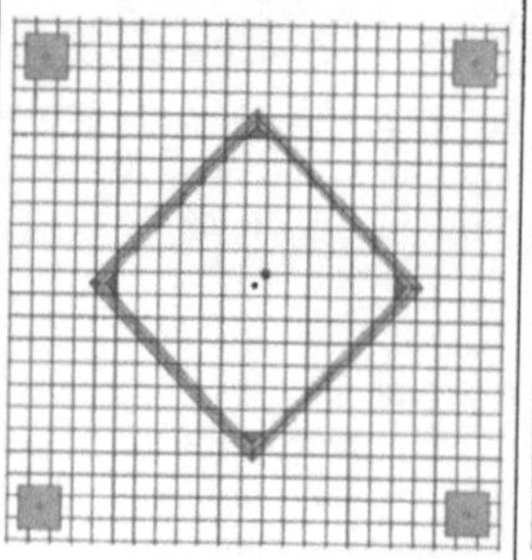	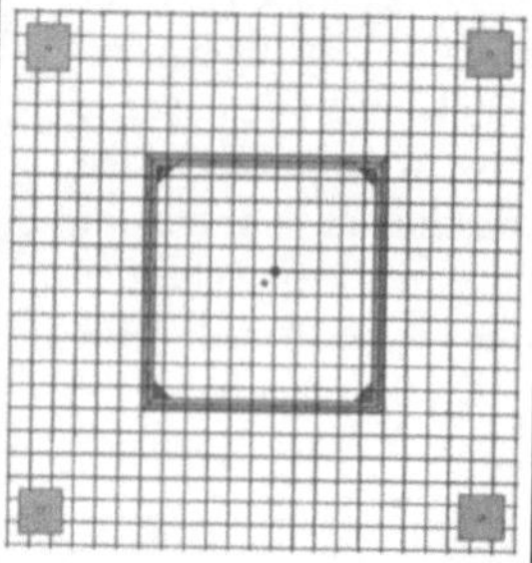	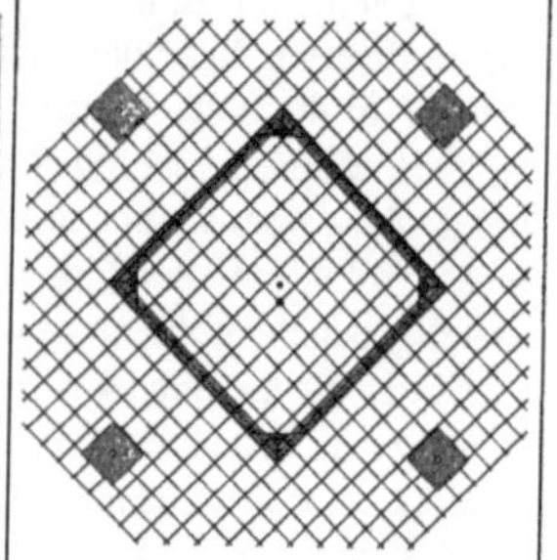

Table 2. Summary of testing program.

Test Designation	Bolt Configuration	Bolt Spacing (m) a	b	Plate Orientation	Plate Size (mm)
$M_sB_s1010P_o075$	Square	1.0	1.0	Oblique	750
$M_sB_s1515P_o075$	Square	1.5	1.5	Oblique	750
$M_sB_s1515P_o105$	Square	1.5	1.5	Oblique	1050
$M_sB_s2020P_o105$	Square	2.0	2.0	Oblique	1050
$M_sB_s2020P_o075$	Square	2.0	2.0	Oblique	750
$M_sB_s1515P_s075$	Square	1.5	1.5	Square	750
$M_oB_o1515P_s075$	Oblique	1.5	1.5	Square	750
$M_sB_s2020P_s075$	Square	2.0	2.0	Square	750
$M_oB_o1520P_s075$	Oblique	1.5	2.0	Square	750
$M_oB_o1515P_o075$	Oblique	1.5	1.5	Oblique	750
$M_sB_s2020P_s105$	Square	2.0	2.0	Square	1050
$M_oB_o1515P_o105$	Oblique	1.5	1.5	Oblique	1050

spacings is given in Figure 4. In this figure, 'a' is the bolt spacing in the longitudinal direction of the mesh and 'b' is the bolt spacing in the cross-wise direction. The bolting patterns are termed either 'rectangular' or 'oblique'. With this terminology, square and diamond bolting patterns result when a = b.

In the testing program, rotating the mesh relative to the fixed location of the bolts effectively resulted in the oblique bolt pattern (i.e. tests with initial designation M_oB_o). The equivalent bolt spacings using the designations given in Table 1 and Figure 4 are a ≈ 1.4 xx and b ≈ 1.4 yy, with the lines of the bolts offset from one another by one half of the spacing. Bolt spacings of 1 m, 1.5 m and 2 m were used in square patterns. One asymmetric, oblique bolt pattern was used ($M_oB_o1520P_s075$). The other oblique arrays effectively resulted in the so-called 'diamond' pattern.

Two different sized square loading frames were used in the test program; 750 mm and 1050 mm. These sizes were deemed to be appropriate for determining the behaviour of mesh subjected to loadings of several tonnes (previous testing has often been conducted with ~300 mm square plates). The actual sizes were also selected to be compatible with the spacing of the wires in the mesh. The bolts which attached the steel frame to the cross-head were located so that they passed through the mesh in a position which did not interfere with the interaction between the frame and the mesh.

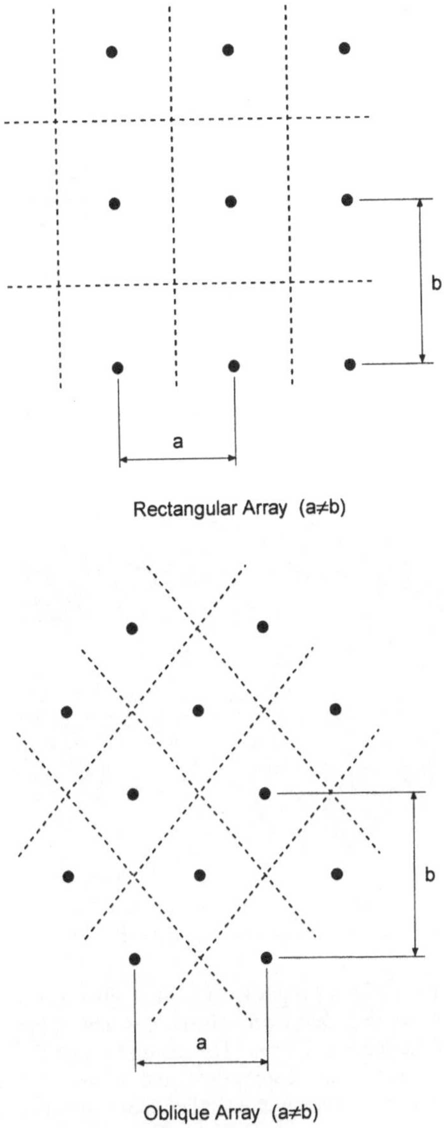

Figure 4. Mesh bolting patterns.

3.3 *Physical observations*

A number of important features associated with the performance of mesh were observed during the course of the testing program. Many of the physical observations can be used to explain the force-displacement responses given in the next section and to identify the essential features that must be included with any mesh analysis methods.

The most obvious feature of the results was the difference in mesh distortion at similar levels of loading as demonstrated by Figures 5 and 6 for

Figure 5. Photograph showing typical large mesh distortion (displacement = 350 mm at load = 18 kN) with 'square' bolt pattern in $M_sB_s2020P_s105$.

Figure 6. Photograph showing typical smaller mesh distortion (displacement = 100 mm at load = 20 kN) with 'diamond' bolt pattern in $M_oB_o1515P_o105$.

square and oblique bolt patterns. It is also worth noting in Figure 5 that the mesh 'bulges' above the plane of the loading frame. This confirms that the use of a frame was appropriate and it was not necessary to have a continuous, plate-like loading to simulate loading from a single rigid block of rock.

The other interesting observations were related to the mesh distortion between the plates and the loaded area. Gross distortion of the mesh was observed in all cases where a 'square' bolt pattern was used, irrespective of whether the loading frame was 'square' or 'oblique'

For example, Figure 7 shows a close up of the mesh distortion in the region between the corner of the loading frame and one of the plates and bolts in test $M_sB_s1515P_s075$. However, beyond this distorted region of approximately 3 'squares' in each direction, the mesh has retained its original shape parallel to the edges of the loaded as shown in Figure 8.

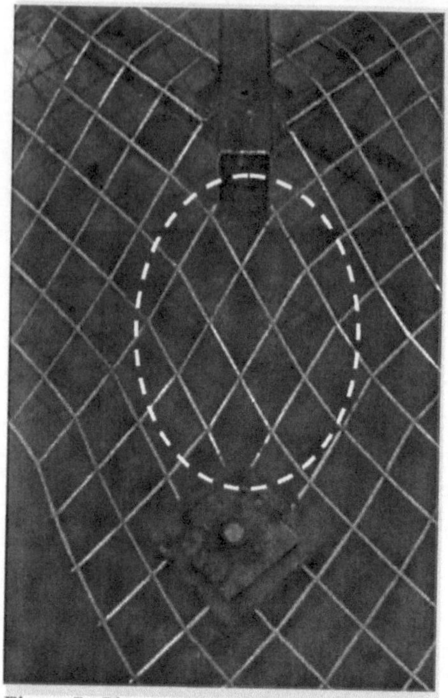

Figure 7. Photograph showing close up of mesh distortion in $M_sB_s1515P_s075$.

Figure 8. Photograph showing deformed mesh in $M_sB_s1515P_s075$.

Figures 9 and 10 show that, for tests in which the mesh and bolting pattern is square and the loading is oblique, the mesh has retained its original shape inside an area delineated by wires (indicated by the dashed lines) passing from the corners of the loading frame towards a plate and bolt. In both cases, the mesh is only severely distorted from the corner of the plate towards the apparently undistorted area .

Figure 11 shows the mesh in the region of one of the plates for an oblique pattern of bolts. In this photograph, the wires located immediately either side

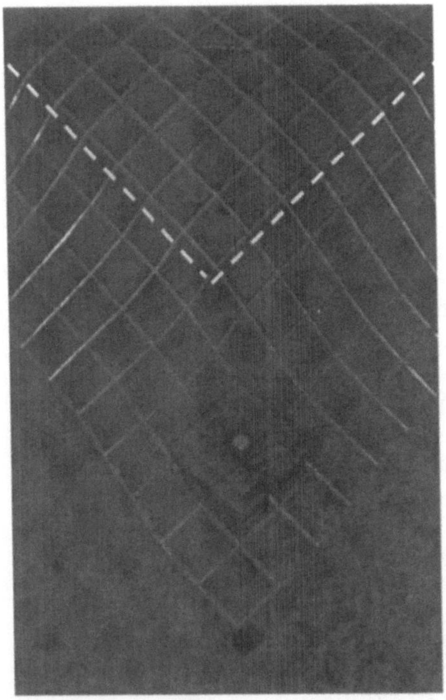

Figure 9. Photograph showing mesh distortion in $M_sB_s2020P_o105$.

Figure 10.Photograph showing mesh distortion in $M_sB_s2020P_o075$.

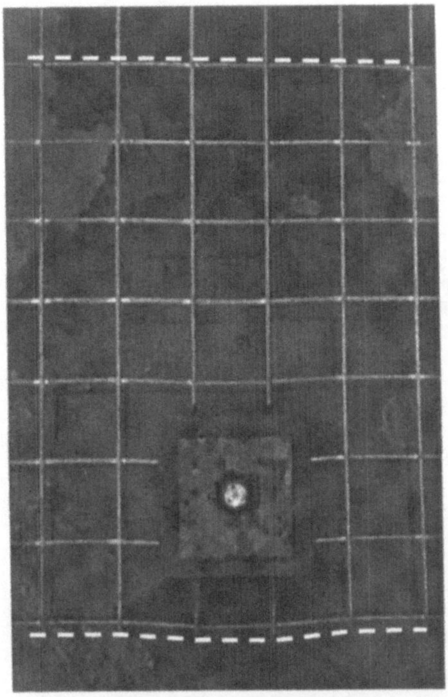

Figure 11. Photograph showing mesh distortion near a plate and bolt in $M_oB_o1520P_s075$.

of the plate appear to have caused distortion of the initially straight, lowermost wire. On the other hand, the uppermost wires are almost straight. This indicates that the wires directly restrained by the plate are not the only wires transferring force from the loaded area of the mesh to the plates. However, the larger strains in the two central wires indicates that the force transmitted through these wires must be significantly larger than that transmitted by the adjacent wires.

Figures 9 and 10 also show distortion of the mesh near the plates but the relative distribution of forces in the wires is not evident.

It is also worth noting that the majority of the mesh distortion shown in Figure 5 remained after removal of the loading and restraint provided by the plates and bolts.

3.4 Mesh testing results

Two tests were performed for each of the test designations. The differences in the load-displacement responses were small and the average responses are reported in Figures 12 to 16. The combination of responses shown in each of these figures was selected in an attempt to graphically show the differences caused by changing one or more of the parameters. For example, Figures 12 and 13 show the

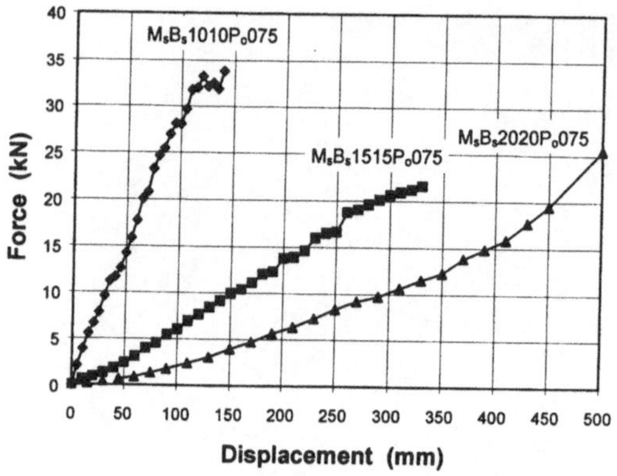

Figure 12. Variation in mesh response as the square bolt spacing was changed for the 750 mm oblique loading.

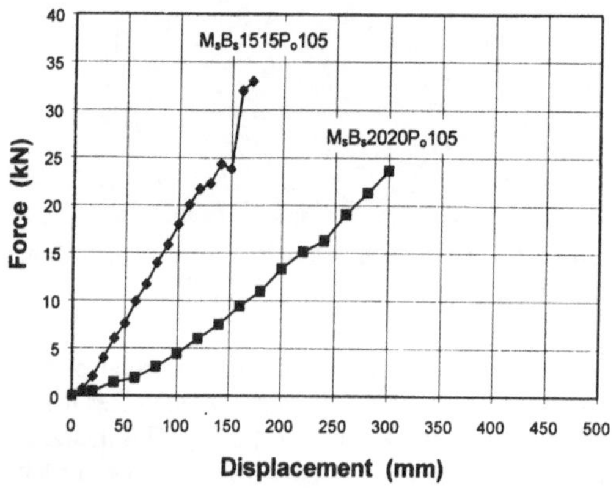

Figure 13. Variation in mesh response as the square bolt spacing was changed for the 1050 mm oblique loading.

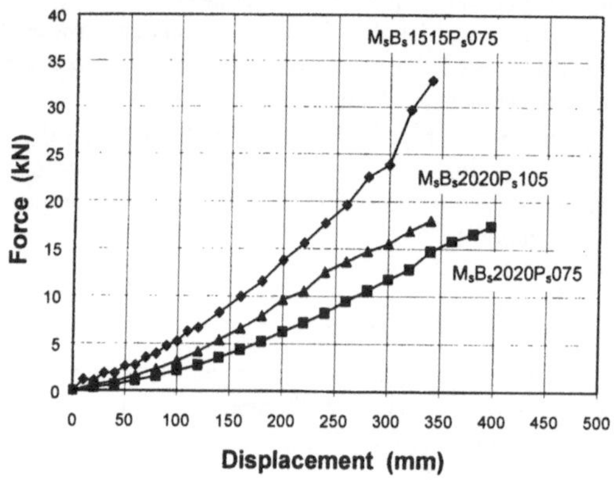

Figure 14. Variation of mesh response for square bolt spacing and square loading orientation.

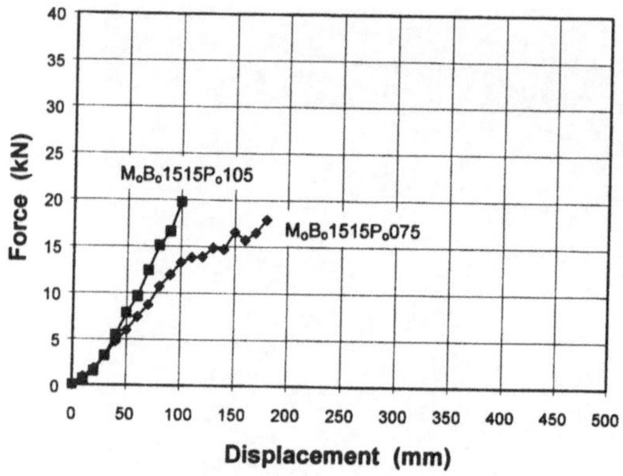

Figure 15. Variation of mesh response with change in oblique bolt pattern for the square orientation of 750 mm square loading.

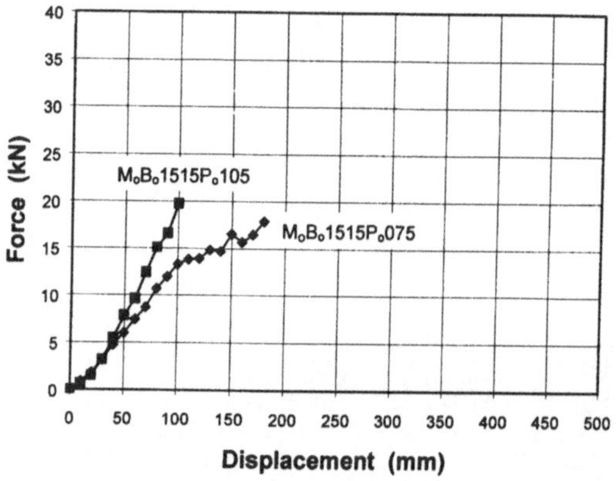

Figure 16. Variation in mesh response for oblique bolt pattern subjected to different sizes of oblique loading.

effect of varying the bolt spacing relative to the small and large loading areas, respectively.

The curves in Figures 12 to 16 do not show the measured reductions of force which accompanied slip of the mesh at one or more of the plates. In some cases, slip of the mesh was prevented by a wire catching on a plate edge.

4 INTERPRETATION OF RESULTS

The major features of the results relate to the stiffness of the force-displacement response, slip of the mesh relative to the plates and bolts, and the deformation of the mesh.

4.1 Stiffness

The stiffness of response is related to both the configuration of the mesh relative to the bolts and the distance of the corners or edges from the plates.

The stiffest response was recorded when the corners of the loading frame were restrained by wires passing directly between the plates (e.g. tests $M_sB_s1010P_0075$ and $M_sB_s1515P_0105$. As the bolt spacing increased for the same type of test, the load was transferred less directly from the loading to the plates and the stiffness decreased, as shown in Figure 12 and Figure 13, respectively. Figures 14 to 16 generally show that the stiffness of response reduces as the distance increases between the load and the plates and bolts.

4.2 Slip at bearing plates

Lateral loading of a restrained, flexible element causes large axial forces (relative to the magnitude of the lateral load) to be developed in the element. Similarly, in the mesh tests, large wire forces were developed and these were sufficient to cause slip of the mesh relative to the plates at relatively low applied loadings. The slip of the mesh at the plates has the effect of reducing the stiffness of response compared with the stiffness that would have resulted if the mesh had been completely restrained.

4.3 Deformation and plastic deformation

It is clear from Figure 17 that the distortion of the mesh involves bending and permanent, irrecoverable plastic deformation of the wires. It also appears that the welds cause some local increase in the bending stiffness of the wires (as highlighted in this figure) and the welds are subjected to twisting forces.

Figure 17 Distortion of the mesh near a plate and bolt showing the local stiffening of the wires due to the welds.

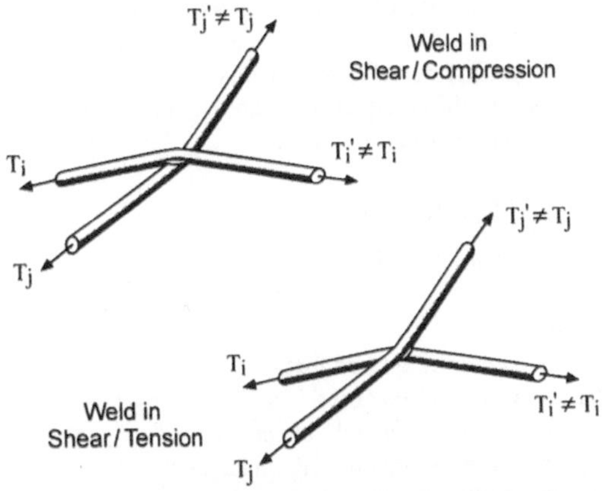

Figure 18 Schematic diagram showing the effects of the relative location of mesh wires at a welded intersection.

5 MESH ANALYSIS REQUIREMENTS

The requirements for mesh analysis, based on the testing program results and other considerations related to the design of alternative mesh configurations, are:

- Variable wire diameters.
- Variable wire spacings.
- Non-linear stress-strain properties for the wire.
- Weld strength.
- Slip of the mesh at the plates and bolts.
- Variable bolt tensions.
- Variable bolt spacings.
- Variable mesh orientation relative to bolt pattern.
- Mesh lay relative to wire loading.
- Variable load types and areas.
- Large mesh displacements.

Most of these requirements are self-explanatory; however, some are not. For example, the mesh lay refers to the location of the cross-wires relative to the longitudinal wires (i.e. above or below as shown in Figure 18). The relative location of the wires will influence whether the forces at a particular intersection in the mesh will produce tension or compression combined with shear in the weld.

Variable load types and areas refers to whether the mesh is loaded at a number of discrete points within the mesh (i.e. as modelled in the test program to simulate a rigid block of rock) or by distributed loading of an area the mesh (i.e. to simulate loading by a number of small blocks formed in closely-spaced, jointed rock). Lastly, large displacements mean that the changes in mesh geometry must be able to be taken into account.

6 THEORY OF MESH BEHAVIOUR

An analysis method incorporating all the requirements listed in the previous section is currently being developed. The precise details of the mesh model are beyond the scope of this paper. However, the basic features of the method can be briefly outlined and some of the important outputs detailed.

The basis of the method is to satisfy equilibrium of forces and moments and compatibility of displacements and rotations at every longitudinal wire and cross-wire intersection or 'node' in the mesh. It is assumed the mesh is restrained at a number of nodes (representing plates and bolts) and subjected to either defined force or displacement loading at a variable number of nodes (representing rock loading). In the general case, there are 6 equations of equilibrium associated with each node in the mesh. This means having to find the solution to a large number of simultaneous equations.

The resulting equations can be summarised in partitioned matrix notation as:

$$\begin{bmatrix} F_a \\ F_u \end{bmatrix} = \begin{bmatrix} K_{aa} & K_{au} \\ K_{ua} & K_{uu} \end{bmatrix} \begin{bmatrix} d_a \\ d_u \end{bmatrix} \tag{1}$$

where:

F_a = vector of applied forces
F_u = vector of unknown forces
d_a = vector of applied displacements
d_u = vector of unknown displacements
K_{ij} = 'stiffness' matrix relating forces and moments to displacements and rotations at each node.

Forces and displacements may be directed in any of the three coordinate directions. Applied displacement may be used to simulate either loading or restraint at the plates and bolts. That is, zero applied displacement at a node implies rigid restraint. Non rigid restraint may be used to allow for modelling the effects of slip of the mesh relative to the plates.

Equation 1 may be assumed to consist of two separate matrix equations:

$$[F_a] = [K_{aa}][d_a] + [K_{au}][d_u] \tag{2}$$

$$[F_u] = [K_{ua}][d_a] + [K_{uu}][d_u] \tag{3}$$

These simultaneous matrix equations are solved in two parts. Firstly, equation 2 is used to solve for the unknown displacements.

$$[d_u] = [K_{au}]^{-1}\{[F_a] - [K_{aa}][d_a]\} \tag{4}$$

where $[K_{au}]^{-1}$ is the inverse of $[K_{au}]$.

Finally, the unknown forces at the nodes may be calculated by substituting $[d_u]$ into equation 3.

In setting up equation 1, relationships are established between mesh nodal displacements and the forces in the wires. These relationships are used to calculate the wire forces which are then used to calculate the weld forces. This process appears to be simple and straight forward. However, in the previous section important requirements were identified involving non-linearities in the material properties and changes in the geometry of the mesh. The need to satisfy these requirements means that a new set of equations need to be developed to account for the material properties and the calculated changes in mesh geometry. Starting with this new geometry, the solution procedure is repeated until equilibrium of forces and compatibility of displacements are satisfied or failure is predicted for one or more of the components (wires or welds). At the time of preparation, the computer program was still being developed and no results are given.

7 SIMPLIFIED ANALYSIS OF MESH BEHAVIOUR

The following analysis method is an extension to the work of Tannant (1995) and is restricted to cases where the wires passing directly between the plates held by the rock bolts carry most of the force imposed by the loading. If it is further assumed that the wires have a low bending stiffness relative to their axial stiffness, then the deformed wire shape and the forces acting in each wire passing directly between the plates are as shown in Figure 19. Vertical equilibrium for the mesh, plate and bolt requires:

$$N_{wp} = T_b / n \tag{5}$$

$$N_{wr} = N_{wp} - T_w \sin\theta \tag{6}$$

where n is the number of wires restrained by the plate. Equation 6 indicates that as the mesh is displaced, the force between the wire and the rock decreases while there is a corresponding increase in the force between the wire and the plate. If the force applied to the wire increases sufficiently, then contact between the wire and rock may be lost at one edge of the plate but, due to plate rotation, contact is maintained at the other edge of the plate.

In practice, the restraining force at the plate will be a function of the friction and mechanical interlock between the mesh and the plate and the mesh and the rock. However, in general, the maximum value of the horizontal wire restraining force (F_{hmax}) at the plate may be assumed to be given by:

$$F_{h\,max} = F_{wp} + F_{wr} \tag{7}$$

where:

$F_{wp} \leq \mu_{wp} N_{wp}$

μ_{wp} = coefficient of friction between wire and plate

$F_{wr} \leq \mu_{wr} N_{wr}$

μ_{wr} = coefficient of friction between wire and rock

$F_{ws} \leq \mu_{wr} N_{ws}$

μ_{ws} = coefficient of friction between wire and load

Subject to not violating the preceding limits of friction load transfer at the plate and at the assumed loading point, the following relationship may be established between the tension in the wire (T_w) and the displacement (d):

$$T_w = \frac{K_w L_0\left(\sqrt{1+\left(\frac{d}{L_0}\right)^2}-1\right)}{1+\frac{1}{\sqrt{1+\left(\frac{d}{L_0}\right)^2}}\left\{\frac{K_w}{K_s}\left(1-\mu_{ws}\frac{d}{L_0}\right)+\frac{K_w}{K_p}\right\}} \tag{8}$$

where:

K_w = stiffness of wire between the plate and load
= AE/L_0

K_s = stiffness of wire between the centre and edge of the load
= $2AE/w_s$

A = cross-sectional area of the wire

E = Elastic Modulus of the wire

K_p = stiffness of restraint against wire sliding at the plate. This resistance is assumed to be relatively stiff up to when sliding initiates, after which, the effective stiffness will be low with sliding at more or less constant force.

To predict the response for a given set of test conditions, a simple computer program was written to calculate (using equation 8) the wire force resulting at different values of the applied displacement. At each step, the limiting conditions for sliding at the plate were checked. If the predicted force exceeded the limiting value, the force in the wire was reduced to be consistent with the limiting value. The overall force at each displacement was then calculated by multiplying the predicted force in one wire by the number of wires that were assumed to be acting for the particular test conditions.

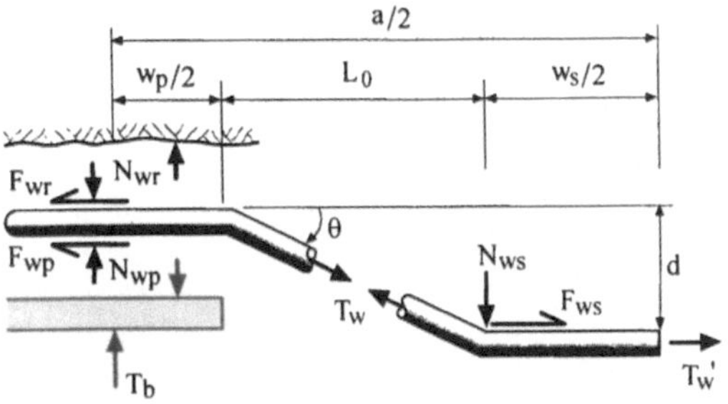

a = effective bolt spacing

w_s = size of loaded area (side / diagonal dimension)

w_p = size of plate

L_0 = initial length of wire between plate and load

d = displacement applied to the wire

θ = artan(d/L_0)

T_w = tension in wire between plate and load

T_w' = tension in wire between loading points

T_b = bolt tension

N_{wp} = average normal force between wire and plate

F_{wp} = friction between the wire and plate

N_{wr} = average normal force between wire and rock

F_{wr} = friction between the wire and rock

N_{ws} = average normal force between wire and load

F_{ws} = friction between the wire and load

Figure 19 Assumed deformation and forces for symmetrically restrained, loaded wire.

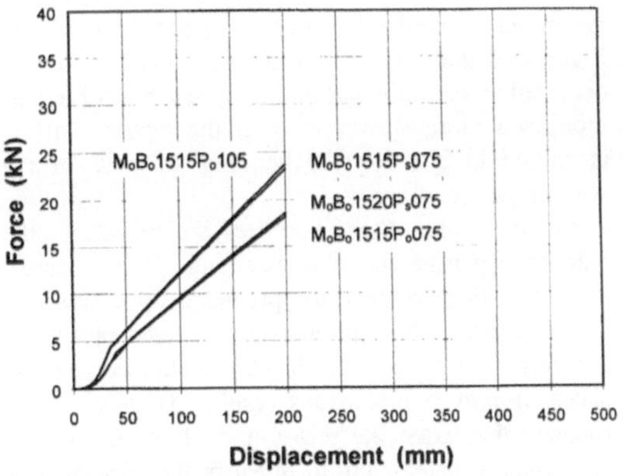

Figure 20 Computer predictions of mesh response based on wire model shown in Figure 19.

This simple wire model for mesh can only be applied to tests $M_oB_o1515P_o075$, $M_oB_o1515P_o105$, $M_sB_s1010P_o075$, $M_sB_s1515P_o105$, $M_oBo_s1515P_s075$ and $M_oB_o1520P_s075$ in which most of the force applied to mesh is transmitted by wires passing directly between plates either side of the mesh (see

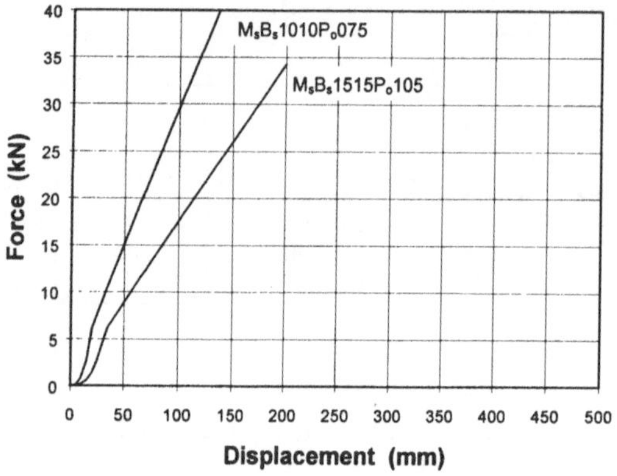

Figure 21 Computer predictions of mesh response based on wire model shown in Figure 19.

Table 1). The results from the analyses are given in Figures 20 and 21. The assumed wire properties are:

Diameter = 5.6 mm
Elastic modulus = 200 GPa
Yield strength = 500 MPa (12.3 kN)
Ultimate strength = 550 MPa (13.5 kN)
Friction coefficients $\mu_{wp} = \mu_{wp} = \mu_{wp} = 0.2$

The wire forces in all cases are predicted to remain well below the wire yield strength due to slip at the plates and bolts. This was consistent with the observations in the corresponding tests.

In some cases, reasonable comparisons were obtained between the test results and the computer generated force-displacement responses shown in Figures 20 and 21. The most obvious differences occurred at small displacements where the force was mobilised more slowly than in the tests. This is attributed to ignoring the bending stiffness of the wire in the model.

In other cases, the differences are related to the number of wires assumed to be acting. In most cases, at a given displacement, the predicted force was less than measured. If additional force is assumed to be transferred in wires not directly restrained by the plates, improved predictions result. However, it is apparent that extending this empirical approach is of little value in attempting to quantify the performance of alternative mesh configurations and the computational approach outlined in Section 6 will be required to optimise mesh for mining applications.

8 CONCLUDING REMARKS

The mesh testing program and associated analysis investigation have identified a number of important features of mesh response to loading. The stiffness of response is a function of the mesh configuration and the bolting pattern. However, the stiffness is also controlled by slip of the mesh relative to the plates and bolts. Typical bolt tensions of about 50 kN will not be sufficient to generate frictional resistance which prevents mesh slip. Alternative mesh configurations for mining applications are being investigated.

ACKNOWLEDGEMENTS

The authors wish to acknowledge WMC Resources Ltd-St Ives Gold for allowing the test program to be undertaken at Junction Gold Mine, Kambalda and the invaluable assistance provided by various mine site personnel – in particular David Finn and Tao Li. The authors also wish to thank BHP Reinforcing Products and Smorgon ARC for providing the mesh used in the testing program.

REFERENCES

Ortlepp, W.D. 1983. Considerations in the design of support for deep hard-rock tunnels. *Proc. 5th Int. Cong. on Rock Mech.*, V2, D179-D817, Balkema: Rotterdam.

Ortlepp, W.D. & T.R. Stacey 1997. Containment of rockburst energy with appropriate tunnel support. *Proc. Int. Symp. on Rock Support* (Eds. E. Broch, A Myrvang & G. Stjern), 609-620, Norwegian Society of Chartered Engineers: Oslo.

Stacey, T.R., W.D. Ortlepp & H.A.D. Kirsten 1998. Practical static and dynamic tests of mesh, mesh/shotcrete and fibre reinforced shotcrete. *SCHOOL – Shotcrete and its application.* SAIMM: Johannesburg, 18p.

Standards Australia 1991. AS 1304 – 1991. *Welded wire reinforcing fabric for concrete.* Standards Australia: Sydney, 12p.

Standards Australia 1998. DR 98119. Steel reinforcing materials. *Draft Australian/New Zealand Standard for Comment*, Standards Australia: Sydney, 34p.

Tannant, D.D. 1995. Load capacity and stiffness of welded-wire mesh. *Proc. 48th Canadian Geotechnical Conference*, 729-736, CIMM: Vancouver.

Wojno, L. 1998. Alternative membranes. *SCHOOL – Shotcrete and its application.* SAIMM: Johannesburg, 12p.

Rock Support and Reinforcement Practice in Mining, Villaescusa, Windsor & Thompson (eds)
© 1999 Balkema, Rotterdam, ISBN 90 5809 045 0

In-situ trials for structural membrane support

L.Wojno & Z.Toper
CSIR Mining Technology Division, Johannesburg, South Africa

ABSTRACT: In a quest for safe and cost efficient support techniques it has become increasingly apparent that non-conventional solutions addressing the problem of insufficient areal coverage provided by many support systems and the labor intensity of support installation, are crucial considerations for the survival of many of the South African mines. One of the strategies to overcome these problems and to achieve associated benefits is to develop support units and systems which are better able to control the fractured rock under slow and rapid loading and also allow for easy installation of support. Accordingly, the CSIR Mining Technology Division, developed and is testing a new membrane type full areal coverage support for tunnels, stopes and pillars. This development addresses the important need for easily applied, areal coverage by mine support systems.

1 INTRODUCTION

In this paper initial in-situ trials of a new membrane type support called Evermine is discussed. This structural membrane support was designed to address the problem of insufficient areal coverage and/or limited yieldability of conventional support systems, with a view to improve the safety and stability of mine excavations and pillars in gold, platinum and coal mines.

Problems with maintaining stability of excavations by preventing falls of ground and/or unravelling of low cohesion or stress fractured rock as well as spalling of pillars are quite common in South African mines and are not always sufficiently addressed by conventional support such as shotcrete. The latter support is widely used internationally in underground mining. However, in addition to many positive features of shotcrete, this type of large areal coverage support also has limitations. Among the more important ones are: brittleness, limited deformability and tensile strength, rebound (during application), problems with transporting and handling of large volumes of material in confined spaces and dust (dry-mix), to mention a few.

Among concerns and challenges that the Evermine structural membrane support could address are the ground control and safety on the one hand, and costs / productivity on the other.

Figures 1 to 4 illustrate some of the excavation instability problems that are related to insufficient areal coverage by conventional supports.

The evaluation of the effectiveness of the Evermine structural membrane in ameliorating the types of damage illustrated in the figures is one of the main objectives of the in-situ trials. Once sufficient, quantified field data on the performance of the membrane is obtained it should be possible to specify Evermine as a new type of fabric support with a potential to either complement or even, in some applications, replace conventional support such as shotcrete or wire mesh and rope lacing.

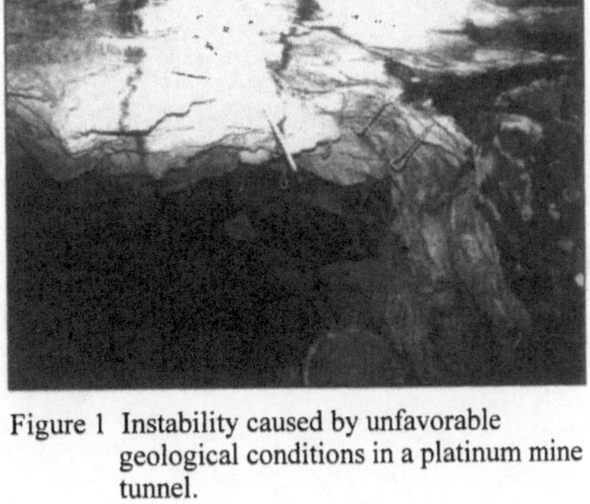

Figure 1 Instability caused by unfavorable geological conditions in a platinum mine tunnel.

Figure 2 Fall out in a deep gold tunnel despite high density conventional support

Figure 3 Coal pillar scaling/unravelling due to a lack of areal coverage support

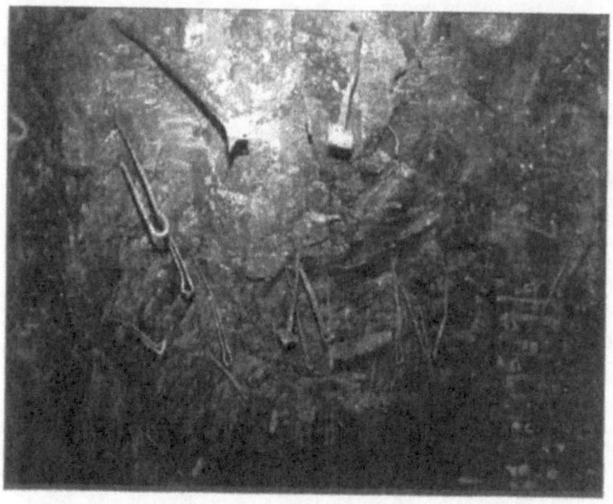

Figure 4 Falls of ground in a stope gully caused by limited areal coverage of the support

2. BRIEF DISCUSSION OF ROCK FAILURE MECHANISMS

For all practical purposes, it can be safely assumed that rock loses strength after failure. In the post failure stage the hard rock material will possess a residual strength which is less than its peak strength. This residual strength is *the* most critical parameter controlling the load bearing capacity of over-stressed abutments such as side walls of deep level excavations and pillars. Barron (1982) investigated the theoretical strength of pillars by means of limit equilibrium analyses and derived the following equation for the vertical stress distribution in the failed zone of a pillar abutment:

$$\text{Vertical Stress} = UCS_B * W \qquad (1)$$

where UCS_B is the uniaxial compressive strength of the *broken* rock,

$$W = \left\{\frac{(1+\sin\phi_B)}{2\sin\phi_B}e^{\chi}\right\} - \frac{1-\sin\phi_B}{2\sin\phi_B},$$

ϕ_B is the friction angle of the *broken* rock

and $\chi = \tan\phi_B\left(\frac{1+\sin\phi_B}{1-\sin\phi_B}\right)\left(\frac{x}{h}\right)$

where: x - distance into abutment ,
h - ½ of height of excavation (h=H)

Although this equation only describes the theoretical stress distribution in the failed material, Figure 5 shows the (expected) stress distribution in the intact material as well. From equation (1) the importance of the residual strength (UCS_B) can be appreciated. However, the calibration of this parameter has not been standardized and its value can only be assumed. At present the uniaxial strength of broken rock is often assumed to be zero in numerical simulations, but this would imply that no stresses can be generated in the crushed zone of an abutment as indicated by equation (1). This is unrealistic and it is clear that some, as yet not well established, value for the residual strength has to be selected.

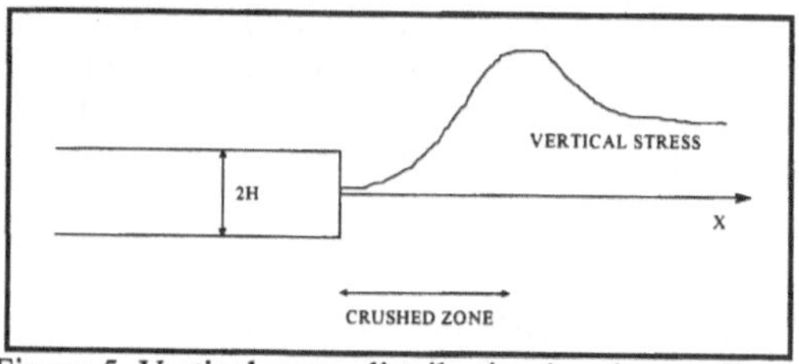

Figure 5 Vertical stress distribution in a longwall abutment (after Barron, 1982)

Although the exact mechanism by which residual strength increases by the application of a limited amount of confinement is debatable, the effects can be substantial and therefore of great practical relevance (Wojno,1997). An increase in residual strength would lead to a decrease in the extent of the fractured/damaged zone and associated dilation. The important requirement for a permanent increase in residual strength is however that such a confinement can be maintained under all circumstances, even if large deformations have to be absorbed. This is where the yielding capability of a structural membrane plays an important role. A support which is too stiff would fail when subjected to large dilatation and associated high stresses which could easily be generated in hard rock environments. This is exactly what the flexible membrane has been designed for. It should provide, under all circumstances, a limited amount of confinement which can result in a substantial increase in the residual strength of brittle rock.

Since in some areas the structural membranes are intended to replace wire mesh, a short comparison of reinforcing mechanisms of these supports could assist in a better understanding of the support action involved (Wojno,1997). With regard to control of the unravelling process of already fractured rock, the support/reinforcement mechanisms which apply to structural membrane systems are very different from those activated by conventional wire mesh. In the first place, the effect of a well adhered membrane is evident from the onset in that it immediately resists differential movement of the fractured skin rock, thus retaining the initial integrity of this rock and improving it's capacity to confine and strengthen the deeper fractured rock i.e. conforming to the adage of assisting the rock to support itself, a support reinforcing type mechanism. With the membrane the forces applied to the rock are more uniform thus better maintaining the overall integrity of the structure and hence increasing its strength. It is realized however, that this superior characteristic of membrane applies only up to some critical deformation limit.

It should be noted though, that a properly designed integrated support system comprising rock bolts (with spacing appropriate for the specific loading conditions and rock structure) and structural membrane should adequately address the problem of the limited deformation capability of the membrane. However, these aspects can only be properly understood and assessed through observational methods and in-situ monitoring.

In the case of wire mesh a degree of unravelling, depending on the stiffness of the mesh, must take place before it's confining effect is evident. The mesh is therefore only playing a containment role, passively supporting skin rock which has been allowed to loosen ("degrade"). The ultimate result with mesh is: a reinforced rock structure around the excavation with very variable width and strength depending on the distance from the nearest rock bolt.

The shell of fractured rock is thus allowed to weaken and importantly, it is believed, the volume of loose rock and thus the dead weight that the mesh has to carry is greater than with the case of the membrane, for the same rockmass conditions and rock bolt spacing. Thus, the dead weight carrying capacity of various fabric supports should not be the only criterion used to differentiate their capabilities as support/reinforcement elements.

3. TEST SITES FOR EVERMINE EVALUATIONS AND APPLICATIONS OF THE PRODUCT IN SOUTH AFRICA

In South Africa, the in-situ evaluations of the effect of Evermine structural membrane support on overall stability of mine excavations is being carried out in 6 sites. It was decided that the investigations should be carried out in two gold, two platinum and two coal mine sites.

3.1 *Gold mines -deep tunnel, Hartebeestfontein GM*

The experimental site was selected where large tunnel deformations are expected in a relatively short time after application of Evermine. Examples of stability problems experienced in the selected tunnel are given in Figures 6a and 6b.
In the experimental tunnel the conventional support comprising either grouted rock tendons, or rock tendons+wire mesh was not sufficient to stabilize the tunnel walls. Further stress change related deterioration of conditions is expected and these are being monitored in the tunnel.

3.2 *Deep stope gully - Western Deep Levels, East*

In the experimental site the newly exposed hangingwall of the strike gully was treated with Evermine and the overall stability of the coated section compared with an adjacent, untreated one.
Examples of serious problems with ensuring stability of the selected gully are shown in Figure 7. The roof strata are highly fractured parallel to bedding, to a

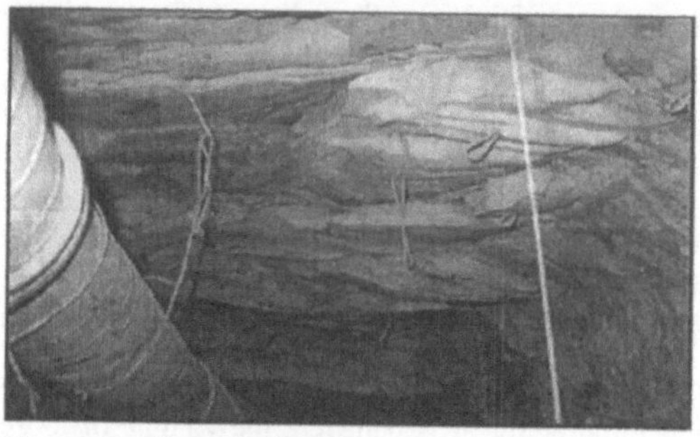

Figure 6a. Unstable blocks supported with fully grouted rock tendons in the Hartebeestfontein tunnel. Note fall out between and around rock tendons.

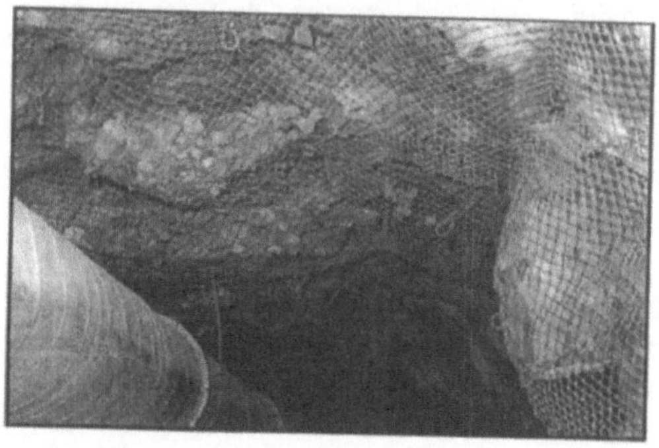

Figure 6 b. Wire mesh containing large volumes of unravelled rock at Hartebeestfontain site.

Figure 7 Stability problems in the WDL East Gold Mine strike gully

height of approximately 2m.

At present the gully is adversely affected by stress concentrations caused by the adjacent stoping. However, it is expected that conditions will deteriorate even further because the selected excavation is approaching a fault zone which caused severe problems in all nearby gullies.

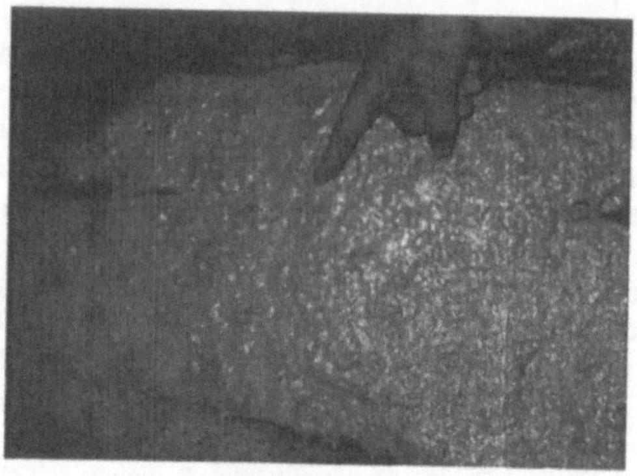

Figure 8. Evermine bridging a 5 mm (+) wide crack in the rockwall

Figure 9 Irregularly shaped quartzite wall treated with Evermine

A more general view of pre-fractured rock that was coated with Evermine is shown in Figure 9.

During the initial in-situ spray tests using a prototype spraying nozzle and mixing equipment, some problems with bridging cracks of a few millimetres wide were experienced. Subsequent improved nozzle/mix design enabled the membrane to bridge these cracks. Figure 8 illustrates the effectiveness of Evermine coating to bridge over a 5 mm wide crack in quartzite by using the latest nozzle design in the spray system

3.3 *Platinum mine tunnel - Impala Platinum Mine.*

In this site, tunnel stability is controlled by the geological discontinuities forming key blocks in the surrounding rock and/or low cohesion environment caused by serpentinization of the host rock. Figure 10 illustrates stability problems caused mostly by unfavourable geometry of intersecting joint sets.

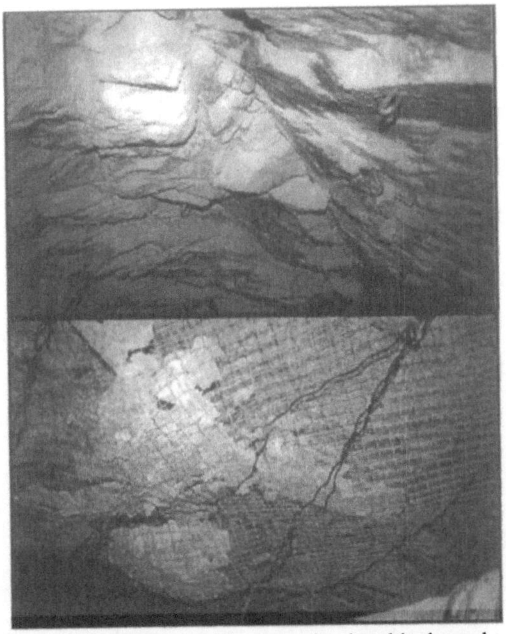

Figure 10 Problems with supporting key blocks and preventing hangingwall rock falls.

Figure 11 Example of low cohesion environment caused by heavily serpentinised rock.

Further problems with ensuring stability of the selected tunnel are related to low cohesion zones associated with highly serpentinized rocks surrounding the excavation, as shown in Figure 11

3.4 *Stope gully sidings - Impala Platinum Mine*

Small, approximately 6m x 3m, chain pillars are used as the main support to prevent large collapses in shallow gold and platinum mines. The pillars are cut immediately below strike gullies about 30m apart on dip. In certain circumstances the pillars become severely fractured causing not only a directsafety

Figure 12a Example of a scaling pillar and a gully siding supported with timber props

Figure 12b Examples of gully siding support by timber packs.

hazard because of the large slabs, but also weakening the pillar. The former hazard is reduced by cutting 2m deep sidings below the gully to distance the pillar from the gully travelling way edge. This operation is both labour intensive and costly. Examples of problems with gullies and siding stability are illustrated in Figures 12a and 12b.

The potential of a structural membrane to eliminate the need the for cutting and supporting gully sidings is being investigated. The membrane is being applied to the gully hangingwall and pillar, shortly after the blast and before cleaning. The effect of the membrane support on excavation / pillar stability is being monitored. If proven, the elimination of a need to cut and support sidings brought about by application of the structural membrane, would not only allow for significant cost savings but also

Figure 13 Coal pillar condition soon after cutting.

improve safety and overall stability of platinum stopes.

3.5 *Coal Mine - stability of coal pillars at New Denmark Colliery*

Two aspects of coal pillar stability are being investigated at the experimental site:

a) selected pillars that are likely to spall are being coated with the membrane soon after they are cut and their condition compared over a period of time with that of adjacent, untreated pillars.

b) reinforcing effect by the coating applied to an already heavily fractured coal pillar.

The coal pillars soon after being cut are usually in a very good condition as is illustrated in the Figure 13. An overall condition of such a pillar rapidly deteriorates for depths of over 650 m below surface, as illustrated in Figure 14.

Fractured coal on the perimeter of such pillars can be removed by hand. Although, in principle, such a fragile and incompetent "rock substrate" is not envisaged for the standard structural membrane applications it was nevertheless decided to apply the coating to the pillar on an experimental basis. During application the Evermine was found to be easily used in terms of handling and it allows for spray rates exceeding 35 m^2 per hour by a team of only 3 semi skilled persons. The product was also found to be economically viable as the total cost of the applied Evermine was not exceeding that of the conventional shotcrete. The product is also environmentally safe.

Figure 14 Severely fractured / spalled pillar before scaling it and spraying with Evermine.

Figure 15 Coal pillar treated with Evermine.

Figure 15 shows the same pillar a few weeks after treating it with the Evermine coating. No further deterioration of the pillar has occurred.
More general view of a pillar that was stabilized with the structural membrane is shown in Figure 16.
It is worth noting that several days after coating the pillar with structural membrane a few attempts to remove previously loose coal on pillar perimeter were made. It was found that a considerable effort was required to remove the coated coal pieces from the pillar wall.

Figure 16 Overall view and close-up of treated coal pillar.

Figure 17a Bord and rib side stability stability problems in burnt coal area

3.6 *Roadway stability in burnt coal at Gloria Colliery*

Maintaining the stability of roadways situated in burnt coal areas is one of the major problems at the Gloria Colliery. Conventional supports used on the mine do not offer sufficient areal coverage with the result that unsafe conditions and falls of ground frequently occur. Evermine was applied in a selected roadway to both, rib side and roof of the excavation with a view to evaluate the support effects of the structural membrane and it's effectiveness in preventing weathering of the low cohesion coal.

Examples of stability problems experienced in the selected burnt coal area are given in Figures 17a 17b.

4. OTHER IN-SITU EVALUATIONS OF EVERMINE STRUCTURAL MEMBRANE

The preliminary tests of the structural membrane support have also bee carried out in Poland and Canada. The field tests were performed at the Rudna Copper Mine of KGHM and the Fraser Mine of Falconbridge Limited, respectively.

In the preliminary evaluations the main intent was to address two concerns: adhesivity of Evermine to the rock and the time required to cure the product in the temperature and humidity conditions specific to the experimental sites. In addition, the user friendliness of the product during it's field applications was evaluated.

The preliminary findings were as follows:

- Evermine adheres very well to clean or even slightly dusty rock surfaces.
- drying time varies from 30 to 100 minutes, depending on the sprayed thickness and the rock surface's proximity to the vent tubing
- the product is safe and easy to work with
- due to the limitations of the Ist version of the spraying equipment (nozzle design) Evermine is not well suited to filling cracks
- improvement in the overall stability of excavations treated with Evermine was observed

Figure 17b Bord and rib side stability problems in burnt coal area

Further small scale field evaluations at the Rudna and Fraser mines revealed more about the product. Among the most important findings were the following:

- no significant convergence was recorded in excavations were the product was applied to the roof and walls
- in a 9 m wide breast excavation good adhesion and coverage were observed. Following the spraying of Evermine membrane it was wire screened using push plates which were subsequently damaged by the blast. In the same time blast related damage to the structural membrane was minimal (about 10%), with most of the damage occurring at or near the face.
- as a result of a significant reduction or even elimination of the falls of ground brought about by the structural membrane it was decided, for one of the experimental areas not to use any wire screen and continue monitoring the area for long term stability
- blinding of the crushing plant screen was considered a possible source of difficulty in dealing with Evermine. This was investigated and no obvious blinding of the screens was noticed in the trial
- in tests to check the adhesion of shotcrete to Evermine sprayed rockwall it was found that adhesion of shotcrete to the structural membrane appeared to be good, with considerable effort being required to remove the shotcrete from the Evermine
- a prototype pre-mix type spraying system which included a pressure pot used in the preliminary field evaluations. It was found that this needed modifications to make it easier to operate and allow for a better control of the spraying process
- problems with inconsistency between different batches sprayed were observed. This indicates the need for good quality control measures
- Evermine was sprayed onto a leaking fill fence in an attempt to stop the leaks. The experiment was unsuccessful. The water based resin component allowed dilution and even washing off of the membrane before it cured. This effectively prevented the product from being used for stopping leaks in the fill fence.

5. CONCLUSIONS

From the preliminary field evaluations (on-going) of the reinforcing effect of the structural membrane support, Evermine, on pillars, in tunnels and stope gullies in gold, platinum,coal, copper and nickel mines, the following conclusions can be drawn:

- The Evermine structural membrane support appears to have a capacity to markedly ameliorate problems with falls of ground, which in turn can improve overall stability of tunnels, stope gullies and pillars
- A better understanding of the effectiveness of Evermine as a support and mechanisms involved was obtained
- The product is easily used in terms of rock surface preparation, handling and application and allows for spray rates of over 35 m^2 of rock wall per hour, by a team of only 3 semi skilled persons
- The product is environmentally safe
- Evermine adheres well to all the rock types tested on gold, coal, platinum , nickel and copper mines
- Some lack of continuity in the structural membrane observed in various underground applications on irregular surfaces does not appear to have a major effect on support / confining capacity of the membrane
- Discontinuous, low cohesion rock strata, which under normal circumstances would not be able to transmit shear and tensile stresses, can be stabilized by a flexible structural coating
- Evermine is cost efficient. In South African conditions it's applied cost does not exceed that of a conventional shotcrete.

ACKNOWLEDGMENTS

A part of the work presented in this paper is supported by the Technical Committee of the Multi-group cost shared R&D, Project number: GE9711. This support for the project is gratefully acknowledged.

REFERENCES

Barron, K. (1982) An analytical approach to the design of pillars in coal. Final Report, CANMET project no. 334-101, Canada.

Wojno, L (1998) Alternative Membranes. School: Shotcrete And Its Applications. The South African Institute of Mining and Metallurgy.

Wojno, L and J.S. Kuijpers (1997). Spray-on, user friendly and flexible support for mine excavations. Proc. of International Symposium on Rock Support, Applied Solutions for Underground Structures, Lillehammer, Norway.

Rock Support and Reinforcement Practice in Mining, Villaescusa, Windsor & Thompson (eds)
© 1999 Balkema, Rotterdam, ISBN 90 5809 045 0

In-situ trials and field testing of two polymer restraint membranes

D.J. Finn
WMC Resources Limited, St. Ives Gold, Kambalda, W.A., Australia

P. Teasdale
WMC Resources Limited, Perth, W.A., Australia

C.R. Windsor
Rock Technology, Perth, W.A., Australia

ABSTRACT: Two spray on polymer membranes were trialed at Junction mine as a potential replacement for mesh. The membranes trialed were a polyurethane material called Mineguard™ and an acrylic based polymer membrane Everbond Mark II. The polymers were trailed underground in a development heading and at the brow of a long hole stope. A series of performance tests were also conducted on surface, which included pull tests on steel plates which had been sprayed over with membrane and strength tests of concrete and rock cores which had been coated with membrane.

The underground trials showed that the membranes were easy to apply and the logistics of spraying could be easily managed. The trials also showed that Mineguard™ had problems adhering to rock if it were wet or dusty and that Everbond II was unable to set in areas where ground water was flowing.

The surface plate tests showed that Mineguard™ could be used to replace mesh as a structural surface restraint but Everbond II was unable to sustain the necessary high loads. The core tests showed that both membranes prevented the explosive failure of the porphyry rock cores and increased the post failure strength of the concrete cores.

1. INTRODUCTION

Traditionally mesh and shotcrete have been used to provide areal support in cases where bolts alone are inadequate to support the rock mass. In recent years spray-on plastic membranes have been developed as alternatives to these traditional methods. These have been developed in Canada and South Africa and are known by their trade names Mineguard™ and Everbond Mark II, the first is a polyurethane membrane and the latter is a water based acrylic system.

The potential for these products to act as a replacement for mesh was recognised and a series of trials and tests were organised at Junction mine which is part of WMC Resources Ltd.-St. Ives Gold operations. The perceived benefits of the membranes were an improved support capability, reduced handling and relatively faster application rates.

The trials and tests were designed to measure these potential benefits and gain first hand knowledge of the products and their application. These trials were conducted underground where development headings and longhole stope brows were sprayed and on the surface, where a series of tests had been devised to measure the membrane performance.

2. JUNCTION MINE

Junction mine is situated 35km south of Kambalda and has been operating as an underground mine since 1989. The deposit is situated in the Norseman Wiluna greenstone belt and gold mineralisation is located in a 5m to 15m wide shear zone, dipping at between 40° to 65° to the north east, the economic part of the shear zone is hosted within a highly competent dolerite. The ore zone is typically an altered dolerite with extensive quartz veining, brecciation and mylonitisation. It is a moderately jointed rock with discreet hangingwall and footwall shears and typically has a rock mass Q rating of between 5 and 15 (Finn 1998).

The measured rock stresses at a depth of 395m are 38.5MPa for σ_1, 22MPa for σ_2 and 17.7MPa for σ_3 (Pascoe 1998). Stress slabbing and strain bursting are common in the ore drives especially where quartz veining or brecciation are present and in areas where porphyry intrusions are encountered.

The ore drives are accessed via a footwall decline and the principle mining method is long hole open stoping. The ore drives are now mined to a 4.7m wide shanty back profile and are currently supported using split sets and mesh, the split sets are grouted after installation to increase their effectiveness (Thompson & Finn 1999).

3. HOW AREAL SUPPORT WORKS

The different types of areal support, shotcrete, mesh and plastic membranes employ different mechanisms to support rock and an understanding of these support mechanisms will help in the understanding of when and where to apply the products. Unfortunately the mechanisms employed by the different types of areal support are not fully understood and there is still some debate on how areal support works. In spite of this some general comments can be made about the behaviour mechanisms of the different areal support types.

Shotcrete will penetrate and fill the wider cracks and fractures increasing the interblock cohesion and producing a wedging effect between blocks. It also provides a surface restraint minimising the movement of keyblocks and wedges into the excavation, preventing them from dislodging, which could lead to the subsequent unravelling of the rock mass. The limitation of this surface restraint action will be the shear strength of the shotcrete layer. Both these factors, crack sealing and restraint, enhance the rock mass strength and its ability to support itself. In thicker layers of shotcrete, (those greater than 100mm) shotcrete can develop high stresses within the shotcrete layer (Myrvang & Stjern 1993), this shares the load in the rock mass and acts as a support arch.

Mesh will provide a restraint for loose rock between the bolts, but the mesh itself does little beyond restraining this loose material.

Spray on plastic membranes are thought to act in a similar way to a layer of shotcrete. The membrane would penetrate cracks and increase the cohesion between the blocks, the membrane also prevents the movement of blocks therefore preventing the rock mass from unravelling. Plastic membranes hold the rock in place because of the increased adhesion between the blocks and the high tensile strength of the surface restraint (Espley et al 1994), not because of their shear strength and their ability to develop high compressive loads within the membrane layer.

Both shotcrete and plastic membranes will protect the rock from exposure to the atmosphere and any deterioration that may be associated with this.

4. DESCRIPTION OF THE MEMBRANE MATERIALS & THEIR APPLICATION METHODS

4.1 *Description of Membrane Materials*

The two polymer membranes trialed at Junction mine were Mineguard™ and Everbond Mark II.
Mineguard™ is a 100% solids polyurethane coating which consists of two components:

Part A: Diphenylmethane Disocyanate
Part B: Polyol Resin

Mineguard was developed by the Mining Industry Research Organisation of Canada in conjunction with Urylon Plastics Inc. The development has been ongoing since 1989 and the product has been trialed in a number of Canadian mines (Archibald et al 1997).

Everbond Mark II is a two part water based acrylic membrane system comprising:

Part A: Acrylic Emulsion
Part B: Cementitious Filler

Everbond has been developed by the CSIR Mining Technology Division of South Africa in conjunction with Stratabond SA Pty Ltd. Everbond has been trialed in a number of South African mines but further refinements are expected (Wonjo & Kuijpers 1997).

4.2 *Storage and Handling*

The two components comprising Mineguard are supplied as a liquid in 200 litre drums. They have a storage life of over 12 months but must be stored at temperatures of greater than 15.5 °C, at temperatures below this, Component A will start to crystallise and will clog up hoses during spraying. If Component A has any contact with water including any moisture in the air, it will start to react. Because of this the drums containing Component A have dry nitrogen squirted into the container before sealing (Mineguard Operations Manual).

Spillages can be absorbed with conventional absorbents and then neutralised with water. Personnel involved in a clean up must wear appropriate respirators.

The liquid component part of Everbond is stored in 200 litre drums which must be sealed and kept at temperatures above 0 °C. Below this temperature coagulation of the fire retardant can occur. During trialing at Junction the liquid component in some of the drums had formed a skin and lumps, this was attributed to the fire retardant not being properly dispersed and was overcome by thoroughly mixing the liquid component. The powder component is also stored in 200 litre drums and must be kept dry. It is supplied in 7kg bags as part of the standard mix.

The fire retardant in the liquid component may cause mild respiratory irritation if inhaled in poorly ventilated conditions due to the fire retardant additive antimony trioxide or the formaldehyde. The powder component should be treated like any other cement product.

4.3 *Application Methods*

Mineguard is sprayed onto the excavation surface using Graco air operated pumps. The two components are pumped from their respective 200 litre drums by air operated Graco Monarch drum pumps to a Graco Bulldog pro-staging pump at a 1:1 ratio. Electric heaters are used to heat the two components to a temperature of between 49°C and 60°C where each of the components has a similar viscosity, which assists in the metering and mixing of the components. The high pressure pro-staging pump delivers the two components at a pressure of between 138 and 172 bar which ensures sufficient mixing of the components in the mixing chamber of the gun.

Everbond is sprayed onto the excavation surface using a batch mixing system involving an air pressure kettle with a venturi effect spray gun. The two components are mixed together in the proportions 21 litres of liquid component A to 7 kg of the cementitous powder component B. An air operated augur is used to mix the two components which are then poured into the pressure kettle for spraying.

Mineguard and Everbond are both sprayed manually. The close proximity of the operator to the face allows the operator to identify and spray up into cracks or spray over a suspect area several times. There is however potential for both products to be applied remotely.

4.4 *Fire Resistance*

During the early development of both products a high priority was given to ensuring that the final membrane was fire resistant.

Flame retardants have been added to Mineguard and tests in Canada indicate that when exposed to a direct flame at temperatures above 480°C the membrane will decompose giving off CO_2. The fire resistance of Mineguard can be improved further by adding fine vermiculite powder onto the sprayed surface (Archibald et al 1997).

The liquid component of Everbond has had fire retardants added and both the liquid and final membrane are not considered flammable. Decomposition from exposure to fire will give off CO and CO_2.

5. INSITU TRIALS AND FIELD TESTS

5.1 *Aims of the trials and tests*

The trials at Junction involving underground in situ trials and controlled tests on surface were to:

- Evaluate the membranes ability to maintain the integrity of a zone of fractured rock surrounding an excavation, by preventing the formation of loose scats between rock bolts.
- Evaluate the membranes as a replacement for mesh at Junction mine.
- Evaluate the adhesion, tensile strength and reaction to the presence of water.
- Evaluate the curing time in the underground environment of the Eastern Goldfields.
- Evaluate the performance when subject to adjacent blasting in longhole brows and development ends.
- Evaluate application methods, preparation, set up and application times.
- Review the practicalities of implementing Occupational Health and Safety requirements.
- Review the practicalities of implementing any special handling characteristics which may be required.
- Review the degree to which the technology could reduce mining cycle times in development headings.

5.2 *Description of the underground trials*

Two underground trials were conducted at Junction both in the ore zone, the first trial was conducted in a typical ore drive. This drive required meshing of the backs and 2m down the side walls to contain strain bursting and stress slabbing of the rock. The trial in this area was to compare the performance of the membranes with that of mesh and to evaluate the effects of blasting a face adjacent to the membranes. The procedure involved mining a 4.2m development advance then, following mucking, the area was scaled and rock bolts installed using a development Jumbo, the usual meshing requirements were omitted and the spraying process was done after bolting. Four development cuts were sprayed using each of the products on two adjacent cuts and between the two areas sprayed with the membranes one advance followed the normal support procedures using mesh. The membranes were sprayed across the backs and 2m down the walls.

The second trial was conducted close to the brow of a stope. The area had been developed to the orebody width of 8m and supported with rock bolts and 6-8m long cable bolts for brow support, no meshing had been used in this area. Experience at Junction showed that the brows of stopes required extensive rehabilitation, following each firing because of blast damage and possibly stress damage resulting in slabbing and loose rock. The purpose of this trial was to evaluate the effectiveness of the membranes in controlling this damage.

Both the underground sites relied upon observations to assess the performance of the membranes relative to mesh. Survey points were installed into the membranes to measure convergence and displacement.

5.3 *Description of the field tests*

The surface trials allowed membrane performance to be assessed in a more controlled environment and measurements to be conducted.

The trials comprised:

- Pull tests of various sized metal plates through a coating of membrane sprayed over the top of the plates and onto a surrounding concrete floor. The plates were 100mm by 100mm, 250mm by 250mm, 500mm by 500mm, 750mm by 750mm and 1000mm by 1000mm and were each covered with membrane extending out one and a quarter of the plate side length beyond the edge of the plate.

- A flexure test where a layer of paving stones were sprayed with membrane then turned over and loaded in the centre.

- Core samples of concrete and porphyry rock were coated with membrane of various thickness and tested in uniaxial compression. Some of these samples were fitted with strain gauges to measure deformation during testing. These samples were load tested at the West Australian School of Mines (WASM).

6. UNDERGROUND TRIALS USING MINEGUARD

6.1 *Site preparation*

Mineguard experiences problems with adhesion onto wet, oily or dusty surfaces. The part A component Diphenylmethane Disocyanate reacts with water to form CO_2 this causes the product to foam which adversely affects strength and the adhesion to the surface. Before spraying, an area must be thoroughly washed down and allowed to dry. Areas of high humidity will also cause problems because of condensation forming on the rock surface. Weak or friable rock will affect the adhesion of the membrane as the rock itself will tend to break away and the resulting membrane will detach from the main body of rock. Before each trial at Junction the rock was washed down with fresh water and then sufficient time was allowed for the area to dry. In the areas where Mineguard was sprayed, ground water was present only in the development ore drive. This occurred around a diamond drill hole which was dripping water and in a damp area along the hangingwall structure. The humidity and temperature of each heading were measured to check the potential of condensation on the rock surface which could adversely affect the adhesion of Mineguard. The results showed that condensation on the rock surface at Junction was minimal. The rock strength in the ore zone at Junction was not considered to be a problem for the creation of the membranes.

6.2 *Spraying*

The spraying equipment was mounted in the basket of an IT which was used as the spraying platform. This was required to access the 6m high backs at Junction.

The spray from the gun comes out in a flat fan. This allows the operator to initially work the spray along a crack which increases the penetration of the spray into the crack before spraying back and forth across the crack. Tests by Urylon Plastics and MIROC indicate penetration depths of 10mm for a 0.9mm crack and 33mm for a 2.2mm wide crack.

To attain a uniform coating of around 2-3mm an area was sprayed with up to 7 or 8 passes.

The skill of the operator is an important factor in the correct application of the membrane for both crack penetration and the uniform coverage of the rock. There was relatively little rebound of material during spraying.

Reaction and curing times could not be measured directly during the trials but discussions with the operators indicated that, the two components react in about 7 seconds to form a 100% solid polyurethane. The resulting membrane is touch dry in 30 seconds, attains 50% of its strength in 3 minutes and reaches full strength after 8-10 minutes.

In total 652 litres of the product was used to spray an area of 230m² at a rate of 58m²/hour. The resulting membrane had an average thickness of 2.8mm.

6.3 *Control of the Airborne Contaminants*

When component A Diphenylmethane Disocyanate is heated and becomes atomised it has a strong irritant affect on the respiratory tract of most people and will cause asthmatic symptoms in some people. The operators must wear full face positive pressure breathing apparatus while spraying and anyone entering this area must be similarly equipped. In theory, if the two components have been mixed correctly at the gun the chemical reaction should take place in 7 seconds and the atomised material is converted into polyurethane and becomes a nuisance dust. During spraying a water curtain was set up at Junction down wind of the spray area, the water neutralises any products in the air stream. If a water curtain is not used an area of 120m down wind has to be declared no entry, unless appropriate protective breathing apparatus is worn. Once spraying has been completed re-entry requirements are that unprotected personnel do not enter the area for 20 minutes. Entry can be allowed 5 minutes after spraying but an organic respirator must be worn. The Draft Code of practice from Worksafe WA indicates the exposure limits for isocyanates are a Time Weighted Average (TWA) exposure of 20 ug/m³ over 8 hours and a Short Term Exposure Limit (STEL) of 70 ug/m³, over a 15 minute time interval. Table 1 shows the results of air quality samples taken in the development heading to test the levels of isocyanates and the effectiveness of the water curtain during spraying operations.

The results show high density levels of isocyanates in the vicinity of the spray gun, the drop off away from the spraying area and the effectiveness of the water curtain in controlling the isocyanates.

The necessity of wearing breathing apparatus in the vicinity of spraying is also evident even if spraying two headings a day under similar ventilation conditions the TWA would be exceeded.

Table 1. Results of Air Quality Tests

SITE	DISTANCE FROM SPRAYING	TIME (mins)	ISOCYANATES ug/m³
Sprayer in Basket	0	60	46
5m from face	0	60	69
15m from face	10m	35	13
10m inside water curtain	35m	48	13
10m outside water curtain	55m	50	<1

6.4 *Observations on the Performance of Mineguard Underground*

6.4.1 *General Observations*

The sprayed membrane is white in colour which improves visibility in the area. The contours of the rock are readily visible in the membrane and any slabs or cracks are easily identified. The spray does not bond across large cracks unless a special effort is made, however it does penetrate up the cracks.

The membrane is opaque and any deterioration of the rock mass can not be seen directly. Mineguard contours the backs well and because it is sprayed in a thin layer any ground movement would be easily noticed.

6.4.2 *Ore Development Drive*

In the development drive spraying was conducted after the face had been bolted with split sets. The spraying was taken to within 0.5m of the face and 2m down either wall in each cut, then another 4.2m development advance taken.

After the first firing about 2.0m² of membrane had peeled off, due to the displacement of a large slab from the hangingwall. Damage was evident along the leading edge of the membrane where the membrane had been detached from the rock and slabs of rock were levered open by the blast. The operator of the bolting jumbo scaled some membrane along this front edge. Fly rock damage was evident with small areas (4cm x 4cm) chipped out of the membrane.

Figure 1. Damage to the leading edge of the Mineguard membrane after a development firing.

The damaged parts of the membrane along the front edge were cut off after the first firing and this area was sprayed over once more. The membrane was then sprayed to within 0.5m of the newly established face. After this face was fired similar damage was observed due to fly rock and along the leading edge of the membrane. In addition the wet area around the diamond drill hole had peeled off over a 0.5 m^2 area.

An inspection after 24 days revealed a strip 2m long and 0.5m wide of membrane that had peeled off the hangingwall. It appears that a rock had dislodged and tore through the membrane peeling the membrane off the wall. As time progressed the areas which had already peeled continued to so and peeling had started at a number of other locations.

Along the ore drive the mesh could be seen to be retaining 10% to 20% scats by area of backs. These were generally minor scats with a size range between 100mm to 200mm thick and 100mm to 500mm in length and breadth. The area sprayed with Mineguard showed no development of any scats in the backs and half barrels could still be seen in the backs where the membrane had been sprayed, half barrels were not visible in the area that had been meshed.

Later on the area was drilled for production stoping and the collars of the holes inspected. The area that had been meshed showed signs of the ground opening up for between 200mm to 250mm up the holes. The area in which Mineguard was applied showed no ground opening up around the collar of the hole or up the hole.

The survey pegs in the membrane surface showed no significant movements during resurveys.

6.4.3 *Stope Brow*

Mineguard was sprayed onto the backs and 2m down the walls, a total of 109m^2 was sprayed. Spraying started 6m from the current brow. This drive was developed 8m wide and had been supported with mechanical anchor bolts and cable bolts but not meshed. The ground had already started to slab open and the area required manual scaling prior to spraying.

Where the production holes were drilled the Mineguard did not bridge the hole and because of the white colour the holes were clearly visible after spraying.

Once blasting started in the area that had been sprayed with Mineguard the membrane was cut using a grinder to detach it from the area being fired, so it

Figure 2. Peeling of the Mineguard membrane along the hangingwall and in the backs.

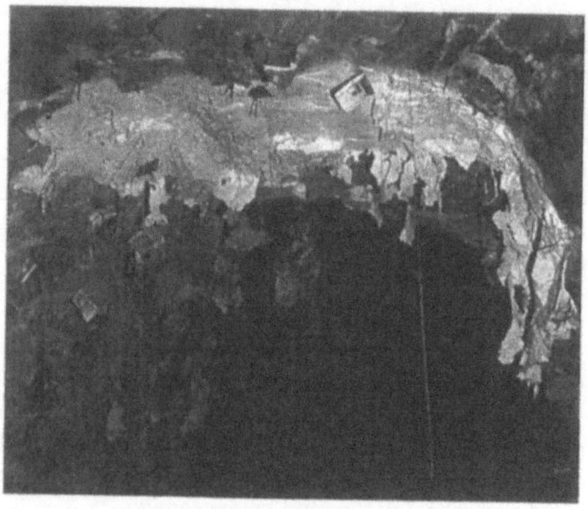

Figure 3. Damage to the Mineguard membrane along the brow following a stope firing.

would not be ripped down with the membrane underneath the blast rings during firing. In spite of this the membrane was extensively damaged by the blast and peeled back from this leading edge to the next ring, a distance of 2.5m from the new brow where it hung in tatters. It was assumed that the gases from the blast got in behind the membrane and detached it from the rock mass.

Along the hangingwall a patch of Mineguard was seen to be holding up a shattered block of rock. The Mineguard was between 4 rock bolts and the block was about 50kg, the block was obviously detached from the rock mass and showed the support capabilities of Mineguard.

After this the area was extensively scaled by a jumbo, this scaling included the remaining Mineguard in the drive. This ruined any further useful observations of the performance of the membrane in this area but did demonstrate the prejudice and distrust associated with the introduction of a new ground support type.

7. UNDERGROUND TRIALS USING EVERBOND

7.1 *Site preparation*

Everbond is not as sensitive to the effects of dust or water as Mineguard. The sites were washed down prior to spraying using the spray equipment. The water used to wash down the surfaces was the water used to clean out the spray equipment prior to the next batch. It was claimed that this water which contains some of the Everbond emulsion was a good primer and bonding liquid when used to wash the surface down. Any large accumulations of dust or rock powder were also removed during washing down.

7.2 *Spraying*

The equipment for spraying is fairly compact and could be easily handled by two people. To access the high backs at Junction it was mounted into the basket of an IT.

The spraying was conducted in batches with each batch consisting of 21 litres of emulsion and 7kg of powder. It took 10 minutes to spray each batch and another 10 minutes to clean the equipment out and mix another batch ready for spraying. A larger mixer and sprayer could be designed to reduce the time of mixing and cleaning.

The Everbond spray is relatively coarse and therefore the penetration into narrow cracks is limited, the spray operator indicated that cracks up to 5mm could be sealed by Everbond. The round nozzle shape is not ideal for spraying into cracks.

A uniform coating of between 3-5mm was achieved by spraying over an area a number of times. During spraying there was relatively little rebound evident. As with Mineguard the final result will depend on the skill and experience of the spray operator.

Reaction and curing times could not be measured directly during the trials but a surface drying time of 30 minutes is quoted in some of the literature and during the trial it was observed that the formation of a skin which was reasonably resistant to breaking occurred in approximately one hour. Literature on tensile tests of earlier versions of the product, Everbond I, indicated that it achieved half its strength in 1 day and its full strength could take up to 7 days to be achieved.

In total 462 litres of the product were used to spray an area of $120m^2$ at a rate of $34m^2$/hour. The resulting membrane had an average thickness of 3.9mm.

7.3 *Control of the Airborne Contaminants*

There are no airborne contaminants associated with the spraying of Everbond. All the components used are considered to be non-toxic.

7.4 *Observations on the performance of Everbond Underground*

7.4.1 *General Observations*

The final membrane is grey in colour. The contours of the rock are still visible behind the membrane but because of the slightly greater thicknesses and grey colour the presence of slabs and cracks do not stand out as well as for Mineguard. The spray does not bond across large cracks unless a special effort is made and it does not penetrate very far up the cracks.

The membrane is opaque and any deterioration of the rock mass can not be seen directly. However as the membrane is stretched it turns a whitish colour which could be used to detect movement behind the membrane.

7.4.2 *Ore Development Drive*

The spraying sequence and process was the same for Everbond as for Mineguard. Two 4.2m advances were sprayed after each advance had been bolted. The spraying was done to within 0.5m of the face and 2m down the side walls. A 4.2m advance which was supported with bolts and mesh separated the two membranes.

A wet area occurred along part of the hangingwall and in the backs with water dripping from the rock, the rest of the section was dry. After the first firing an area had peeled back 1.5m from the front edge of the membrane in the wet area and it was observed that the membrane had failed to cure properly. Here it formed a crumbly paste at worst or where partially cured it could be easily detached from the rock surface.

After the face was fired it was observed that fly rock caused general damage to the membrane with (4cm x 4cm) pieces being chipped from the membrane.

The areas of fly rock damage and damage to the leading edge were resprayed and the final cut of the drive was sprayed to the face.

The membrane around the area of dripping water never cured properly and the water could be observed dripping some months later when the heading was revisited, here the membrane remained crumbly when rolled between the fingers.

It was also observed in other areas that with time the membrane had become more brittle and could be torn easily or broken by flexing it back and forth. No stress slabbing was observed in the backs, where Everbond was sprayed and an inspection of the production holes showed that there had been no opening of the ground as happened in the meshed areas of the same drive. As in the area sprayed with Mineguard half barrels were visible in the backs where Everbond had been sprayed.

7.4.3 *Stope Brow*

Everbond was sprayed onto the backs and 2m down the walls, a total of $40m^2$ was coated with membrane. The drive is 8m wide in this area and is supported with mechanical anchor bolts and 6m to 8m long cable bolts, there is no mesh in the area.

Where the production holes were drilled the Everbond did not bridge the holes.

Figure 4. Damage to the Everbond membrane following a stope firing.

Before each firing the Everbond was cut at the start of the next ring to prevent it being ripped down with the blast. A large slab of rock failed on the footwall side of the drive and around this area scaling and rebolting were required.

8. RESULTS OF SURFACE TESTS

8.1 *Mineguard Plate Tests*

These tests were conducted in the surface concrete storage bays at Junction. The area was cleaned out, washed and the concrete slabs allowed to dry before spraying started.

The plates were 100mm x 100mm, 250mm x 250mm, 500mm x 500mm, 750mm x 750mm and 1000mm x 1000mm in size. They were made from 10mm thick steel and the corners were rounded to prevent them tearing through the membrane cover. A nut was welded into a recess in the middle of each plate to allow a bar for pull testing to be attached.

The plates were arranged within the concrete bay and membrane was sprayed over the top of the plate and onto the concrete surface around each plate. The membrane was extended out about one and a quarter of the plate length beyond the plate edge and over the top of the plate.

A rock bolt was threaded into the attachment nut and the pull tests were done using a steel beam resting on the walls of the concrete bays. A hydraulic pump and jack were used to pull up the rock bolt and plate. The load was measured using a load cell mounted on the rock bolt and displacements were measured using a digital micrometer initially or a tape measure when the displacement became too large.

Two 250mm plates were also placed on rocks from the low grade ore pad and these were then sprayed over with membrane and the plates loaded using the same system.

The pull test results are plotted in Figure 5 for the 250mm, 500mm, 750mm and 1000mm plates. The response was similar in each case: an initial stiff loading with little displacement, reaching a maximum of 25KN for the largest plate. At this point a crack was heard, followed by a small drop in the load then deformation at a fairly constant load before the load began to increase to a point when one of the sides tore off the concrete and there was a sudden drop in load. Even at this point there was still residual strength in the membrane before the other sides were detached.

The analysis of the pull tests is very complex: The membrane is being stretched, the bond to the concrete is being broken by a low angle pulling action and there is probably a suction effect as the air could only enter at the rock bolt attachment.

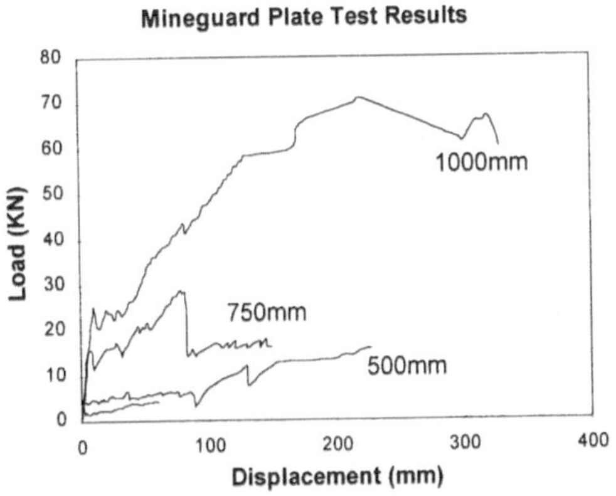

Figure 5. Results of Mineguard Pull Tests on different size plates

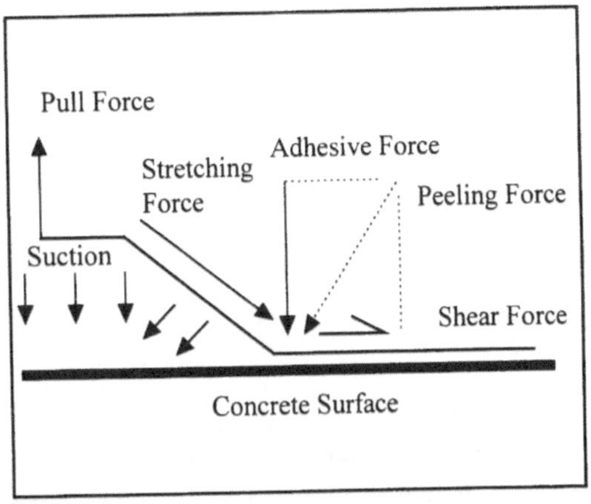

Figure 6. Forces in the system during stage 3 of plate tests.

To analyse the results the graphs were broken up into a number of stages which allows results from the two membranes to be compared. These stages are shown in Figures 10 to 13 and described below.

Stage 1 is characterised by an increase in load with only a minor increase in displacement. This is attributed to a combination of the adhesive strength of the Mineguard which had penetrated underneath the plates and the membrane being loaded to the point where it started to stretch around the perimeter of the plate. At the maximum load a crack was heard as the membrane/concrete bond was broken and there was a drop in the load, in the case of the 1000mm plate the load dropped from 25KN to 23KN.

At stage 2 there was no increase in load but the membrane continued to deform plastically, this was attributed to the membrane being stretched around the perimeter of the plate. The stretching load on the membrane was less than the tensile strength of the membrane so the plate did not tear through the membrane but the membrane started to peel off the concrete surface.

The peeling of the membrane off the concrete surface was a complex interaction of many factors which is seen in stage 3 of the graph. The membrane was plastically deforming while also being decoupled from the concrete surface. The decoupling behaviour occurs by a combination of adhesive tensile failure and shear bond failure. In addition to these forces there is a suction effect as the membrane had effectively sealed the surface and air could only enter where the rock bolt had been attached to the plate . Figure 6 shows this complex interaction of forces. The increasing load during this time was due to the increasing perimeter which increased the adhesive and shear forces and the increase in membrane area over which stretching could take place.

Figure 7. Mineguard 1000mm plate test at stage 3.

Failure occurred at stage 4 as the membrane was detached along its edge and there was a sudden loss in load capacity.

During the load tests on the plates sprayed onto the rocks from the ore pad, the rocks were lifted off the ground at loads of about 15 KN and the testing was stopped.

8.2 *Everbond Plate Tests*

The Everbond plate tests were set up in the same way as those for Mineguard using the same concrete bay and plates and the same pull test equipment.

Two 250mm plates were also placed on rocks from the low grade ore pad and these were then sprayed over with membrane and the plates loaded using the same system.

The results of the tests for the 500mm, 750mm and

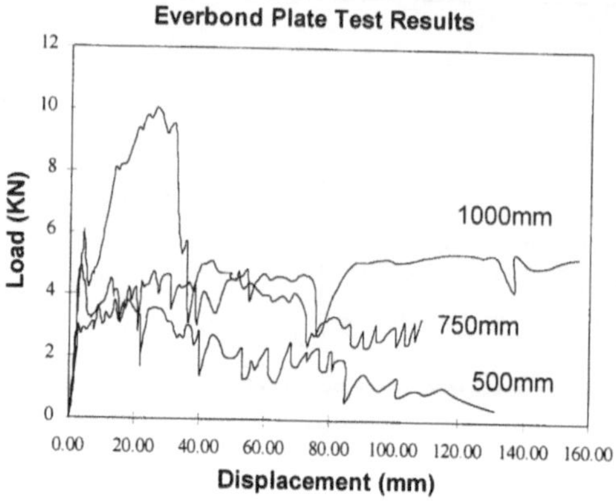

Figure 8. Results of Everbond pull tests on different size plates.

Figure 9. Everbond 1000mm plate test at failure.

1000mm plates are shown in Figure 8. As with the Mineguard tests the results are complex because of the different forces that are acting on the system. Again the results were graphed and broken up into different stages and compared to Mineguard in Figures 10 to 13.

Everbond achieved much lower pull off loads than Mineguard in these tests. The maximum load recorded for the 1000mm plate was 10 KN, the displacements were also much lower as can be seen from Figure 8.

Stage 1 is similar to that of Mineguard, the build up of load is again attributed to the adhesion and tensile loading of the membrane around the edge of the plate but now at lower loads.

The plate then started to lift off after the adhesive strength to the concrete was reached. At stage 2 the membrane was stretched to its limit and the graph shows displacement with decreasing or constant loads. At stage 3 failure occurred as the load exceeded the tensile strength of the Everbond and the plate tore through the membrane.

It was observed that the corners of the membrane started to turn a whitish colour and cracks opened up in the membrane at the corners, as the plate started to tear through.

During the load tests on the plates sprayed onto the rocks from the ore pad the Everbond membrane was torn through at relatively low loads.

8.3 *Comparison of Mineguard and Everbond Performances during Plate Tests*

Stage 1: System loaded until the adhesion of the membrane is broken and the tensile yield reached.

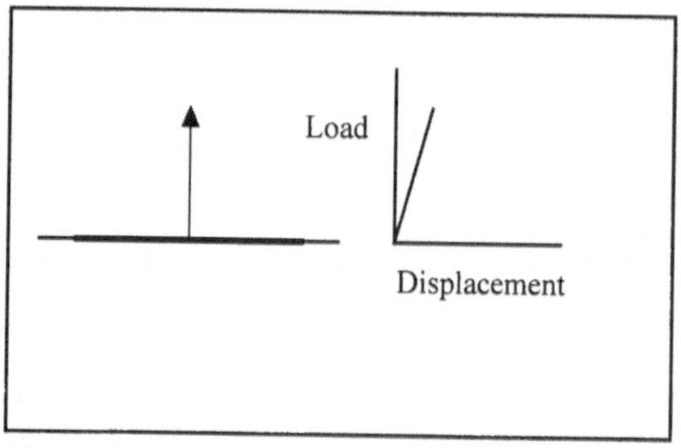

Figure 10. Stage 1 of Plate Loading Mineguard and Everbond.

Stage 2: Membranes continue to stretch plastically without an increase in the load, a gradual loss in load occurred in some of the Everbond samples.

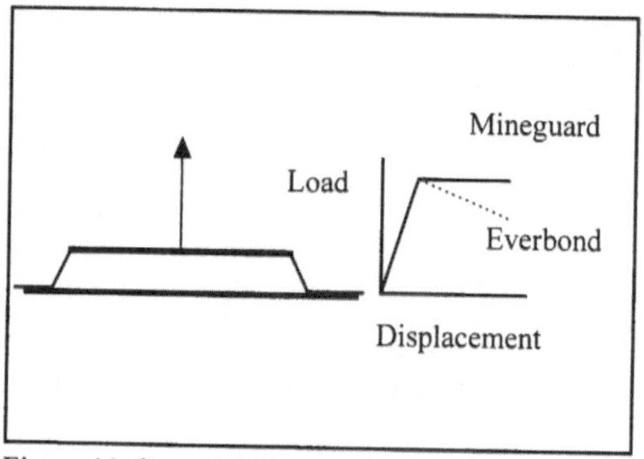

Figure 11. Stage 2 Mineguard and Everbond.

Stage 3: Mineguard continues to stretch and peel off the concrete surface, the tensile load has not

exceeded the tensile strength of the membrane.
The tensile load has exceeded the tensile strength of Everbond and failure occurs.

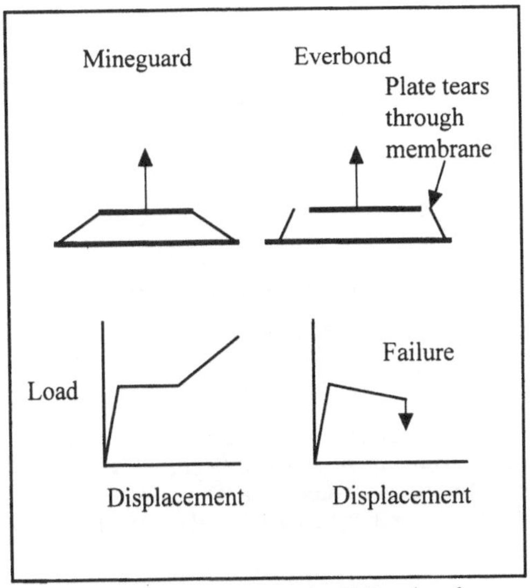

Figure 12. Stage 3 Mineguard and Everbond.

Stage 4: The membrane is pulled off the surface at the edges and load carrying capacity is lost.

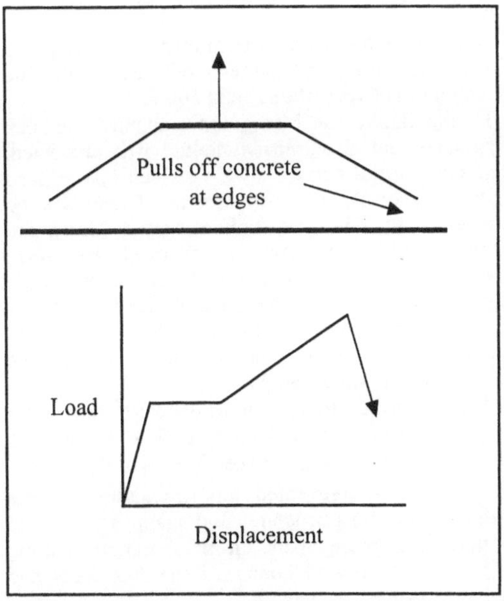

Figure 13. Stage 4 Mineguard only.

8.4 *Relevance of Plate Tests to the Application of Membranes Underground*

The tests at Junction were used to assess whether the membranes could act as a replacement for mesh.

With regard to the Mineguard tests only the largest plate is judged to be representative of the insitu loading conditions underground. The loads applied by this plate would be similar to what a block or wedge could develop on the membrane underground and the area covered by the membrane would be similar to the areas between bolts in the current support designs. The loads developed by the smaller plates are much higher than what would be expected from a wedge or block of similar surface area. However these smaller plate tests do show that smaller blocks would definitely be prevented from moving by the membrane. Beyond stage 1 the membrane begins to debond and stretch plastically and peel off the surface. Stages 2 and 3 of the graph show that this type of response could continue with no failure to the membrane.

The results of all the plate tests indicate that Mineguard would be an effective structural replacement for mesh supporting large blocks and easily supporting small blocks.

The adherence to a clean and dry rock surface would be expected to be better than to the concrete test surface because even when washed there was a residue of dust in the pores and along this surface. This would improve the expected performance in dry underground conditions.

However in jointed rocks as the backs deform sharp edges are formed between blocks and as the membrane is stretched there is potential for the sharp edges to slice through the membrane causing it to fail prematurely. This may have caused the peeling failure described in the ore drive.

The Everbond plate tests showed that this membrane would be unable to retain large blocks or wedges of ground. The 1000mm plate attained a maximum load of 10 KN before it failed. This is much less than the loads that the mesh currently used at Junction is able to restrain.

The results of the plate tests indicate that Everbond at this stage of its development would not be an effective structural replacement for mesh.

8.5 *Flexure Tests of Mineguard*

Two square boxes were built of wood, 1075mm x 1075mm x 50mm and the second was 2148mm x 2148mm x 50mm. These were filled with paving stones of 153mm x 153mm x 60mm dimensions. The slab of paving stones was then sprayed with Mineguard.

The boxes were then turned over and the slab supported at each corner on a stack of paving stones.

Figure 14. Flexure tests with Mineguard on the large slab just before failure.

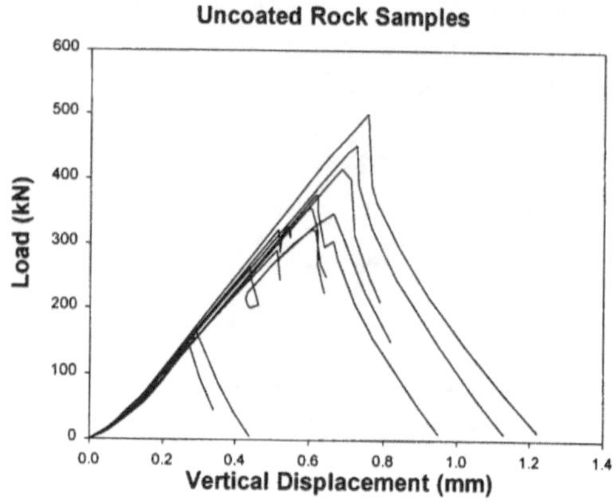

Figure 15. Uncoated porphyry core tests.

The system was then loaded at the centre using the same apparatus as for the plate tests but using it to push now rather than pull as shown in Figure 14.

The larger slab was loaded until the membrane began to tear. The smaller slab was loaded until it started to lift the steel beam off the walls of the concrete bays and the testing was stopped as the system was no longer in balance.

8.6 *Flexure Tests of Everbond*

The flexure tests were set up in the same way as those for Mineguard but the tests were never completed because the membrane tore as the boxes were being turned over. This was a significant outcome and supports the findings of the plate tests.

8.7 *Uniaxial Compressive Tests of Mineguard and Everbond*

The rock core samples were microgranites/ microgranodiorites with a porphyrtic texture which were collected from exploration diamond drill holes at Junction. The concrete samples were made up from a 32MPa batch of concrete and a 28MPa batch of concrete these were poured into polypipe to produce 50mm diameter core samples. Strain gauges were attached to the concrete samples prior to spraying.

The samples were coated with Mineguard by spraying the cores of concrete and porphyry and by dipping the sample into a mixed batch of Everbond. It was extremely difficult to control the thicknesses of the membrane being applied and because of this some of the porphyry core samples were sent to Canada and South Africa to have a more even coating of the membranes applied.

In all 8 samples of porphyry core and 8 samples of concrete with Everbond were tested and 5 samples of porphyry core and 4 samples of concrete were tested with Mineguard. Three concrete samples and 11 samples of porphyry core were also tested with no membrane.

8.7.1 *Testing of Porphyry Cores*

The tests on the uncoated porphyry samples resulted in explosive failure at between 300 to 500 KN, a concave plot characterises the post failure curve. Two of the samples failed prematurely along pre-existing structures in shear at loads of less than 200 KN.

None of the porphyry core sprayed with Mineguard failed prematurely, all the tests reached an ultimate strength of between 300KN and 500 KN.

During failure the Mineguard contained the rock fragments and an explosive failure was prevented. The post failure curve shows a sudden loss in load with no increase in displacement. There was no concavity to this post failure curve. Mineguard provides little if any improvement to intact specimens at failure or immediately after failure because there is no change to the ultimate strength or any significant change to the immediate post failure curve. Mineguard does not provide any effective confinement to these samples.

As the tensile strength of Mineguard comes into play the specimens achieve a residual strength of between 32-105 KN or between 7-29% of the UCS.

Everbond also prevented any premature failures along pre-existing structures and again there was no noticeable increase in the ultimate strength. During failure the Everbond contained the fragments and prevented an explosive failure.

There was no change in the shape of the post failure curve and the Everbond membrane tore as the deformation continued so no residual strength

Figure 16. Unsprayed porphyry core tests.

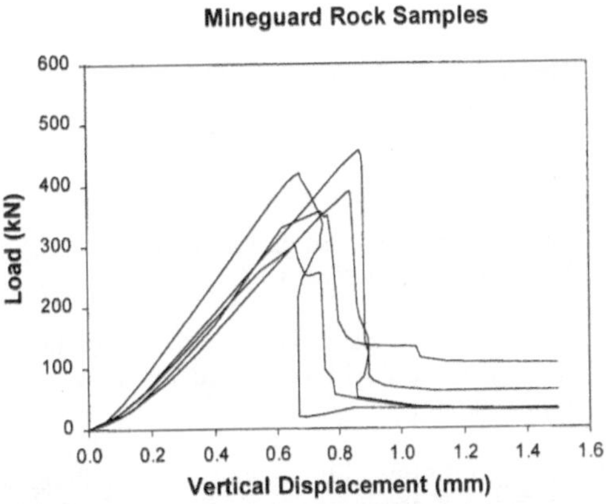

Figure 17. Porphyry core samples coated with Mineguard.

developed in the sample. Everbond provided no effective confinement to these specimens.

8.7.2 *Testing of Concrete Samples*

Aggregated concrete was used to represent a softer rock mass which on failing would have rough irregular contacts along the failure surface.
The ultimate loads of the uncoated concrete were between 40 KN and 60 KN.

The concrete sprayed with Mineguard showed no increase in the ultimate strength of the sample. However the confinement provided by Mineguard increased the frictional resistance along the rough irregular failure contacts in the failed concrete increasing the post failure strength to about 70% of the UCS. There was a significant increase in the post failure toughness of the coated concrete samples.

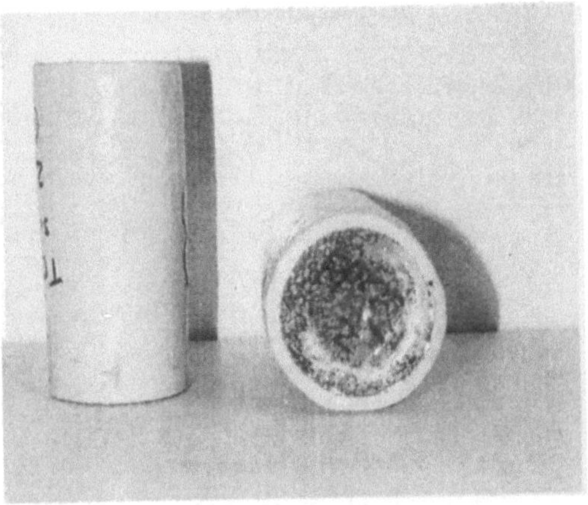

Figure 18. Porphyry core samples coated with Mineguard.

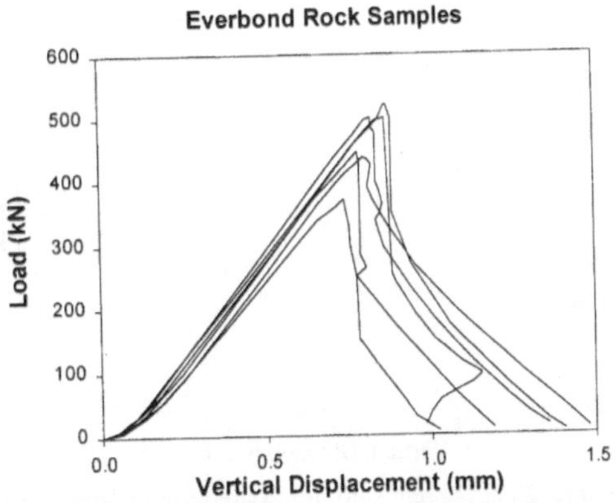

Figure 19. Porphyry core samples coated with Everbond.

Figure 20. Porphyry core samples coated with Everbond

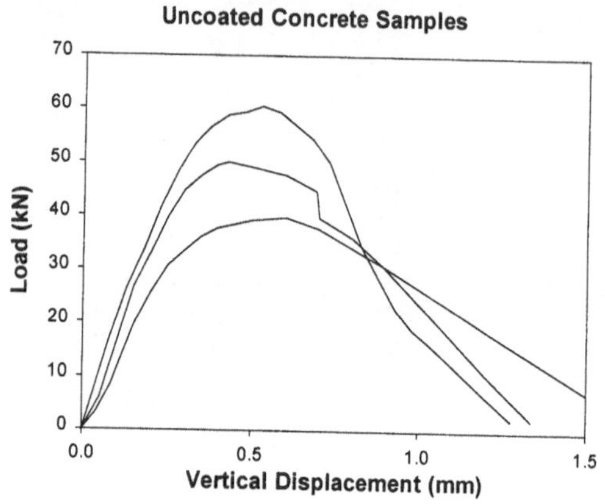

Figure 21. Uncoated Concrete Samples.

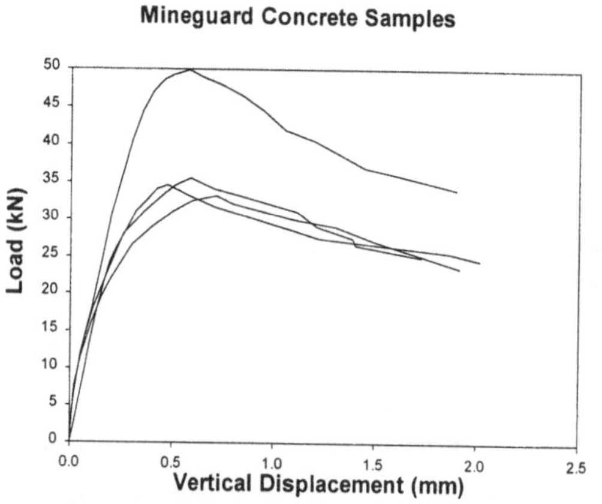

Figure 22. Concrete samples coated with Mineguard.

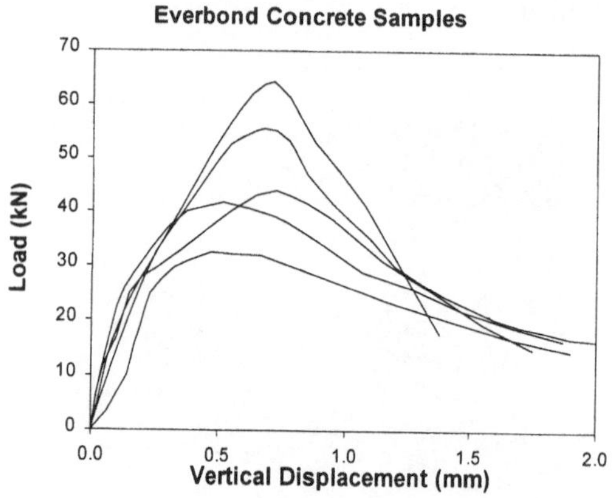

Figure 23. Concrete samples coated with Everbond.

The results for concrete samples which had been sprayed with Everbond were similar to Mineguard. There was no increase in the ultimate strength but the confinement increased the post failure toughness as the frictional resistance along the contacts was increased. As deformation continued the Everbond membrane tore and the increase in toughness was not as significant as that for Mineguard.

9. CONCLUSIONS

9.1 *Underground Trials*

- Both membranes prevented the development of scats and maintained the integrity of the ground.
- Both the membranes can be applied relatively easily and there are no logistical problems associated with them. The current spraying systems could be improved to increase the rate of application and allow them to be applied remotely.
- Damage from fly rock and blast gases could be managed in development headings by maintaining a critical distance from the face to minimise these effects.
- Damage from the blast gases on membranes at the brows of stopes is significant.
- Isocyanates associated with the spraying of Mineguard could be controlled by the use of a water curtain.
- Mineguard's inability to adhere in anything but dry pristine conditions make it difficult to apply in production headings and areas of ground water inflow even where the amounts are minor.
- Everbond's inability to be used in areas of flowing water (even minor amounts) where poor curing and adhesion were observed, make it unsuitable for use in wet conditions
- Both products are unsuitable for use in weak incompetent rock as the surface layer of rock may fail detaching the membrane from the main part of the rock mass.

9.2 *Plate Tests*

- Mineguard was able to achieve high loads during plate tests due to its high tensile strength and elongation capabilities.
- Mineguard could replace mesh as a restraint structural system underground.

- Everbond was unable to support loads of greater than 10KN in the plate tests suggesting that if used underground it will tear through when loaded.

- Everbond is currently not an acceptable structural replacement for mesh underground.

9.3 *Uniaxial Compression Tests*

- Both products prevented the porphyry rock from exploding during core tests. The Mineguard also maintained some residual strength after failure.

- Neither product significantly altered the post failure behaviour of the porphyry rock. This is the anticipated result when used as a membrane on a strong stiff rock mass.

- Both products increased the post failure toughness of the concrete core samples. This is the anticipated result when used as a membrane on a jointed or soft rock mass.

9.4 *General*

- Both products have potential and it is envisaged that the use of spray-on plastic membranes will become common place in the mining industry in a few years. But a number of improvements are necessary to both products before their use becomes widespread.

- At this stage the use of mesh is adequate for the current restraint requirements at Junction mine. It is relatively easy to handle, flexible and relatively cheap and quick to apply

ACKNOWLEDGEMENTS

The authors would like to thank WMC Resources Ltd. for permission to publish this paper. The operators from ECL and Stratabond who sprayed the membranes during underground and surface trials. Prof. E. Villescusa and J. Piang for conducting the uniaxial compressive tests. We express a special thanks to Danny Watts for his help with the test work on site and to the other operators and staff at Junction mine who were involved in the test work.

REFERENCES

Archibald, J.F., Lausch, P., Espley, S.J. 1997. Field and Laboratory Support Response of Mineguard ™ Spray-on Polyurethane Liners. *Proc. Of International Symposium on Rock Support, Applied Solutions for Underground Structures, Lillehammer, Norway.*

Espley, S.J., Langille, C. and McCreath, D.R. 1994. Innovative liner support trials in underground hardrock mines at INCO. *Technical Paper presented at CIM, AGM, Toronto.*

Finn D.J. 1998. Junction mine Ground Control Plan. *Internal WMC Resources St. Ives Gold Operations Document.*

The MIROC Mineguard Operations Manual. Urylon™ Plastics Inc.

Myrvang, A and Stjern, G. 1993. Determination of in-situ stresses and adhesivity of sprayed concrete. *Sprayed Concrete. R. Kompen, O.A. Opsahal and K.R. Berg (ed.) Proc. Int. Symp. On Sprayed Concrete. pp 424-438. Fagernes, Norway.*

Pascoe, M.J. 1998. Junction Mine 603.0 Access Virgin Stress Measurement. *Australian Mining Consultants Pty Ltd., report to WMC Resources St. Ives Gold Operations.*

Thompson A. & Finn D.J. 1999. The Performance of grouted split tube bolts. *This publication.*

Wojno, L and Kuipjers, J.S. 1997. Spray-on, user friendly and flexible support for mine excavations. *Proc. Of International Symposium on Rock Support, Applied Solutions for Underground Structures, Lillehammer, Norway.*

Rock Support and Reinforcement Practice in Mining, Villaescusa, Windsor & Thompson (eds)
© 1999 Balkema, Rotterdam, ISBN 90 5809 045 0

Laboratory testing of weld mesh for rock support

E. Villaescusa
Western Australian School of Mines, Kalgoorlie, W.A., Australia

ABSTRACT: Laboratory test results have been used to determine the failure mode and failure strength for weld mesh. The preferred mode of failure for underground support purposes is identified. A minimum breaking load for the weld strength is recommended. Similar studies can be used by mesh manufacturing companies to carry out quality control and quality assurance programs. Mining companies can use the results to award contracts during a tendering process.

1 INTRODUCTION

The practice of meshing is an essential part in providing personnel safe access to underground excavations. It is a form of passive support required to provide surface restraint at an exposed excavation boundary. Rockbolts are likely to control the overall excavation stability through keying, arching or composite beam reinforcing actions, while mesh is installed to support the small, loose pieces of rock detached within the bolting pattern. The mesh loading mechanism can either be uniformly distributed loading forces such as in rock bulking or point loading by loose individual rock blocks.

The aim of meshing is to prevent small pieces of loose rock from falling, particularly in rock masses having small block sizes or undergoing stress changes. The mesh is not designed to carry excessive loads of broken rock without failure and it can be easily damaged by flyrock from blasting when installed very close to an active face. The amount of loose rock on the mesh is usually a function of the in-situ block size distribution and the degree of damage to the excavation due to blasting or stress re-distributions.

Figure 1 shows a conceptual situation in which loose rocks or scats are being supported by the mesh installed within a rock bolt pattern. Although the principal stabilizing mechanism is by load transfer between the reinforcing element and the rock mass, mesh can be used to contain small rock pieces between the bolts and to build up a back pressure to inhibit further upward slabbing. The depth of disturbance along the bolt axis is controlled by the in-situ block sizes, the subsequent level of damage following mesh installation and the effectiveness of the reinforcing elements. Under no circumstances should the main stabilizing action depend mainly upon mesh strength.

2 LABORATORY TESTING

A series of shear strength tests have been carried out at the Western Australian School of Mines (WASM) in order to determine the suitability of weld mesh for the long term support of underground excavations. The tests were based on the Australian Standard (AS 1304 - 1991). This standard was designed to test mesh for concrete reinforcing, and a modification is required when testing mesh for underground support.

This lack of an accepted Standard Test for underground rock support membranes, suggests that a rock mechanics engineer must rely on experience, laboratory testing of weld shear strength and field observations to select

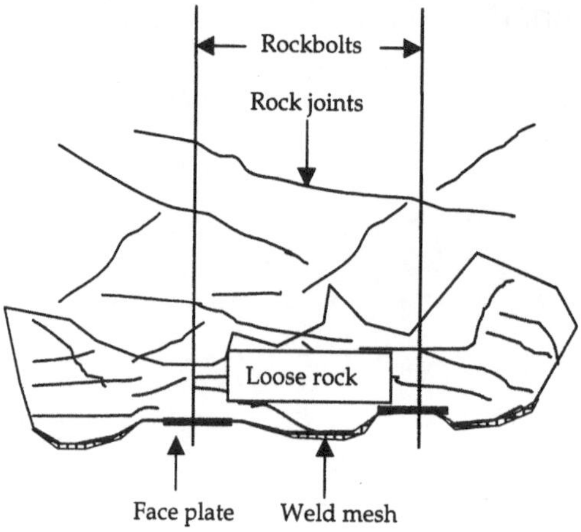

Figure 1. Loose rock supported by a combination of rockbolts and mesh.

the best mesh product to match the expected ground behaviour on a particular mine site.

The current Australian Standard test on weld shear strength for concrete reinforcing (AS 1304 - 1991) requires that a pulling force be applied to the longitudinal mesh wires. However, the actual mesh loading during underground support can be complex and a mesh weld shear strength specification must be established by pulling on both longitudinal and cross wires.

3 FAILURE MODES

In rock support applications, failure of the mesh within the wires rather than at the weld itself is preferred, and under no circumstances should the weld or wires fail during mesh handling and installation. Failure within the wires can be related to the inherent wire strength, while failure at the weld points can occur at any mesh loading level. In extreme cases of poor quality control, the wires can be separated by hand. In other cases the weld strength equals the tensile strength of the wire. Failure at the weld points at low strength levels is an undesirable property from a safety point of view, especially when small overlaps of one or two wires are used between individual sheets during meshing operations.

Three failure modes have been identified during laboratory testing of weld mesh at WASM. These can be described as shear failure at the weld points, failure on the Heat Affected Zone (HAZ) and tensile failure of the wire. Failure at the weld is an indication of the weld technology and quality control (dirty electrodes or dirty wire) during mesh manufacturing. Failure at the HAZ is caused by weakening of the wire during the welding process due to excessive weld head pressure and temperature, while tensile failure of the wire is controlled by the wire manufacturing process. Figure 2 shows a schematic of the laboratory samples used for testing indicating the different failure modes.

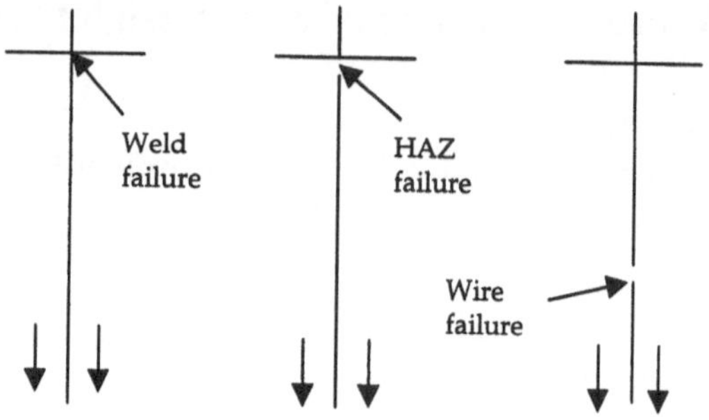

Figure 2. Modes of failure during laboratory testing of weld mesh.

4 FAILURE STRENGTH

To date there is no published standard covering weld strength for underground support. However, the Australian Standard (AS 1304 - 1991) for concrete reinforcing coupled with many years of meshing experience at Mount Isa Mines can be used as the guidelines for design. The Australian Standard AS 1304-1991 9.2a states that "the minimum allowable average shear value in newtons (N) shall not be less than 250 multiplied by the nominal area of the longitudinal wire in square millimetres (mm^2)".

As an example, for a 5.6mm thick galvanized wire the minimum breaking force should be approximately 6.2kN. However, this standard is for concrete reinforcement and it is not directly applicable to rock support applications, where the preferred mode of failure is at the HAZ or through the wire. Consequently, the weld strength must be designed to have a strength equal to that of the line wire strength. The formula in terms similar to the Australian Standard AS 1304-1991 9.2a would read, "the minimum breaking load in newtons (N) shall not be less than 500 multiplied by the nominal area of the longitudinal wire in square millimetres (mm^2)". Experience at WASM indicates that a 5.6mm thick galvanized wire can indeed achieve a minimum weld shear strength ranging from 11to 12 kN.

5 CASE STUDIES

5.1 Mine Site A

Underground observations at this site suggested that weld shear failure was the most common mesh failure mode. Mine personnel raised concerns regarding the actual weld shear strength as this could be critical to the performance of their support schemes, especially in highly stressed (burst prone) areas. Consequently, laboratory testing of a representative number of weld points has been carried out to determine the suitability of the current mesh product being used at this site.

The laboratory results are shown in Figures 3 and 4. In addition, several welded wires were collected from an area that had undergone a seismic event (See Table 1). Galvanized weld mesh having a 5.6mm thick wire on a 100 x100mm square pattern is used at this site.

The results suggested that the limiting strength was controlled by the weld shear strength for the mesh product used at this site (See Figures 3 and 4). Different average weld shear strength values were determined for each of the 4 sheets tested. These results appear to indicate quality control problems during the welding process.

In addition, a series of tensile tests were carried out on the wire to determine the extent of strength reduction experienced by having failures at the weld points. The average tensile strength of the wire was found to be 15.2 kN (See Figure 5).

The laboratory results indicate that on average, a 21% reduction on the nominal wire capacity was found due to failures at the weld points. The average failure load at the weld points was found to be 12 kN. However, a 60% reduction on the nominal wire capacity was found for the lowest individual weld shear strength result (6 kN). This strength reduction could be critical especially in highly stressed areas, where a significant amount of loading on the mesh is expected.

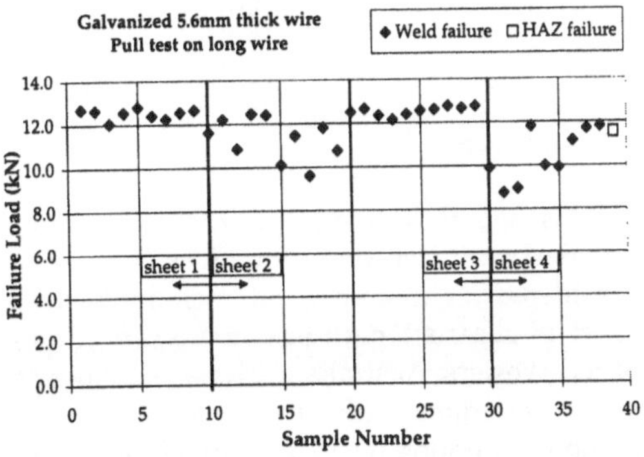

Figure 3. Maximum failure strength by pulling on long wires.

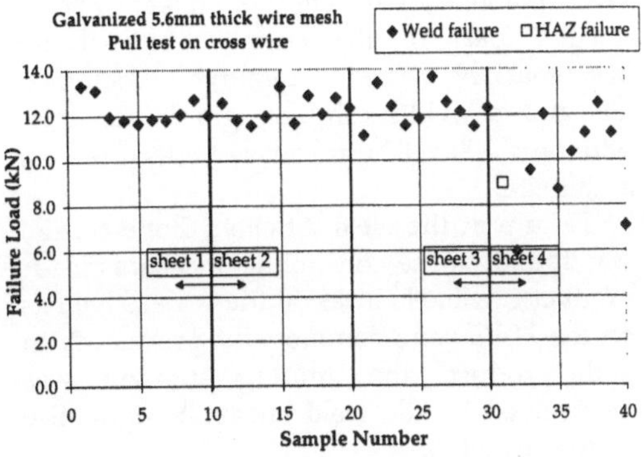

Figure 4. Maximum failure strength by pulling on cross wires.

Table 1. Weld shear strength from a seismically loaded wire mesh.

Sample No	Failure Load (kN)	Failure Mode
1	11.40	Weld
2	9.55	Weld
3	13.50	Weld
4	8.50	HAZ*
5	6.30	HAZ*

* Damaged by the rockburst.

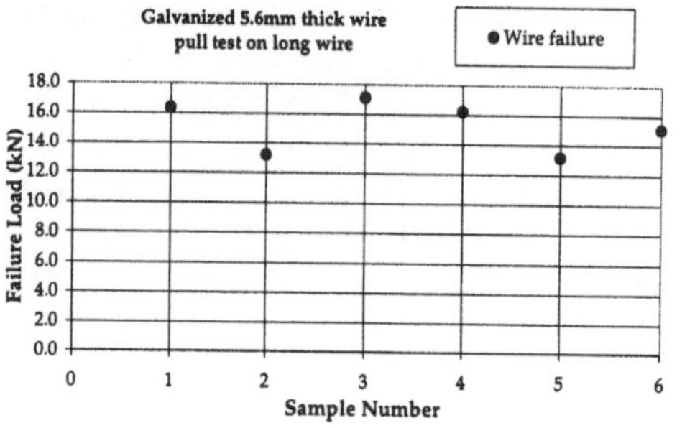

Figure 5. Maximum tensile failure of the long wire.

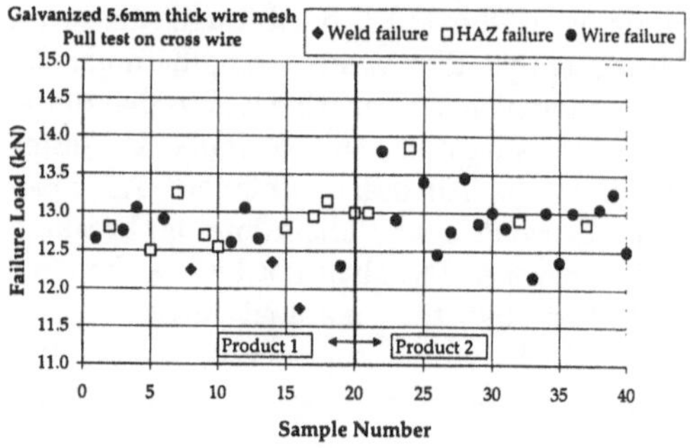

Figure 6. Maximum failure strength by pulling on cross wires.

5.2 Mine Site B

At this site weld mesh products from two different manufacturing companies were being considered for underground rock support. A series of shear strength tests were carried out at the Western Australian School of Mines to help determine the most suitable and economical option for long term support. This company was in the process of tendering and awarding a contract for the supply of galvanized weld mesh and the results of the WASM laboratory testing were likely to be an integral part of the supply contract for benchmarking and quality assurance. Galvanized weld mesh having a 5.6mm thick wire on a 100 x100mm square pattern is used at this site.

Failure at the Heat Affected Zone (HAZ) was found as the controlling mechanism for product 1, while failures on the wire as well as on the HAZ were common for product 2. In both products the limiting strength was determined by the weld shear strength (See Figures 6 and 7).

Slightly higher strength was determined by pulling on the cross wires compared to pulling on the longitudinal wires for both mesh products. Consequently, the mesh design specification was established by pulling on longitudinal wires as shown in Figure 7.

A large dispersion on the failure load values was found for mesh product 2 as shown in Figure 7. In general, the weld shear strength of product 2 appeared to be slightly higher than that determined for product 1. The experimental results indicate a minimum

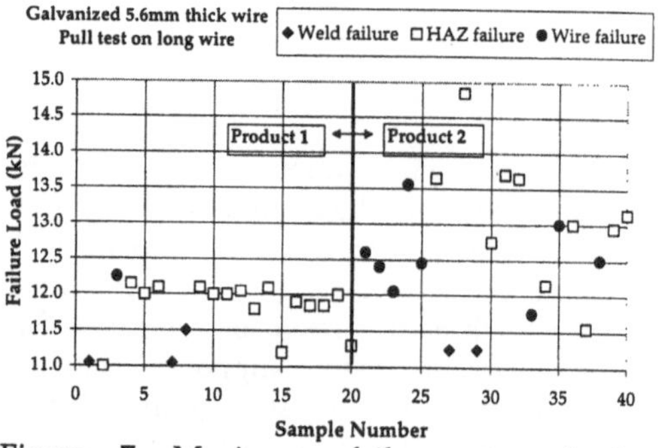

Figure 7. Maximum failure strength by pulling on long wires.

breaking load value of 11 kN for both meshing products.

Post-galvanizing weld procedures were used by both weld mesh manufacturers in order to minimize corrosion damage to the mesh products. However, visual inspection of the sample sheets provided indicated that the corrosion process had already started at some weld points. Consequently, although not all galvanizing is lost, it can be argued that use of welded galvanized mesh is of limited benefit over non-galvanized welded mesh, due to the galvanizing layer being removed in the area most susceptible to failure.

Significant savings can be achieved by using standard non-galvanized wire in non-critical areas such as stope drive support. These are temporary areas that are consumed by the stoping process itself, and where corrosion deterioration is not usually an issue. This practice has been successfully implemented at the McArthur River Mine in the Northern Territory.

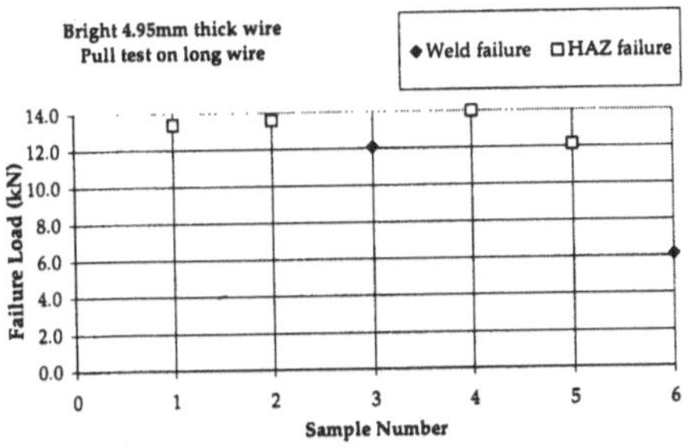

Figure 8. Maximum failure strength by pulling on long wires .

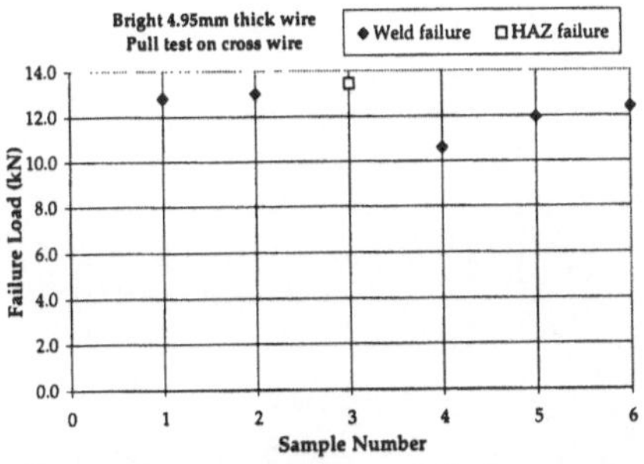

Figure 9. Maximum failure strength by pulling on cross wires.

5.3 Mine Site C

A series of shear strength tests have been carried out to help determine the most suitable and economical options for the long term support at this site. The results of the WASM laboratory testing are likely to be an integral part of a quality assurance program. Bright mesh (black, non-galvanized) having a 4.95mm thick wire on a 100 x 100 mm square pattern is used at this site.

The results shown in Figures 8 and 9 suggest that the pulling direction did not appear to control the resulting failure strength. In some cases, pulling on the longitudinal wires produced a slightly higher strength than that determined by pulling on the cross wires. The opposite was also true. Consequently, the mesh design specification at this site must be established by pulling on both longitudinal and cross wires.

Failures at the Heat Affected Zone (HAZ) and at the weld points were found as the controlling mechanisms, while the limiting strength was determined by the weld shear strength.

6 CONCLUSIONS

Meshing is likely to be used routinely in most underground support operations in Western Australia. Laboratory testing of weld shear strength can be used for quality control and quality assurance. A new standard has been recommended to ensure adequate weld mesh integrity in underground support installations.

7 ACKNOWLEDGEMENTS

The author acknowledges the financial support from the Australian Centre for Geomechanics and Curtin University of Technology for his position as Professor of Mining Geomechanics at WASM. The Australian Centre for Geomechanics has received funding for this position from the Government of Western Australia, Centres of Excellence Program.

8 REFERENCES

Standards Australia, 1991. Welded wire reinforcing fabric for concrete. AS 1304-1991, 12p.

3 Shotcrete

Rock Support and Reinforcement Practice in Mining, Villaescusa, Windsor & Thompson (eds)
© 1999 Balkema, Rotterdam, ISBN 90 5809 045 0

Keynote lecture: International practices and trends in sprayed concrete

Tom A. Melbye
MBT International Underground Construction Group, Zürich, Switzerland

ABSTRACT: Over the years since the start of sprayed concrete application, one fact cannot be denied: There has been a rapid development within all aspects of the method, from concrete technology and equipment to working environment and capacity. It is also clear that the development continues at a rapid pace.
Sprayed concrete is but one of several ways to cast concrete. As with traditional methods of casting, sprayed concrete also makes its special demands on the characteristics of the concrete during casting. At the same time all normal concrete technological demands, such as w/c ratio, amount of cement, correct consistency and curing must be complied with and followed. The paper will deal with some of these requirements and gives an update in latest development of special admixture systems for sprayed concrete (like hydration control system, internal curing, alkali-free accelerators, etc.).

1. INTRODUCTION

A well known fact about the underground construction industry and mining in particular is that all projects are unique. The degree of complexity due to the intertwining of the variety of project related parameters is higher than in many other industries, thus forcing contractors, mining companies and suppliers to be adaptable and flexible.

The enormous advantages of sprayed concrete as a construction and rock support process and the improvements of materials, equipment and application know-how have made it very important and a necessary tool for modern underground construction works as well as mining.

2. WHAT IS SPRAYED CONCRETE?

Sprayed concrete or Gunite is not a new invention. Sprayed concrete (mortar) has been known for more than 80 years. The first sprayed concrete jobs were done in the United States by the Cement Gun Company Allentown as early as 1907. The sprayed mortar was named "Gunite". This mortar contained fine aggregates and a rather high percentage of cement. The name Gunite is still being used. In some classifications Gunite stands for sprayed mortar but the limits for the grain size are not consistent. Depending on the country the limit for the maximum aggregates is defined by 4, 5 or even 8 mm.

In order to avoid this confusion between sprayed mortar or sprayed concrete / shotcrete / Gunite, EFNARC defined the new expression "Sprayed concrete" for every sprayed mixture of cement and aggregates placed with high velocity. Later, this will also be used in the new European Norms and Standard for Sprayed Concrete (CEN) which is under preparation.

3. TWO METHODS - WHAT IS THE DIFFERENCE?

Today there are two application methods for sprayed concrete: The dry mix and the wet mix procedure. In the beginning, there was only dry mix sprayed concrete.

Both methods have their advantages and disadvantages. Depending on the project requirements and the experience of people the best suited method should be chosen. There will continue to be a need for both methods in the future.

Until the mid 1990's, dry spraying has been the dominant method. However, the wet mix method is being used more and more now, especially for rock support works. In the future the wet spraying will be the dominant method. This in terms of m^3 sprayed concrete as this method offers a substantially better working environment, higher and more consistent quality and a much higher production.

The future developments within the sprayed concrete technology, products and equipment will be mainly made in connection with the wet mix process. Good examples for recent developments of sprayed concrete are the new ad-vanced admixtures (Total Consistency Control System, Hydration Control System, Concrete Improver [Internal Curing], micro silica and fibres).

The situation nowadays is that world-wide more than 60% of the sprayed concrete is being applied by the wet mix method and with a rapidly increasing tendency.

In some areas, however, the wet mix method is already dominating (Scandinavia, Italy, Switzerland with almost 100%). Within the next five years the wet mix method could be used for more than 70 - 80% of all sprayed concrete works world-wide. Today, more than 8 million m^3 of sprayed concrete are being applied world-wide every year.

4. WHY THIS DEVELOPMENT?

A brief presentation of the sprayed concrete development in Scandinavia and why it happened will illustrate why countries like Italy, Switzerland, France, UK, Spain, Greece, Australia, Korea, Brazil and some other European countries have been moving in the same direction.

Between 1971 and 1980, the sprayed concrete application in Norwegian tunnelling turned from 100% dry mix method to 100% wet mix. A similar change took place in Sweden and Finland.

The next dramatic change was from mesh reinforcement to steel fibres. Again, the fastest turn-around happened in Norway. During the same period a similar change from manual to robot application took place. Since 1976 - 78 steel fibres and micro silica have been added to wet mix sprayed concrete in rapidly increasing volumes.

It is not unfair to say that Norway leads the way into real wet mix sprayed concrete. They have definitely the longest experience and know most about wet spraying.

The main reasons for this change are clear:

- rapid increase in shotcrete output per pump hour - from 3 - 5 m^3/hour to 15 - 20 m^3/hour
- rebound reduction from typically 30 - 40% down to 10% and less
- substantially reduced transport and set-up time per application
- reduced number of people per equipment
- less compressed air and total energy for sprayed concrete on the wall
- as a direct consequence, the cost reduction was dramatic. In fixed value, the cost of sprayed concrete on the wall was reduced by 75% from 1972 to 1986.

Such a cost reduction may seem unrealistic at first sight. But if:

- the pump output increases from less than 5 to about 18 - 20 m^3/hour,
- the rebound reduces by 66%,
- robot application and integrated self-sufficient equipment units cut the set-up time from an average of two hours to 20 minutes, and the
- necessary staff is reduced by 75% (from 3 - 4 men down to 1 man),

it may all look a bit more plausible.

4.1 Some highlights from other countries

The data presented in Table 1 are not 100% accurate but still reasonably correct and can be used for the purpose of showing today's situation and tendencies. Some comments are given as a supplement.

When the construction period started in London Underground for the Jubilee Line and Heathrow Express and other well known projects in the area, the market was practically 100% dry mix. Within slightly more than a year this was turned around to wet mix. This is illustrated very well by the fact that MBT Intl. UGC alone had 17 MEYCO® Suprema wet mix pumps in operation (see Figure 1).

Sometimes it is also commented or claimed that in the "Third World Countries" only the simple, low cost dry mix method can work properly. The reality is different since there is no relation between the choice of method, local development tendency and such popular regional tags. This is well illustrated by the figures in Table 1 for China, Brazil, Colombia, India/Nepal compared to Austria and Germany.

Table 1: Shotcrete methods and volumes per year (approximate)

Country	Dry %	Wet %	m^3/year	Tendency
Japan	10	90	2-3 mio.	wet
China	60	40	> 1 mio.	wet
Hong Kong	30	70	100.000	wet
Asia/Pacific	40	60	> 1 mio.	wet
USA	50	50	500.000	wet
Brazil	20	80	400.000	wet
Colombia	40	60	200.000	wet
Rest of Latin America	40	60	> 300.000	wet
UK	20	80	100.000	wet
Switzerland	10	90	150.000	wet
France	10	90	250.000	wet
Italy	0	100	700.000	wet
Greece	30	70	200.000	wet
Spain	20	80	300.000	wet
Scandinavia	0	100	250.000	wet
India/Nepal	50	50	300.000	wet
Austria	98	2	150.000	???
Germany	70	30	600.000	wet???

Austria and partly Germany have been traditionally very strong dry mix areas as Switzerland was some years ago. Switzerland turned to wet mix in less than one to two years whilst Austria and Germany have developed new systems within the dry mix method, using extremely quick setting cement types (without gypsum). This allows spraying of thick layers in one go without the use of any admixtures or accelerators. This new alternative with quick setting cement appears to be favourable at first sight because of lower material costs per m^3 mixed. But there are some serious negative aspects to be considered, such as high rebound (> 30%), dust (above all limits), sensitivity to cement quality, high to very high energy cost, low capacity, complicated equipment and difficult handling of the material as well as labour intensive. Furthermore, there is no realistic chance of using steel fibres because of high fibre rebound (50 - 70%).

The development of quick setting cement is driven by the cement lobby in Austria and Germany in order to protect themselves from the import of cement, especially from East European countries. However, the method is regarded by the author as a blind track led by people who have not a high understanding of sprayed concrete and who have no interest in the end user. There are clear signs that a change in the direction of the wet mix method will also happen in Germany and Austria in the near future. Some first big projects have been completed successfully with the wet mix method both in Germany and Austria (Königshainer Berge, Ditschhardt Tunnel, Irlahüll-Ingolstadt [Highspeed Link, 400.00 m^3], Sieberg Tunnel). Another interesting area may be Australia which has not been mentioned in Table 1. Here the mining industry was ahead of civil construction in switching from dry to wet mix and partly to steel fibres. This has led to a substantial market share for wet mix in both mining and civil construction works in Australia (e.g. Pasminco mine, Melbourne City Link).

5. DRY MIX METHOD

5.1 Fibres

Steel or synthetic fibres should never be used in the dry mix because of the high fibre rebound. The loss of fibres in dry mix is measured in the range of 40 - 80%. Therefore, the cost/performance becomes critical. In order to meet the normal specified requirements for toughness and ductility and due to the high rebound, you have to add 80 - 100 kg of steel fibres.

5.2 Equipment for dry spraying

Mostly used are different types of rotary machines which operate at relatively high costs due to wear and tear (10 - 12 SFR/m^3).

5.3 The future of dry mix

In the future it is to be expected that the main application for dry mix will be in projects with relatively small volumes and/or high requirements regarding flexibility and/or limited space (e.g. repair) or long conveying distances (> 400 m).

6. WET MIX METHOD

Wet mix is similar to ordinary concrete in that a ready mixed concrete from a concrete batching plant is used. It is possible to check and control the w/c-ratio and thus the quality at any time. The consistency can be adjusted, e.g. by means of admixtures.

Figure 1: MEYCO® Suprema

With the wet mix method it is easier to produce a uniform quality throughout the spraying process. The ready mix is emptied into a pump and forwarded through the hose by pressure (thick stream transport). Today, piston pumps are dominating.
At the nozzle, at the end of a hose, air is added to the concrete at a rate of 7 - 15 m^3/min. and at a pressure of 7 bar, depending on whether the spraying is performed manually or by robot. The air is added to increase the speed of the concrete so that good compaction is achieved as well as adherence to the substrate/surface. A mistake often made with the wet spraying is that not enough air is used. In most cases only 4 - 8 m^3/min. are added which gives a bad result in compressive strength, bond and rebound. For robot spraying, a minimum of 12 m^3/min. is required. In addition to the air, liquid set accelerators are added at the nozzle.

6.1 Advantages of wet mix

The advantages of wet mix are:
- low rebound (5 - 10%)
- better working environment, low dust
- thick layers because of effective admixing material
- controlled w/c-ratio and quality
- higher capacity (60 m^3 in 8 hours with one robot)
- use of steel/synthetic fibres and new advanced admixtures/additives possible
- better total economy on applied sprayed concrete

6.2 Disadvantages of wet mix

The disadvantages of the wet mix are:
- limited conveying distance (max. 300 m)
- higher demands to mix design
- cleaning costs (can be solved by using hydration control admixtures)
- limited open time/workability (can be solved by using hydration control admixtures)

6.3 Sum-up of the wet spraying method

The wet spraying method will dominate worldwide for all types of rock support in civil construction as well as in mining because of its advantages. The wet method should mainly be applied as robotic spraying.

7. CONCRETE TECHNOLOGY

Some people maintain that sprayed concrete is a special concrete. However, sprayed concrete is but one of several ways to cast concrete. As with traditional methods of casting, sprayed concrete also makes its special demands to the characteristics of the concrete during casting. At the same time, all usual demands on the concrete technology (such as w/c-ratio, amount of cement, correct consistency and curing) have to be complied with. The reason why so much concrete of poor quality has been applied in many parts of the world is because one seems to forget that sprayed concrete is a way of casting and that all concrete technological requirements have to be fulfilled.

Available aids to get a good quality sprayed concrete with the wet mix method are:

- cement
- microsilica/additives
- aggregates
- admixtures
- set accelerators
- fibres
- curing
- correct equipment
- correct execution

The individual areas will be dealt with in order to evaluate the advantages but also point out drawbacks due to the use of improver. As mentioned earlier the same requirements are made on sprayed concrete as on normal concrete, namely:
- low w/c-ratio
- less water
- less cement
- good casting ability

The conflict between the properties of fresh concrete and set concrete is particularly strong for sprayed concrete and used to reduce the quality of wet mix sprayed concrete. New superplasticizers as well as micro silica and alkali-free set accelerators have, however, altered this.

7.1 Cement type and quality

Before the spraying is started, a test for the compatibility of the cement and accelerator has to be made. This is conducted in order to check the reactivity/setting of the accelerator.

For wet mix, a minimum of 450 kg of binder should always be used. This is to ensure proper workability (no slump loss) and quality (w/c-ratio). A wet mix concrete needs a minimum of 200 - 210 litres of water to maintain workability and to avoid slump loss. In order to keep the w/c-ratio below 0.45, you will need a minimum binder content (200 litres/0.45 = 444 kg binder).

It is possible to spray with a low cement or binder content but this will result in a high w/c-ratio > 0.5, reduced quality, lower early and final strength. A lower binder content also increases the rebound dramatically (below 425 kg will increase the rebound with 30 - 40% compared to 450 kg of binder). Usually the cement/ binder content in the placed concrete is even higher than only 450 kg.

It is wrong to reduce the binder content as low as 400 kg and even less for economical reasons. In reality, this sprayed concrete is much more expensive in place due to the following parameters:

- higher rebound (7.5 SFR/m^3 extra costs in rebound compared to higher binder content of 450 kg/m^3)
- higher consumption of set accelerators because of a higher w/c-ratio (1 - 2% => 4 - 8 kg/m^3 at 1 SFR/m^3)
- lower production because of poor consistency and higher volume required because of higher rebound

Considering these points, it is easy to understand that it pays to add 50 kg more cement/binder in order to get better economy and quality and problem-free sprayed concrete. The costs for 50 kg of cement are approx. 0.12 SFR/kg x 50 kg = 6 SFR/m^3, compared to extra rebound costs which are considerably higher.

Binder materials include:

- cement
- high quality fly-ash
- microsilica

It is recommended to use a cement type 42.5 or 52.5 because of a better reactivity with set accelerators, better early and final strength results and lower accelerator consumption.

7.2 Microsilica

Silica fume, or microsilica, is considered to be a very reactive pozzolan. Microsilica has a definite filler effect in that it is believed to distribute the hydration products in a more homogeneous fashion in the available space. That leads to a concrete with reduced permeability, increased sulphate resistance and improved freezing and thawing durability.

When considering the properties of microsilica concrete, it is important to keep in mind that microsilica can be used in two ways:

- as a cement replacement, in order to obtain reduction in the cement content - usually for economic reasons.
- as an addition to improve concrete properties - both in fresh and hardened state.

In sprayed concrete, microsilica has to be used as an additive rather than as a substitute for cement to improve the concrete and spraying properties.

7.3 Special advantages of sprayed concrete with microsilica

Normal sprayed concrete qualities, i.e. 20 - 30 MPa cube strengths, can be produced without micro-silica, whereas a practical and economical production of higher strengths is more or less dependent on the use of microsilica. It seems favourable from a technical point of view to use 5 - 10% (by cement weight) of microsilica. The correct use of micro-silica can provide the sprayed concrete with the following properties:

- Better pumpability (lubricates and prevents bleeding and segregation)
- Reduced wear on the pumps
- Increased cohesiveness of the fresh concrete and therefore reduced consumption of accelerator which is positive for the final strength

- Increased bonding strength to various substrates and between sprayed concrete layers
- Improved strengths
- Improved resistance against alkali aggregate reaction
- Improved permeability resistance
- Reduced rebound
- Improved sulphate resistance

In fibre reinforced sprayed concrete it provides:

- Easier mixing and distributing of fibres
- Reduced fibre rebound
- Improved bonding between cement matrix and fibres

Because of these positive effects it is recommended that microsilica should always be added to sprayed concrete in order to obtain the best possible quality.

When adding microsilica to concrete because of its fineness, it is necessary to always add a high rate of plasticizers/superplasticizers to disperse the microsilica. The dosage of admixtures increases by approximately 20 %, compared to sprayed concrete without microsilica.

7.4 Aggregates

As for all special concrete, the aggregate quality is of major importance for the fresh and hardened concrete. It is particularly important that the grain size distribution and other characteristics show only small variations. Of particular importance are the amount and characteristics of fines, i.e. the grain size distribution. However, it is not relevant to talk about choice of aggregate, as normally the available material must be used and the prescription has to be adapted to it. Nevertheless, for wet-mix spraying, the following criteria have to be observed:

- Maximum diameter: 8 - 10 mm. This is because of limitations in the pumping equipment and in order to avoid too much rebound loss. From a technological point of view, one should wish for a larger maximum diameter.
- The granule distribution basket is also very important, particularly its lower part. The fine material content in sieve no. 0.125 mm should be min. 4 - 5% and not higher than 8 - 9%.
- Too little fine material gives segregation, bad lubrication and risk of clogging. However, in the case of fibre concrete the surplus of fine material is important, both for pumping and compaction. A high fine material content will give a viscous concrete.

As the margins in the sieve basket are relatively small, it may often be convenient to combine two or more fractions, e.g. 0 - 2, 2 - 4 and 4 - 8 mm, by adjusting the proportion between them, to make a sieve curve that fits within the ideal curve limits (see Table 2). Too little fine material will be compensated by using more cement or microsilica. Too much fine material is primarily compensated by increasing the dosage of water-reducing admixtures.

The quantity of 8 mm particles should preferably not exceed 10%. The larger particles will rebound when spraying on a hard surface (when starting the application) or penetrate already placed concrete producing craters difficult to fill. During screening, storing and handling of the aggregates, measures should be taken to prevent the presence of particles in excess of 8 mm. Coarse particles may block the nozzle and subsequent cleaning can be very time consuming.

Table2: Grain size

SIEVE mm	MIN%	MAX%
0.125	4	12
0.25	11	26
0.50	22	50
1.0	37	72
2.0	55	90
4.0	73	100
8.0	90	100
16.0	100	100

An improvement of the grain size curve for a natural sand by the use of crushed materials often results in an increased water demand and poorer pumpability and compaction. Before crushed materials are employed as part of the aggregates, tests for comparison should be done to establish whether the addition of crushed material gives an improved result.

7.5 Admixtures (plasticizers/superplasticizers)

In order to obtain specific properties in the fresh and hardened concrete, concrete admixtures should always be used in the wet-mix spraying method. Concrete admixtures are no new inventions. The old Romans used different types of admixing material in their masonry, such as goat blood and pig fat in order to make it more mouldable. The effect was good, since the constructions are still standing.

The fact is that concrete admixtures are older than PC-cement, but it is only during the last 30 years that more stringent requirements for higher quality and production have sped up development, research and utilisation of admixing materials. Water reducers are used to improve concrete workability and cohesiveness in the plastic state. The water reducer can give a significant increase in slump with the same w/c ratio, or the w/c ratio can be reduced to achieve the same slump as for a mix not containing the water reducer. The reduced w/c ratio relates to a direct increase in strength. The higher slump adds to an increased pumpability.

The wet-mix method is attractive as the concrete is mixed and water is added under controlled and reproducible conditions, for instance at a concrete plant. The w/c ratio, one of the fundamental factors in the concrete technology, is under control. One often forgets, however, that the equipment makes heavy demands on the fresh concrete first of all in terms of pumpability. Furthermore, the method requires a larger amount of fast setting admixing materials, which may lead to loss of strengths in the final product.

Today, combinations of lignosulphonate, naphthalene and melamine are often used. This is to obtain the best possible and production-friendly concrete. Naphthalenes/melamines (superplasticizers) are chemically distinct from lignosulphonates (plasticizers/water reducers). They are better known as high range water reducers since they can be used at high dosages without the problems of set retardation or excessive air entrainment often associated with high rates of addition of conventional water reducers. The effect of superplasticizers/plasticizers to disperse "fines" makes them perfect and needed admixtures for sprayed concrete. The slump increase achieved by adding conventional superplasticizers is time and temperature dependent. However, pumpability can only be maintained for a limited time (20 - 90 min.) after mixing, and excessive dosages of admixtures can result in a total loss of cohesiveness and in segregation. Normal dosage is from 4 - 10 kg/in^3 depending on the quality requirements, w/c ratio, required consistency, as well as cement and aggregate type.

7.6 A new generation of special superplasticizers

A new generation of high performance superplasticizers has entered the market during the last two years. They are based on modified polycarboxylic ether and they have a much higher water reduction than traditional superplasticizers.

- Glenium™: > 40% water reduction
- BNS/melamine: 20 - 30 % water reduction

This opens up new possibilities for sprayed concrete. With these types of admixtures you can produce a sprayed concrete with a w/c-ratio of ≤ 0.38 and with a slump of 15 - 20 cm. Advantages are:

- Longer / better slump retention
- Faster setting
- Higher early strength
- Higher final strength
- Reduced rebound
- Possibility to handle water ingress or to lower the dosage of accelerators (2 - 3%)

Glenium™ (polycarboxylic) is already being used widely in Europe in combination with alkali-free accelerators. This is the future generation of sprayed concrete admixtures

7.7 Set accelerators

The wet-mix method needs the addition of fast setting admixtures at the nozzle. The primary effect of the material is to reduce the slump (consistency) at the moment of spraying from liquid to paste while the concrete is still in the open air, so that it will adhere to the surface when the layer thickness is increased.

With the use of set accelerators, effective spraying on vertical surfaces and overhead becomes possible. The setting effect allows to apply sprayed concrete for initial support - an important function in the New Austrian Tunnelling Method (NATM). Water inflow (e.g. from the rock behind) usually calls for a higher proportion of admixtures to accelerate the setting of the sprayed concrete.

Accelerators are added in liquid form via a pressure tank or a special dosage pump (piston or worm pumps). The dosage of accelerator will vary, depending on operator's ability, the surface and the w/c ratio (high w/c ratios will increase the need for accelerators in order to reduce consistency).

Every coin has two sides. A side effect of the accelerators is the decrease of final strength. Compared to plain concrete (without accelerators), the 28-day strength can be reduced significantly. Therefore, the accelerator consumption should be kept at a minimum at all times (lower consumptions on walls than in the roof).

The types of accelerators used include:

- waterglass
- modified silicate
- consistency activators (MEYCO® TCC)
- aluminates (sodium or potassium or a mix)
- alkali-free

Accelerators containing aluminates should not be used because of their strong negative influence on working conditions and environment. Due to their high pH (> 13) they are aggressive to skin and eyes.

7.8 Modified sodium silicates

Modified sodium silicates give only momentarily a glueing effect (<10 sec.) of the sprayed concrete mix (loss of slump) and take no part in the hydration process like alkali-free based accelerators (if dosages do not exceed 20% of b.w.) Modified sodium silicates bind the water in the mix. Dosage is therefore depending on the w/c ratio: The higher the w/c ratio the more modified sodium silicate/ water glass is required in order to glue the water and the mix.

Modified sodium silicates do not give very high strength within the first 2 - 4 hours. Normal final setting is from >30min. (depending on cement type and temperature).

Advantages of modified silicates:
- Work with all types of cement
- Less decrease in final strengths than with aluminate based accelerators at normal dosages (4 - 6 %)
- Very good gluing effect
- Environmentally friendly, not so aggressive for skin. The pH is <11.5, but still direct skin contact has to be avoided and gloves and goggles should always be used.
- Much lower alkali content than aluminate based products (Na_2O < 8.5%)

Disadvantages of modified silicates:
- Temperature depending (cannot be used at temperatures below + 5°C).
- Limited thickness: max. 8 - 15 cm

Typical dosages:
- Modified sodium silicates: 3 - 6 % by weight of binder.
- Water glass should normally not be used because high dosages (>10 - 12%, normally 20%) are needed that decrease strengths and give a very bad quality and a false security. Normal water glass is more or less banned in Europe.

The European Sprayed Concrete Specification (EFNARC) allows a maximum dosage of 8% by weight of the cementitious material for the use of liquid accelerators.

In most applications with a reasonable modified silicate dosage (3 - 6 %) and a good quality control, not more than 20 % strength loss is acceptable. In practice the loss is between 10 - 15%.

Note that an 18 year old wet-mix sprayed concrete recently tested in Norway has the same strength today as after 28 days. This is in conflict to what some people claim.

7.9 Alkali-free sprayed concrete accelerators

Of late, safety and ecological concerns have become dominant in the sprayed concrete accelerator market and applicators have started to be reluctant to apply aggressive products. In addition, requirements for reliability and durability of concrete structures are increasing. Strength loss or leaching effects suspected to be caused by strong alkaline accelerators (aluminates) have forced our industry to provide answers and to develop products with better performances.

Due to their complex chemistry, alkali-free accelerators are legitimately much more expensive than traditional accelerators. However, accelerator prices have very little influence on the total cost of in-place sprayed concrete. Of much larger consequence are the time and rebound savings achieved, the enhancement of the quality and the safe working environment.

7.10 Liquid non caustic alkali-free accelerators

The increasing demand for accelerators for sprayed

concrete termed "alkali-free" always contains one or more of the following issues:

1) Reduction of risk of alkali-aggregate reaction, by removing the alkali content arising from the use of the common caustic aluminate based accelerators.
2) Improvement of working safety by reduced aggressiveness of the accelerator in order to avoid skin burns, loss of eyesight and respiratory health problems.
3) Environmental protection by reducing the amount of released aggressive and other harmful components to ground water, from sprayed concrete and its rebound.
4) Reduced loss of sprayed concrete final strength, normally in the range of 15 to 50% with older accelerator products.

The focus within different markets, regarding the above points, is variable. Where most sprayed concrete is used for primary lining (in design considered temporary and not load bearing), points 2 and 3 are the most important. When sprayed concrete is used for permanent structures, items 1 and 4 become equally important. This variation in basic reason why to require a new accelerator technology has caused some confusion.

7.11 Choice of accelerator

The most important rule is: Carry out a compatibility test with the cement(s) to be used for every sprayed concrete application with alkali-free accelerators and before any practical concrete spraying. MBT has developed in-house standard tests that are very simple and quick to execute.

7.12 Alkali-free accelerators in powder form

The approach with powder products involves numerous practical limitations and constraints:

- Costs for an additional dosing unit
- One additional man for the dispensing of the accelerator into the dosing/dry spraying machine
- Higher dosages: approx. 7-10 % by weight of binder
- More air required (4 - 5 m^3/min.)
- Higher rebound. Experience from practical tests has shown that the rebound is 10-15 % higher as compared to liquid alkali-free accelerators.
- The approach with powder products is not practical in modern rapid tunnelling where wet-mix high performance steel fibre reinforced sprayed concrete plays an important role: The set-up of the equipment between each spraying cycle is too complicated and too time consuming.
- In addition to all these practical and economical limitations and constraints, powder products also raise environmental concerns:They create a lot of dust and therefore cause a very bad working environment.

7.13 Type of dosing system for alkali-free accelerators

Alkali-free accelerators are usually suspensions and therefore not all types of dosing pumps will work properly. In order to achieve a good result, it is of the utmost importance to secure a constant and adequate dosage. We recommend mono pumps and peristaltic pumps.

We believe the following should not to be used:

- Piston pumps
- All systems with seat valve system
- Pressure tank
- Gear pumps
- Membrane pumps

8 HYDRATION CONTROL SYSTEM

The supply and utilization of sprayed concrete mixes for infrastructure projects in congested environments creates problems for both the contractor and ready-mixed concrete supplier.

Sprayed concrete mixes, wet or dry, only have a useful pot-life of 1.5 to 2 hours and even less at temperatures above +20°C. Material sprayed after this time will exhibit lower strengths and increased rebound, due to the commencement of the cement hydration.

Long trucking distances from the batching plant to the site, delays in construction sequences as well as plant and equipment breakdowns ensure that much of the concrete actually sprayed is beyond its pot-life.In addition to this, environmental regulations may well impose restrictions upon the working hours of batching plants in urban areas, meaning that a contractor who requires sprayed concrete mixes to be supplied 24 hours per day, may only be able to obtain material for 12 hours each day. Problems such as these create unnecessary additional costs of construction for both the contractor and the client. A chemical system for the controlling of cement hydration in both wet and dry sprayed concrete mixes has been developed by MBT, enabling the working life of such mixes to be substantially increased. The first component of this system is a stabilizer which is capable of inhibiting the hydration of Portland cements for periods of up to 72 hours. The second component of the system is an activator, which is a hydration accelerator that is added to the stabilized concrete before placing.

The hydration control system is able to give flexibility to the production and spraying of concrete mixes in large underground projects and, at the same time, offers considerable cost savings to contractors, owners and concrete producers.

It also ensures that all sprayed concrete which is sprayed through the nozzle contains a fresh cement that has undergone little or no hydration reactions.

The system brings revolutionary benefits to sprayed concrete (in particular wet-mix sprayed concrete) and is currently being used on a lot of big projects in Europe, America, the Middle East and Far East.

8.1 Selected case study - Wet-mix spraying with Delvo®crete in Athens Metro

The civil construction part of the project comprises 20 stations and 18 km of tunnels. In the tender documents, the dry-mix method was specified. It was, however, possible to convince the contractors that wet-mix spraying is beneficial in all aspects and the method is now being used.

From a central mixing plant the concrete is distributed by truck mixers to a number of sites. The individual sites have a buffer storage for concrete, holding about 12 m^3 maximum. The buffer is also an agitator that can be operated when necessary. The concrete goes from the agitator into a concrete pump, delivering through a pipe system down the shaft (typically 20 m deep) into the tunnel, ending in the spraying pump. To the spraying nozzle it is normally 100 to 150 m.

A normal work sequence means application of 3 to 4 m^3 of sprayed concrete and then a full stop of 3 to 4 hours until the next application. During this time, the whole system from day-buffer to the sprayed concrete nozzle is left untouched with concrete inside. A full cleaning of the delivery system is carried out about once a week.

This logistics system is only possible due to the Delvo®crete Stabilizer. By adding a maximum of 2% drawn on the cement weight, it is possible to prevent any hydration for up to 72 hours. When Delvo®crete is used for sprayed concrete, an activator must be used in the nozzle to start the hydration process.

Some key data of the sprayed concrete mix:

Cement	400 kg
Rheobuild® 716	1.2%
Delvo®crete Stabilizer	1%
w/c ratio	<0.45
Slump (at the batching plant)	18 - 20 cm
Delvo®crete Activator S71, added at the spraying nozzle	5 - 6%

This produced a rebound rate < 10%, early strength development better than the class J3 according to the Austrian Sprayed Concrete Norm, 24-hour strength of 13 - 17 MPa and 28-day strength of 30 MPa.

9. THE TCC SYSTEM - THE SOLUTION TO CRITICAL MIX DESIGNS

The TCC system provides Total Consistency Control of wet-mix sprayed concrete from the moment of mixing until the application.

The TCC system is a two-component admixture system. Its first component acts as a consistency stabilizer and provides good workability and pumpability, also with difficult aggregates and critical mix designs. Its second component is a special set accelerator which causes an instant loss of slump of the concrete, with adjustable setting within a period of one minute to one hour.

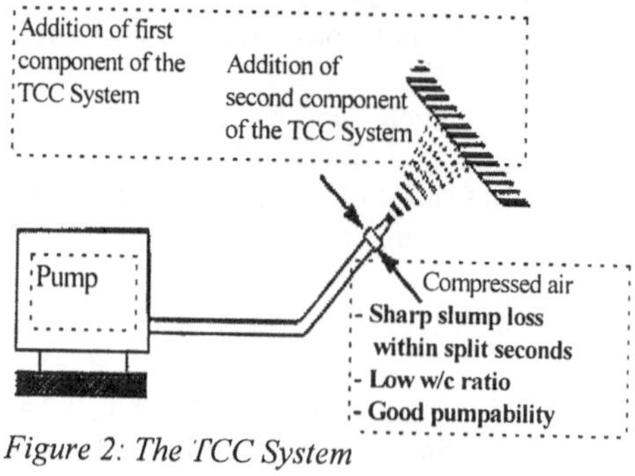

Figure 2: The TCC System

9.1 The single components of the TCC system

Component A - the consistency stabilizers

Component A is added to the concrete during batching, together with the other admixtures used for the mix. It is added just as a normal superplasticizer. In most cases, Component A of the TCC system makes the addition of a further superplasticizer unnecessary.

The main advantages of the consistency stabilizer as compared to ordinary superplasticizers are:

- Longer open time
- Improved pumpability
- More homogeneous concrete
- Improved pumpability and workability with concrete with a low cement content, with aggregates of lacking fineness or deficient grading or with fibre reinforced concrete
- Higher reliability and durability
- Free of chlorides

Component B - the consistency activators

Component B is added to the concrete mix at the nozzle just as a set accelerator or activator. In most cases, Component B of the TCC system makes the addition of a set accelerator or activator unnecessary as it works both as a consistency activator ("slump killer“) and as a set accelerator.

The TCC activators combine the latest modified silicate technology with modern concrete polymers and improve viscosity, reduce the alkali content and work with moderate dosages. They can also be used in combination with hydration controlled concrete mixes. They work as activators and replace the use of the hydration control activator.

TCC activators are compatible with all types of cement. Dosage: 2 - 12% by weight of binder.

9.2 A proven technology

The TCC system is intensively used in different big projects around the world; e.g. in Sweden, Norway, Saudi Arabia, U.K., Hong Kong, Spain, Italy, Mexico, Venezuela, India, Malaysia, Australia, Belgium, China, France, USA, South Africa, etc., with excellent results everywhere. In fact, the TCC system has already set higher standards in steel fibre reinforced sprayed concrete for rock support.

10. CONCRETE IMPROVING (INTERNAL CURING)

Tunnels and other underground construction projects have some of the worst conditions for curing due to the ventilation that blows continuously dry (cold or hot) air into the tunnel. It can be compared with concrete exposed to a windy area. One would think that tunnels have ideal curing conditions with high humidity (water leakage), no wind and no sun exposure. However, this is not the case.

10.1 Background

Curing is one of the basic and most important jobs in sprayed concrete because of the large cement and water content of the mix and the consequent high shrinkage and cracking potential of the applied concrete. Another reason is the danger of rapid drying out due to the heavy ventilation as is usual in tunnels, the fast hydration of accelerated sprayed concrete and the application in thin layers. Therefore, sprayed concrete should always be cured properly by means of an efficient curing agent. However, the use of curing agents involves several restrictions: They must be solvent-free (use in closed rooms), they must have no negative influence on the bonding between layers and they must be applied immediately after placing of the sprayed concrete. Most of the in-place sprayed concrete around the world has poor bonding and many cracks, due to the fact that no curing is applied.

With the use of sprayed concrete as permanent final lining, long-term quality and performance requirements have built up significantly. These requirements are: good bonding, high final density and compressive strengths to ensure freeze/thaw and chemical resistances, watertightness and a high degree of safety.

When curing sprayed concrete with a curing agent, one has to be very careful with the cleaning procedure of the substrate before applying a subsequent layer. Cleaning must be done with high pressure air and a lot of water (use spraying pump and nozzle, adding air at the nozzle). Another problem with curing agents is to be able to apply them quickly enough after finishing of spraying. To secure proper curing of sprayed con-crete, the curing agents must be applied within 15 to 20 minutes after spraying. Due to the use of set accelerators, the hydration of sprayed concrete takes place a very short time after spraying (5 to 15 minutes). The hydration and temperature are most lively during the first minutes and hours after the application of the sprayed concrete and it is of great importance to protect the sprayed concrete at this critical stage.

The application of curing agents requires consuming working operations twice: Application of curing agent and cleaning/removal of the curing agent from the sprayed concrete surface between the layers in the case of multiple layers.

In many countries with experience in wet-mix sprayed concrete like in Norway and Sweden and in big projects world-wide, there is an obligation to cure the sprayed concrete with a curing agent. Very good experiences have been made with the use of a special curing agent for sprayed concrete. It is used in many big projects and in different countries, everywhere with very good results. The use of specially designed curing agents for sprayed concrete improves bonding by 30 - 40% compared to no curing (air curing), reduces shrinkage and cracking and also gives a slightly higher density and compressive strength (at 28 days). These results are confirmed by several laboratory tests and field trials. However, in order to achieve these results, proper cleaning is required before subsequent layers of sprayed concrete can be applied. Even with easy-to-apply products, curing of sprayed concrete remains a time consuming job and is often felt as a hindrance to other tunnelling operations.

10.2 Concrete improving

MBT has developed a new system for more efficient and secure curing of wet-mix sprayed concrete, repair mortars as well as concrete.

Concrete improving (internal curing) means that a special admixture is added to the concrete/mortar during batching as a normal admixture. This admixture produces an internal barrier in the concrete which secures safer hydration and better resistances than the application of conventional curing agents. The benefits resulting from this new technology are impressive:

- The time consuming application and, in the case of various sprayed concrete layers, removal of curing agents are no longer necessary
- Curing is guaranteed from the very beginning of hydration
- There is no negative influence on bonding between layers

As a consequence of this optimum curing effect, all other sprayed concrete characteristics are improved: density, final strengths, freeze/thaw and chemical resistances, watertightness, less cracking and shrinkage. In addition, it also improves pumpability and workability of sprayed concrete, even with low-grade aggregates. It particularly improves the pumpability of steel fibre reinforced sprayed concrete mixes.

10.3 A proven technology

The concrete improving system has been tested with good results both in laboratories and on big jobsites. Comprehensive investigation programmes were carried out in Norway (SINTEF), in Switzerland (LPM Institute) and in Austria (University of Innsbruck).(See Table 4)

Bond strengths were higher than 2.0 MPa with failures discovered in the concrete only and not in the bonding area. Density and mechanical strengths at 28 days were more than 10% higher than in conventionally cured reference sprayed concrete.

Example results from a large Middle East project:

- Increased bonding compared to no curing: 100% (from 0.5 - 0.7 to >2.0 MPa)
- Increased bonding compared to curing with special curing agent: >30 - 50 % (from 0.7 - 1.2 to >2.0 MPa)
- All cores of sprayed concrete treated with concrete improver show bond >2.0 MPa. Failures were discovered only in the concrete and not in the bonding area.
- Increased density (>15 %) compared to sprayed concrete treated with external curing agents
- Increased strength (28 days) compared to air cured sprayed concrete or treated with external curing agent (>10 %). No signs of cracking.

10.4 Benefits of concrete improving

- No influence on bonding between the layers. Always good bonding, high security
- No additional work operation for the application of curing agents or other curing methods
- No need for additional work operation for cleaning and removal of curing agents
- Curing from the first second and therefore during the critical time
- Less cracking
- Better chemical resistance

Table3: Cost comparison per m^3 of concrete improver, external curing and water curing

		Water curing	External curing	Concrete improver (internal curing)
Material		--	SFr 14.--	SFr 15.--
Application:	- man hours	SFr 25.20	SFr 1.--	--
	- machine costs	SFr 280.--	SFr 18.--	--
Removal:	- man hours	--	SFr 10.80	--
	- machine costs	--	SFr 80.--	--
Total costs m^3		**SFr 305.20**	**SFr 123.80**	**SFr 15.--**

Table 4: Mechanical performances of the three mixes

	Reference (no curing)	External curing	Concrete improver (internal curing)
Flexural strength tests on concrete beams (10 x 10 x 40 cm), UNI 5133, MPa:			
7 days	3.8	--	5.9 - 6.1 = **6.0**
28 days	5.5 - 4.5 = **5**	4.5 - 4.5 = **4.5**	6.4 - 6.8 = **6.6**
Pull-out test Rilem/CEB/FIP RC6, MPa:			
7 days	--	--	2.1 - 1.9 = **2.0**
28 days	**1.5**	2.0 - 1.8 = **1.9**	2.4 - 2.2 = **2.3**
Adhesion on concrete (*), MPa:			
7 days	0.92 (P)	0.9 (P)	1.5 (P)
28 days	1.02 (I)	1.5 (I)	2.8 (P)
Cracks on beams:			
1 day	cracks	no cracks	no cracks
7 days	cracks	no cracks	no cracks
14 days	breakings	superficial cracks	no cracks
28 days	breakings	cracks	no cracks
Static modulus of elasticity, UNI 6556, MPa:			
7 days	17150	--	19100
28 days	21650	--	22400
Dynamic modulus of elasticity, MPa:			
7 days	28500	28000	39400
28 days	36600	37300	39600

- Improved watertightness (less cracks)
- Improved freeze/thaw resistance
- Improved workability and pumpability
- Works independently from aggregate quality, grading and lack of fineness
- Works particularly well with steel fibre reinforced sprayed concrete; better fibre orientation, re-duced fibre rebound and increased toughness values
- Less time per m^3 due to increased production and less working operations. Time is money!
- Increased density
- Improved final compressive strengths

10.5 A safer and cheaper solution

- The concrete improving system, whilst guaranteeing safer curing, provides a new state of the art application procedure of curing agents in the easy-to-apply form of a concrete admixture.
- With the concrete improving system, overall savings of the spraying job are achieved: the elimination of additional work operations for application of curing compounds and preparation of substrate, as well as the reduced rebound and fibre rebound more than offset the extra material cost. (See Table 3)

10.6 Results from some spraying tests

In the tests (Table 4) a great number of parameters have been fixed in order to evaluate the real performance differences of the three mixes and systems.

(*): The values are the average of two tests

P: The breakings have occured in the application, i.e. in the product

I: The breakings have occured at the interface between the application and the concrete slab.

10.7 Selected case studies

Midmar potable water tunnel, South Africa

Midmar tunnel will ultimately convey 1 million m3 of potable water a day from Midmar dam to Pietermaritzburg in South Africa. The water comes from Midmar dam and runs through a system consisting of a pump station, water treatment plant, pipelines and a tunnel with a length of 6.5 km, all forming part of the Midmar project. The sprayed concrete specifications called for a sprayed concrete lining with a strength of 40 MPa at 28 days. A total of 45.000 m^2 of lining had to be applied as post excavation operation.

Mix design (per m^3):

Cement	450 kg
Gravel (max. particle size 10 mm)	470 kg
Fines	865 kg
Rheobuild® 3510	4.5 kg
Pozzolith® LD	4 kg
Water content	220 l
MEYCO®TCC 735	
(concrete improver)	5 kg
Delvo®crete Stabilizer	2.5 kg
(used only sporadically)	
Dramix Steel fibres (30/50)	30 kg
Addition at the nozzle:	
MEYCO®TCC766 (2.2%)	10 kg
Application thickness, variable	50, 100 or 150 mm
Slump	15 - 18 cm
Compressive strength:	
7 days	33 MPa
28 days	46 MPa
Rebound	5 - 9%

North Cape Subsea Tunnel, Norway

At the North Cape subsea road tunnel (6.8 km length), sprayed concrete for preliminary support and final lining was demanded by the client. The project put the contractor into a difficult situation as the bad ground conditions cut back the blast rounds to 1.5 m instead of the planned and normal 5 m. In addition to this, they had to follow with the insitu concrete lining all the way to the face for each blast round, which dramatically affected the advance of the tunnel.

First, the concrete improver was not included in the mix design. The sprayed concrete was being cured partly with water and partly with membranes. When after a short while the first cracks were found in the concrete structure, the contractor decided to use the concrete improver. The concrete improver not only solved the problem of the cracking but also increased both compressive and flexural strength as well as flexural toughness by 15 to 20%.

Lærdal road tunnel, Norway

The Lærdal tunnel is today the world's longest road tunnel with a final length of 24 km. At present, approximately 19 km have been excavated.

The tunnel has a very high overburden and some substancial rock bursting problems occured. In order to overcome these and other problems and to reduce cracks in the sprayed concrete, the mix design was adjusted for the project. The mix now includes Glenium™ 51, concrete improver and the hydration control system as well as an alkali-free accelerator, and it has proven to give significant better results, such as:

- Early strength increase
- No problems to build up thicker layers in one pass
- Very little problems with rock burstings
- No cracks in the sprayed concrete structure
- Significant increase in final compressive and flexural strength as well as flexural toughness
- Rebound reduced to a minimum (approx. 2 - 3%)
- Very low dust development during spraying

11. FIBRES IN SPRAYED CONCRETE

Fibre concrete is a new material undergoing fast development with new and better fibres, hand in hand with improved concrete technology and application techniques.

The use of steel fibre reinforced sprayed concrete has advanced substantially in the last few years. It has been accepted for rock support by engineers, specifiers, owners and contractors around the world.

11.1 Types of fibres

11.1.1 Glass fibres

Glass fibres cannot be used as a permanent material

because, after some time, they will become brittle and be destroyed by the basic part of the concrete matrix. Therefore, they have to be avoided in all types of concrete, sprayed concrete and cement based mortars.

11.1.2 Plastic fibres (polypropylene fibres)

Normal short plastic fibres are resistant and durable in the concrete environment. Yet, their mechanical properties are similar to those of concrete and therefore cannot improve them or make concrete more viscous. This makes plastic fibres unsuitable for the use in rock support.

However, for applications where only reinforcement against shrinkage, and in particular plastic shrinkage, is asked for, as in sprayed concrete repair, plastic fibres are well suited: They are very efficient at distributing microcracks during the plastic phase of hardening and they also help reduce rebound in wet-mix spraying.

Recently, developments in the US (Synthetic Industries) have come up with a new type of plactic fibre which resembles more a steel fibre in terms of shape and form. The so called new HPP plastic fibres are made of high quality materials and are delivered in a length of 35 mm. Different test results from Australia and Europe show that this type of fibre can reach a suitable toughness if dosed moderately (10 - 13 kg/m^3). The tests show that these fibres reach about 700 - 900 Joules according to the EFNARC plate test. This result is more or less equal to the result achieved with 30 kg/m^3 of high quality steel fibres. This new type of fibres is of interest for the industry and can be an important addition to sprayed concrete where steel fibres cannot be used because of several reasons (e.g. surface correction, fibres in the surface and where it is required to have an efficient reinforcement to improve the ductility of the sprayed concrete). One of the main problems we are facing today is the high fibre loss with the use of these new HPP fibres (the mix design has to be redesigned, higher slump and the request for a different spray pattern closer to the substrate and less air).

It could be interesting to combine a low dosage of the new HPP fibres together with steel fibres. You will get an excellent ductility, less cracks, low rebound and economical savings due to the use of less fibres per m^3.

11.1.3 Carbon fibres

From a technical point of view the mechanical properties of carbon fibres should be ideal for rock support, but in practice they are not used because of their high price.

11.1.4 Steel fibres

Steel fibres are the most commonly used fibres in sprayed concrete. There are several types and qualities available on the market, but only a few types meet the requirements set for fibre reinforced sprayed concrete. Critical and important parameters of the steel fibres are:

- Geometry
- Length
- Length/thickness ratio (L/D)
- Steel quality

In practice, we are looking for a thin and long fibre with high steel quality (same or higher than ordinary reinforcement). Most of the steel fibres available on the market have an insufficient steel quality. Typical fibres that meet the requirements for steel fibre reinforced sprayed concrete are Dramix fibres 30/50 and 40/50, NOVOTEX 0730 (0.7 x 30 mm) and Harex CF 30/0.5.

11.2 Technical advantages of steel fibres

Rock support includes the constant risk of unexpected loads and deformation. The best possible safety margin is achieved by the highest possible fracture energy (ductility) of the sprayed concrete layer.

Whereas the addition of ordinary steel fibres doubles the fracture energy of unreinforced sprayed concrete, modern steel fibre technology improves it 50 - 200 times, see Figure 3. In practical terms this

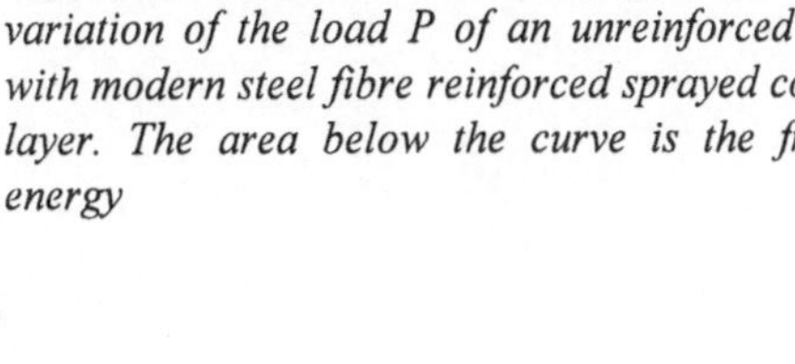

Figure 3: The 2 curves show the deformation under variation of the load P of an unreinforced and a with modern steel fibre reinforced sprayed concrete layer. The area below the curve is the fracture energy

means that with modern steel fibre technology a sprayed concrete layer may crack and deform and still have a lot of bearing capacity left, so that under normal circumstances there is ample time for cracks/deformations to be noticed and measures taken.

The fracture energy of steel fibres is also higher than of wire mesh. This has been proven by a large scale test run at the beginning of the eighties by the independent Norwegian Technical Research Association (NTNF), see Diagram 1.

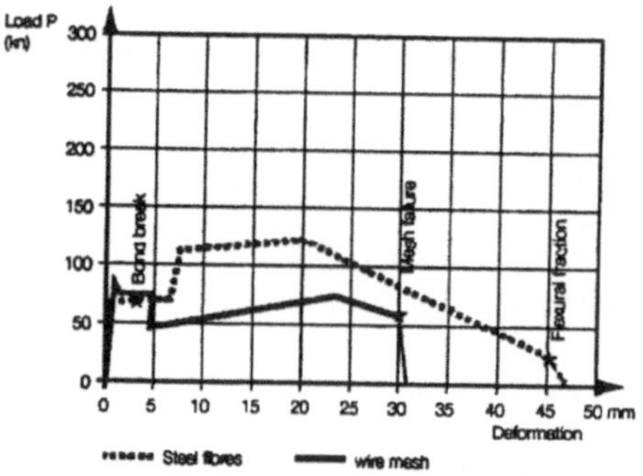

Diagram 1: Fracture energy of steel fibres vs. wire mesh.

The test simulates a block falling on a 10 cm sprayed concrete layer.

a) Sprayed concrete with 1 % steel fibres

b) Sprayed concrete with centric applied wire mesh

Both types of reinforced sprayed concrete layers were applied with a 10 cm thickness on three granite stone blocks (see Figure 4). After 28 days the middle block was exposed to various loads (P). The resulting deformation was measured.

The test shows that the fracture energy of the steel fibre reinforced sprayed concrete is much higher than that of traditional wire mesh reinforced sprayed concrete.

Theoretically, wire mesh reinforced sprayed concrete may produce similar results if the layer thickness is above 15 cm and the steel quality is good. However, commonly used wire mesh is produced from cold drawn wire. This mesh will break already under very small deformation and is therefore dangerous since in rock support deformation has to be taken into consideration constantly.

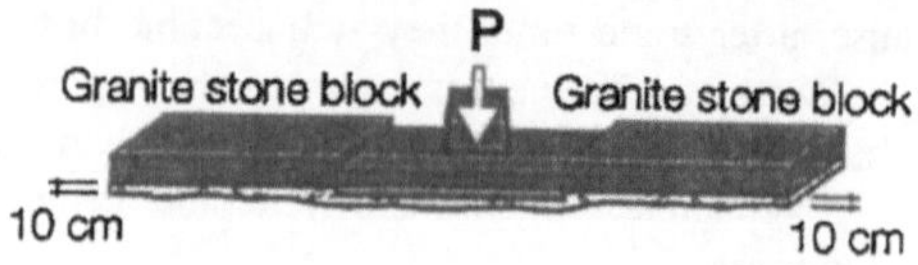

Figure 4: Test simulation

Reinforcing sprayed concrete by WWF also creates a quality problem. The shadow effect may produce voids behind the bars. This is often a serious problem, because it will eventually cause reinforcement corrosion and concrete spalling.

The danger arising from the insecurity about the wire mesh quality actually used and the problem of the shadow effect can be easily avoided by using steel fibre reinforcement which lends itself so well to wet-mix sprayed concrete, and at a lower cost, too. In rock support where one always has to allow for deformation, this feature is a very strong quality asset of the wet-mix method.

11.3 Economical advantages of steel fibres

By replacing welded wire mesh with steel fibres a time-consuming and dangerous operation can be avoided. This makes fibre concrete able to compete with traditional wire mesh. Steel fibres save money and time:

- Savings on direct costs: Direct cost of steel fibres is 50 to 60 % of the direct cost of wire mesh (labour plus material).
- Savings on indirect costs: Indirect costs due to the application of sprayed concrete in two layers which the use of wire mesh makes necessary, can be avoided, and no delay is caused in other tunnelling operations.
- Savings on sprayed concrete used: With steel fibres the required thickness of sprayed concrete can be applied over the whole surface, independent of the irregularity of the substrate.
- The increased rebound caused by wire mesh as well as the effect of "shadows" behind the mesh are avoided.

11.4 Mix design for steel fibre reinforced sprayed concrete

Steel fibres require the knowledge and skill of practical mix design.

Figure 5: Modern sprayed concrete: Robotic spraying with advanced admixtures and steel fibres at Melbourne City Link

- Fibre reinforced sprayed concrete requires the use of microsilica and admixtures, in order to cancel the negative effects of the fibres on pumping and spraying. Furthermore, it is important that the bonding (adherence) between steel and concrete matrix is optimal, and this is achieved by the addition of microsilica and with a maximum aggregate grain size of 8 mm.
- A higher content of fine material (min. 400 kg) is required.
- The slump has to be increased to a minimum of 10-14 cm. This means that fibre reinforced sprayed concrete requires a higher dosage of superplasticizers.
- For anchoring reasons the fibres should be at least twice as large as the largest aggregate granule.
- The fibre length should not exceed 50 to 60 % of the pumping hose diameter. This means that for manual spraying the normal maximum fibre length is 25 mm and for robots with 65 mm hoses, it is possible to spray with a fibre length of up to 40 mm.
- Steel fibres can be added before, after or during batching of the concrete materials. If balling occurs, it is usually eliminated by altering the batching sequence.

12. SPRAYED CONCRETE EQUIPMENT

Typical for the underground environment are numerous technical solutions, high risks and time pressure. Consequently, the contractor needs a competent and reliable partner. However, quality products alone are not enough. Only with a balanced utilisation of reliable equipment, high performance products and competent service can the required quality and efficiency be achieved.

Parallel to the devolopment in material technology there has been a constant innovative development in the equipment sector to produce ma-chines suited for the new products and that are adaptable to the everchanging conditions in the construction business. The result is a wide range of sys-tems that cover all sprayed concrete works: from huge tunnelling jobs with large quantities of con-crete mixes to be sprayed down to small volume repair works. Common to all developments in equip-ment is the tendency toward integrated and automated systems which ensure higher production output, consistent and controllable quality, as well as safer and more operator-friendly working conditions. Machines that work on the rotor principle are the type most generally used nowadays for dry-mix spraying. In wet-mix spraying the professional applicators trust in double-piston pumps.

12.1 Developments for wet-mix spraying

To ensure even spraying, the latest equipment developments aim at realizing a pulsation-free conveyance of the wet-mix from the pump to the nozzle.

This is put into practice with MEYCO® Suprema: The electronically controlled push-over system that is integrated into the output adjustment brings the pulsation of the material flow to a minimum which is hardly noticeable at the nozzle. An integrated memory programmable control system (PLC) supervises, coordinates and controls all functions of the machine. The PLC system allows checking and controlling of data which can also be printed out, e.g. dosing quantity of admixtures, output capacity etc. A dosing unit for liquid admixtures is integrated into the drive system of the machine and connected to the PLC system. This guarantees forcible regulation of the dosing analogous to the spraying capacity.

12.2 Spraying manipulators

Spraying manipulators or robots are suitable for use wherever large quantities of sprayed concrete are applied, especially in tunnel and gallery constructions or for protection of building pits and slopes. Thanks to mechanized and automated

Figure 6: MEYCO® Roadrunner: This mobile spraying system contains all specified equipment of a spraying mobile, mounted on a roadworthy truck. All equipment can be powered either from an external source or by its own Diesel motor.

equipment even large volumes of sprayed concrete - dry-mix and wet-mix - can be applied under constantly optimum conditions and without fatigue for the nozzleman who also profits from higher safety and improved general working conditions. The spraying robots typically consist of:

- Lance-mounting with nozzle
- Boom
- Remote control
- Drive unit
- Turntable or adapter-console (for different mounting versions)

12.3 Spraying mobiles

Many suppliers also offer complete mobile systems with integrated equipment for the complete spraying job. A spraying mobile typically consists of:

- Wet-mix spraying machine/pump
- Spraying manipulator
- Accelerator storage tank
- Dosing unit for accelerator
- Cable-reel with hydraulic drive
- Air compressor, capacity 12 m^3/min
- Central connection and control system for external power
- High pressure water cleaner with water tank
- Working lights

12.4 Benefits of mechanized spraying

- Reduced spraying cycles due to higher output capacity and the elimination of time-consuming installation and removal of the scaffolding, particularly in tunnels with variable profiles.
- Cost savings thanks to reduced rebound and labour savings
- Improved quality of the in-place sprayed concrete thanks to even spraying
- Improved working conditions for the nozzleman thanks to protection from cave-ins, rebound, dust and accelerators.

12.5 State of the art of concrete spraying robots

Spraying manipulators or robots are suitable for use wherever large quantities of sprayed concrete are applied. A new machine, based on the world-wide well known kinematic principle of the MEYCO® Robojet has now been developed. In co-operation with Industry and University for this manipulator with 8 degrees of freedom a new automatic and human oriented control system has been developed. The new tool enables the operator to manipulate the spraying jet in various modes, from purely manual to semi automatic and fully automatic, within selected tunnel areas. In one of the modes the operator uses a 6-D joystick (spacemouse). The calculation of the kinematics is done by the control system. A laser scanner sensor measures the tunnel geometry and this information is used to control automatically the distance and the angle of the spraying jet.

The aim of this control is not to automate the whole job of spraying but to simplify the task and enable the operator to use the robot as an intelligent tool and to work in an efficient way with a high level of quality. With the correct angle and continuously constant spraying distance a remarkable reduction in rebound and therefore savings in cost are achieved. Furthermore, if the tunnel profile is measured after spraying, too, the system will give information of the thickness of the applied shotcrete layer which was up to today only possible with core drilling and measurement. If an exact final shape of a tunnel profile is required, the control system manages the robot to spray to these defined limits automatically.

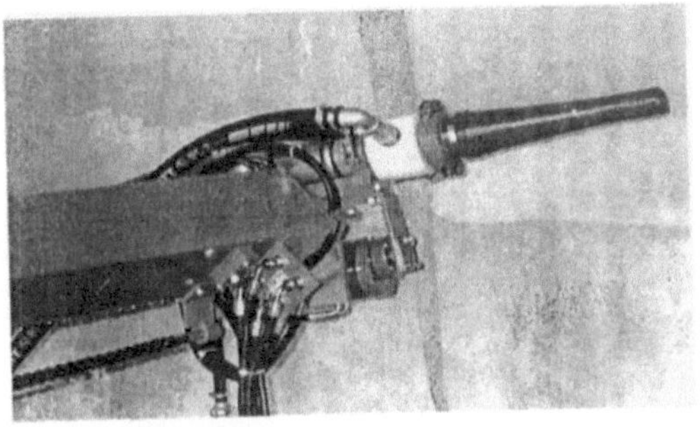

Figure 7: Nozzle for wet-mix spraying

12.6 Dosing systems

When using liquid admixtures it is important to ensure that dosing is constant and uniform in relation to the weight of the binder. To achieve this, the use of an appropriate dosing pump is necessary.

12.7 Nozzle systems

Nozzle systems are an important part of the spraying equipment. Nozzles essentially contribute to providing:

- Lower rebound
- Improved bonding
- Improved compaction by proper mixing of accelerators/activators and air in the case of the wet-mix spraying method, and by proper mixing of accelerators/activators and water in the case of the dry-mix spraying method

Only with the correct nozzle system - adapted to the type of application (wet-mix/dry-mix method, robot/hand application) and the accelerator/activator used - can low wear and outstanding quality of the in-place sprayed concrete be obtained.

13. SPRAYED CONCRETE KNOW-HOW

There are a few major sprayed concrete consumers who from practical experience, research and development have acquired know-how. Equipment and control methods have also gone through a development which has led to a rational production as well as a more uniform quality of the final product. From an international point of view it is safe to say that we have come a long way from when sprayed concrete was used for securing rock, but it is also fair to say that we are lagging behind when using sprayed concrete for building and repair works. It is not easy to find a reason. The know-how exists, however, it is not fully utilised.

Prevailing regulations make special demands on the concrete technological know-how of the people doing the spraying work. Present requirements have led to a better training of involved personnel. The result of this is an improved quality of the work. The number of special contractors who are working with sprayed concrete has increased over the last few years, which improved the quality of the application. However, there is a risk of getting badly executed work by less serious contractors. This is particularly the case with smaller jobs where the contractor often lacks knowledge about sprayed concrete. These are, however, things that can be eliminated if the contractor makes more stringent demands for his competence, experience, trained personnel, knowledge about concrete and authorisation.

The contractors should demand an authorisation arrangement for sprayed concrete with general validity as it exists e.g. for casting and sheathing (like the Sprayed Concrete Association in UK).

14. CONCLUSIONS

The wet-mix sprayed concrete technology has reached a stage of development which allows the production of high quality, durable sprayed concrete for permanent support. Further substantial technical advantages are available when reinforcement is required and steel fibres are used.

The time and cost saving potential in application of wet-mix SFRS as permanent support is in most cases substantial and sometimes dramatic. It must be ensured that the design method allows the use of such permanent support measures. Also, contract terms that are counter-productive for the utilisation of modern support techniques must be avoided.

Other important advantages like totally flexible logistics, very good working safety and good environmental conditions complete the range of reasons in favour of using wet-mix sprayed concrete technology. It is not an experiment any more, the solutions are well proven.

Sprayed concrete is mostly used today in rock support where it has solved many difficult problems and has become a necessary aid. There is a clear tendency that more and more sprayed concrete is being applied in international tunnelling for rock support. The total volume alone in Europe alone is more than 3 million cubic metres per year. In my opinion, this increasing trend will continue for several years to come.

Sprayed concrete as a building method would have a much larger field of applications; however, until today the degree of utilisation is unfortunately still rather limited. One of the advantages of the sprayed concrete is its flexibility and speed. Concrete which is to be applied simply with a hose against form work, rock surface or concrete surface, may architecturally and constructively be varied. The only limit is imagination and the desire for experimentation.

I therefore call upon all contractors, architects, authorities and consultants: Concrete technology, know-how, equipment and materials exist and may be mobilised to increase the range of our building activities as soon as someone plucks up courage to utilise the building method of the future: Sprayed concrete.

15. REFERENCES

Melbye, T.A., *Sprayed Concrete for Rock Support (6th edition, 1997*)

Garshol, K.F., 1998, *International practices and trends.* School on Shotcrete and it's Application. SAIMM - Randburg, South Africa.

Annett, M.F. and Varley, N., 1996 *High Performance Wet Shotcrete in London Clay.* Second International Symposium on Sprayed Concrete, Gol, Norway

Rock Support and Reinforcement Practice in Mining, Villaescusa, Windsor & Thompson (eds)
© 1999 Balkema, Rotterdam, ISBN 90 5809 045 0

An investigation into the behaviour of shotcrete pillars

J. Hadjigeorgiou & F. Habiyaremye
Department of Mining and Metallurgy, Université Laval, Ste-Foy, Que., Canada

J. S. Lessard & P. Andrieux
Noranda Mining and Exploration Incorporated, Brunswick Mining Division, Bathurst, N.B., Canada

ABSTRACT: This paper reports on recent work on the behaviour of shotcrete pillars. The use of shotcrete pillars is becoming more and more popular in Canadian underground mines as an alternative means of support. This paper addresses design considerations based on a design methodology derived from traditional structural analysis as dictated by the Canadian code. The latter part of the paper looks at the results of two and three-dimensional numerical analysis with reference to pillar capacity as well as foundation capacity.

1.0 INTRODUCTION

This paper discusses the introduction of shotcrete pillars at the Brunswick Mining Division of Noranda Mining & Exploration and their relative success in providing adequate support for different mining and ground control conditions.

The Brunswick Mining Division of Noranda Mining & Exploration is a 9,300 tonnes/day operation. The mine is located near Bathurst, New Brunswick, about 800 km Northeast of Montreal, and has been in operation since 1964. The orebody contains zinc, lead, copper and silver. The principal extraction methods are primary-secondary open stopes, pillar-less pyramidal open stopes, modified AVOCA and mechanised cut & fill. Two shafts (#2 & #3) provide underground access. They reach respective depths of 900 m and 1,375 m.

The orebody is composed of a series of 10 parallel, mineralised lenses along a north-south orientation. The arrangement of these lenses is typically complex and usually only three to six lenses are present and extracted on a given level. The orebody is approximately 1.2 km long with a maximum width of 200 m. It is mined to a depth of 1,125 m. The main rock types are massive sulphides (ore and waste), metamorphic sediments and porphiric dyke.

Ground control is an integral element of mine design and worker safety. Problems at Brunswick Mining have been linked to a high stress environment as well as the presence of several geological features. Stress related problems have been linked to high extraction rates (more than 70%), high in-situ ground stresses (sub-horizontal virgin principal stress σ_1, following an east-west direction, with a gradient of 0.052 MPa per meter of depth) and deep mining. Successive folding phases undergone by the orebody over geological times, have resulted in the development of several features that have an adverse influence on the stability of the excavations.

In an effort to manage the resulting induced stresses the mine considers different mining sequencing and varied stope dimensions. Parameters that cannot be managed include: the location of existing openings, previous loading-unloading stress cycles, pillar geometry, location of mined out stopes, location of old falls of ground. Consequently, the performance of the installed ground control support is of paramount importance under these conditions.

2.0 SHOTCRETE PILLARS AT THE BRUNSWICK MINE

2.1 Background

Since 1995 Falconbridge pioneered the use of shotcrete pillars, Beaudry (1997). At the Brunswick Mining Division of Noranda Mining & Exploration, shotcrete pillars have been in use since June 1996. Based on the success of the first applications at the mine they have since come to be accepted by the underground workforce.

Between June 1996 and November 1998, a total of 102 shotcrete pillars were built. During the same period, only one wooden crib was erected, that in a zone not accessible to shotcrete equipment. In the course of a few months, shotcrete pillars have completely replaced wooden cribs to become an integral part of the ground support arsenal at Brunswick Mining.

2.2 Applications

Shotcrete pillars are used at Brunswick as a support tool for quite a range of under different conditions.

Pre-support of Brows

Shotcrete pillars have occasionally been used to support planned brows. The most successful application has been the 56 Stope. The planning for this stope involved blasting phases that would result in long brows that would require support. Constructing a row of pillars, 2 meters in advance of the brow insured the stability of the mucking horizon. Shotcrete pillars have proven not only stable but also quite resilient. They were often subjected to blows from the operating 8 yd^3 scooptrams as well as being exposed to blasting. Nevertheless they did not show any significant deterioration.

Sill pillar Recovery

Probably one of the earlier and most successful applications for shotcrete pillars at Brunswick was for sill pillar recovery. In the 580-1 stope the pillars were used to provide a drilling, loading, blasting and mucking horizon. Several ground movement monitors had been previously installed in the stope back to detect any movement. The dimensions of the back were 80 by 50 meters. Eleven pillars were constructed on cemented rockfill. Mining was completed without any significant problems. This success greatly contributed to acceptance of shotcrete pillars at the mine.

Reduction of large stope spans

At the Brunswick Mine it has been observed that mining induced stress changes quite often result in the formation of sub-horizontal fractures in the stope backs. These fractures are in addition to the pre-existing sub-horizontal joint set typically present in the ore and can be spaced anywhere from a few centimetres to several meters.

Furthermore, mining of nearby parallel lenses results in stress shadowing. A potential for gravity failure is then developed in the massive sulphides because of the fractures parallel to the back. Since it is necessary to maintain access to several of these stopes, it is important to select a type of ground support that will prevent this kind of gravity failure from occurring. In small or mid-sized stopes, cablebolts are generally adequate for back support. In larger stopes it may be necessary or expedient to reduce the effective span of the excavation by introducing shotcrete pillars.

The support used to prevent gravity failures in wide spans needs to possess a great loading capacity, a high stiffness and needs to start acting immediately. Shotcrete pillars fulfil all those restrictions and have been erected over rockfill and solid rock.

Support of large size wedges

The presence of major discontinuities at the larger openings can at times result in the formation of very large wedges in the back. Scaling down these blocks can result in a long, costly and tedious reconditioning job. If local operation constraints permit shotcrete pillars can provide an effective and economic option to cablebolts to support these wedges.

Advantages and drawbacks of shotcrete pillars

Shotcrete pillars are built mostly in areas where wooden cribs would have been used in the past. Shotcrete pillars can also be used has an alternative to cablebolting although, most of the time, these two types of ground support are used in conjunction.

Loading capacity

The greatest asset of shotcrete pillars is their high loading capacity as shown in Section 4. Wooden cribs, on the other hand, have a capacity that varies a lot according to their shape and filling ratios. Also their resistance is directly related to the displacement of the back. Their ultimate resistance is only achieved after displacements measured in inches, if not in feet. For example, an 8" x 8" x 36" wooden crib can support 65 tonnes after a displacement of 1 inch in the back. Its resistance will increase to 170 tonnes after a displacement of 10 inches, Barczack & Gearhart (1994). This type of support is very soft and allows displacements at which the back can come down and the area can become inaccessible.

Resilience

The resilience of shotcrete pillars has been proven on several occasions. They were hit by heavy equipment, particularly around drawpoints, and no significant damage was recorded. Under the same circumstances, wooden cribs are liable to break apart, thus loosing a great portion, if not all, of their loading capacity.

Construction time

Under favourable conditions, two workers can build a shotcrete pillar in little more than one regular shift. Building the form only takes a couple of hours. When pillars are built on non-cemented rockfill, a foundation base must be built. This slows down the whole process. Wooden cribs have a wider base and as a rule of thumb do not require a foundation base. Nevertheless, the overall construction time for both shotcrete pillars and wooden cribs is comparable. Shotcrete pillars, however, must be allowed time for curing before they develop their full strength.

Worker's safety

When building either a shotcrete pillar or a wooden crib, drilling the back is unnecessary. Disturbance of the back caused by such drilling is avoided. This is one of the main advantages of this type of support. Also, work is being carried out to develop techniques for building shotcrete pillars entirely remotely.

Rigidity

Shotcrete pillars are quite stiff. If they are resting on intact rock, their maximum loading capacity is achieved after a very short displacement. On the other hand, if the pillars rest on rockfill, the pillar-base ensemble is a lot less rigid and allows greater displacements in the back (sinks in the floor).

In most applications, the rigidity of the pillars is an asset and translates into an increase in the general stability of the supported ground. Wooden cribs, on the other hand, allow significant displacements of the rock mass before reaching their maximum loading capacity. Quite often this is inadequate for the sulphides.

Nevertheless, on some occasions, high rigidity turns out to be a drawback. One of these situations which occurs regularly at Brunswick is floor heaving. Most pillar failures are related to this problem. Floor heaving is often caused by inverted raise blasting or high seismic activity. This is a violent phenomenon that creates significant dynamic loading, which these pillars were not designed to withstand.

Another drawback to shotcrete pillars that is linked to their rigidity is the lack of progressing visual indications of loading. On the other hand, visual inspections of a wooden crib will reveal any variations in load.

Spatial limitations

At times shotcrete pillars and wooden cribs cannot be used because of spatial limitations in the openings to be supported. Heavy traffic requires a certain clearance and mucking operations (drawpoints for example) require that the operator have an unobstructed view. However, the high loading capacity of shotcrete pillars, as opposed to wooden cribs, reduces the number of units that need to be built to support a given weight.

Load distribution

Shotcrete pillars, just as wooden cribs, provide only punctual support. There is a quickly reached limit beyond which the pattern cannot be tightened further without seriously hindering travelling of heavy equipment in the supported area. On the other hand, cablebolts can be installed in very tight patterns thus producing quasi-uniform back support. At Brunswick, cablebolt patterns as dense as 3' x 3' (with 2 cables per hole) have already been used in extreme cases.

3.0 DESIGN OF SHOTCRETE PILLARS

3.1 Construction

The construction of shotcrete pillars at Brunswick is very simple. When built on solid rock, they are built as a column of shotcrete with a diameter varying between 1.3 and 1.5 meters, Figures 1 & 2. When casing is used, it is made of #9 screen cylinder of slightly smaller diameter. The contacts at the floor and back are slightly expanded to give it an hourglass shape.

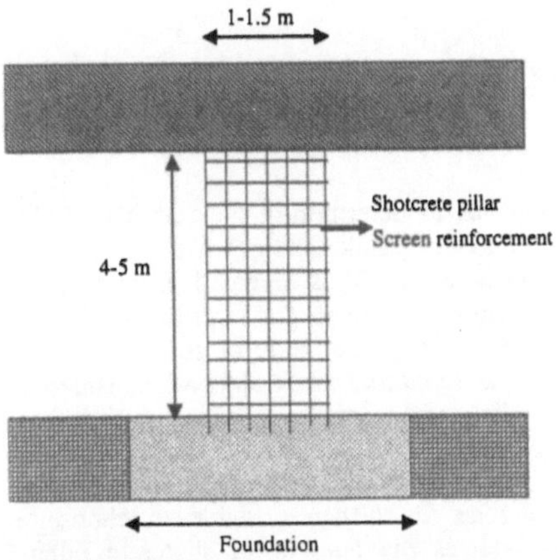

Figure 1. Construction of a shotcrete pillar.

Cleaning the floor is the first action to be taken when installing a shotcrete pillar. The form is made either of fiberglass or of screen and burlap, with a half cylinder shape. Two previously installed posts brace it. Once the form is attached to the posts, a casing cylinder can be installed. Twenty to thirty bags of shotcrete are then used to complete the pillar. When it is necessary to build a base, it consists in a ditch approximately 1.5 meter in depth dug with a scooptram. This ditch is then filled with cemented rockfill or cement with scrap metal for reinforcement. The pillar is then built on the base as presented above.

Despite the relative success of shotcrete pillars there are some quality control concerns of problems that arise when shooting pillars of diameter greater than 1 meter. These problems include trapping of the shotcrete rebound within the pillar by the screen reinforcement and lack of control due to the greater distance of shooting. Consequently the mine is now investigating different alternatives including pouring rather than shooting, and eliminating the screen.

Figure 2. Construction of a shotcrete pillar.

3.2 Design Of Shotcrete Pillars

Preliminary considerations

The first step lies in determining the stability of the excavation. This can be undertaken by use of any of the traditional methods. If it is deemed necessary to support the excavation then shotcrete pillars should be considered. At this stage the mode of possible failure, access as well as economic constraints are considered. If shotcrete pillars are deemed as a viable option then the design process shown in Figure 3 can be applied. This involves the determination of the load caused by the unstable zone. It is then possible to choose a diameter as well as the number of shotcrete pillars required. The resulting design should then be checked and see if it is stable but also whether it is operationally feasible and economically competitive.

Evaluation of the unstable zone and the determination of the applied loads

The determination of the extent of the unstable zone is undertaken by standard rock mechanics methods, deflexion theory in a laminated rock mass, structural analysis, empirical methods or by numerical modelling.

Shotcrete pillar design

3.2.1 Determination of shotcrete pillar strength using the Canadian standard (CSA, 1994)

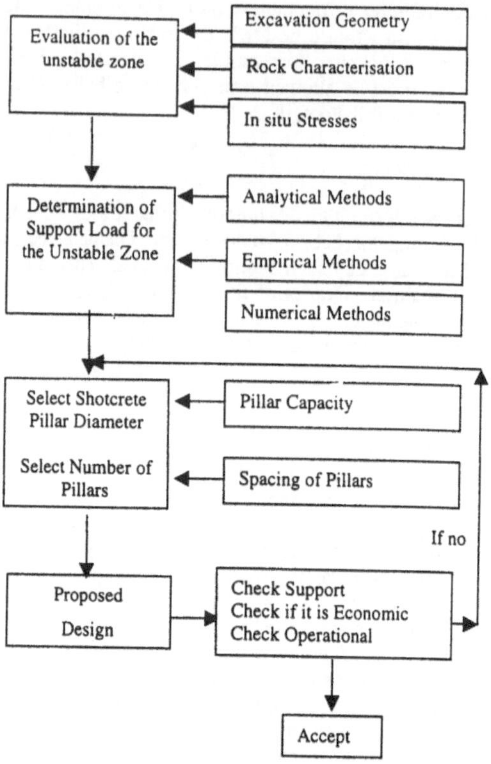

Figure 3. A design methodology for shotcrete pillars.

a) Basic assumptions

In the absence of a code for shotcrete, the pillar strength was calculated by using the Canadian standard CSA A 23.3-94, "Design of Concrete Structures". The code treats in detail of concrete columns as used in civil engineering. The design is done using the unified limit state philosophy. In this method, the design is aimed at avoiding the attainment of a Limit State by underestimating the resistance of the pillar and/or overestimating the load effect. Applying partial safety factors does this.

The pillar is taken as an axially loaded short column reinforced by a screen that is converted into longitudinal and tied reinforcements, Figure 4.

The effects of slenderness are ignored, according to the Clause 10.15.2 of CSA standard A23.3-94 for columns subjected to axial load. In fact, this code defines a limiting slenderness ratio kl_u/r below which slenderness effect is insignificant.

$$\frac{kl_u}{r} \le \frac{25-10(M_1/M_2)}{\sqrt{P_f/(f'_c A_g)}} \qquad (1)$$

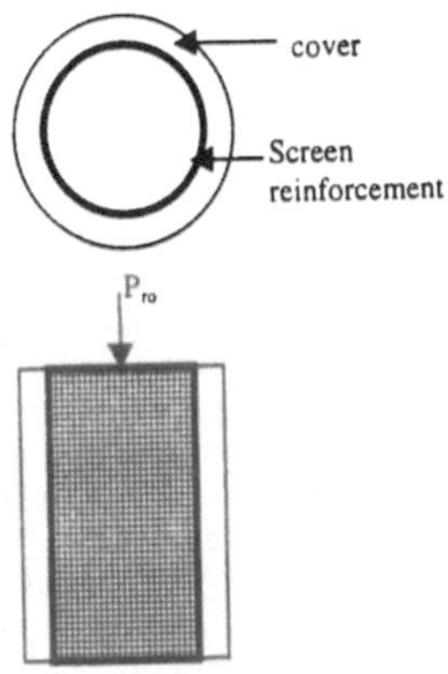

Figure 4. A screen reinforced shotcrete pillar.

where

l_u= unsupported column length,
k= effective length factor reflecting end restraint and lateral bracing conditions of a column (k = 1)
r= radius of gyration reflecting the size and shape of a column cross section (r = 0.25 d);
M_1/M_2 ratio of smaller to larger end moments. By considering that the load eccentricity is insignificant (e = 0) the ratio is equal to zero.

b) Determination of the shotcrete pillar strength under axial loads

According to the Canadian Standard CSA A23-3.94, the axial load resistance for reinforced concrete columns is given by

$$P_{ro}= 0.85\phi_c f'_c (A_g - A_{st})+ \phi_s f_y A_{st} \quad (2)$$

where
P_{ro} = factored axial load resistance at zero eccentricity
f'_c = specified compressive strength of concrete
f_y = specified yield strength of reinforcement
A_g = gross area of section
A_{st} = total area for longitudinal reinforcement
ϕ_c = resistance factor for concrete (=0.60)
ϕ_s = resistance factor for reinforcing bar (=0.85)

c) Design of the reinforcement for a shotcrete pillar

In underground mines, the majority of shotcrete pillars are reinforced by grid. This is quite different from the traditional reinforcement of concrete columns.

A concern is whether to employ traditional longitudinal and tied reinforcements and decrease the design diameter of a shotcrete pillar. Previous work by the authors has demonstrated that the advantage obtained from the addition of longitudinal rebars in the shotcrete pillar becomes of less importance as the nominal pillar is increased. Furthermore recent studies, by Mau et al. (1998) on columns confined by Welded Wire Fabric (WWF) revealed that the confinement provided by the use of screen is so uniformly distributed along the column, that the grid constitutes an integral part of the pillar. It has been concluded that the screen provides a better reinforcement.

To determine this reinforcement of shotcrete pillar, the screen is converted into longitudinal and tied bars that are calculated by dividing respectively the screen length and the width by the wide.

i) Ratio of longitudinal reinforcement

According to the Canadian standard CSA A 23.3-94, the minimum number of longitudinal reinforcing bars, in a column, should be 4 for bars within circular ties, and 6 for bars enclosed by spirals. The area of a longitudinal bar (A_{st}) shall be greater than 0.01 times the gross area of the section, A_g. In other words the following condition must be satisfied:

$$\rho_t = \frac{A_{st}}{A_g} \geq 0.01 \quad (3)$$

For a column with ρ_t smaller than 0.01 but larger than 0.005 the value for ρ_t, the factored axial and moment resistance should be multiplied by the ratio $\rho_t/0.01$.

ii) Ratio of the confining reinforcement

The degree of confinement depends on the configuration, the size and the longitudinal spacing of the transverse reinforcement. The confinement is given either by the spiral reinforcement or by the transverse bars. According to the Canadian standard CSA A23-94, for the circular column confined by spirals, the ratio ρ_s of spiral reinforcement is given by the following formula:

$$\rho_s = 0.45\left(\frac{A_g}{A_c} - 1\right)\frac{f'_c}{f_y} \quad (4)$$

where

A_c= area of core of spirally reinforced compression member measured to outside diameter of spiral;
A_g= gross area of column section;
f_c = compressive strength of concrete;
f_y = yield strength of lateral steel.

iii) Diagram of shotcrete pillar capacity

Figure 5 was developed to provide a preliminary design for shotcrete pillars for a range of pillar diameter and shotcrete strength resistance assuming several strength and pillar diameters.

Several assumptions have been undertaken in this work. In the first place the use of methodology design for concrete structures with the inherent safety factors is indeed somewhat arbitrary and may not necessarily be the most appropriate for shotcrete pillars. In the second place while there is good control over the shotcrete quality the design of the excavation and the rock mass characteristics is subject to the usual uncertainties. Consequently the proposed design methodology is only a starting point. It is assumed that a mine would develop a detailed record of every shotcrete pillar constructed and its subsequent behaviour so as to develop a design methodology based on its acquired experience. Under this light the mine is currently investigating the use of shotcrete pillars which are poured without the benefit of screen reinforcement. In certain cases it would appear that the reduction in pillar capacity may be tolerable and in cases even compensated for by improved quality control. A further consideration is that in practice the mine still uses rockbolts and quite often cablebolts between pillars.

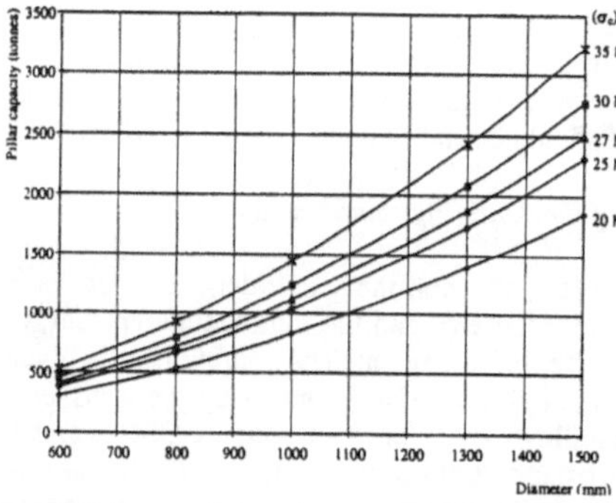

Figure 5. Determining the shotcrete pillar capacity.

A further consideration is that any design should be operationally feasible with good quality control. Under these conditions it is evident that shotcrete pillars do not provide the optimal support choice for every scenario.

The choice of a shotcrete pillar diameter

Figure 5, provides a means to select the appropriate pillar diameter for a given shotcrete strength for a given required capacity.

Spacing of shotcrete pillars

In this work a series of parametric studies were undertaken to develop guidelines in selecting an appropriate spacing. To these purposes a series of numerical models were run for situations typically encountered at Brunswick. The inherent assumption has been that the introduction of shotcrete pillars will control the resulting deformations of the excavation and will apply a load on the excavation floor. Consequently the foundation of the pillars is of importance, solid rock or backfill, as it will have different deformations. For the parametric runs a finite difference code, FLAC^{2D}, Itasca (1991), and its equivalent three-dimensional finite difference code, FLAC ^{3D} were used, Itasca (1997).

a) Two dimensional numerical model

If the pillars are resting on a solid rock foundation the pillar properties are the controlling parameter. Furthermore the pillars act as a local support of the span by generating a zone of influence in the rock mass. Thus, for a given excavation, and shotcrete pillar dimensions the zone of influence of an individual pillar as well as the overlapping of these zones by an array of pillar is conceptually shown in Figure 6. This was obtained by varying the spacing of the pillars and by tracing the plasticity plots for a stable system. An operation constraint is that the spacing of pillars should be adequate to allow the circulation of underground equipment. It is then possible to arrive at an optimum pillar dimension. In practice it may also be advantageous to provide local rock bolt support between the pillars.

Figure 6. Selecting an appropriate spacing for the pillars.

Following a series of simulations an empirical relationship was derived linking spacing of pillars and optimum diameter. Figure 7 was determined for shotcrete having a 28-day compressive strength of 27 MPa.

$$\text{Diameter} = 0.8222 + 0.0028\, e^{(\text{spacing})}$$

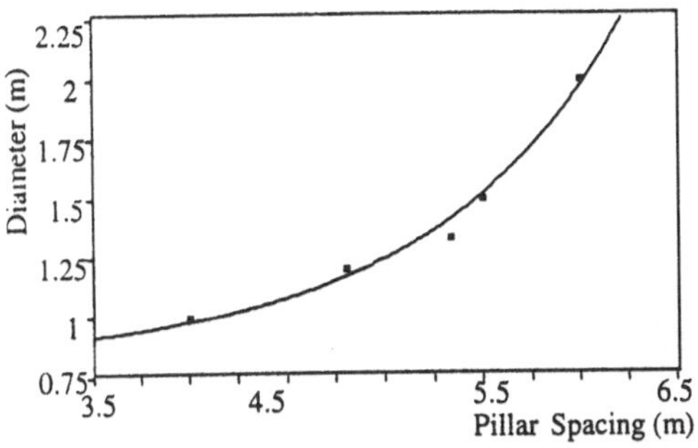

Figure 7. Shotcrete pillar spacing vs pillar diameter for a 27 MPa shotcrete compressive strength.

b) Three dimensional stability assessment for shotcrete pillars footed on cemented rockfill floor

The analysis presented in the preceding paragraph is valid when the pillar is resting on solid rock or on cemented rockfill. If the pillar is resting in non-cemented rockfill then it is necessary to account for the resulting higher deformations in the ground. This necessitates a design for a foundation that would define an acceptable settlement. Under these conditions the relatively stiff shotcrete pillar is pushed into the soft and compressible fill floor. Problems arise under high loads and with an inadequate foundation. This can be manifested as a general, localised or punching shear foundation failure, Hunt (1986). In practice any of the bearing capacity theories can be used. Numerical modelling provides an interesting tool to complement such analyses. The acceptable capacity is given by the relationship between the bearing capacity and the acceptable indentation. As this indentation is a function of the applied loads and of the fill compressibility, it was possible to employ numerical modelling to evaluate the stability of the system roof-pillar-floor. In summary, the shotcrete pillar design procedure has four steps;

- Evaluation of the loads applied to the foundation (pillar dead weight and the loads caused by roof deflection)
- Selection of the thickness of foundation;
- Running numerical models to determine the bearing capacity of the foundation
- Verify if the settlement is acceptable.
- Iterate the design process until an acceptable design is reached.

4. 0 EVALUATION OF STRENGTH, NUMBER AND SPACING OF SHOTCRETE PILLARS

The proposed methodology was used to estimate the strength of existing shotcrete pillars at Brunswick

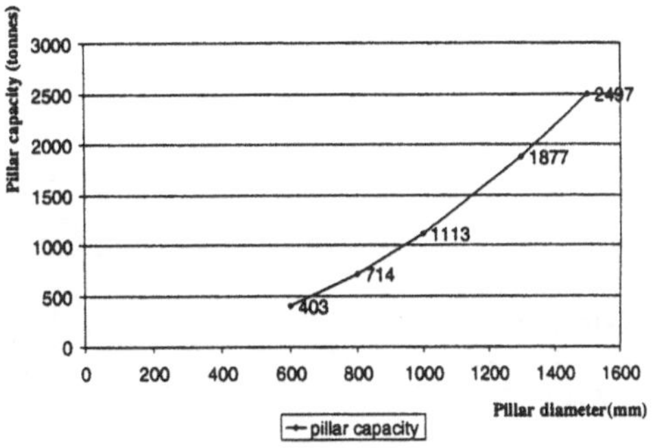

Figure 8. Shotcrete pillar strength for different diameters.

Table 1: Design of reinforced shotcrete pillars using a. screen.

a) Shotcrete Pillar					
Pillar diameter (mm)	600	800	1000	1300	1500
Cover (mm)	40	40	40	40	40
Diameter of shotcrete core (mm)	520	720	920	1220	1420
Height (mm) (lu)	4000	4000	4000	4000	4000
Slenderness (klu) limit (mm)	34000	34000	34000	34000	34000
b) Grid					
Length needed (mm)	4000	4000	4000	4000	4000
Width needed (mm)	1634	2262	2890	3832	4461
Bar spacing (4"X4") (mm)	102	102	102	102	102
Bar diameter (mm)	3.8	3.8	3.8	3.8	3.8
Calculation of longitudinal reinforcement					
Number of bars	16	22	28	38	44
Reinforcement area (mm^2)	182	252	321	426	496
Cross sectional area of pillar (mm^2)	282743.1	502654.4	785397.5	1327321.8	1767144.4
Ratio of longitudinal reinforcement (%)	0.06	0.05	0.04	0.03	0.03
Calculation of confinement					
Number of bars	39	39	39	39	39
Area of confinement bars (mm2)	445	445	445	445	445
Core section (mm2)	212371.4	407150.0	664760.4	1168985.6	1583675.5
• • s (%)	0.21	0.11	0.07	0.04	0.03
Strength of shotcrete pillar					
Resistance factor of concrete• • c):	0.6	0.6	0.6	0.6	0.6
Resistance factor of reinforcing bars(• s):	0.85	0.85	0.85	0.85	0.85
f'c (MPa)	27	27	27	27	27
fy MPa)	400	400	400	400	400
Strength (KN)	3953	7004	10920	18416	24495
Strength (t)	403	714	1113	1877	2497

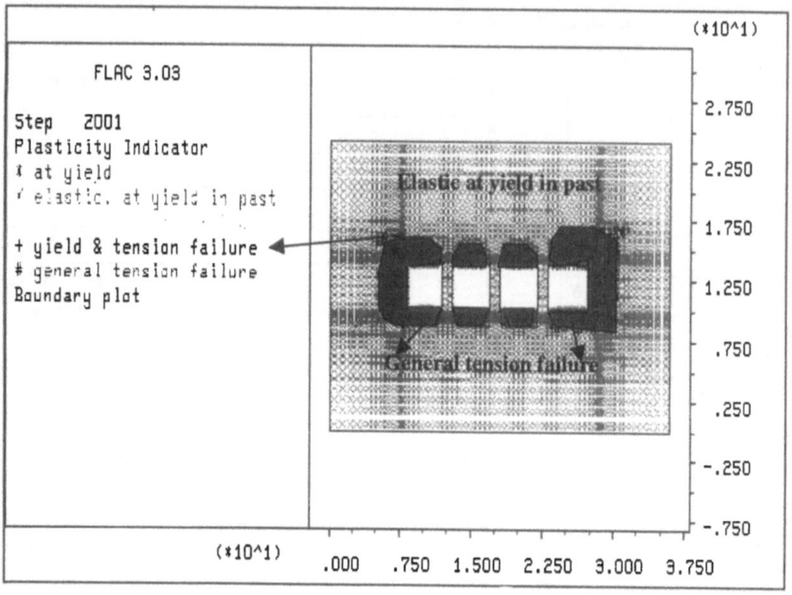

Figure 9. Plasticity indicator for the excavation supported by three shotcrete pillars.

mine. Most existing pillars are formed from a cylinder of standard welded-wire screen. A typical shotcrete pillar is about 1.3 m in diameter. One half of the pillar is covered either by burlap or steel cover. To evaluate the influence of the screen on the strength of the existing shotcrete pillars, we converted the screen into longitudinal and tied reinforcements. The number of longitudinal and tied bars is calculated by dividing respectively the screen length and width.

Table 1 summarises the proposed approach. The pillar capacity is calculated by assuming a 28 days shotcrete compressive strength of 27 MPa. Five different pillar sections were calculated to examine the influence of the size on the shotcrete pillar strength. For each pillar, equation (2) was used. As shown in Table 1, the capacity of the existing 1.3 to 1.5 meter diameter pillars built with screen reinforcement at Brunswick Mine varies between 1800 and 2500 tonnes. The influence of pillar diameter on its resulting capacity is shown in Figure 8.

To determine the spacing and the number of pillars needed for an overall span performance for a given stope, different models have been constructed. The constructed model consists of an opening in rock mass 20-m of wide and of 4-m height excavated at 580 m deep. In order to point out the number of pillars required for a given span length, a total of three models of shotcrete pillars inserted in the excavation were considered. Each pillar has a diameter from 1 to 1.5 m and a height that varies between 4 and 5 m. The influence of the in situ stresses was taken into account by using a domain in which, the rock mass surrounding the excavation is two times the height of the opening. Figure 10 shows a plasticity graph for an excavation supported by three shotcrete pillars. The zone of influence can be schematically traced as in figure 6 in order to select a suitable spacing.

As shown in Figure 6, the elastic zones above the pillar overlap. For this scenario spacing the pillars at a distance of 4 m was acceptable. This assuming 1-m diameter pillars each with a capacity of 1100 tonnes. If the load applied is 2372 tonnes, the safety factor is 1.4. However, as shown on Figure 9, there is an unstable zone that remains between two contiguous pillars. Thus its effectiveness will be maximal if the shotcrete pillar is used along with another back support.

5.0 ASSESMENT PILLAR STABILITY WHEN FOOTED ON CEMENTED ROCKFILL

At Brunswick Mine the fill floors are often layered in unconsolidated rockfill and 5 % cemented rockfill layers. In order to point out the influence of pillar size and the foundation thickness on the roof-shotcrete pillar-fill floor stability, a total of six models were considered in which parameters such as the pillar diameter and the foundations depth were varied. The three-dimensional explicit finite difference program FLAC^{3D} was used to investigate this stability. Figure 10 and Table 2 show the different configurations considered.

Table 2. Different configurations considered.

Model type	CRF thickness (m)	Pillar diameter (m)
A	0.5	1
B	0.5	1.5
C	1	1
D	1	1.5
E	1.5	1
F	1.5	1.5

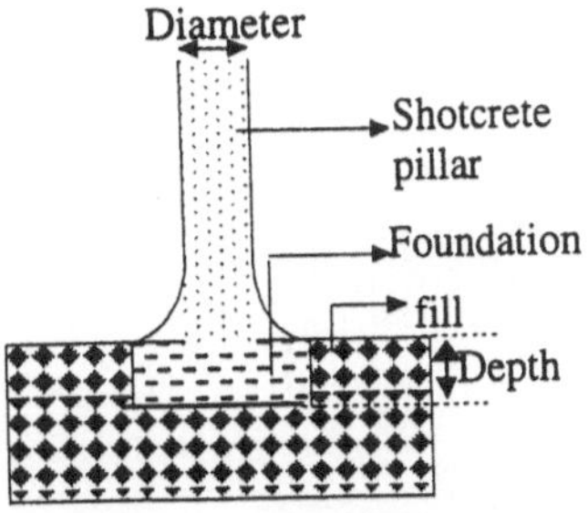

Figure 10. Model configuration for shotcrete pillar erected on cemented rockfill.

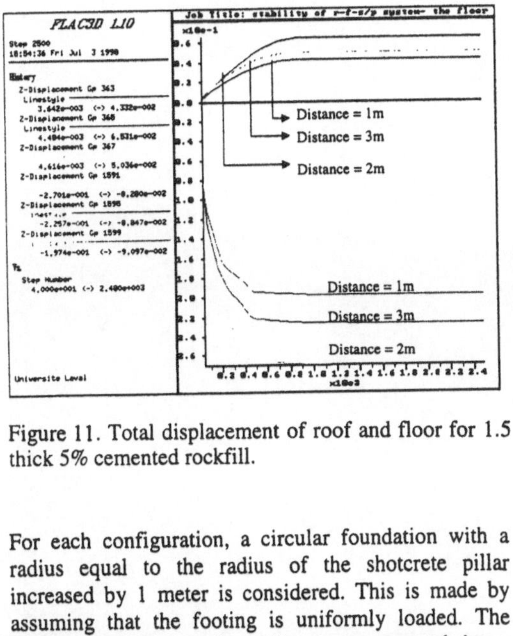

Figure 11. Total displacement of roof and floor for 1.5 thick 5% cemented rockfill.

For each configuration, a circular foundation with a radius equal to the radius of the shotcrete pillar increased by 1 meter is considered. This is made by assuming that the footing is uniformly loaded. The loading applied by the pillar is represented by a constant, slow vertical velocity applied to all of the nodes of elements, which lie beneath the footing. As the velocity is applied, the program is used to sample the reaction forces beneath the footing.

Figure 11 summarises some of the results obtained by the three-D numerical model for shotcrete pillars footed on 1.5 m thick of 5% cemented rockfill. The pillar has a diameter of 1-m. Referring to Figure 11, the pillar is loaded based on a vertical deformation from the roof of 8 cm. This load, as well as the loads due to the weight of the pillar result in a localised settlement of the fill of 26 cm.

CONCLUSIONS

The use of shotcrete pillars has been shown advantageous for certain situations at Brunswick Mining. The proposed methodology can be used to provide preliminary design specifications. As more empirical data become available and based on operational and quality control considerations the mine modifies its design strategy.

In practice it has been demonstrated that the diameter of the pillar is the critical design parameter. Based on the undertaken analysis a pillar of 1.3 to 1.5 m diameter should be adequate for most situations. It is always possible to increase the pillar capacity by introducing more reinforcement. This however should be examined further based on economic as well as operational and controls considerations. Barring operational constraints the use of grid reinforcement is an attractive option.

The results of the two dimensional modelling for pillars footed on solid rock provide a preliminary answer as to the number of pillars required. This is possible by assuming that each pillar has its own zone of influence, which should overlap, with the adjacent ones. At different areas there is however the need for bolts between pillars. The shotcrete pillar remains a local support mode. It persists indeed a zone of collapse between two pillars, which suggests that, the pillar will be effective for the reconditioning operations or when the pillar is combined with other support as bolts or screen, as it is always the case at Brunswick Mine. If adequate support as cable bolts is installed it maybe possible to increase the spacing of the pillars.

Based on the results of a three dimensional model assuming that the pillars are placed on a 5% cemented rockfill floor it would appear that the pillar should require a foundation of 1.5-m layer thickness the 5% cemented rockfill for the pillar diameter varying between 1 and 1.5 m.

ACKNOWLEDGEMENTS

The authors would like to thank the direction of Noranda's Brunswick Mining Division for granting permission to publish this work.

REFERENCE

Barczack T. M. & Gearhart, D. F. (1994). *Optimizing Wood Crib Design.* E&MJ, December 1994, pp 23-30.

Beaudry, M. (1997). *Shotcrete Pillar for Ground Support Use at Falconbridge Ltd.* 12th Ground Control colloquia, Section 9, Val d'Or, Québec, Canada.

Board M., 1996, "*Consulting Report From the Ground Control Review*". Report to Ground Control group of Brunswick Mining Division, may 13-17.

Brummer, R K, 1997, "*Structural behaviour of shotcrete in underground construction*", 12th Ground Control colloquia, Val d' or, Quebec.

Canadian Portland Cement Association, 1995, "*Concrete Design Handbook*". 2nd ed.

Canadian Standard Association, 1994, CSA A23.3-94, *Design of Concrete Structures,* Rexdale(Toronto), Ontario, Canada.

Habiyaremye F., 1998, *Les applications du béton projeté dans les mines souterraines.* M.Sc. thesis Université Laval.

Hunt R.E., 1986. "*Geotechnical engineering analysis and evaluation*". McGraw Hill 729 p

Itasca Consulting Group inc., 1991, "*FLAC Fast lagrangian analysis of continua. Version 3.0*", User manual.

Itasca Consulting Group Inc., 1997, "*FLAC3D: Fast lagrangian analysis of continua in 3 Dimension*". Version 1.1 User manual Vol 1.

Mau S., Holland J et Hong L, 1998, "*Small column compression tests on concrete confined by WWF*". Journal of structural engineering vol 124 n°3 pp252-261.

Rock Support and Reinforcement Practice in Mining, Villaescusa, Windsor & Thompson (eds)
© 1999 Balkema, Rotterdam, ISBN 90 5809 045 0

The rationale, design and implementation of steel fibre reinforced wet shotcrete, in highly stressed hard rock tunnels, where rapid development advance is needed, in South Africa

A.J.S.Spearing
MBT International, Underground Construction Group, Zürich, Switzerland

W.A.Naismith
AngloGold, South Africa

ABSTRACT: The paper outlines current support requirements in hardrock, highly stressed (deep) tunnels. The traditional practice of using meshing (screening) and lacing is impractical, for deep level orebodies, where rapid access is required, in order to meet financial targets. Quality fibre reinforced shotcrete can offer a viable support alternative, that readily meets the support requirements and overcomes the time constraints. The design of tunnels at depth is also discussed, particularly methods to minimise the effects of the stress regime on rock failure and deformation around the excavation periphery. This must be minimised, not only to reduce the cost of the overbreak, but also to allow the temporary (primary) support to be combined with the permanent (secondary) support, to save time and money. Practical results to date are also discussed, including the logistic problems, that need to be overcome.

1. INTRODUCTION

Deep level hard rock mines, such as those mining to depths close to 4000 m below surface, in the Witwatersrand Basin in South Africa,are subject to exceptionally high rock stresses. Mines at depth need to counter additional problems, not usually apparent in shallower, or soft rock mines. These problems include:

- high stress regimes, fractured rock and significant seismic activity
- increased ventilation and cooling requirements
- complex and long infrastructure to reach the orebody
- high capital costs
- long time periods before stoping (ore production) can commence on a new project

Rapid tunnel access development is therefore essential at depth, in order to reduce overall costs.

This paper discusses this potential problem, and concludes that steel fibre reinforced wet shotcrete can be an essential support component, under such circumstances, to resolve the issue, and has been successfully applied in practice.

2. SHOTCRETE

Shotcrete (sprayed concrete) is defined by the American Concrete Institute, in ACI 506R-85, as mortar or concrete pneumatically projected at high velocity onto a surface.

In mining, a key issue that is frequently ignored (or overlooked) during the design phase of a mining project, involving the use of shotcrete, is the infrastructure to deliver the material from the storage facility to the shotcrete mixer and pump. In most projects, particularly at depth, this aspect requires the much attention.

Sprayed concrete can be applied using either the wet or the dry process. The selection is totally dependent on the local conditions. Table 1 gives a brief comparison of the methods. It has been estimated (Garshol 1998) that over 8 million m^3 of sprayed concrete are placed annually. At least 50% is now using the wet process, and this figure is rising steadily.

The main operational difference between the methods is where the water is added to the mix. In the dry process it is added near the nozzle, and the mix is transported from the pump pneumatically in a dry (or near dry) form. In the wet process water is

Table 1. Comparison between the shotcrete processes

Wet process	Dry process
Little dust	Considerable dust
Low maintenance	High maintenance
High capital	Low capital
Low rebound (typically about 10%)	High rebound (usually more than 25%)
Moderate to high placement rate (between 3.0 and $20m^3/hr$)	Low to moderate placement rate (up to about $4.0m^3/hr$)
Low transport distance (up to about 200m)	High transport distance
Moderate to high placed quality	Moderate placed quality

added in the mixer, prior to pumping.

The quality of the sprayed concrete is therefore very reliant on the skills of the crew (especially the nozzleman), when using the dry process.

3. HIGH SPEED TUNNEL NEEDS

High tunnel development rates are particularly essential at depth, in order to open up ore reserves as rapidly as possible. Up-front capital expenditure is enormous (up to $500 million), when opening up new mining areas at depth, and the first significant revenue from mining, can take up to 10 years to realise. Every effort to speed up ore reserve access and infrastrucure is therefore essential.

Technologies exist to drill, blast and clean a tunnel every shift (ie: 3 times per day), using proven mechanised equipment such as drill rigs and LHDs (load, haul and dump vehicles). This has the added benefit of reducing the manpower level working directly on the face (an increased potential hazard in hard rock mines at depth). The serious bottleneck is perceived to be the support installation portion of the development cycle.

In large cross sectional area tunnels (typically $+20m^2$), it is possible to install some of the support (such as shotcrete) during the cleaning phase. A typical cycle time is given in Table 2 (Spearing 1992).

The most common form of (permanent) support in hard rock tunnels at depth, is meshing (screening) and lacing with grouted shepherd's crooks (looped rebar). This is usually installed some distance from the tunnel face, in relatively long tunnel lengths, to obtain reasonable labour efficiency. The tunnels

Table 2. Typical tunnel development cycle

Activity	**Time (minutes)**
Water down and make safe	15
Cleaning (mucking-out)	150
Marking-off and drill preparation	30
Drilling the round & install support	150
Clean-up and drill removal	15
Charge-up	75
Remove all equipment	15
Blast and re-entry	30
TOTAL	**8 hours**

therefore usually need some form of temporary support, until this permanent support is in place. When the permanent support is installed, the bulk of the rock deformation has already taken place, and only further mining induced or seismic deformation needs to be accomodated.

Traditional meshing and lacing cannot be installed close to the advancing face, as it is too time consuming and is subject to significant blast damage. In high deformation environments, the system is too soft to contain the rock movements, and may need to be replaced several times during the operational life of a tunnel. In addition, the support system is also prone to corrosion, in the hot and humid conditions. Care must be taken when comparing results from mesh and fibres in shotcrete, that are not conducted in the field.

4. TUNNEL SHAPE DESIGN AND SUPPORT AT DEPTH

4.1 Tunnel Shape

Mining engineers, unlike civil engineers still tend to develop tunnels with square or rectangular cross sections. Such profiles are largely governed by the prevailing geological conditions, but may not be appropriate for high stress conditions.

Under a high stress regime in hard rock, the tunnel is developed within an envelope of failed rock. This presents a serious problem in the tunnel sidewalls, where intense sloughing (scaling) takes place. This makes it virtually impossible to combine the temporary and permanent supports.

A ready solution (Spearing 1992,1994 and 1997) is to round (make concave) the tunnel sidewalls (at least), to make the tunnel shape more compatible to the effects of the stress regime, and hence reduce the unplanned dilution (sloughing). The use of yielding tendons, such as the cone bolts (Wojno 1987), are also required, in order to accomodate the inevitable and significant deformation. The practicalities of creating a profiled tunnel need to be balanced against the advantages to be gained.

4.2 Tunnel Support Design

Many tunnel support design methodologies exist, such as:

- previous experience and local knowledge
- Rock Quality Designation - RQD - (Deere et al 1969)
- Q system (Barton et al 1974)
- Mine Rock Mass Rating System - MRMR (Laubscher 1990)
- Ground Characteristic Curves using FLAC (Speers et al 1996)
- analytical and/or numerical calculations
- observational methods

The methodology adopted is mainly dependent on the strategic importance of the tunnel, the rock mass information available, the anticipated stress regime over the planned life of the tunnel and the available rock mechanics manpower and infrastructure.

The tunnel support however, can be designed with a relative confidence, using a number (or combination) of guidelines.

4.3 Tunnel Support

Tunnels in hard rock at depth, tend to be rockburst prone, and hence the following support requirements are generally needed:

- an active support
- ability to sustain rapid deformation (ie: absorb energy)
- ability to deform in a stable manner under quasi-static deformation
- high percentage areal cover
- corrosion resistant
- limit unplanned overbreak
- maintain the overall tunnel cross section, over the life of the tunnel

In addition, as shown in Table 2, the support must be typically installed in 150 minutes (as an absolute maximum).

Steel fibre reinforced shotcrete is therefore an ideal solution because:

- it can be applied quickly (at over 20 m^3/hour if required)
- with correct accelerators, it gains strength rapidly (over 2 MPa after 2 hours, depending on the temperature)
- the fibre addition gives it good durability (and therefore good deformation characteristics)
- it is relatively blast resistant, and any damage can be readily repaired
- it can be kept up to the advancing face, after each blast
- it can be remotely applied, if a (robot) manipulator is used
- the drilling round length can be substantially increased if SFRS is applied using a manipulator

The wet application process generally is preferred because:

- the dust is negligible and hence other operations can continue at the same time in the vicinity of the spraying
- the application rates are much faster
- rebound is significantly reduced and hence less waste material must be handled
- it is more cost effective

Recent developments in the field of shotcreting (and in particular wet shotcreting), have also made the system even practical:

- the development of safe and environmentally alkali free accelerators, that also have little adverse effect on the long term strength (unlike silicates and aluminates)
- hydration control admixtures that increase the open time of the mix from about 3 hours to up to 72 hours
- new generation superplasticizers that reduce the water to cement ratio very effectively
- concrete improvers (internal curing compounds)

that are added to the mix and significantly increase the shotcrete properties

A perceived drawback to the increased use of steel fibre reinforced shotcrete (SFRS) is the logistical problem of an adequate material supply. This can be overcome by the use of pipelines and/or agitator vehicles.

5. THE GROWTH OF QUALITY WET SHOTCRETING IN HARDROCK MINES

Wet shotcreting has been applied in small quantities on hard rock mines, in South Africa, for many years (eg: Thabazimbi Iron Mine), but it has only really started to grow, and gain acceptance in the last few years.

Anglogold has helped to pioneer the introduction, when in 1996, wet shotcreting was started at 11 Shaft, Vaal River Operations. This project used a robojet (articulated application arm, using a remote control) for shotcreting the shaft infrastructure and nearby major excavations.

Subsequent to this, the use of wet shotcreting has been expanded on Vaal River Operations, and also introduced onto other mines, administered by Anglogold, including Western Deep Levels, Elandsrand and Freddies Gold Mines. Other hardrock mines have also implemented wet shotcreting, such as JCI's South Deep Project, Anglovaal's Target Project, West Diefontein and Palabora Copper Mine.

6. PRACTICAL RESULTS

The following case studies, show the dramatic results achieved in practice to date:

6.1 11 Shaft Moab Khotsong - Vaal River Operations

11 Shaft is a new shaft system being developed to exploit reserves of the gold bearing reef at depths from about 2200 m to 3500 m.

Wet shotcreting was initially applied on the 73 Level station development, at a depth of 2300 m below surface. Being a sinking shaft, hoisting time was very limited, and the shotcrete could only be transported in a 150 mm diameter pipeline already installed in the shaft.

On the surface, a basic concrete batching plant was available, with crushed aggregate, of a relatively poor quality. A shotcrete mix was designed, that needed to have:

- unusual cohesiveness to limit segregation in the vertical pipeline
- a long open time, even at elevated temperatures
- relatively rapid strength gain after placement
- high durability

The mix design selected, after laboratory and field trials, was as follows:

• cement	450 kg
• micro-silica	40 kg
• -8 mm aggregate	1400 kg
• sand	320 kg
• water reducers	10.7 kg
• internal curer	5.0 kg
• hydration controller	2.8 kg
• accelerator	5 %
• steel fibre	50 kg

The relatively high cement content was mainly needed to improve the cohesiveness of the mix.

The mix was sent down the shaft pipe, and into an energy dissipating kettle at the bottom. From the kettle, the shotcrete passed directly into a conventional concrete pump, that conveyed the material to the shotcrete pump. A MEYCO Suprema shotcrete pump with an integrated dosing pump was used, and a MEYCO Robojet manipulator mounted on an existing chassis was used to spray the shotcrete, rapidly and safely.

The wet shotcrete application was very successful, and no pipe blockages were experienced. A total of about 400 m^3 of shotcrete was placed on the level, and the client estimated less than 5 % rebound on average.

6.2 Elandsrand Gold Mine

Elandsrand Gold Mine extracts gold from the Ventersdorp Contact Reef at typical depths between 2000m and 2700m.

Prior to 1991, the mine used very little shotcrete in tunnels, and the most common type of support was meshing and lacing. Around 1991, the mine decided to implement dry shotcreting, because it was a proven technology and could offer support benefits such as improved safety and higher productivity.

The strategy was to apply a skin of around 25 mm of shotcrete, concurrent with the tunnel face advance. This method was successful in reducing strain bursting on the face, but had limitations.

In 1997, a decision was made to change to wet shotcreting, in most applications, because:

- it can be placed faster
- a superior placed product is possible
- rebound is reduced
- dust is virtually eliminated

At present the mine uses 2 hand sprayed wet shotcreting machines (the 1 operated by a contractor).

Experience to date has shown that the wet shotcreting has been able to meet the mine's needs, and the necessary infrastructure is being planned and installed. This will consist of a surface batching plant, with a pipeline networks down the shafts, with the final horizontal delivery to the shotcrete machines being via railed mounted agitator cars. The wet shotcreting will then be expanded considerably, and 2 additional shotcrete machines (both with manipulators) have been ordered.

6.3 Western Deep Levels - South Mine

As part of a shaft sinking programme, that is needed to access ore reserves lying in excess of 4000m below surface, a replacement ventilation shaft has been sunk from 84 to 109 Level. This is about from 2580m to 3300m below surface. A novel method of raise boring, using a "V" Mole was used to develop the 7.0m diameter shaft (Moll 1998).

Sidewall support was installed concurrently with the sinking, and comprised split sets and SFRS applied using a manipulator, attached to the sinking stage in the shaft.

The shotcrete was mixed at a batching plant on 84 Level, and fed via a 150mm diameter pipeline into a 1.0m³ remixing tank, located on the second deck of the sinking stage.

Admixtures were used in the mix to aid pumpability, consistency and strength gain. The mine estimated the overall rebound to be around 5%.

The shaft development located in the Ventersdorp lava sequence, that has an unconfined compressive strength (UCS) of about 500 MPa, proceeded according to the planned schedule. In the relatively weaker Witwatersrand quartzites, with a UCS of about 200 MPa, excessive sidewall slabbing occurred behind the cutter head, before the shotcrete could be practically applied. This tended to delay the sinking, and it was concluded by the project team, that under such conditions, the SFRS should have been applied right up to the face.

7. CONCLUSIONS

Existing deep level and hard rock mining operations have proved that SFRS can provide a technically and economically viable tunnel support system.

For the successful implementation of such a system, the following factors should be addressed early in the planning phase:

- defining the tasks to be done by the support (particularly in terms of energy absorption)
- designing a mix to meet the performance requirements (using locally available materials wherever possible)
- the infrastructure to deliver adequate shotcrete to the faces within the time constraints
- matching the shotcrete pump to the required delivery and supply infrastructure
- appropriate quality management systems and procedures should non-compliance occur

ACKNOWLEDGEMENTS

The authors gratefully acknowledge the Managements of Anglogold and MBT for permission to publish and present this paper.

REFERENCES

Garshol, K 1998. *International practices and trends.* School on Shotcrete and it's Application. SAIMM - Randburg, South Africa.

Spearing,AJS 1992. *A possible fast track tunneling method at depth.* Tuncon 1992 - Maseru, Lesotho.

Spearing,AJS, Speers,CR, Forsyth,MI 1994. *Design concepts and support considerations for highly stressed tunnels in deep hard-rock mines.* J.S.Afr.Inst.Min.Metall. Vol 94, No.5.

Spearing,AJS 1997. *Support methodology and design for rapid access at depth.* Technology for Rapidly Accessing the Orebody - SAIMM - Randburg, South Africa.

Wojno,L et al 1987. *Recommended performance requirements for yielding rock tendons.* Design of Rock Reinforcing - SANGORM - South Africa

Deere,DU et al 1969. *Engineering classification of in situ rock.* AFWL Report TR67-144 - Washington, USA.

Barton,N et al 1974. *Engineering classification of rock masses for the design of tunnel support.* Rock Mechanics, Vol 6, No 4.

Laubscher,DH 1990. *Geomechanic classification of jointed rock masses - missing applications.* Trans.Instn.Min. Metall.ppA1-8

Speers,CR, Spearing,AJS 1996. *The design of tunnel support in deep hard-rock mines under quasistatic conditions.* J.S.Afr.Inst.Min.Metall. Vol 94.No 2.

Collett,K, van Rooyen,B 1998. *The past, present and future of shotcrete at Elandsrand Gold Mine.* J.S.Afr.Inst. Min.Metall. Vol 98, No 3.

Moll, I 1998. *Shotcreting at Western Deep Levels South Deepening Project.* School on Shotcrete and it's Applications - SAIMM - Randburg, South Africa.

Rock Support and Reinforcement Practice in Mining, Villaescusa, Windsor & Thompson (eds)
© 1999 Balkema, Rotterdam, ISBN 90 5809 045 0

The use of shotcrete in Australian underground mining – A contractors perspective

M.J.K.Clements
Jetcrete Australia, East Freemantle, W.A., Australia

ABSTRACT: The underground mining of metals in Australia is developing rapidly as large near-surface deposits reach maturity. Although Australian mines are largely characterised by hard rock conditions, ground support conditions in underground mines are becoming increasingly difficult as depths and stresses increase. Against this background, the use of shotcrete as a ground support element has gained an increasing amount of attention. This paper discusses the issues concerning a contractor in the Australian shotcrete industry. These issues include design, specification and project quality control. The concept of a Shotcrete Management System is discussed and practical examples given of its application. Performance testing, especially the measure of toughness, is also discussed and a new type of toughness test devised in Australia is described.

KEYWORDS
Shotcrete, spraying, toughness, quality, shotcrete management system, in-cycle, absorption, thickness control, roughness factor, steel fibre

1 INTRODUCTION

The use of shotcrete in Australian underground mining is almost exclusively by wet mix application. Modern techniques owe more to adaptations of piston-type concrete pumps used on building sites than to evolution of old mining practices using dry spraying equipment.

Australian underground mines are typified by large decline access and diesel powered equipment for development. Typical decline sizes are 6m x 6m and level development 4.5m x 4.5m.

The most common method of shotcrete production is to use a nearby commercial plant to produce the shotcrete mix. This is transported underground in standard five cubic metre agitator trucks and transferred directly to a robotically controlled shotcrete spraying machine. These arrangements generally allow for a good standard of mix production and a high volume of delivery where required.

2 THE SHOTCRETE CONTRACTOR

Many Australian mines dabbled with the shotcrete process in the 1980's. One of the first uses by a mine of a specialist shotcrete contractor was at Broken Hill in 1992 when Pasminco lined their new pump chamber using high quality steel fibre reinforced shotcrete. They also chose to line their raise bore sump with 50MPa shotcrete in preference to a steel liner on the basis of cost and the speed of installation.

From this point, using specialist shotcrete contractors became more common for campaign style works. Work was generally carried out using trailer mounted shotcrete pumps and hand spraying techniques.

At the Perseverance Mine at Leinster W.A. ground conditions in the underground nickel mine made development slow and hazardous. Jetcrete Australia was awarded a contract in 1993 to rehabilitate large areas of the disseminated orebody with shotcrete. This proved to be successful and the mine decided to try using shotcrete in the new development headings. This saw the beginning of the first in-cycle application of shotcrete in an Australian mine.

In order to undertake this work safely, Jetcrete adapted a drilling jumbo rig by building a shotcrete pump on the rear and pumping the shotcrete to a remotely operated nozzle on the jumbo boom. The nozzle could travel along the boom on the drill feed. This was the first documented use of a remote spraying shotcrete rig in Australia.

Steel fibre reinforced shotcrete became an integral part of the Perseverance Mine development cycle and remains so to this day.

The NorthParkes Mine was a greenfield development in NSW. The mine elected to support most of their development for the block cave copper mine with steel fibre reinforced shotcrete. In 1996 a contract for the shotcreting work was let to Boral Shotcrete and a shotcreting industry became a reality in Australia. Ironically, Pasminco at Broken Hill, who started the trend towards specialist shotcrete applicators, are now the only major mine who routinely produce and spray their own shotcrete.

But for most mines, using a specialist contractor eliminates the need to purchase expensive equipment, which would often not be well utilised. There is no learning curve, and with a well written specification they can insist on a quality which they might not be able to produce themselves.

There are now more than a dozen specialist shotcrete contractors operating in Australia. Several of the larger mining contractors have also started specialised shotcreting divisions. It is estimated that shotcrete consumption in underground mining has increased from 40,000m3 in 1995 to 100,000m3 in 1998. It is further estimated that this volume will triple in the next five years.

The rapid growth in demand and correspondingly rapid growth in number of shotcreting contractors has had a positive impact on shotcreting costs. Typical steel fibre reinforced shotcrete prices have dropped from an average of A$900.00 per cubic metre applied in 1995 to less than A$600.00 per cubic metre in 1998. This saving has been achieved by more efficient and highly productive equipment, often operating with one man crews.

However, the entry of a large number of new contractors into a young industry has also brought its fair share of problems, mostly associated with regulation, specification and quality.

3 IN-CYCLE APPLICATION

Rockbolts, mesh and shotcrete, in descending order, are the major support elements available to miners for ground support. It is fair to say that the order of descension reflects the installed cost per lineal metre of support.

Mines have traditionally used shotcrete as a last resort measure to manage ground that has not responded to control by other means. Firstly, this means that shotcrete is the third attempt at ground control (after bolting, then meshing) making for a very expensive total solution. Secondly, the shotcrete now has to cover the previously installed mesh. This may double the volume that may otherwise have been required. Thirdly, the shotcrete is applied over the top of the previously installed rockbolts, minimising any structural connection between the two elements.

Ideally, the shotcrete element should be installed as the first pass support. This has many advantages. Immediately after blasting the first application can be quickly executed by robot over the muck pile, with the reinforcement included. The first stage can be a full thickness shotcrete layer, and any required rock bolts can be drilled through as a second stage. The ground support cycle is then complete.

The volume of shotcrete is minimised by incorporating steel fibre into the shotcrete mix and thereby eliminating mesh. The shotcrete has maximum possible adhesion to the ground, maximising initial load transfer. The bolt bearing plate is placed on the shotcrete surface, producing maximum load transfer between the two support elements. Additionally, the bolt installation takes place under a properly supported area.

4 SPECIFICATIONS AND QUALITY

The area of specification of quality in mining contracts is perhaps the area that requires greatest attention. It is unfortunately still common to see specifications of one paragraph with no mention of even a compressive strength requirement.

There is no Australian Standard for shotcrete but there are several excellent publications, which can be readily referred to.

4.1 Quality Specifications

The Australian Standard for Concrete Structures AS3600 specifies testing and assessment for compliance of concrete specified by compressive strength in Section 20. These rules and especially the requirements for tests on cores taken from the structure in Section 21.4.3 can and should be applied to shotcrete.

Other Australian Standards delineating methods of testing concrete that apply to shotcrete include;

AS 1012.1 *Methods of testing concrete.* This standard sets out the method for obtaining samples of freshly-mix concrete.

AS 1012.3 *Methods for the determination of properties related to the consistency of concrete.* This section of the standard sets out the method for determining the slump of concrete, when the nominal size of aggregate does not exceed 40mm.

AS 1012.8 *Methods for making and curing concrete compression, indirect tensile and flexure test specimens, in the laboratory or in the field.* This section sets out the method for the making and curing of compression test specimens of concrete sampled in the laboratory or in the field. Tensile and flexure tests are rarely specified for shotcrete with toughness now being preferred as a measure of flexural performance.

With reference to standards that are written specifically for shotcrete, there are several in the international arena. American publications include;

ACI 506R-90 *Guide to Shotcrete* also incorporates ACI 506.3R-91 *Guide to Certification of Shotcrete Nozzlemen.*

C-1018 1994b American Society for Testing and Materials Standard. "*Test Method for Obtaining Flexural Load vs. Deflection Behaviour of Fiber-Reinforced Concrete*", ASTM, Philadelphia, 1994.

The Norwegian "*Sprayed Concrete for Rock Support - Technical Specification and Guidelines*", published by the Norwegian Concrete Association contains excellent practical advice for setting up shotcrete contracts. This standard sets out the use of roughness factors which have become a worldwide industry standard. Roughness factors are explained further in Section 5.

The EFNARC "*European Specification for Sprayed Concrete*" is a recent addition. This standard has gained worldwide attention for the specification of the EFNARC panel test for determining toughness. This aspect of testing is discussed further in Section 6.

However possibly the most recent effort to write a specification for shotcrete has been undertaken by the NSW Roads and Traffic Authority who have issued QA Specification B82 "*Shotcrete Work*" in draft form. This specification deals with steel fibre reinforcement in a comprehensive fashion and is the first standard to incorporate the round determinate panel as the means for estimating toughness. This is discussed further in Section 6.

4.2 Performance based specifications

Once sufficient data for writing a proper specification has been obtained, there is sometimes a tendency to over specify the product. For instance, if a minimum compressive strength is specified, there is little need to specify a minimum cement content, since the compressive strength grade will already determine that factor.

The specifier must concentrate on specifying the performance required from the mix. For shotcrete, this will normally require a minimum compressive strength and a minimum toughness level. Items such as cement content, water/cement ratio and slump should be determined by the supplier and contractor to best suit their equipment and project. This is not to say these aspects do not require control. The levels should be set by the contractor prior to commencement and quality procedures to ensure compliance set in place. For instance, if the contractor decides 100mm slump is correct for project in hand, then the slump for each batch should be measured and recorded as satisfying that requirement.

The danger for specifiers is that if they overspecify the product, they may inadvertently assume responsibility for the performance. For instance, if a specification requires a cement content of 400kg per m3 and the resultant mix fails some other criteria, the contractor may have a case for claiming to have satisfied the contract specification despite the fact that the mix has failed.

With regard to steel fibre reinforced shotcrete (SFRS), Bernard 1998 makes the point that five years ago the range of fibres available and knowledge about mix design and its influence on performance was limited. Specifications for FRS were predicated on the belief that fibre dosage was the primary determinant of post-cracking performance and no distinction was made between competing fibre types . Recent years have seen the introduction of many new and varied types of fibres, and experience in producing and testing this material has demonstrated that mix design has a significant effect on performance. The result is that *prescriptive* specifications based on minimum fibre dosages, or any other mix parameter, are archaic and discourage innovation. Specifications for post-cracking capacity should therefore be *performance* based, which requires that performance must be assessed by some means. Testing to determine post-cracking performance therefore plays a central role in ascertaining the suitability of a suggested mix.

4.3 Specifying compaction

One of the most troublesome areas for engineers specifying shotcrete is how to ensure that the insitu material is well compacted.

Most specifications for shotcrete projects call for the spraying of test panels during shotcrete placement from which cores are drilled and sent to a laboratory for compressive strength testing. Panels must be sized to allow for standard 75mm diameter x 150mm long cores to be taken.

Cores from test panels can also be visually inspected and graded on quality. The panels must be sprayed at the work face using the same machinery, operators, mix and accelerator settings used to spray the ground. In this way, core compressive strengths offer a reasonably accurate indication of the in-situ quality of the shotcrete product.

The density of the cores can be measured before testing. Typically, shotcrete cores measure about 2300kg per m3. However, specifying this as a minimum density in the hope of ensuring adequate compaction may not achieve the required ends. Suppliers frustrated with attempting to meet an artificially imposed minimum density start looking for denser aggregates in order to comply with the specification. This can result in poorly compacted samples containing very dense aggregate passing tests aimed at achieving good compaction.

A much better method of determining compaction is to check the absorption and volume of permeable voids. This can be done in accordance with ASTM C642-90. Morgan,1998 suggests a limit of 8% for maximum boiled absorption and a limit of 17% for maximum volume of permeable voids.

5 PROJECT CONTROL

A detailed inspection and testing plan (ITP) is a basic requirement for project control. The quality of the final in-situ material must be controlled from the production of the source material through to the site. The ITP contains two major elements. The first is a batch plant ITP and the second is an application ITP.

5.1 Batch plant inspection and testing plan

It is necessary to control the quality of the shotcrete material from the quarry for aggregates, suppliers of admixtures etc. This is achieved via a Batch Plant QA Inspection and Testing Plan. For each process incorporated into the work the ITP must specify the applicable standard, the test and test method required, the frequency of testing and the form on which the information is recorded. A suitable plan is illustrated in Appendix 1 .

5.2 Application inspection and testing plan

An application ITP monitors the processes used in applying the shotcrete to the wall. These typically include surface preparation, precise location recording, thickness and testing carried out. A typical application ITP is illustrated in Appendix 1 .

5.3 Thickness

One of the most difficult and crucial areas to control is thickness. It is normal for the design engineer to specify the minimum thickness assumed in the structural calculations.

At the Perseverance Mine site at Leinster, a shotcrete management system (SMS) was introduced to ensure minimum thickness were obtained whilst limiting overuse of shotcrete. Clements and Carpenter 1998 describe how the SMS works at the Perseverance Mine site. As work proceeds, a 75mm probe is used during the placement of the first layer of shotcrete in the cut. Each 2.5m cut is probed approximately 10 times. In addition, the SMS was introduced to control shotcrete usage and ensure correct thickness was developed. This method relies on the “Dale Scale”; a table calculated using the Norwegian concept of roughness factors agreed on between the shotcrete contractor and client.

A pocket sized laminated card is provided to all crews and to shift bosses. It calculates the exact amount to be placed in the development cuts for a given length of cut. Any deviation from this scale is noted on the shift report and signed by shift boss and crew leader. A deviation might occur due to water control problems or excessive overbreak.

The roughness actor is a factor that describes the ground conditions after blasting. Before minimum thickness can be achieved, all the minor irregularities have to be filled. The volume of shotcrete consumed by these irregularities is determined from the roughness factor.

The theoretical volume is calculated by multiplying the profile size by the minimum thickness, then this number must then be multiplied by the roughness factor to obtain the required shotcrete volume. The Norwegian standard then proposes a 10% allowance for rebound and waste.

The formula for calculating the required

Table 1 The "Dale Scale" shotcrete management system.

75mm cover directly to rock at face			
Depth of cut metres	Good profile Factor 1.5	Ave profile Factor 1.7	Poor profile Factor 1.9
1.5	2.6	3.0	3.4
2.0	3.6	4.0	4.6
2.5	4.4	**5.0**	5.6
2.6	4.6	5.2	5.8
2.7	4.8	5.4	6.2
2.8	5.0	5.6	6.4
2.9	5.2	5.8	6.6
3.0	5.4	6.0	6.8

volume of shotcrete consumption then becomes;
(length of profile) x (minimum thickness) x (roughness factor) x 1.1 (10% rebound allowance)

The Norwegian standard quotes roughness factors varying from 1.3 to 1.8 times the theoretical volume. The Norwegian standard relates closely to tunnelling applications. Experience in Australian mines has shown that roughness factors are generally in the range of 1.5 to 2.0. The roughness factor can be strongly influenced by blasting pattern designs. Any shotcrete quality system that does not record blasted profiles is ignoring the largest influence on shotcrete usage.

Example of calculation of shotcrete volume in a 4.8 x 4.8m drive, 75mm thickness;
(4.8m x 3) x 0.075 x 1.7 x 1.1 x 2.5 = 5.05m3
drive size x thickness x roughness factor x rebound allowance x depth of cut
= total volume of shotcrete required

This Shotcrete Management System ensures even coverage of shotcrete. It minimises waste from overspraying and eliminates underspraying. A 17% saving in materials was attributed to the use of this system.

A key performance indicator can be used to assess compliance. At the Perseverance site, the shotcrete contractor was contracted to place the exact amount of shotcrete planned per lineal metre of development. A tolerance of +/- 0.2m3 per lineal metre was deemed allowable. A substantial financial penalty was attached to non-compliance.

6 EVALUATING TOUGHNESS

Steel fibre reinforced shotcrete has opened up new horizons for the use of the material. Many of these applications are in tunnel linings and embankment stabilisation projects.

The ability of steel fibre reinforced shotcrete to deform and carry load after cracking is the key to the usefulness of the product. "Toughness" is the term given to describe the post-crack performance of the material.

6.1 The EFNARC panel test

Toughness estimates can be derived using the standard flexural beam in a modified test described in the American C1018 standard. In recent years, panel tests have become popular as the test specimen is a plate and is seen as being more representative of thin shotcrete structures. The European Specification for Sprayed Concrete outlines a method of determining toughness using a 600mm x 600mm x 100mm thick panel. The testing arrangement is illustrated in figure 1.

In the test, the square shotcrete plate is deflected through 25mm and the load carrying ability plotted against deflection. Typical

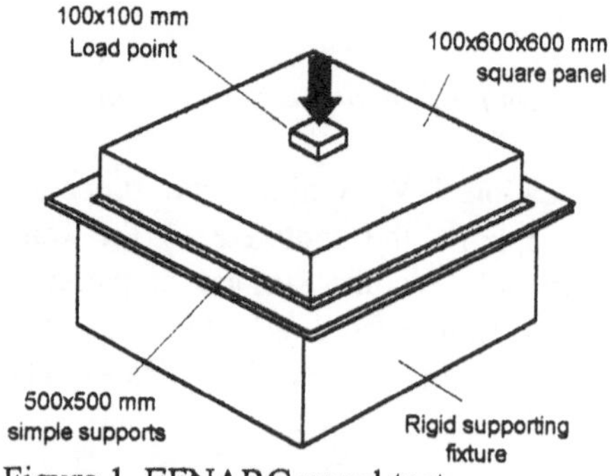

Figure 1. EFNARC panel test.

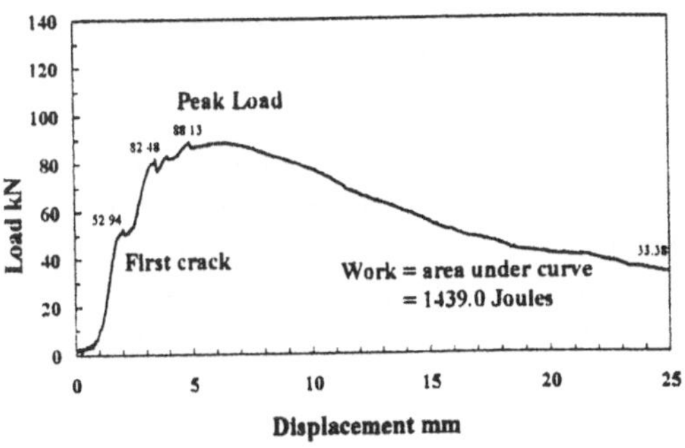

Figure 2. EFNARC panel test output.

output from a panel toughness test is shown in figure 2.

As the output demonstrates, the square panel is capable of sustaining higher loads than those creating the first crack. The panel continues to carry load throughout the 25mm of deflection. The important feature is the *total energy* absorbed during the test. This is given by the area under the graph and is expressed in Joules.

6.2 The round determinate panel

An alternative panel test has been developed in Australia by Dr Stephen Bernard at the University of Western Sydney. The *round determinate panel* test uses a circular panel 75mm thick with a supported span of 750mm. This gives an aspect ratio of twice that of the EFNARC panel and ensures bending failure.

Other advantages of the round determinate panel include;

- panel 30% lighter than the EFNARC panel
- no saw cutting required
- low co-efficient of variation of results
- less expensive than the EFNARC to test
- more slender specimen (aspect ratio 10 cf 5 for EFNARC) ensuring bending failure
- statically determinate

This test is gaining international acceptance and has already been included in the draft NSW Roads and Traffic Authority shotcrete specification in Australia.

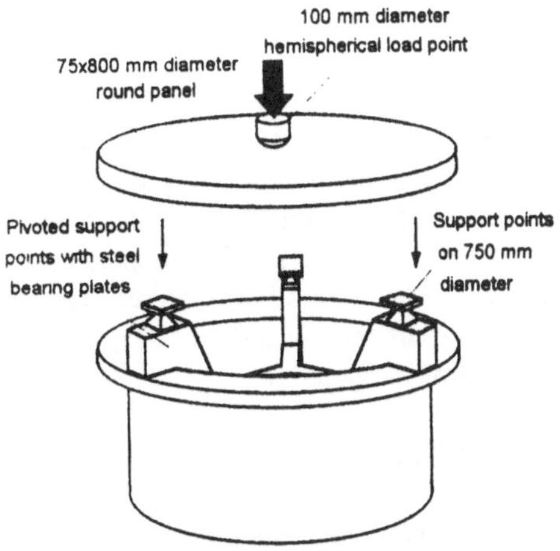

Figure 3. Round determinate panel test.

7 CONCLUSIONS

In Australia, a growing and strengthening demand for shotcrete for ground support is being met by a fast growing shotcrete industry. Concerns over quality can be overcome by providing a well written specification at time of tender. To check the quality of applied shotcrete, the client can ask for;

- compressive strength cores to be taken from sprayed test panels at the rate of at least 4 cores per 50m3
- boiled absorption and permeable voids tests can be carried out on the cores to check the degree of compaction
- rebound must be kept under 10%
- an Inspection and Test Plan must be in place both at the batching plant and underground
- a Shotcrete Management System should be put in place to control the volume usage.
- Toughness can be specified to ensure performance criteria are met

Ongoing education including conferences and workshops need to be available to enable the necessary engineering skills and confidence to be passed on. Good shotcrete practice needs to be defined in a continually updated standard, which can be used widely for technical specifications.

Recent improvements in evaluating the performance of steel fibre reinforced shotcrete have been developed in Australia. The round determinate panel test for estimating toughness is gaining international recognition.

The recent formation of the Australian Shotcrete Group will assist in the pursuit of these goals. The Australian Shotcrete Group is a sub-group of the Australian Underground and Tunnelling Association.

8 REFERENCES

ACI 506R-90 Guide to Shotcrete.

American Society for Testing and Materials Standard C642-90 "Standard Test Method for Specific Gravity, Absorption and Voids in Hardened Concrete", ASTM, 1990.

American Society for Testing and Materials Standard C-1018 "Test Method for Obtaining Flexural Load vs. Deflection Behavior of Fiber-Reinforced Concrete", ASTM, Philadelphia, 1994.

Application Inspection and Testing Plan, Jetcrete Australia P/L, Darlot Mine site, 1998.

Australian Standard 3600-1988 Concrete Structures.

Australian Standard AS 1012.1 Methods of testing concrete, 1990.

Bernard, E.S. "Point Load Capacity in Round Steel Fibre Reinforced Concrete Panels", *Civil Engineering Report CE8*, School of Civic Engineering and Environment, UWS Nepean, July 1998.

Bernard, E.S., "Measurement of post-cracking performance in fibre reinforced shotcrete", *Australian Shotcrete Conference*, Sydney, October 8-9, 1998.

Bernard, E.S., and Clements, M.J.K., "Variability in Toughness Estimates Derived Using Beam and Panel Tests", *Australian Shotcrete Conference*, Sydney, October 8-9, 1998.

Clements, M.J.K., and Carpenter D. "Trends in the Use of Shotcrete in Underground Mines", *Safety in Underground Mining Conference*, Kalgoorlie, October 19-20, 1998.

Clements, M.J.K., "The Performance of Steel Fibre Reinforced Shotcrete in Mining", *3rd National Shotcrete Workshops*, Kalgoorlie, November 18, 1997.

European Specification for Sprayed Concrete, European Federation of National Associations of Specialist Contractors and Material Suppliers for the Construction Industry (EFNARC), 1996.

"Guide to good shotcrete practice" Concrete Institute of Australia, 1973

Morgan, D.R., "Steel fibre reinforced shotcrete for underground support", *Australian Shotcrete Conference*, Sydney, October 8-9, 1998.

QA Specification B82, draft issue, 30 December 1997, NSW Roads and Traffic Authority.

Shotcrete Inspection and Testing Plan, CSR Readymix, Perseverance Mine site, 1997.

Sprayed Concrete for Rock Support - Technical Specification and Guidelines, Norwegian Concrete Association, Publication No. 7, 1993.

APPENDIX 1

INSPECTION AND TESTING PLAN

BATCH PLANT QUALITY CONTROL

	Inspection and Test Plan	Criteria	Frequency	Procedure	Form
1.	Cement compliance	AS 3972	Monthly	AS 2350	Certificate
2.	Silica fume compliance	AS 3582.3	6 Monthly	AS 3583	Certificate
3.	Water compliance	AS 1379	6 Monthly		Lab Report
4.	Source Rock Analysis	AS 2758	Annual/changes	AS 1141.	Certificate
5.	Course aggregate grading	AS 2758.1	Monthly	AS 1141.11	Lab Report
6.	Course aggregate Bulk density and water absorption	AS 2758.1	6 Monthly	AS 1141.5	Lab Report
7.	Fine aggregate grading	AS 2758.1	Monthly	AS 1141.11 AS 1141.12	Lab Report
8.	Fine aggregate Bulk density and water absorption	AS 2758.1	6 Monthly	AS 1141.5	Lab Report
9.	Fine aggregate Clay fine silt	Max 8%	Monthly	AS 1141.33	Lab Report
10.	Fine aggregate Organic impurities	AS 2758.1	Monthly	AS 1141.34	Lab Report
11.	Mix Design Letter		Initial / changes		Letter
12.	Use of admixtures		Initial / changes		Letter
13.	Compressive Strengths	AS 1379	1 test / 50m^3	AS 1012	Lab Reports
14.	Agitators Cleanliness of interior Condition of blades		Fortnightly		Inspection records
15.	Test personnel		Trained before commence work		
16.	Batch plant scales calibration	AS 1379	6 Monthly	CAP11.01	Report
17.	Batch plant liquid metering equipment	AS 1379	6 Monthly	CAP11.01	Report
18.	Moisture content of aggregates		Weekly or changes	AS 1012.3	Reports and batch records
19.	Slump		Every load	AS 1012.3	Batch records
20.	Water content		Every load		Batch records
21.	Steel fibres Type and amount used		Every load		Batch records

INSPECTION AND TESTING PLAN

SHOTCRETE APPLICATION

	Inspection and Test Plan	Criteria	Frequency	Procedure	Form
1.	Daily shift report	JA	Every shift	JA	JA form
2.	Compressive strength	32MPa	1 test/50m3	AS1012	NATA cert
3.	Dramix Steel Fibre	30kg	Per m3	RC65/35	Shift report
4.	Slump		Every load	AS 1012.3	Batch records
5.	Sprayed test panel	32MPa	1 test/50m3	AS1012	NATA cert
6.	Wash down face	Water spray	Each cut		Shift report
7.	Thickness applied	Remote probe	Each layer		Shift report
8.	Quality Assurance Report	Darlot Inspection and Test Plans	Monthly	JA	Report to site
9.	Batch Plant Audit	Batch Plant ITP	2 Monthly	Batch Plant ITP	JA Audit

Rock Support and Reinforcement Practice in Mining, Villaescusa, Windsor & Thompson (eds)
© 1999 Balkema, Rotterdam, ISBN 90 5809 045 0

The important properties of steel fibre reinforced shotcrete (SFRC) used in mining

R. Ratcliffe
Scancem Materials (Australia) Pty Limited, Osborne Park, W.A., Australia

ABSTRACT: The reasons for using SFRS, including the way a SFRS performs in comparison to plain shotcrete as a support element in mining tunnels, are discussed. The mechanisms by which steel fibres reinforce shotcrete and the need for specifying performance criteria are highlighted. A description of current testing procedures along with a critique of their suitability are included and are accompanied by a summary of international recommendations on suitable performance classifications. Finally recommendations on the practical considerations necessary to ensure good quality in-place SFRS are made including a calculation method for determining minimum steel fibre dosages.

1 INTRODUCTION

Shotcrete is becoming more and more popular as a means of ground support in the mining industry for a number of very important reasons:-

- Flexibility - it can offer a cost-effective solution over a range of rock types from jointed massive (blocky) to squeezing ground in terms of both temporary and permanent support.
- Safety - With the remote robot arms now being used to apply shotcrete it is possible to apply support early without personnel needing to be beneath the area to be supported. Shotcrete also offers a level of protection against rock burst activity, where meshing will not.
- Maintenance - The ability of shotcrete to seal the rock face overcomes problems with slaking as well as ravelling between rock bolts thus having the potential to significantly reduce rehabilitation costs.

The level of interest that has been generated in replacing conventional mesh reinforcement with steel fibres was driven primarily by the cost savings that are achievable in both time and materials. Another very important consideration, some say the most important, is the improvement in operating safety afforded by removing the need to fix reinforcement to a partially supported or fully unsupported rock face prior to shotcreting.

It must be borne in mind that shotcrete is simply concrete that has been sprayed (propelled at high speed using compressed air) into position and as such exhibits all the plastic and hardened properties of a conventionally placed concrete, albeit with a higher fines content than a similar grade of conventionally placed concrete would have. The reason for the higher fines content is practical, in that the larger aggregates will initially rebound from the substrate and fall to the ground until a sufficient thickness of mortar has been deposited to absorb the energy of the aggregates and retain them.

The strength and quality of an in-place hardened shotcrete are determined by the water/cement ratio and the level of compaction achieved. The important plastic properties of pumpability, rebound and achievable build thickness are determined by the binder components, aggregate grading, water content and admixtures.

Many references are now available on how to make and place good quality shotcrete and they nearly all give recommendations on the proportions of the base ingredients as follows for either dry or wet process normal weight shotcrete with a wet mass of 2300-2400kg per cubic metre:-

Cementitious: Total binder content should be in the range of 350-500kg/m^3 and will typically contain between 35-70kg/m^3 of silica fume.

Aggregates: Total aggregate content for normal weight aggregates is typically in the range of 1600-1800kg/m^3. The important criterion with aggregates however, is that the overall grading curve falls within the envelope given in ACI 506R-85[1] Table 2.1, in order to ensure minimum rebound, permit good compaction and generally give a high durability shotcrete. With a maximum recommended aggregate size of 8-10mm for most applications this gives the grading shown in Table 1 below.

Table 1 Shotcrete aggregate grading curve.

Sieve Size	Percent by weight passing
12.7mm	100
9.5mm	90-100
4.75mm	70-85
2.36mm	50-70
1.18mm	35-55
600μm	20-35
300μm	8-20
150μm	2-10

The important point in terms of steel fibre reinforced shotcrete (SFRS) is that *no special changes need to be made to conventional shotcrete mix designs in order to incorporate steel fibres.*

2 WHAT IS THE ROLE OF SHOTCRETE?

Shotcrete used as a ground support element does not work in isolation from the ground that requires support. The technology has come a long way from the concept that the support structure has to be strong enough and stiff enough to support the overburden loads by itself with the realisation that when a tunnel is excavated there is a local disruption of the stresses in the ground surrounding the tunnel that results in a new set of stresses being induced. The reaction of the ground to these stresses and the resultant stresses and deformations needed to be carried by the support structure will then differ with the nature of the ground.

The major problem in designing support to underground openings is associated with determining the strength and deformation properties of the ground and matching it with the chosen support structure. Two distinctly different types of failure mechanism can be identified dependent on the nature of the supported ground as follows:-

2.1 Competent blocky rock

The behaviour of blocky rock is directly related to the nature of its jointing, with failure being dependent on the ability of individual blocks to fall, topple or slide into the tunnel opening. In blocky rock the role of a shotcrete lining will therefore be to

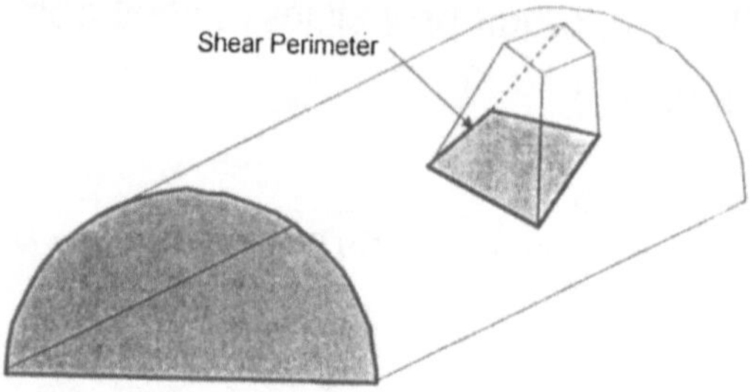

Figure 1. Plain Shotcrete in Blocky Rock

effectively hold back individual blocks and in a shotcrete that is effectively bonded to the rock it does this by virtue of the shear resistance offered by the shotcrete at the perimeter of the block (Fig.1). This support mechanism is therefore totally dependent on achieving and maintaining an adequate level of adhesion to the rock and where this is managed no flexural stresses will result in the shotcrete and it is perfectly legitimate to use a plain unreinforced shotcrete.

Where there is an adhesion failure to the rock however, the shotcrete will then have to carry the loads from a moving block back into the nearest adjacent point of support (typically rock anchors), hence putting the shotcrete layer into flexure.

2.2 Soft and/or fractured rock

In heavily jointed or soft rock the failure mechanism is a much more ductile one, with the total support load being shared between the ground and the support structure in proportion to their relative stiffnesses.

In a tunnel the face initially provides partial support to the ground and this support has to be replaced by other methods of ground support as the face progresses. Because the face must advance before support can be installed there will already be a reduction in the support stresses required accompanied by an inward movement of the ground. The stresses carried in the support structure will be determined from both the stiffness of the support structure as well as the time of installation. This effect can be seen in Figure 2.

The important points to note in Figure 2 are that the Rock Mass Displacement Curve will result in a total radial displacement that is determined from the displacement prior to the installation of the support system(d_r) plus the displacement of the combined rock plus support system (d_s), which is determined by both the rock characteristics and the stiffness of the support system.

The stiffness of the support system is determined from the slope of the support system load versus deflection curve. Effective support is achieved only if these two curves intersect, with timing being critical. The support system will be ineffective if the support is installed too late to prevent loosening and collapse or so early that the support system yields under the applied pressures.

The complexity of this kind of analysis, coupled with the difficulty of defining the rock parameters, typically means that designers rely on empirical design approaches, such as the NMT's (Norwegian Method of Tunnelling) Q-system of rock classification to specify support systems.

*The most important issue in all of this in terms of shotcrete support systems is that in anything other than competent blocky rock that the shotcrete **must** be capable of accommodating movement.*

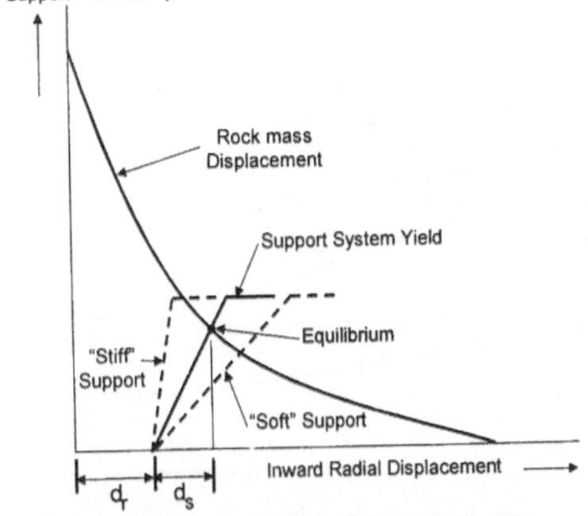

Figure 2 Support pressure versus displacement

3. WHEN SHOULD REINFORCEMENT BE USED IN SHOTCRETE

Shotcrete is made from concrete and concrete is a brittle material with low tensile capacity. The strength related properties of a commonly specified 32MPa shotcrete are typically:-

Compressive Strength at 28 days	32MPa
Flexural Strength at 28 days (taken as $0.4 \times f'_c{}^{2/3}$)	4MPa
Youngs Modulus "E"	31000MPa

For a layer of this shotcrete spanning continuously between rockbolts at 2 metre centres and supporting a pressure "p"(Fig. 3) it is possible to estimate the pressure and hence deflection at which cracking will occur using standard engineering theory:-

Cracking Bending Moment "M" = $pl^2/12$
$= 4N/mm^2 \times t^2/6000$ kNm/m

Solving this equation for p gives

$$\mathbf{p = .008t^2/l^2\ kN/m^2}$$

where t = thickness in mm
l = span in m.

Similarly the deflection "d" is given by the equation:-

$$\mathbf{d = p \times l^4/(384 \times E \times I)}$$

where I = Concrete Moment of Inertia
$= t^3/12\ mm^4/mm$

and using the value of "p" causing a $4N/mm^2$ cracking stress gives:-

$$\mathbf{d = 8.06 \times l^2/t\ mm}$$

Therefore for a shotcrete thickness of 100 mm continuously spanning 2 metres the calculations in Figure 3 show that the pressure needed to cause cracking is $20kN/m^2$ (equivalent to a 2 metre pressure head of water) and the mid point deflection at which cracking occurs is 0.32mm.

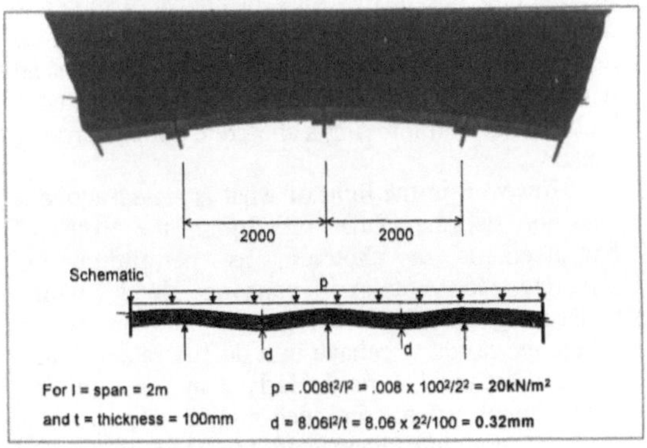

Figure 3 Shotcrete supported ground

These two values stress some very important aspects of thin shotcrete linings:-

1. A thin shotcrete layer has very little capacity in terms of its ability to carry any significant overburden loads. Its true value lies in its ability to act as a membrane that can deflect with the rock mass as the overburden stresses are redistributed within the rock to accommodate the opening.

2. Extremely small deflections, in terms of ground movement, will cause cracking in shotcrete. In unreinforced shotcrete a complete loss of load carrying capacity will occur at deflections that are imperceptible to the human eye. It is also very important to note that the effect of thickening up the shotcrete layer is to actually **reduce** the deflection at which cracking will occur. This is an important lesson to be learned because it means plain, unreinforced shotcrete, is unsuitable for supporting moving ground and the incorporation of thicker and thicker layers can also be self defeating.

In moving ground, in order to optimise the support solution, it is necessary to design the support system on the basis of the grounds stiffness properties with any shotcrete membrane needing to be reinforced if a sudden brittle failure is to be avoided.

Because reinforced shotcrete fails in a ductile way ie it is accompanied by significant deflections and cracking, there is normally adequate warning to either undertake remedial works, such as adding extra shotcrete or rockbolts, or at the worst run to safety.

4 WHAT ARE THE PROPERTIES OF FIBRE REINFORCED SHOTCRETE?

In all but the best quality rock the choice of shotcrete for support means the incorporation of reinforcement. Typically the incorporation of steel reinforcement, usually in the form of mesh, has been on the basis of improving a shotcrete's load carrying capacity.

However, in the light of what is stated above it may be worth while redefining the role of reinforcement in shotcrete as permitting the shotcrete to continue carrying load whilst undergoing significant deflections. In other words "shotcrete needs to behave in a ductile rather than a brittle manner", particularly in a mining environment, where continued mining activity tends to cause ongoing changes in the required ground support pressures.

The ability to sustain load whilst undergoing significant deflections needs to be an intrinsic property of the chosen support system.

The presence of steel fibres in a concrete matrix, with the potential to bridge and apply tension across any cracks that form, has the ability to change an inherently brittle material into a ductile one. There are several international standard test methods available to measure the ductility of SFRS, although the more commonly used term is "toughness".

Toughness testing to the American ASTM C1018[2] and the Japanese JSCE-SF4[3] is performed on the basis of third point loaded beam tests as shown in Figure 4 and some typical results are shown in Figure 5.

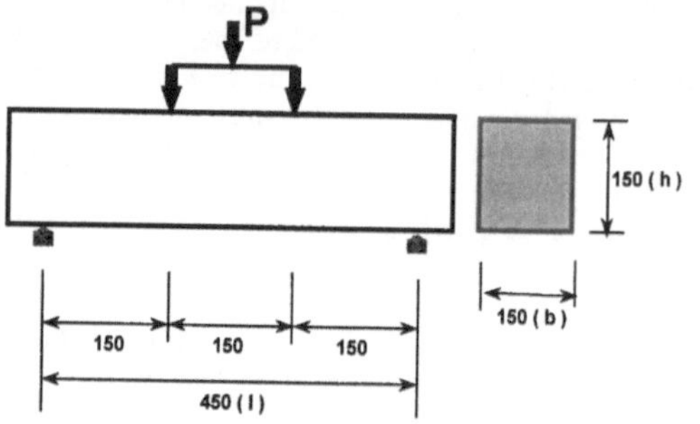

Figure 4 Toughness testing for beams

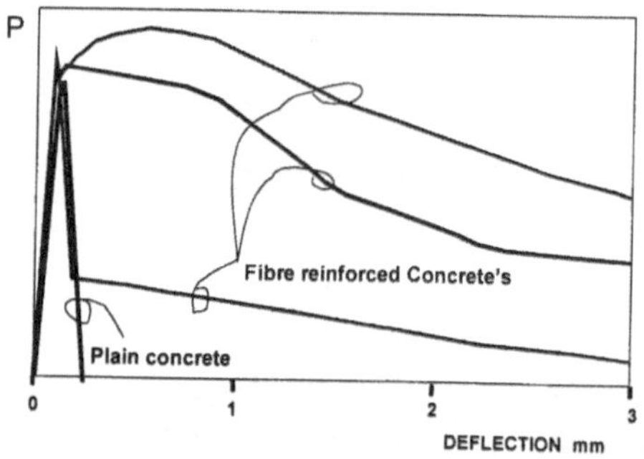

Figure 5 Toughness test results

Toughness is actually measured in terms of the area under the load deflection graph to a predetermined value of deflection and has the units of load times deflection (kNm), more commonly being referred to as work or energy. The deflection value nominated for the area evaluation may be different depending on the standard test method followed, but the approach is universal.

Another way to look at these results is that once a plain shotcrete cracks it losses all load carrying capacity whereas a fibre reinforced shotcrete will continue to support load after cracking, even up to very significant deflections.

Figure 5 shows that the post crack loads a SFRS will carry will be different for different fibre types and dosages even if the same concrete flexural strength is adopted. The fact that all fibres do not perform the same has been confirmed from a significant amount of testing both in Australia and overseas on the large range of fibre types that are available in the market place.

In order to reflect the known variability in fibre types and their performance ***it becomes absolutely mandatory that SFRS is specified in terms of a performance requirement.***

Any specification requesting a minimum fibre dosage will automatically ensure that pricing will be based on the minimum dosage of the cheapest fibre available in the market. This approach is usually followed by the grim realisation (for the contractor) that the cost of achieving the required performance will be significantly higher than was priced for.

If a suitable performance standard is specified the problem of having to specify minimum dosages is overcome and the urge to include it should be strenuously resisted. The reason for resisting it is that contractors that do not understand SFRS will simply quote on the basis of the minimum dosage rate, whereas contractors that do understand SFRS will have to quote similarly simply to avoid putting in too high a price. The result is usually pressure being applied to reduce any performance criteria requirements once the contract is awarded.

5. WHAT PERFORMANCE IS REQUIRED FROM FIBRE REINFORCEMENT?

A complete design approach for steel fibre reinforced concrete and shotcrete has been detailed by the German Concrete Association[4] that is suitable for combined flexure and compression stresses and outlines the toughness measuring procedure it is based on.

The only problem with this, and most other design approaches, is the difficulty of evaluating the loads that will be applied to the lining by the ground. Given this problem and the nature of the mining environment, it seems justified to use a qualitative rather than a quantitative approach to shotcrete specification and then monitor the performance insitu.

The suitability of this "suck it and see" approach should be evaluated whilst bearing in mind that there is always an opportunity to add more shotcrete or ground anchors to obvious trouble areas as long as the safety of a mine is not compromised.

The qualitative approach is one that has been addressed by several specifying bodies and each seems to have a different recommendation.

All the recommendations are however, still in terms of the load carrying capacity after cracking and therefore are essentially variations on the theme of specifying the toughness performance of the fibre shotcrete. The following summarise some proposed values and the source:-

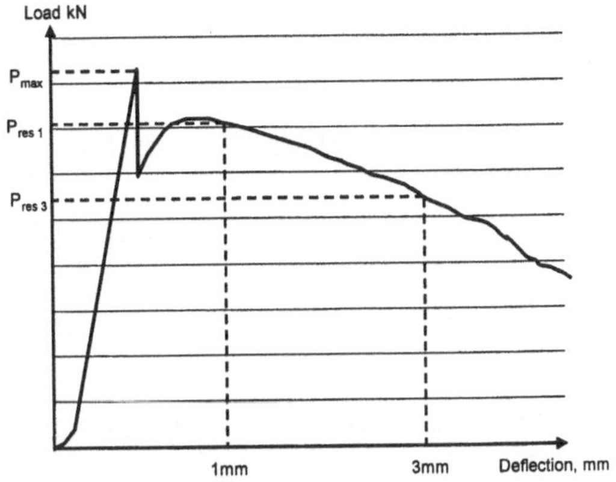

Figure 6 Load-deflection curve

5.1 Norwegian Concrete Association[6]:

Beams that are 75mm deep and 125mm wide are cut from shotcrete panels and made to span 450mm with load applied at third points. The residual stress in the beam is calculated from the load versus deflection graph at specified deflections (Fig.6) and forms the basis for the specification of a toughness class as shown in Table 2.

Residual stress values are calculated from the beam section modulus ($bd^2/6$) and the load value at the designated deflection expressed as a bending moment

Eg. Residual stress at 1mm deflection

$$= \frac{0.5P_{res1} \times 150 \times 10^3}{125 \times 75^2/6}$$

$$= 0.64\ P_{res\,1}\ N/mm^2$$

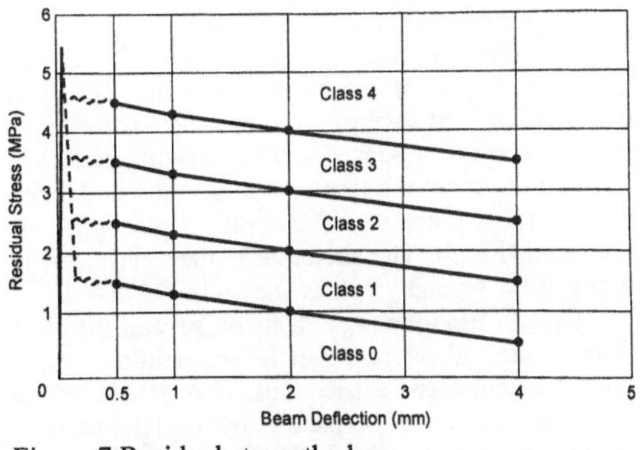

Figure 7 Residual strength class

Table 2. Residual Strength Value Toughness Classes

Toughness Class	Deflection 1mm	Deflection 3mm
0	Non reinforced concrete	
1	Type and amount of fibre specified	
2	2.0MPa	1.5MPa
3	3.5MPa	3.0MPa

Table 3 Residual Strength Class Definition Points

Deformation Class	Beam Deflection (mm)	Residual Stress (MPa) for strength class 1	2	3	4
	0.5	1.5	2.5	3.5	4.5
Low	1	1.3	2.3	3.3	4.3
Normal	2	1.0	2.0	3.0	4.0
High	4	0.5	1.5	2.5	3.5

5.2 EFNARC "European Specification for Sprayed Concrete"[5]:

Gives the option of specifying the toughness of the shotcrete using either a *Residual Strength Class* (from a beam test) or an *Energy Absorption Class* (from a plate test).

5.2.1 Residual Strength Class

Is based on the shape of the beam stress/deflection curve and is defined in terms of the five different classes summarised in Figure 7 and Table 3. The test beams are cut from sprayed panels to measure 125 wide and 75 deep and are loaded at third points while spanning 450mm.

5.2.2 Energy Absorption Class

Is determined from tests on shotcrete plates supported on four sides and loaded in the centre. In this test plates measuring 600 x 600mm in plan and 100 –0/+10mm in thickness are continuously supported around the four edges and centrally loaded through a contact point of 100 x 100mm.

The plate is loaded to a central deflection of 25mm at a rate of 1.5mm per minute. The load deformation curve (Fig. 8) is produced and then from this curve a second curve showing absorbed energy(area under the first curve up to the deflection value) as a function of central displacement is produced (Fig.9). The shotcrete is then classified in terms of the second graph as shown in Table 4

Results from the panel tests on several different fibres and also on panels containing mesh reinforcement were carried out in Australia at the University of Sydney on panels shot by Jetcrete and the results published by MJK Clements[7] at the IX Australian Tunnelling Conference in 1997.

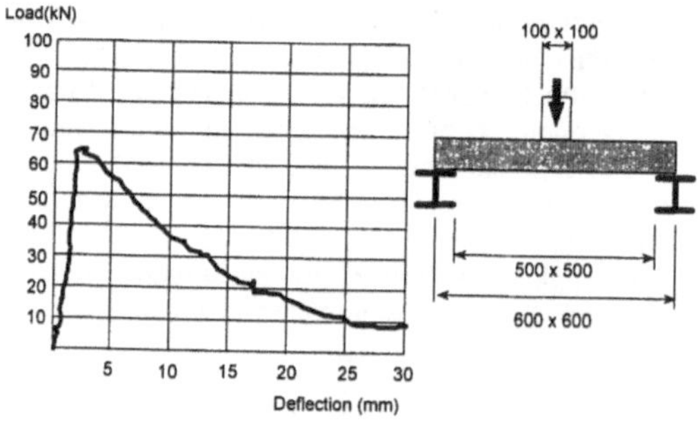

Figure 8 Load/displacement graph & Panel elevation

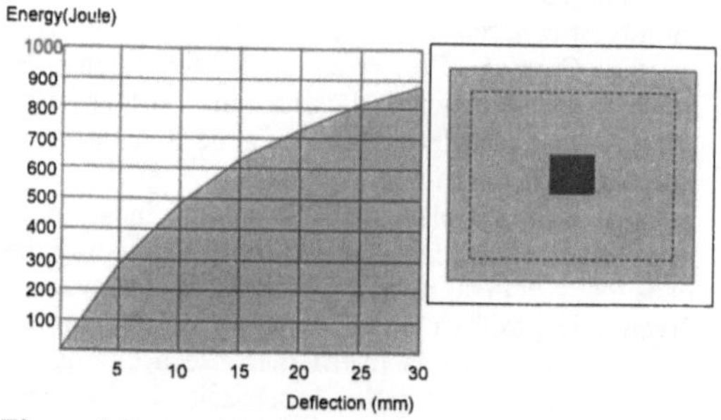

Figure 9 Energy/displacement graph & Panel plan

5.2.3 ASTM-C1018 Toughness Indices

This standard defines toughness in terms of toughness indices, defined in Fig. 10, and residual strength factor.

The Residual Strength Factor "R" is calculated from the toughness indices and is expressed as a percentage as follows ⇒

$$R_{20-50} = \frac{100(I_{50}-I_{20})}{50-20}$$

$$R_{10-20} = \frac{100(I_{20}-I_{10})}{20-10}$$

The standard gives no guidance for the classification of fibre reinforced concrete based on either toughness indices or residual strength factors.

6 WHICH PERFORMANCE TEST FOR SFRS?

It has already been stated that the usefulness of a quantitative test for SFRS performance is of dubious value in a mining environment due to the difficulty in assessing the loads that will be applied and the changing nature of these loads as mining activity proceeds. For these reasons it is considered that a suitable qualitative test is to be preferred and some of the authors experiences with those listed above are worth mentioning.

6.1 ASTM C1018

It is an acknowledged fact by several authors(8)(9) that there is significant room for error involved in performing this test. The basic problem is that the results depend totally on the ability of the equipment to measure and report the areas beneath the load/deflection graph with a high level of accuracy.

The accurate determination of the first crack deflection "δ" is absolutely critical to the reported results. If the area under the load deflection curve to first crack is incorrectly evaluated all of the ensuing results will be incorrect.

Reasons for an incorrect evaluation of the load-deflection areas are due to poor deflection measuring technique, inclusion of testing equipment "slack" in the deflection measurement, difficulty in pinpointing the first crack and using a machine that is too slow in its response or not stiff enough relative to the tested beam to correctly reflect the beams behaviour.

The author was personally involved in the commissioning of this test on two occasions in Australia when the first crack deflection, which should be measured at around 0.06mm was reported

Table 4 Energy Absorption Toughness Classes

TOUGHNESS CLASSIFICATION	ENERGY ABSORPTION IN JOULE FOR DEFLECTION UNTIL 25MM
a	500
b	700
c	1000

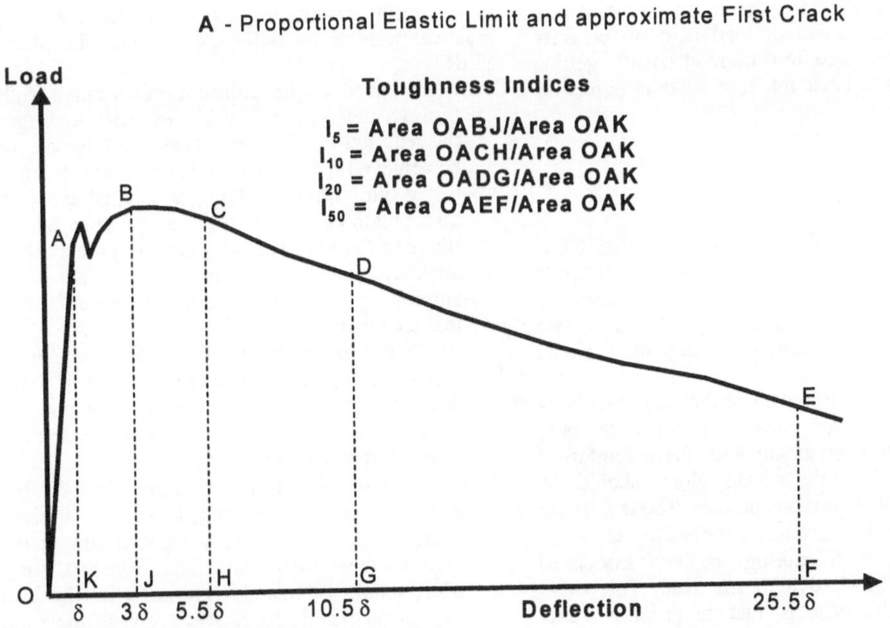

Figure 10 Toughness indices to ASTM C1018

as in excess of 1mm. The result of this is that all the reported results were incorrect and were only picked up as being so from observation of the load-deflection graphs.

This is not an easy test to perform correctly and has not commonly been performed in Australia until relatively recently. Test results should only be trusted if accompanied by the load–deflection plots and evaluated by a suitably experienced person.

6.2 Residual Strength Tests

The use of residual strength values are a deliberate attempt to overcome the problems experienced with the ASTM C1018 approach of having to effectively determine the first crack deflection.

This is achieved by completely ignoring the early, uncracked portion of the graph and only determining the load carrying capacity, expressed as a residual stress, of a fully cracked section. The results are deliberately determined away from the first crack deflection to ensure the results are being measured in the more stable part of the load-deflection curve.

The only problem the author has with this test, apart from the fact that the test beam dimensions are different to those in C1018, making comparison impossible, is that it is carried out on a one dimensional beam element that takes no account of the fact that shotcrete is typically applied as a thin two dimensional element with the ability to redistribute stresses.

The main advantage a fibre reinforced shotcrete has over plain shotcrete is the ability to maintain a two dimensional spread of load carrying capacity even after it is cracked, thus more efficiently getting the applied loads back into the support points(eg. rock anchors).

6.3 Panel Test

The panel test is the only test presently available that goes some way to reproducing how a shotcrete actually works in the field, which it does by measuring the load carrying capacity of a two dimensionally slab instead of a one dimensional beam.

In terms of a qualitative test that can effectively be used for comparison purposes it is also the only test that permits comparison with mesh reinforced shotcrete and for this reason alone should be considered in any evaluation process. The test is also relatively simple to carry out and does not have any of the reporting or measuring problems associated with the ASTM C1018 beam test. The other advantage of this test is that it gives a better indication of how a SFRS will crack and deform in comparison to mesh reinforced shotcrete in actual use.

The one problem the author is aware of with this test is that it is possible to develop different crack patterns in panels dependent on how flat the panel is and hence how well it actually seats on the supporting frame. Different crack patterns, in terms of number and length of cracks, affect the energy absorption capacity of the panel and can lead to significantly different results in otherwise identical panels.

7 WHAT ARE THE PRACTICAL CHARACTERISTICS OF A GOOD SFRS?

To be an attractive proposition in a mining environment a SFRS needs to be able to cope with dynamic as well as static loads, whether from impact from equipment or ore loading out activity or perhaps due to blasting and seismic activity. Whatever the cause of a dynamic load the ability of a shotcrete to cope with it will be determined by its ability to absorb energy, which in turn is directly measured in toughness testing.

The higher a particular SFRS rates in terms of its toughness the more suitable it will be for coping with the sort of duty that will typically be placed on it in a mining environment. For this reason it seems fair to say that a high toughness requirement is an investment in achieving a shotcrete with a low maintenance requirement for a mine.

As fibres are what impart toughness to shotcrete it is worth considering some of the following practical aspects to achieving a good "in place" SFRS:-

1. Fibres need to be uniformly distributed right through the concrete matrix in order to impart the required level of toughness and this will not be achieved if fibre balling is occurring. In order to avoid fibre balling and ensure good distribution care needs to be taken when adding fibres to the mix, either by ensuring loose fibres are individually rained into the concrete or by employing the collated and glued Dramix® fibres that are available.
2. Blockages to shotcrete delivery lines can be an expensive nightmare underground and to avoid them it is necessary to avoid balling and use fibres that are no longer than 70% of the delivery lines internal diameter.
3. Rebound of steel fibres is higher than for the shotcrete, typically resulting in a lower in place dosage than was originally batched and as the fibres are a very expensive part of the mix every effort should be made to minimise it. The best way to do this is to correctly design the mix,

preferably incorporating silica fume and use the wet process in preference to the dry process to take advantage of the lower inherent rebound values.

4. Shotcrete as a material is dependent to a large extent for its in place quality on the ability of the nozzleman and for this reason it is recommended that prior to any nozzleman being employed he is suitably trained and prequalified. The training and/or prequalification period is also an excellent opportunity to establish the performance characteristics of the SFRS to be used and finalise through testing the mix design, fibre type and dosage required to meet the specified toughness and other performance requirements.
5. Once the toughness characteristics of a particular SFRS are established it only remains necessary to ensure the correct quality and thickness of shotcrete are being placed. Shotcrete quality can be established from crushing cores taken from test panels and the thickness of a plastic shotcrete can be easily established by measuring the penetration thickness to the substrate rock. Where adhesion of the shotcrete is critical some form of pulloff test will also be required.
6. In terms of practicalities it is also preferable if anchors are placed through the shotcrete rather than being placed under the shotcrete. There are several reasons for this:-
   - The anchor load plate can more effectively spread anchor loads into the shotcrete if it is placed at the shotcrete surface.
   - Shotcrete placed over the load plate is subject to debonding, which can quickly be followed by cracking and spalling.
   - Anchors that are visible at the shotcrete surface can be visually observed and are accessible for inspection and repair.
7. SFRS should never be specified by dosage rate alone. The reason for this is that not all fibres perform the same, coming as they do in different lengths, aspect ratios (length/diameter), tensile strengths and with differing anchorage details. The only time a dosage rate should be specified is after toughness testing has been carried out and the dosage rates of individual fibres needed to meet the toughness criteria have been established. The dosage rate must then always be related to the fibre type it is applicable to.
8. Minimum dosage rates, where specified, *must* be related to the aspect ratio of a fibre. The reason for specifying a minimum dosage rate in SFRS is to ensure that after placing there are a sufficient number of fibres within the matrix, allowing for rebound and variations in the fibre distribution, to effectively reinforce the shotcrete. What in effect is the aim of this approach is to ensure a certain maximum average 3-dimensional fibre spacing (Fig. 11).

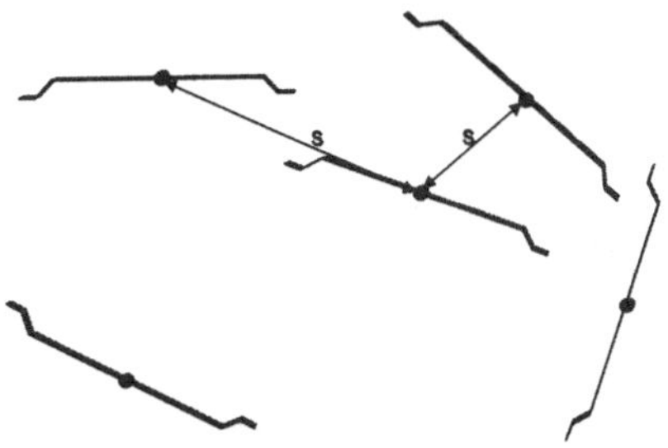

Figure 11 Spacing factor "s" for fibres

It is in fact possible to calculate this "spacing factor" using the "Spacing Theory of McKee[10]" as represented in Equation 1 below. Using Equation 1 for steel fibres and the recommendation that the spacing factor in the placed shotcrete should not be greater than 45%[11] of the fibre length it is then possible to specify a minimum fibre dosage related to any particular fibre aspect ratio as in Equation 2.

Equation 1:

Spacing factor "s" $= \sqrt[3]{\text{(Volume of one fibre/Fibre Volume Fraction)}}$

Substituting into this equation the maximum recommended spacing factor of 0.45l and rearranging it makes it possible to solve for the minimum recommended fibre volume fraction.

i.e. Min. Fibre Volume Fraction = Vol. of 1 fibre/s^3
$= \pi d^2 l/4(0.45l)^3$
$= 8.62d^2/l^2$

where d = fibre diameter
l = fibre length

For steel fibres with a density of 78.5kg/m^3

Fibre Dosage = Fibre Volume Fraction x 78.5 x10^2

Substituting the minimum fibre volume fraction into this equation makes it possible to solve for the minimum recommended fibre dosage:-
i.e. Minimum Fibre Dosage = $8.62d^2/l^2 \times 78.5 \times 10^2$
$= 67658/(\text{Aspect ratio})^2$

Equation 2

Minimum Fibre Dosage = = $67658/(\text{Aspect ratio})^2$

Example:

For a 30mm long fibre 0.4mm x 0.6mm in x-section the equivalent fibre diameter is $(0.4 \times 0.6 \times 4/\pi)^{0.5}$ = 0.55mm

⇒ Fibre aspect ratio = 30/0.55 = 54.5

the minimum fibre dosage $= 67658/54.5^2$
$= \mathbf{23kg/m^3}$

Similarly a 30mm long fibre with an aspect ratio of 25 will have a minimum fibre dosage of $67658/25^2$ = **108kg/m³**.

In both cases a factoring up of these minimum dosages is appropriate to allow for rebound and variations in the as-placed fibre distribution.

REFERENCES:

1. ACI 506R-85 Guide to Shotcrete, *ACI Standard*, ACI, USA 1985
2. ASTM C1018, Standard Test Method for Flexural Toughness and First Crack Strength of Fibre Reinforced Concrete (Using Beam with Third Point Loading), *Annual Book of ASTM Standards V04.02*
3. JSCE-SF4, Method of Test for Flexural Strength and Flexural Toughness of Steel Fibre Reinforced Concrete, *Concrete Library of JSCE*, Japan Society of Civil Engineers.
4. DBV-Recommendation: DBV: Deutscher Beton-Verein E.V. *German Concrete Association.* "Design Principles of Steel Fibre Reinforced Concrete for Tunnelling Works"
5. EUROPEAN SPECIFICATION FOR SPRAYED CONCRETE, *The European Federation of National Associations of Specialist Contractors and Material Suppliers for the Construction Industry (EFNARC)*, 1st Floor, 241 High Street Aldershot, Hampshire GU11 1TJ UK.
6. Sprayed Concrete for Rock Support-Technical Specification and Guidelines, *Norwegian Concrete Association, Publication Number 7 Committee Sprayed Concrete*, 1993.
7. Clements MJK, Measuring the Performance of Steel Fibre Reinforced Shotcrete, *IX Australian Tunnelling Conference, Sydney, Australia, 27-29 August 1996* Published by Institution of Engineers, Australia.
8. Banthia N & Trottier J-F Test Methods for Flexural Toughness Characterization of Fiber Reinforced Concrete: Some Concerns and a Proposition *ACI Materials Journal*, Jan-Feb. 1995.
9. Morgan DR, Chen L & Beaupré D,Toughness of Fibre Reinforced Shotcrete.
10. McKee DC The properties of expansive cement mortar reinforced with random wire fibres PhD Thesis, University of Illinois, Urbana, 1969. Documented in *ACI committee 544 report State of the art report on fibre reinforced concrete 1982.*
11. INFRASTRUCTUUR IN HET LEEFMILIEU Nr. 4 - 1995 *Ministerie van de Vlamse Gemeenschap(Ministry of the Flemish Community), Departement Leefmilieu en Infrastructuur (Department for the Environment and Infrastructure):* Design of concrete structures - Steel wire fibre reinforced concrete structures with or without ordinary reinforcement.

Rock Support and Reinforcement Practice in Mining, Villaescusa, Windsor & Thompson (eds)
© 1999 Balkema, Rotterdam, ISBN 90 5809 045 0

Shotcrete trials and applications in poor ground conditions at Mount Isa Mines

T.Li
WMC Resources Limited, St.Ives Gold, Kambalda, W.A., Australia

B.Cribb
Copper Mine, Mount Isa Mines, Qld, Australia

ABSTRACT: Mount Isa Mines conducted a series of shotcrete trials and used shotcrete for poor ground conditions from 1994 to 1997. The trials and application were extensively monitored and the monitoring results provided considerable information on the quality and performance of shotcrete. The shotcrete trials and the applications have proved that shotcrete is a very effective ground support method in very poor ground conditions and under severe mining induced dynamic and static loadings (mass blasts and stress changes). A number of logistic and quality control issues were identified to improve the quality and performance of shotcrete.

1 INTRODUCTION

Mount Isa Mines have four operating mines and two others under development. The history of mining at Mount Isa, the geology and the mining methods have been extensively described by many authors (Stewart 1980, Hall 1992). The rock masses are highly variable even within the same stoping area, ranging from very competent silicified shales to very weak mylonites and buck quartz. The complex rock masses and the large scale open stoping or bench stoping methods used have presented challenges for mine design, rock mechanics design and mining practices (Matthews 1972, Brady 1977, Harris & Li 1995, Villaescusa et al 1997). Ground support optimisation is one of the areas that continued efforts have been directed to in order to achieve safe and cost-effective support methods for the different ground conditions encountered (Bywater & Fuller 1983, Villaescusa et al 1992).

Poor to very poor ground conditions are often encountered in some areas at Mount Isa Mines. The areas with poor to very poor ground conditions are generally within or below the Basement Contact Zone (BCZ), a fault zone where slaty shales, buck quartz, and carbonaceous mylonites are the dominant rock types. Rapid deterioration of ground conditions, even in the initially competent rock masses can be experienced due to stress changes and blasting, especially near draw points. The conventional ground support methods, such as cables and rockbolts in conjunction with mesh, are sometimes found to be ineffective because of the broken nature of the rock masses or the progressive deterioration of the ground conditions. Time consuming and costly rehabilitation, involving barring down, re-bolting/cabling and meshing, are typical in these areas. Sometimes the conventional support methods can not be used, such as in the very weak backfill masses.

Shotcrete has been recognised as a viable ground support method for poor ground conditions at Mount Isa Mines. The early use of shotcrete dates back to 1964 when gunite, a form of dry mix shotcrete, was used to stabilise pillars. Since then a number of trials and applications of shotcrete were undertaken. However, the use of shotcrete was sporadic and the results were mixed and a large section of the underground workforce developed some scepticism towards the support ability of shotcrete (Li 1995).

In early 1994 when mining commenced in the deeper and more complex areas such as the southern 1100 Orebody, the 3000 Orebody and the 3500 Orebody, shotcrete technology was re-evaluated in searching for viable ground support methods for the poor ground conditions.

The review evaluated the shotcrete technology (Daffy 1971, Brekke et al 1976, American Concrete Institute 1990), and the research and applications within and outside Mount Isa Mines (Petrina 1968, Kaiser 1989, Wood 1989, Zhao et al 1992, Hoek et al 1993). Shotcrete trials were first initiated at the Copper Mine following the review in 1994 and in the following three years, further trials and applications of shotcrete were carried out under a wide range of ground conditions. The objectives for the trials were to:

- Determine the cost effectiveness of shotcrete for ground support in poor ground conditions;
- Test shotcrete specifications under different conditions (eg. draw points, backfill); and
- Develop design criteria and guideline.

An extensive testing and monitoring program was implemented during and after the trials. The quality of shotcrete placement was measured and tested. The performance of shotcrete was monitored and observed regularly. The technical aspects of shotcrete as a ground support method were assessed based on the measurements, observations and instrumentation during the trials.

The following sections describe the shotcrete trials and applications, the monitoring program implemented, the monitoring and testing results, the observations and performance assessment, and the shotcrete guideline developed.

2 ROCK MASS CHARACTERISATION

Two sites were selected for the initial trials, a main decline, W37 Decline, and an access drive with draw points, 18B FWDR. Both sites were situated within the BCZ. The fault zone consists of interbedded slaty shales, buck quartz and carbonaceous mylonites.

The rock mass characterisation at the shotcrete trial sites were assessed using the NGI Tunnelling Quality Index (Q) method (Barton et al 1974). Examination and logging of diamond drill cores were conducted. However, mapping of the exposed rock masses provided more reliable data. Estimates of Rock Mass Rating (RMR) (Bieniawski 1976) were also made for the sites. The intact uniaxial compresive strengths (UCS) were tested at about 60Mpa for the slaty shales; and 80-100Mpa for the buck quartz. The intact UCS for mylonites were estimated between 5-20Mpa. The RQDs for the three dominant rock types were between 0-40 for slaty shales; 20-40 for buck quartz and 0-20 for mylonites. The rock masses in the BCZ were assessed between 0.1 and 0.01 using Q, and 10-25 using RMR.

Figure 1. The unravelled walls and backs after re-bolting at 18B FWDR.

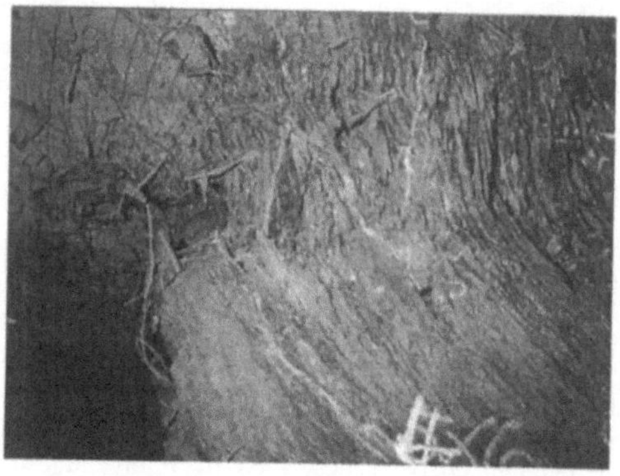

Figure 2. Buckling of rock mass after rehabilitation.

Blast damage and stress changes had caused considerable deterioration of the ground conditions at the trial sites prior to shotcreting (Figure 1). Up to 500mm "squeezing" type of deformation was observed previously in similar conditions.

Figure 1 shows the unravelled walls and backs at 18B FWDR. Three draw points are at the left hand side of the drive and the three stopes that had been extracted and backfilled were 15m above the drive. The drive had been rehabilitated twice before and the size of the drive had increased by 1 to 2 metres.

18B FWDR was rehabilitated to establish access to two stopes to be extracted from the draw points on the right. The ground conditions continued to deteriorate after the rehabilitation and before stope production. Figure 2 illustrates the buckling failure which occurred after rehabilitation.

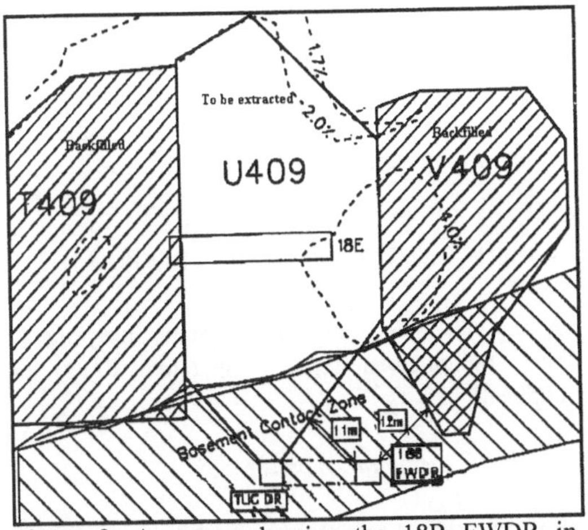

Figure 3. A secton showing the 18B FWDR in relation to the backfilled stopes and the production stope.

More significant deterioraton of the ground conditions were expected during stope producton if there was no additional support. Often the rock mass and support system at draw points deteriorates to the extent that re-support has to be carried out, causing considerable production delays.

Alternative ground support methods were reviewed, such as meshing plus cable lacing, concrete pillars and shotcrete. Shotcreting was considered as a potentially cost-effective method for such difficult ground conditions.

3 SHOTCRETE TRIALS AND APPLICATIONS

3.1 *The initial shotcrete trials*

The initial trials focused on 18B FWDR and W37 Decline. The shotcrete design followed the empirical approaches (Hoek & Brown 1980, Barton 1989, Hoek et al 1993, Clements 1994). 18B FWDR was located at the extraction level of the twin stopes U405/U409 (Figure 4). Figure 3 shows the drive location in relation to the production stopes (U405/U409) and the filled stopes (V409 and T409).

A total of 180.3 m^3 fibrecrete and 42.4 m^3 plain shotcrete were sprayed along an 85 metres drive and in the draw points. The drive size was very irregular ranging from 4 x 4.2m to 6 x 5.5m.

The thickness was designed to be average 150mm within and near the draw points, and 75mm to

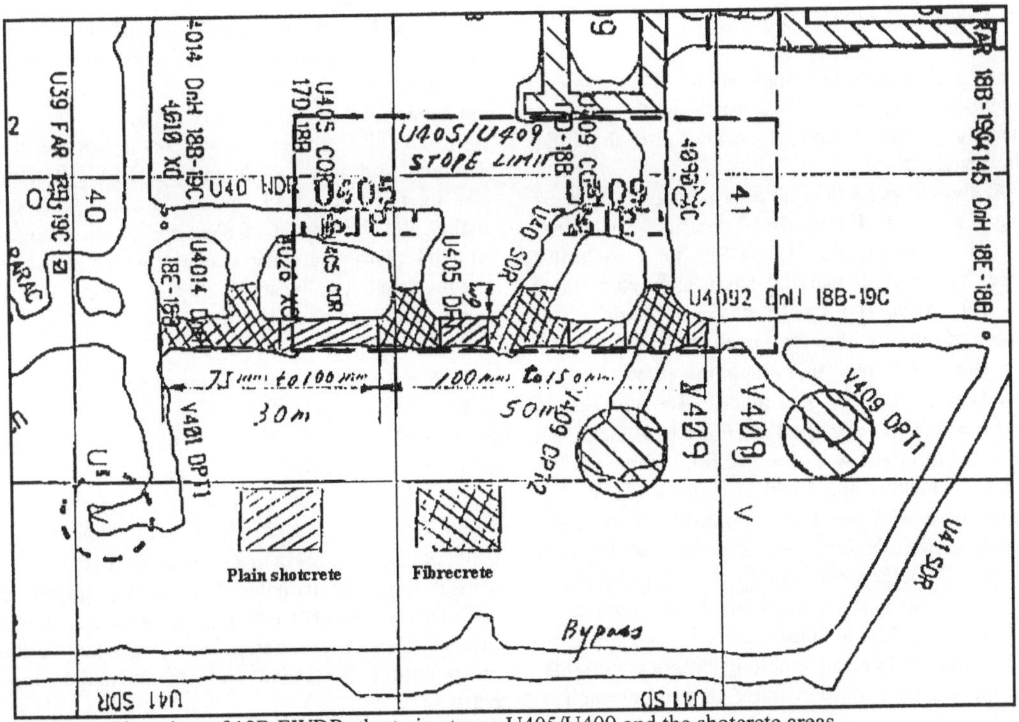

Figure 4. Plan view of 18B FWDR, the twin stopes U405/U409 and the shotcrete areas.

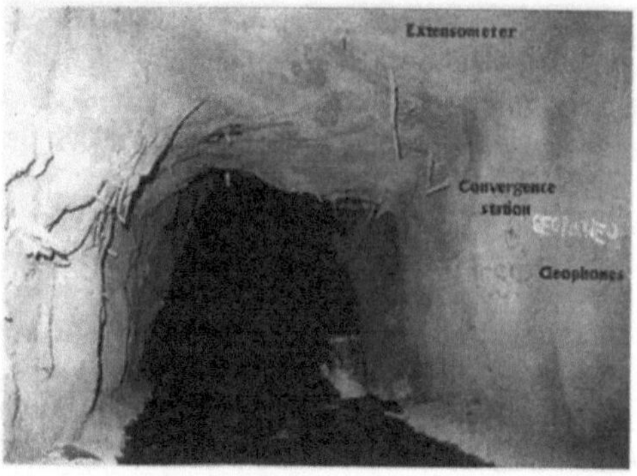

Figure 5. 18B FWDR, after shotcreting and before U405/U409 stope production.

100mm in the drive. The average thickness sprayed was estimated at 140mm and 90mm based on measurements using nails immediately after spraying, the reconciliation of the cubic metres sprayed and the observed rebound. Fibrecrete was sprayed to the walls and the back at two turnouts. Plain shotcrete was sprayed to the back where 150mm x 150mm aperture welded mesh was placed prior to shotcreting. Figure 5 shows the southern end of 18B FWDR after shotcreting and before production.

3.2 *Shotcrete for backfill support*

A number of trials and applications of shotcrete to stabilise and support exposed weak backfill materials were carried out at Mount Isa Mines after the initial trial's success.

At the Copper Mine, an approximately 20 metre long drive with 4m by 4m dimension was mined through cemented slag fill (CSF) in a backfilled stope. The backfill material has a UCS strength of less than 1Mpa. The first 15 metres of the drive was supported only by fibre reinforced shotcrete. The last 5 metres was not supported to allow for the observation of backfill performance. The thickness of the fibrecrete was specified at 70mm to 100mm, but the actual thickness was observed from 10mm to over 150mm. Cracking and buckling failures occurred at the locations with very thin (<25mm) layer shotcrete. Otherwise, the fibrecrete performed very well against the backfill deformation during the excavation, and the stress and blast loadings due to the stoping of nearby stopes (25m to 70m).

At Hilton Mine, an 8m long cross-cut through uncemented rock fill was fibrecreted to re-establish access. Although some cracks developed and small fall-offs occurred at the rock-backfill contact areas, the cross-cut remained stable for its designed life.

Fibrecrete was also used to stablise the exposed low strength backfill in cut-off drives to allow for safety driling and charging operations.

4 SHOTCRETE MONITORING PROGRAM

A comprehensive monitoring and testing program was implemented for the shotcrete trials and applications.

4.1 *Shotcrete quality monitoring*

The mix designs were provided by the contractor. Limited mix quality testing and monitoring, such as the size distribution of aggregate and the percentage of additives in the mix, were conducted.

Slump tests and water/cement ratio checks were carried out by the contractor as a quality control measure.

Rubber rings were placed onto the tails of the protruding bolts and cables to indicate the thickness required. However, these did not work because they were not visible during spraying. Post spraying thickness checks were made using nails and diamond coring.

Limited flexural and tensile strength tests were conducted. A large number of UCS test results were obtained from both a NATA certified laboratory and Mining Research, Mount Isa Mines.

A Schmidt Hammer test was conducted after the shotcrete work was completed on 18B FWDR. This was intended to check the bond quality. Bond quality and the impact of blast and stress changes on bond quality were also made by way of hammer tapping after the blasts at five locations where convergence stations were installed.

4.2 *Shotcrete performance monitoring*

Photographs were taken before and after shotcreting, and before and after major blasts.

The blast vibration near draw points was monitored using a GPX Blastronics monitor and geophones. OYO 8Hz geophones were cast into triaxial geophone modules. Four triaxial geophone modules (G1 to G4) were sprayed into the shotcrete layers. Vibration monitoring was conducted for the initial cut-off blasts and some of the major stope blasts.

Five convergence stations (CS1 to CS5), each

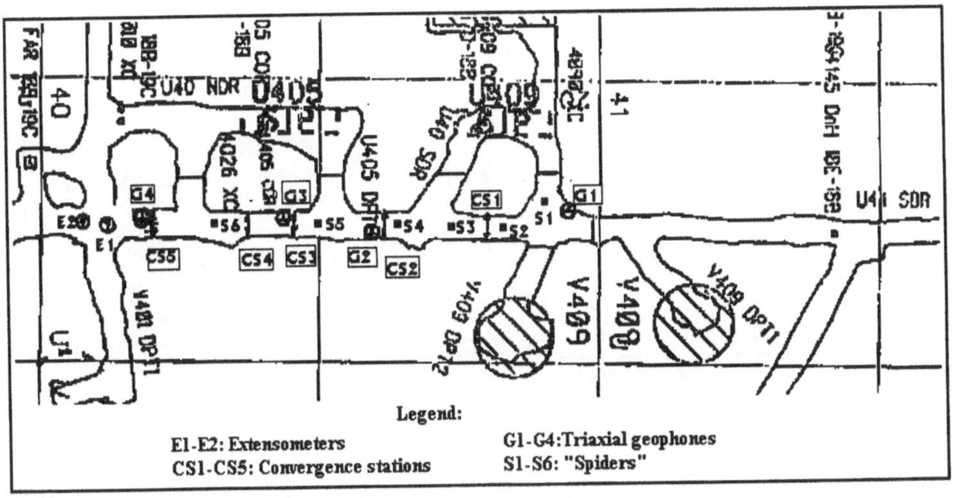

Figure 6. Monitoring locations at 18B FWDR.

consisting of three pins, were installed at 18B FWDR. Regular readings were taken using a tape extensometer. Two short extensometers (E1 and E2) were installed in the turn-out.

Six 'spider' gauges (S1 to S6) were installed to measure shotcrete/rock separation. The monitoring station locations are shown in Figure 6.

5 MONITORING AND TESTING RESULTS

5.1 *Shotcrete strength tests*

Some samples for UCS tests were sourced at the headframe of the shaft before shotcrete was dropped down the pipeline inside the shaft. A large number of samples were obtained from the shotcrete panels cast underground before or during spraying. Figure 7 shows the results from underground shotcrete core tests.

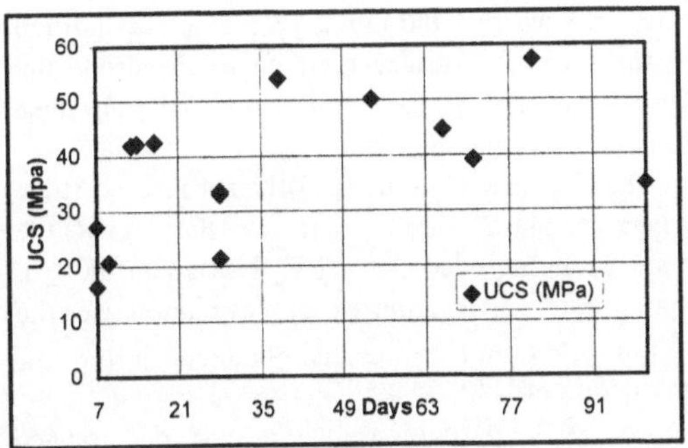

Figure 7. UCS test results.

The results show that there was a large scatter in the UCS strengths. However, the average UCS was very close to the specified strength.

Brazilian tests were done using the core samples obtained from diamond drilling of the cured shotcrete on the walls and backs. The average tensile strength as determined by Brazilian tests was 3.16MPa with standard deviation of 0.64MPa.
An attempt was made to use Schmidt Hammer for in situ strength testing of the cured shotcrete. It was found, however, the reliability of the test results were influenced by a number of factors. The major factors include surface roughness, bond quality, thickness, and aggregate size.

5.2 *Bond quality test and monitoring results*

Bond quality was tested using a hammer to tap the shotcrete surface. A drummy sound indicated poor bond and solid a sound indicated good bond. Grids were painted on the shotcrete surface at five locations where convergence stations were installed. Tapping was done near the grid and repeated after major blasts. The poor bond was detected mainly in the lower part of the walls and was primarily due to greasy surface that could not be cleaned thoroughly prior to shotcreting. Generally, blasting did not seem to affect the bond before any visible cracking, however, cracking occurred and considerable movement was measured at the convergence stations, the bond deteriorated.

Shotcrete layer and rock surface separation was monitored using the 'spiders'. The 'spider' gauges

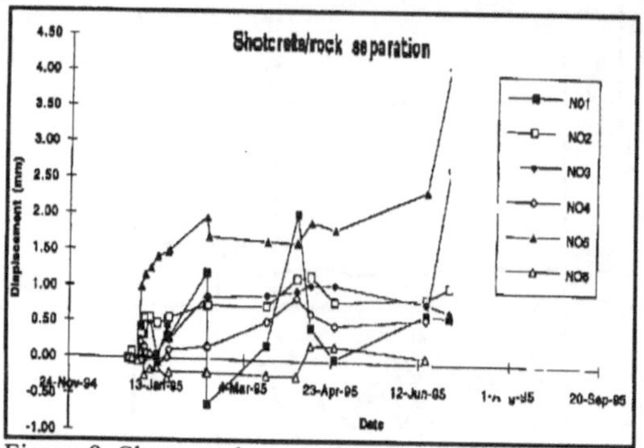

Figure 8. Shotcrete layer and rock surface separation monitoring using 'Spiders'.

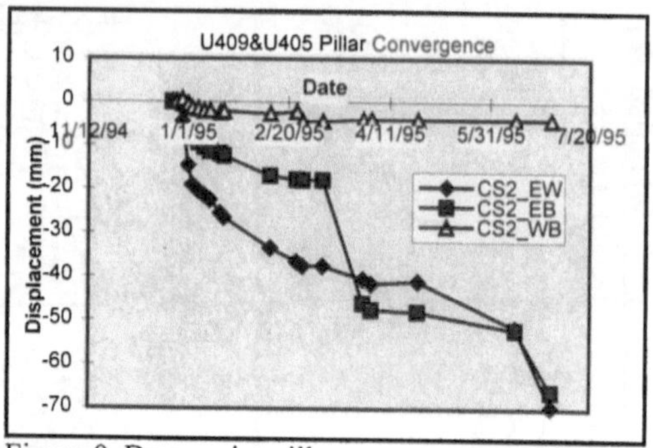

Figure 9. Draw point pillar convergence profiles.

have an inner bar and an outer tube, which is free to slide along the bar. The inner bar was grouted to rock and the outer tube grouted to the shotcrete layer through three 'legs'. Figure 8 shows that the separation between the shotcrete layer and the rock surface might have occurred after the initial trough under cut (TUC) blasts. However, there was evidence that the two distinct increases in separation coincided with the two mass blasts.

5.3 *Thickness measurement results*

Several indicators for shotcrete thickness control were investigated during the trials. Both rubber shrouts and painted nuts were trialed as thickness indicators. They were installed at the protruding bolts and cables prior to spraying. The orange coloured rubber shrouts could not be seen when spraying started due to over spill and poor lighting.

Shotcrete layer thickness was regularly checked soon after spraying using steel nails. Diamond coring of the shotcrete layer was also carried out to check the thickness. The thickness was also estimated from the observed fall-offs, where cross sections of shotcrete layers were exposed.

Generally, a large variation in shotcrete thickness was noticed. For instance, at a few sections, shotcrete layers were only 20mm to 30mm thick compared to the design average of 100 mm. On the other hand, very thick shotcrete layers were exposed after the misfiring of a drainage hole in a draw point, where 250mm thickness was estimated compared to the design average of 150mm.

The major factors influencing thickness control are surface unevenness, nozzleman's skill and meshing quality.

5.4 *Deformation monitoring*

Five convergence stations were installed in 18B FWDR between the draw points. Each station consisted of three pins grouted 500mm into the rock and the shotcrete layer. Convergence measurements were made before and after major stope blasts. Figure 9 shows the typical convergence profiles. Convergences ranging from 15mm to 30mm were measured during the cut-off blast period for the two stations closest to the U409 stope cut-off (CO) slot. The deformation continued with the firing and extraction of the two stopes. However, the two mass blasts did not appear to induce greater movement than the CO blasts. The observed initial cracking occurred when 10mm to 20mm of convergence was measured.

5.5 *Vibration monitoring results*

Vibration monitoring was intended to quantify the dynamic loading due tothe blasting vibration being transfered from the rock mass to the shotcrete layer.

The geophone modules were sprayed into the shotcrete layers on the walls approximately 1.5 metres above the floor. Three geophones, radial (Ri), transverse (Ti), and vertical (Vi), were oriented such that the radial geophone was normal to the shotcrete surface pointing to the drive, the vertical parallel to the surface, and the transverse orthogonal to the other two.

The first few Cut-Off (CO) blasts in each stope where the blasts were colse to the 18B FWDR, the main Trough Under Cut (TUC) blasts and the two mass blasts were monitored. Table1 summarises the vibration monitoring results obtained during the production of U409/U405 twin stopes.

The peak particle velocities recorded ranged between 189mm/s and 1,500mm/s in the shotcrete

Table 1. Vibration monitoring results

Date	Event	Holes or Tonnage (t)	Geophone stations	Distance to source	Maximum ppv(mm/s)
30/12/94	COB1	6 holes	G1&G2	23m	R1:-316
31/12/94	COB2	6 holes	G1&G2	24m	N/A
4/1/95	COB4		G1&G2	30m	R1:-291
11/2/95	MR3-8	115,000	G1&G2	11m-80m	R1:>1500
9/3/95	COB3	6 holes	G3&G4	20m	R3:189
16/3/95	COB6		G3&G4	35m	N/A
17/3/95	TUC1		G3&G4	12m-20m	R3:534
26/4/95	MR8-10		G3&G2	20m-80m	T2:-220
21/6/95	Pillar	108,000	G3&G2	30m-80m	V2:-219

layers. The charge per delay varied from 150kg/delay (TUC blast) to 1,558kg/delay (mass blast). The average charge per delay was about 200kg/delay for CO and TUC blasts, and 600 kg/delay of ANVO (used for its for longer sleeping time than ANFO), and 420 kg/delay of ISANOL50 for the main ring blasts and mass blasts. The maximum particle velocities generated in shotcrete layers by the blasts were estimated from the monitoring results at between 200mm/s and 2,000mm/s in the shotcrete layers.

The measured vibration levels exceeded the damage threshold for concrete structures (Dowding 1985). However, damage to the shotcrete layers was confined to cracking and the limited opening of the cracks.

5.6 *Stress analysis results*

A simplified three dimensional stress analysis using MAP3D was carried out to determine the stress changes at the 18B FWDR due to the extraction of U405/U409 at several stages. An approximately 20MPa stress change was predicted by the stress analysis in the 18B FWDR, while the major principal stress was determined at between 60MPa and 70MPa. However, the actual stress magnitude and stress change would be much lower, given that the rock mass at 18B sub-level is predominantly slatey shales and has a very low rock mass strength and a lower modulus than that used for the modelling (80Gpa).

Figure 10. Shotcrete layer fell off from the lower wall after damage by mucking units, showing layer separation.

6 SHOTCRETE PERFORMANCE ASSESSMENT

6.1 *Overall performance*

Shrinkage cracks developed at limited locations mainly due to poor quality mix. Generally, plain shotcrete tended to develop these shrinkage cracks. Minor cracks (<2mm opening and up to two metre long) appeared after a few cut-off blasts.

Production mucking units then knocked off the lower part of the shotcrete layers near the two draw points in U409 and then the draw point in U405 (Figure 10). More fall-offs of shotcrete scats from the lower part of the walls followed, however, the back of the drive did not show any visible cracks. Poor bond between two layers was evident at the north side of the drive. The poor bond could be due to the fact that the area was sprayed in two passes 24 hours apart and no wetting was done prior to spraying the second pass.

A 140mm drainage hole drilled from 18E intersected the shotcreted north draw point at U405. The hole was mistakenly charged and fired. The shotcrete layer at the back in the draw point was blown out over a 3 metre diameter area. Cracks extended 2 to 3 metres away around the crater. However, the back remained stable throughout the production period. The mesh in the shotcrete layer helped to hold the broken shotcrete layer together.

No production delays were experienced due to any ground condition problem. A total of 530,000 tonnes of ore were extracted from the twin stopes U405/U409 over an 11 months.

In contrast, a hangingwall draw point at the southern end of U405 was not shocreted but was in good conditions prior to stoping. Minor but continuous fall-offs occurred during stoping. Fall-offs measuring 0.5m to 1.5 m near the draw point were observed as indicated by the stripped cables and bolts. Some delays in production from this draw point were experienced.

6.2 *The impact of blast loading*

Vibration levels generated in the shotcrete layers due to major blasts and two mass blasts were recorded (Table 1). Generally, the maximum peak particle velocity (ppv) was the radial component, which was oriented perpendicular to the shotcrete layers and often in the direction pointing to the source. It could be argued that the dynamic stresses induced in shotcrete layers were dominated by shear stress.

The vibration levels expressed in peak particle velocities (ppv) were converted into dynamic shear stress, σ_d, using the following equation:

$$\sigma_d = \rho\, C_s\, ppv \qquad (1)$$

where ρ is the density of shotcrete; and C_s is the S wave velocity.

Assuming that the density of the shotcrete is approximately 2,200kg/m3 and the S wave velocity is 2,000m/s, the dynamic stresses induced in shotcrete layers can be estimated to be between 0.88MPa (for ppv at 200mm/s) and 6.6MPa (for ppv at 1,500mm/s). The shear stress would have been transformed into tensile stress at the surface of the shotcrete layers (Rossmanith 1983). As indicated by the test results, the tensile strength of the shotcrete was about 3MPa. This implies a critical vibration level of 680mm/s, at which cracking due to blasting would occur.

The observations and the monitoring results described in the previous sections support the threshold value for the onset of shotcrete cracking. Since cracking was observed during CO blasts when only a small opening was created, the stress change was negligible. This suggests that the threshold can be taken as a dynamic threshold.

The effects of blasting appeared to diminish in the late stages of the stope extraction, as indicated by a low level of vibration recorded in the shotcrete layers. This could be explained by the deformation and shotcrete/rock separation measurements. As deformation progressed (Figure 9), there was an increasing separation of shotcrete layers and rock surfaces (Figure 8). This separation of the rock mass and shotcrete layer would have reduced the transfer of blast loading from the rock mass to the shotcrete.

6.3 *The impact of stress changes*

Stresses were not monitored for the trial. The stress analysis results indicated high level of stress changes (20 Mpa) and relatively high induced stresses (maximum principal stress of 60-70Mpa). However, such levels of stresses could not be possible due to the low rock mass strength, which would be at 35-40 Mpa. The induced stresses would have dissipated gradually during the stope extraction through deformation. Generally, the tangential stress component is much higher than the normal component in a shotcrete layer, as expected in a rock mass surrounding an opening. Therefore, crushing and buckling of shotcrete layers would be the indicators for the presence of high stresses.

There was no observed crushing or buckling of shotcrete layers in the early stage of the stoping, which could mean that flexural strength of the shotcrete was not exceeded. Tensile cracks were observed after blasts, and these cracks continued to grow in length and opening as stope extraction progressed.

After the stope production and during the backfilling of the stopes, crushing and buckling of shotcrete layers at the top corners in the 18B FWDR drive occurred. Large deformations up to 100 mm were measured during this period. This indicated that there was a stress build up in the shotcrete layers due to large deformations, which were caused by the removal of the support from the broken ore after stope production and by drainage water from backfill.

The measured convergence indicated that the deformation of the drive was mainly due to the deformation of the hangingwall of the drive. The hangingwall side (close to stope void) of the drive was 'pushed' into the drive, while the footwall side (close to filled voids) and the back deformed little. The deformation mode suggests that the pillar between the drive and the stope void was failing. The hangingwall movement typically coincided with the CO blasts and main ring blasts (at convergence stations CS1 and CS2).

It could be argued that low stresses existed in the shotcrete layers, and that the support function of the shotcrete layers was primarily in retaining a broken rock mass. This observation seems to confirm the conclusion from the stress monitoring in shotcrete layers conducted by Myrvang and Stjern (1993). Their monitoring results suggested that very low level stress (<1MPa) is likely to be induced in a thin shotcrete layer (50mm to 75mm) and low level stress (up to 4MPa) in a thick shotcrete layer (>100mm).

6.4 *Fibre reinforced shotcrete vs mesh reinforced shotcrete*

The standard 150mmX150mm and 4mm diameter welded mesh was used. The straight steel fibres used were measured 18mm long, 0.7mm wide and 0.3mm thick with little 'knobs' at each end. The backs at the two trial sites were meshed prior to shotcreting. Plain shotcrete was used on W37 Decline on the backs and walls. Plain shotcrete was used on the backs at 18B FWDR between the draw points and fibre reinforced shotcrete was used on the walls and on the back in the draw points.

It was difficult to assess the performance of fibre reinforced and mesh reinforced shotcrete. However, mesh reinforced shotcrete appeared to have more retaining capacity compared with fibre reinforced shotcrete. Where large deformation and extensive cracking occurred fibre reinforced shotcrete seemed unable to prevent the opening of the cracks. Although other factors might have contributed to the observed behaviour, mesh reinforced shotcrete performed better than fibrecrete under the large deformation conditions.

Shrinkage cracks appeared soon after spraying plain shotcrete. However, no such cracks were observed when reinforced shotcrete was sprayed. Fibrecrete did reduce or eliminate shrinkage cracks.

6.5 *Shotcrete strength, bond and thickness*

Shotcrete strength of 30MPa to 35MPa UCS is normally specified for underground applications, although no rigorous research or trial-and-error tests were conducted. Generally, the actual in-situ strength of the shotcrete layers was determined to be lower than specified. Large variation of the in situ strength could be due to the inconsistency in shotcrete quality. The shotcrete strength within the measured 30-40Mpa range did not seem to affect the performance of shotcrete.

Bond strength did not appear to affect the structural stability and integrity of thick shotcrete (>70mm) when supporting heavily jointed rock masses such as slaty shales. Examination of the shotcrete layers that fell off from the walls indicated that very small pieces of broken rocks adhered to the shotcrete layers. For thin layers of shotcrete, however, bond strength could play a role for the integrity and stability of the support system.

Very significant variation in thickness was observed from the post spraying measurement and from fall-offs. For instance, 10mm to 300mm thick shotcrete layers were observed at 18B FWDR, where 100mm to 150mm thick shotcrete was specified. In the southern end of the 18B FWDR, only 10mm to 30mm fibre reinforced shotcrete was sprayed on broken buck quartz at the lower part of the western wall, as indicated by the spalling. Although the poor bond between the quartz and shotcrete layer could have been a cause of the fall-off, the inadequate thickness would have been the main cause of the premature failure of the shotcrete.

Thickness of shotcrete is the most difficult parameter to determine in shotcrete design and it is also very difficult to control in practice.

6.6 *The cost-effectiveness of shotcrete*

The total costs of shotcreting the draw points at U405/U409 stopes were approximately 2% of the total mining costs and was similar to the total costs of the initial rehabilitations (barring down, bolting and meshing).

Shotcrete costs was slightly lower than the total costs of building concrete pillars. However, time resuired for building concrete pillars would be longer than that required for shotcreting.

The major benefits of shotcreting compared to rehabilitations and other alternative methods were the savings in opportunity costs due to the reduced delays to production, increased availability of access drives, and reduced workforce level.

CONCLUSIONS

The shotcrete trials have proved that shotcrete is a very effective ground support method in poor ground conditions and under severe mining induced dynamic and static loadings (mass blasts and stress changes).

No production delays were experienced at the three draw points that were shotcreted for the stopes U409/U405 extraction (11 months and 530,000t extracted).

Observations and monitoring indicated that thin layers of shotcrete less than 30mm did not provide adequate support as found in many locations and even under low stresses and small deformations.

The specified strength (UCS 35MPa) and the thickness (100mm to 150mm for draw point support, and 50mm to 75mm for drive and decline support) are considered to be adequate for the ground conditions.

The assessment of fibre reinforced shotcrete versus mesh reinforced shotcrete was not conclusive due to the limited trial sites and the uncontrolled conditions. However, it appeared that mesh reinforced shotcrete performed better than fibre reinforced shotcrete when large deformations (>50mm convergence) occurred.

Shotcrete placement quality, in particular thickness control, is of vital importance to achieve the desired performance and low costs. In the absence of viable technology for thickness control, the nozzlemen's skills play a key role.

The experience gained from the trials and applications of shotcrete indicates that the performance of shotcrete is ultimately dependent upon the quality of the shotcrete sprayed. There are opportunities for improvement in the quality control measures and cost reduction.

ACKNOWLEDGEMENTS

The authors would like to thank Mount Isa Mines for permission to publish the materials contained in this paper.

REFERENCES

American Concrete Institute. 1990. Guide to Shotcrete. ACI 506 R-90.

Barton, N. Lien, R., and Lunde, J. 1974. Engineering classification of rock masses for the design of tunnel support, *Rock Mechanics, No. 6*. pp189-236.

Barton, N. 1989. Cavern design for Hong Kong rocks. Proc. Rock Cavern Seminar - Hong Kong. (eds. A.W. Malone and P.G.D. Whiteside). IMM. London. pp179-202.

Bieniawski, Z.T. 1976. Rockmass classification in rock engineering. *Proc Exploration for Rock Engineering*. Cape Town. pp97-106.

Brady. B.H.G. 1977. An analysis of rock behaviour in an experimental stoping block at the Mount Isa Mine, Queensland, Australia. *Int. J. Rock Mech. Min. Sci. & Geomech. Abstr.*. **14**:59-66.

Brekke, T.L., Einstein, H.H. and Mason, R.E. 1976. State-of-the-Art Review of Shotcrete. Final Report to Army Engineer Waterways Experiment Station. U.S. Department of Commence. Rpt. No. AD-A028 031.

Bywater, S. and Fuller, P.G. 1983. Cable support for Lead Open Stope hangingwalls at Mount Isa Mines Limited. *Proc. Int. Symp. On Rock Bolting*. Abisko. Sweden. Rotterdam. Balkema. pp539-556.

Clements, M. 1994. Personal communication.

Daffy, J. 1971. Shotcrete in Engineering construction. Cement and Concrete Association of Australia. Technical Report No. 29.

Hall, B.E. 1992. Mining practices in the 1100 Orebody at Mount Isa. *Proc. Underground Mass Mining Conference.* SAIMM.

Harris, A. and Li, T. 1995. Mining of the Southern 1900 Orebody at Mount Isa. *Proc. 6th Underground Operators' Conf.* Kalgoorlie, WA. pp141-145.

Hoek, E. and Brown, E.T. 1980. *Underground Excavations in Rock.* IMM. London.

Hoek, E., Kaiser, P.K. and Bawden, W.F. 1993. *Design of Support for Underground Hard Rock Mines.* MRD. Sudbury.

Kaiser, P.K. 1989. Use of shotcrete in highly deforming ground. *Proc. of the Seminar on Shotcrete Technology for the Mining Industry.* April 1989. Sudbury, Canada.

Li, T. 1995. Shotcrete trial at the Copper Mine. MIM Technica Report. Ref.:RES MIN 75. 1995.

Matthews, K. 1972. Excavation design in hard and fractured rock at Mount Isa Mines Limited, Australia. *Proc. 8th Canadian Symp. On Rock Mechanics*. pp211-230.

Myrvang, A. and Stjern, G. 1993. Determination of in-situ stresses and adhesivity of sprayed concrete. *Sprayed Concrete. Proc. Int. Symp. on Sprayed Concrete.* (eds. R. Kompen, O.A. Opsahal and K.R. Berg) pp424-438. Fagernes, Norway.

Petrina, A.J. 1968. Ground support at Craigmont. 70th Annual General Meeting of the Institute, Vancouver, April 1968.

Rossmanith, H.P. (ed.). 1983. *Rock Fracture Mechanics*. Spinger Verlag Wien, New York.

Stewart, J.E. 1980. Mining a 100 million tonne orebody without subsidence. *Proc. AusIMM Conference.* Melbourne. pp111-123.

Valencia, F.E. 1973. Practical aspects of shotcrete application. *Use of shotcrete for underground*

structural support. Proc. of the Eng. Foundation Conf. Berwick Academy. Maine. pp114-129.

Villaescusa, E., Sandy, M.P. and Bywater, S. 1992. Ground support investigations and practices at Mount Isa. *Proc. Int. Symp. on Rock Support,* Sudbury Ontario. pp185-193.

Villaescusa, E. Tyler, D. and Scott, C. 1997. Predicting underground stability using a Hangingwall Stability Rating. *Proc. 1st Asian Rock Mechanics Symposium.* (eds. H.K. Lee, H.S. Yang and S.K. Chung). Seoul. pp171-176.

Wood, D.F. 1989. Introduction to shotcrete in mining. *Proc. of the Seminar on Shotcrete Technology for the Mining Industry.* April 1989. Sudbury, Canada.

Wood, D.F. 1992. Specification and application of fibre reinforced shotcrete. *Rock support in mining and underground construction. Proc. Int. Symp. on rock support.* (eds. P.K. Kaiser and D.R. Mcgreath) Rotterdam. Balkema. pp349-376.

Zhao, G., Ren, T. and Wang, C. 1993. Application of monolithic support theory in bolting and sprayed concrete lining of soft rock underground openings. *Sprayed Concrete: Proc. Int. Symp. on Sprayed Concrete.* (eds. R. Kompen, O.A. Opsahal and K.R. Berg). Fagernes, Norway. pp142-152.

Rock Support and Reinforcement Practice in Mining, Villaescusa, Windsor & Thompson (eds)
© 1999 Balkema, Rotterdam, ISBN 90 5809 045 0

The design of shotcrete linings for excavations created by drill and blast methods

Christopher R.Windsor & Alan G.Thompson
Rock Technology, Perth, W.A., Australia

ABSTRACT: The structural design of shotcrete linings has conventionally resulted in the specification of a shotcrete 'thickness' with the implicit assumption that the shotcrete is a prismatic layer of uniform thickness. A new concept of 'shotcrete cover technique' is proposed. This concept abandons the so-called 'thickness' in favour of a more formal specification of 'shotcrete cover dimensions' which allows rock surface irregularities to be explicitly accounted for in the design of shotcrete linings.

1 INTRODUCTION

Shotcrete is an important component in the range of rock improvement techniques used to stabilise civil and mining engineering excavations. Current shotcrete design procedures result in a specification of a so-called 'thickness'. Furthermore, thickness verification is usually attempted by way of shotcrete depth measurement or back calculation involving the volume of shotcrete used. It is suggested here that the assumptions implicit in conventional design procedures and in the verification of in situ dimensions are inappropriate in all cases other than simple smooth, prismatic excavations. In this discussion, thickness is replaced by a new concept of 'shotcrete cover technique' which involves a more formal description of the shotcrete cover dimensions. The concept forms the basis of a series of new structural design procedures being developed for shotcrete, one of which will be discussed here.

2 SHOTCRETE SYSTEM SPECIFICATION

In a previous discussion Windsor (1998) defined the concepts of:

- Shotcrete support systems.
- Shotcrete support schemes.
- Rock surface profiles.
- Cover techniques for shotcrete and mesh.
- Cover dimensions for shotcrete and mesh.
- Cover sequences for shotcrete and mesh.

These concepts are thought to allow formal specification of the thirteen possible variant support schemes involving combinations of mesh, plain shotcrete and fibre-reinforced shotcrete. In this discussion, the structural consequences of the different types of cover technique will be explored for plain and fibre reinforced shotcrete. To facilitate discussion, the concepts of rock surface profiles, shotcrete cover techniques and cover dimensions will be briefly reviewed.

2.1 Rock surface profiles

One of the principal components of a shotcrete system is the rock surface. Consider Figure 1 which shows horse-shoe shaped tunnels created in massive rock, stratified rock and jointed rock. Figure 1a shows the typical result when smooth-wall blasting techniques are employed in massive rock. The other cross-sections show typical results when conventional blasting cuts are employed in the different rock types. In each case, the excavation has a surface roughness which is dependent on the excavation technique and the rock structure. What then is the shotcrete dimensional specification, if in each case the surface needs to be covered with shotcrete? The answer is clearly related to the geometry of the rock surface.

Windsor (1998) has previously described three rock surface profile models (continuous regular, continuous irregular and discontinuous irregular), the two- and three-dimensional methods for measuring rock surfaces by laser and conventional survey methods and the mathematics for defining

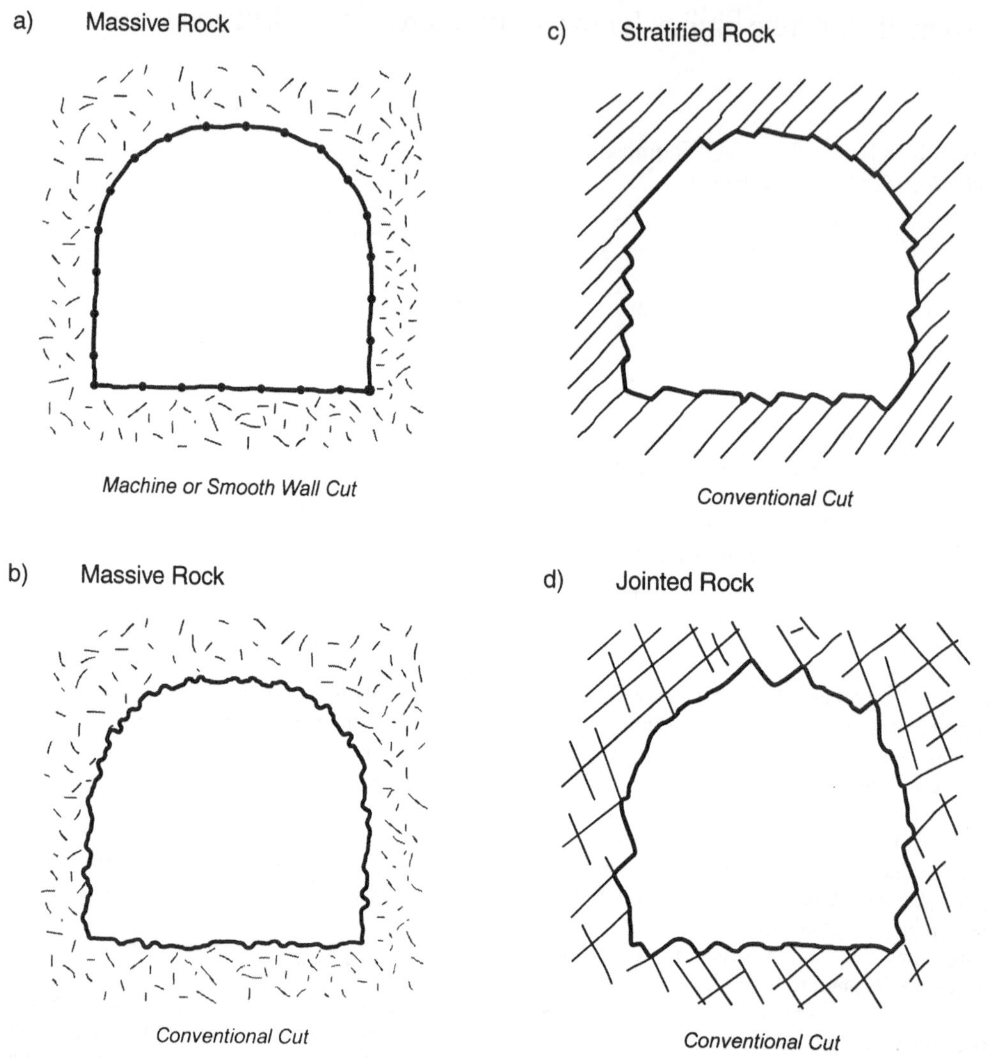

Figure 1. Surface irregularities in a horse-shoe shaped tunnel created in massive, stratified and jointed rock.

the surface and calculating volumes of shotcrete for standard excavation shapes. Of importance here is the general, discontinuous, irregular model for surface roughness. In summary, the surface is assumed to comprise a proportion of irregularities (from here on termed asperities) and a proportion of smooth areas (from here on termed flats).

Any given line in the rock surface (planar or curviplanar) is assumed to sample a discontinuous series of saw-toothed asperities and flats. The asperities are defined by the average depth, width and asperity angle of the notch between them and the flats are simply defined by their length. The profile on a length of line L_s is determined by measuring each asperity 'i' and flat 'j' sampled by the line. This results in an assessment of:

n = The number of asperities sampled in L_s.

d_{av} = The average depth of the ith asperity.

L_{ai} = The length or pitch of the ith asperity.

α_{avi} = The average angle of the ith asperity.

m = The number of flats sampled in L_s.

L_{fj} = The length of the jth flat.

Profiles are sampled in various directions on the rock surface. An average measure of roughness is given by the mean of all the measured profiles.

2.2 *Shotcrete cover technique and dimensions*

The interaction of the shotcrete with the rock surface is critical to shotcrete performance. A definition of the 'Shotcrete Cover Technique' and the 'Shotcrete Cover Dimensions' are required in order for the rock surface to be properly accounted for in design. There are five possible cover techniques for coating and filling a rock surface and these are defined in conjunction with five surface profiles:

- Type 1. Coat the rough rock excavation surface (RE profile) with a minimum thickness to produce a rough coated surface (C profile).
- Type 2. Coat the rough rock surface with an even minimum thickness and partially fill the irregularities or 'notches' to produce a rough coated and filled surface (CF profile).
- Type 3. Fill the rough rock surface to a smooth surface defined by the 'tips' of the rock projections to produce a relatively smooth filled surface (F profile).
- Type 4. Fill the irregularities to an F profile and then apply a minimum thickness cover over the tips to produce a smooth, filled and covered surface (FC profile).
- Type 5. Fill all overbreak and irregularities and continue with covering until a given excavation design surface is achieved to a specified excavation geometry (SE profile).

The five cover techniques and the two-dimensional profiles are shown schematically for a horse-shoe shaped tunnel in Figure 2 together with a magnification of part of the rock surface which shows the associated cover dimensions.

3 SHOTCRETE MECHANICS

3.1 *Shotcrete demand, capacity and equivalence*

The structural design of shotcrete requires that consideration be given to shotcrete 'demand', 'capacity' and 'equivalence'. Demand is defined here as the force-displacement response required to equilibrate a rock mass instability mechanism. Capacity is defined here as the force-displacement response of a shotcrete system to a given arrangement of loadings and boundary configurations. This definition of capacity leads directly to the recognition that there are in fact two forms of capacity:

- Nominal capacity - that associated with contrived testing arrangements of loadings and boundary configurations.
- In situ capacity - that associated with the in situ arrangements of loadings and boundary configurations.

Unfortunately, most of the literature and almost all international codes of practice concentrate on concepts, procedures and methods associated with nominal capacity. This discussion is concerned with balancing the in situ demand with the in situ capacity. This requires that proper attention be given to in situ capacity and the concept of equivalence when representing the actual in situ conditions of the shotcrete. Many excellent texts exist that describe the procedures for the design of shotcrete (e.g. Hoek & Brown, 1980, Brady & Brown, 1985 and Hoek et al., 1995). However, a key to successful implementation of these methods concerns ensuring:

- Equivalence of materials.
- Equivalence of geometry.
- Equivalence of boundary conditions.
- Equivalence of loading conditions.

The equivalence of the materials assumed in design to those actually existing in situ is relatively easy to account for by using quality control and assurance programs during shotcreting. The last three requirements of equivalence are more difficult to satisfy and have not been given sufficient treatment in the literature or in the recommended codes of practice. The effect of incorporating equivalence of geometry into the design of shotcrete linings will be explored in the balance of this limited discussion. How then do the rock surface geometry and the cover technique affect the design of shotcrete linings?

3.2 *Cover technique and shotcrete response*

Two findings result from recognising that the shotcrete system includes the rock surface as a principal component. Firstly, the rock surface geometry and the selected cover technique control the required volume of shotcrete to cover the surface. Thus, the surface geometry must be defined and the relationships connecting the cover dimensions with the required shotcrete volumes need to be determined. The procedures required to conduct these calculations have been given by Windsor (1998).

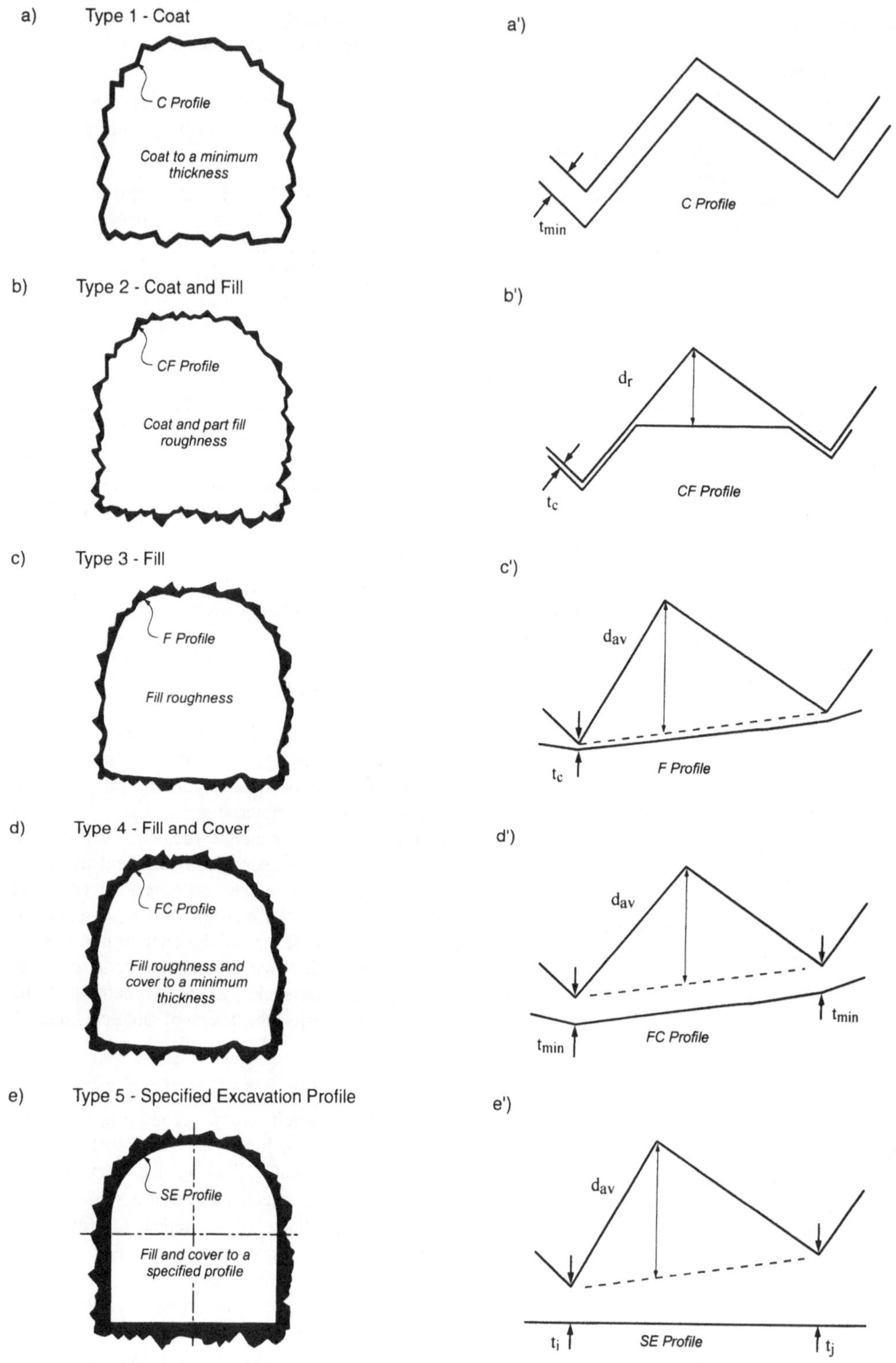

Figure 2. The five shotcrete cover techniques (a to e) and the associated cover dimensions (a' to e').

Secondly, the relationship between the rock surface and the shotcrete affect the loading conditions, the boundary conditions and thus the structural behaviour of the shotcrete/rock composite. Consider, collectively, Figure 2 which shows the five shotcrete cover techniques and Figure 3 which gives a sectional representation through a rough, curviplanar, shotcreted surface under circumferential stress. Now, consider the shear, compression, tension and flexure response of the linings in this stress controlled deformation mechanism.

Radial shear response may occur in the shotcrete for all shotcrete cover techniques. Circumferential compressive and circumferential tensile response in the shotcrete may occur for each cover techniques by stress transfer through the rock to the shotcrete, and vice versa. However, flexural response of the shotcrete can only occur for shotcrete cover techniques 4 and 5 where the shotcrete forms a continuous structural member. Similar observations can be made for structurally controlled deformation mechanisms (Windsor, 1998) where the shotcrete response mechanisms for cover techniques 1, 2 and 3 are suggested to predominantly comprise shear and tension.

Regardless of the so-called thickness specification, it is suggested that in reality most mining and many civil shotcrete projects employ cover techniques 2 and 3. The reader is asked to consider the validity of this assertion and the response mechanisms. If the claims are correct, then the implications are far-reaching in the design, specification, application, testing and verification of shotcrete dimensions.

An example of a stress-controlled deformation mechanism will now be used to demonstrate how the shotcrete cover technique and cover dimensions affect the structural design of a shotcrete lining. The consequences of the cover technique concept in structurally-controlled deformation mechanisms are to be explored in a future publication.

4 ROCK SUPPORT INTERACTION ANALYSIS

The conventional rock support interaction analysis can be used to study the interaction of shotcrete linings with the rock mass for cylindrical excavations created in massive rock under hydrostatic stress conditions. The analysis will be modified here to study what happens when the equivalence of geometry for rough excavation surfaces is taken into account.

4.1 The Rock Mass Characteristic

The calculation scheme chosen here follows that given by Ladanyi (1974) but incorporates the Hoek and Brown rock strength failure criterion. The derivation of the technique and all of the necessary equations are given by Hoek & Brown (1980). Useful extensions and further discussion have been given by Brown et al. (1983).

The method involves solving the two-dimensional differential equation of equilibrium for the instantaneous creation of a long, prismatic cylindrical hole in an elastic continuum under conditions of hydrostatic stress. The solution to the differential equation of equilibrium results in a relationship between the radial stress and radial displacement which, when plotted in radial stress – radial displacement space, produces a relation termed a 'rock mass characteristic'. The design aim is to intersect this rock mass characteristic with a radial stress - radial displacement response characteristic for the shotcrete.

4.2 The Shotcrete Response Characteristic

The rock mass characteristic is dependent on the rock mass properties, the in situ stress and the excavation and may be thought of in terms of 'rock mass demand'. Similarly, the response characteristic for the shotcrete (which is defined by the radial stiffness, stress capacity and displacement capacity of the lining) may be thought of in terms of 'capacity'. In a conventional analysis, the capacity is calculated on the implicit assumption that the shotcrete lining is a cylindrical prismatic annulus. However, previous discussion has suggested that the in situ capacity of a shotcrete lining in a rough excavation is dependent on the rock surface roughness. Thus, the first objective is to find expressions for the radial stiffness, stress capacity and the displacement capacity of the lining taking into account the rock surface roughness, the cover technique and the cover dimensions.

The rock surface and its shotcrete cover will be replaced by an equivalent shotcrete/rock lining. The surface roughness will be assumed to be a continuous, saw-toothed profile with a uniform asperity depth and pitch equal to the mean asperity depth and pitch obtained from measured profiles. The compressive thrust line is assumed to pass mid-way through the shotcrete at the deepest asperity notch. The deepest asperity and the shotcrete cover at the critical point (shown schematically in Figure 3) dictate the radii of the extrados and intrados of the equivalent lining, respectively.

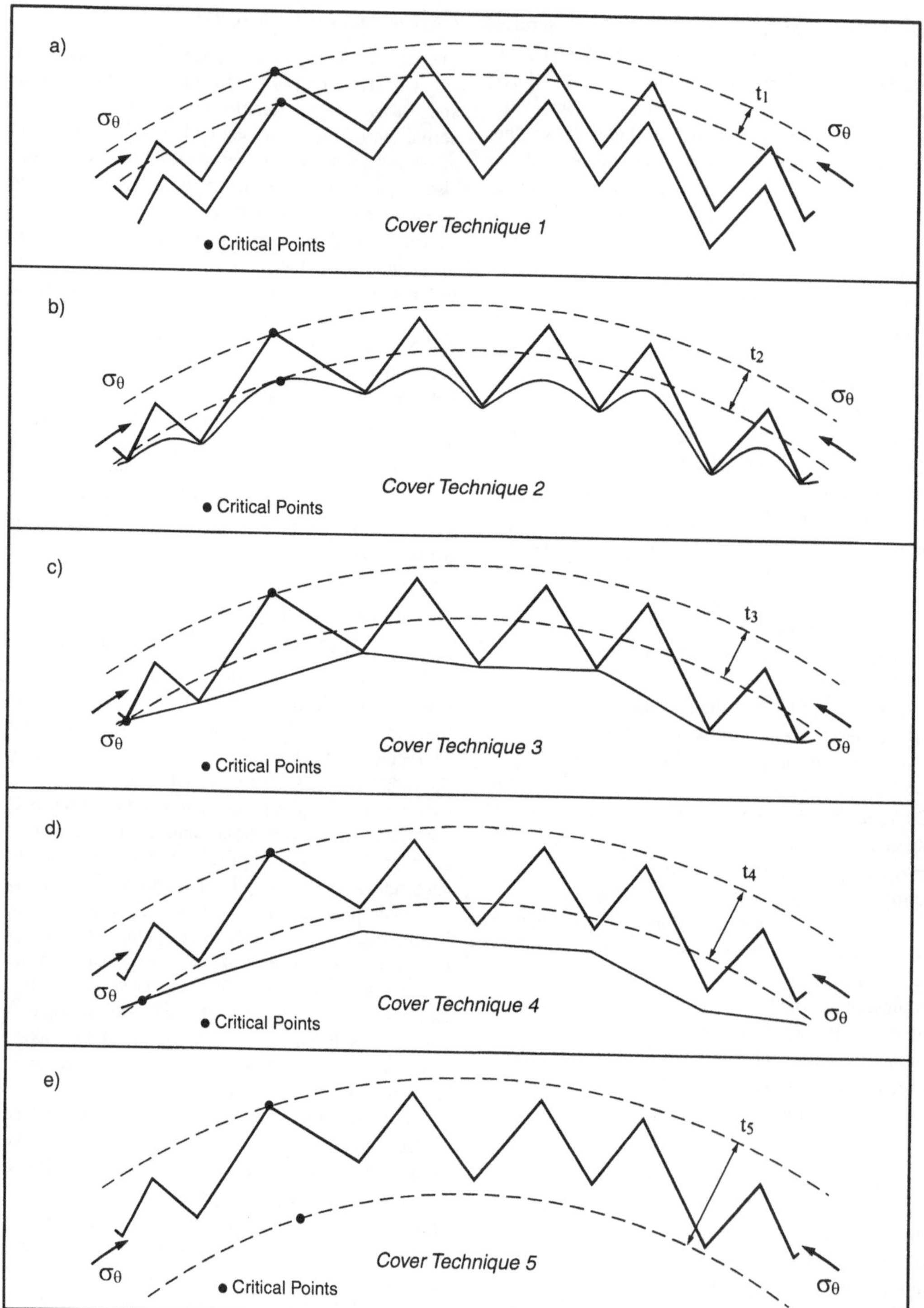

Figure 3. A section through a curved excavation surface shotcreted using the five different cover techniques.

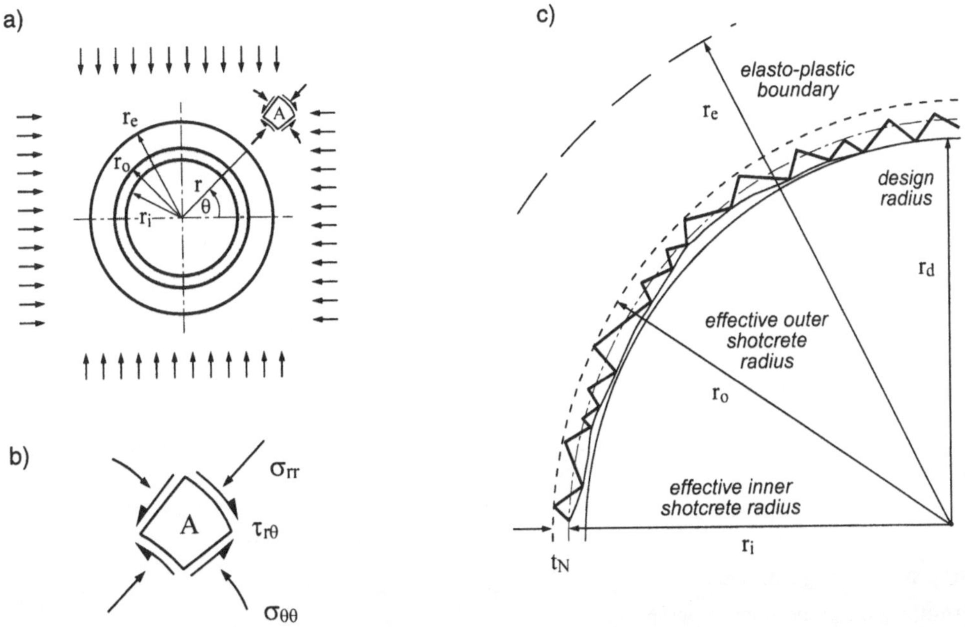

Figure 4. A rough cylindrical excavation undergoing elasto-plastic deformation. (a) A cross-section showing the equivalent shotcrete/rock annulus. (b) The stresses acting on an element of material. (c) Radii of the equivalent shotcrete/rock annulus.

The shotcrete and rock materials are assumed to form a series of sectors around the circumference with their sector widths equal to their width at the thrust line. An excavation undergoing elasto-plastic deformation is given in Figure 4a with the stresses acting on a small element of material shown in Figure 4b. The thickness and radii of the equivalent material annulus are shown in Figure 4c.

4.2.1 *Modulus of Elasticity and Poisson's ratio for an equivalent shotcrete/rock lining*

Simple expressions for the Modulus of Elasticity and Poisson's ratio may be obtained for a lining of equivalent shotcrete/rock material. Firstly, the thickness of the equivalent lining (t_N) is given according to each shotcrete cover technique N:

$$t_1 = t_c/\cos\alpha$$

$$t_2 = d_r$$

$$t_3 = d \qquad (1)$$

$$t_4 = d + t_M$$

$$t_5 = d + t_{SE}$$

where:

t_c = Average shotcrete coat thickness.
α = Average asperity angle.
d_r = Average depth of shotcrete at asperity root.
d = Average asperity depth.
t_M = Average shotcrete cover above asperity tips.
t_{SE} = Average shotcrete cover above asperity tips.

From considering equilibrium of circumferential stress through the assumed line of thrust:

$$E_{EQ} = \frac{1}{\left(\dfrac{u}{E_r} + \dfrac{v}{E_s}\right)} \qquad (2)$$

$$\nu_{EQ} = \frac{u\nu_r + v\nu_s}{(u + v)} \qquad (3)$$

where:

E_{EQ} = Modulus of Elasticity of equivalent material.
E_r = Modulus of Elasticity of rock.
E_s = Modulus of Elasticity of shotcrete.
ν_{EQ} = Poisson's ratio of equivalent material.
ν_r = Poisson's ratio of rock.
ν_s = Poisson's ratio of shotcrete.
u = Proportion of thrust line comprising rock.
v = Proportion of thrust line comprising shotcrete.

The proportion of thrust line comprising shotcrete (v_N) and that comprising rock (u_N) can be determined for each of the N cover techniques according to the cover dimensions:

$$v_1 = \frac{t_c}{2d}\sqrt{1+\left(\frac{2d}{L}\right)^2} \quad (4)$$

$$v_2 = \frac{d_r}{2d} \quad (5)$$

$$v_3 = \frac{1}{2} \quad (6)$$

$$v_4 = \frac{1}{2}\left(1+\frac{t_M}{d}\right) \quad (7)$$

$$v_5 = \frac{1}{2}\left(1+\frac{t_{SE}}{d}\right) \quad (8)$$

where:

L = Average asperity length or pitch.

Now, substituting the proportional relationship

$$u_N = 1 - v_N \quad (9)$$

enables equations 2 and 3 to be used to estimate the equivalent material Modulus of Elasticity (E_{NEQ}) and Poisson's ratio (ν_{NEQ}) for each cover technique N:

$$E_{NEQ} = \frac{1}{\left(\frac{1}{E_r} + v_N\left(\frac{1}{E_s} - \frac{1}{E_r}\right)\right)} \quad (10)$$

$$\nu_{NEQ} = \nu_r + v_N(\nu_s - \nu_r) \quad (11)$$

4.2.2 Radial stiffness of an equivalent shotcrete/rock lining

The radial stiffness of the shotcrete lining for each of the five cover techniques is found by substituting the expressions for t_N, v_N, E_{NEQ} and ν_{NEQ} into an equation for the radial stiffness of a cylindrical lining in a smooth cylindrical excavation subjected to radial compressive stress. That is:

$$k_{NEQ} = \frac{E_{NEQ}(r_N^2 - (r_N - t_N)^2)}{(1+\nu_{NEQ})\{(1-2\nu_{NEQ})r_N^2 + (r_N - t_N)^2\}} \quad (12)$$

where:

k_{NEQ}= Stiffness of shotcrete/rock equivalent lining.

$$r_N = r_i + d \quad (13)$$

= Radius of shotcrete/rock extrados.

and:

r_i = Radius of shotcrete intrados.

4.2.3 Radial stress capacity of an equivalent shotcrete/rock lining

The radial stress capacity (σ_{NEQ}) of an equivalent shotcrete/rock lining of cover technique N, in a rough cylindrical excavation is given by:

$$\sigma_{NEQ} = \tfrac{1}{2}\sigma_c\left[1 - \frac{(r_N - t_N)^2}{r_N^2}\right] \quad (14)$$

where σ_c = Compressive strength of shotcrete/rock lining.

4.2.4 Radial displacement capacity of an equivalent shotcrete/rock lining

The radial displacement capacity (δ_{NEQ}) of a shotcrete lining, of cover technique N, in a rough cylindrical excavation is given by:

$$\delta_{NEQ} = \frac{\sigma_c r_N}{k_{NEQ}} \quad (15)$$

4.2.5 Circumferential strength of an equivalent shotcrete/rock lining

The circumferential compressive strength of the equivalent lining (σ_c) may be found by considering how circumferential stress is transferred though the composite shotcrete/rock ring. The compressive strength may be controlled by either failure though the rock, failure through the shotcrete or shear failure at the shotcrete/rock interface. For example, consider a lining of cover technique number 3 (i.e. the notches are completely filled with shotcrete) under a circumferential compressive stress (σ_θ).

Firstly, failure may occur in the rock adjacent to the shotcrete, in which case:

$$\sigma_c = \sigma_{rc} \quad (16)$$

where σ_{rc} = Compressive strength of rock.

Alternatively, failure may occur in the shotcrete, in which case:

$$\sigma_c = \sigma_{sc} \quad (17)$$

where σ_{sc} = Compressive strength of shotcrete.

Thirdly, failure may occur as a shear failure at the shotcrete/rock interface on the asperity wall. From equilibrium considerations, the shear stress acting on the shotcrete/rock interface (S_c) is given by

$$S_c = \sigma_\theta \cos\alpha \sin\alpha \quad (18)$$

and the normal stress acting on the interface (N_c) is given by:

$$N_c = \sigma_\theta \sin^2 \alpha \quad (19)$$

Shear bond failure occurs when:

$$S_c > \tau_{srb} \quad (20)$$

where τ_{srb} = Shotcrete/rock shear bond strength. (Note that τ_{srb} will be a function of the bond between the rock and shotcrete, the local surface roughness and the normal stress).

Substituting equation 18 into equation 20 gives:

$$\sigma_\theta > \tau_{srb} / \cos \alpha \sin \alpha \quad (21)$$

Setting σ_c to the minimum value of σ_θ required to cause failure results in:

$$\sigma_c = \tau_{srb} / \cos \alpha \sin \alpha \quad (22)$$

In summary, the compressive strength of the composite shotcrete/rock lining is given by the minimum of equations 16, 17 and 22.

5 AN EXAMPLE OF SHOTCRETE DESIGN FOR A ROUGH CYLINDRICAL EXCAVATION

Consider the problem of estimating the volume of shotcrete needed to support a 500 m deep, vertical, cylindrical shaft. The shaft is excavated by drill and blast methods with a required minimum clearance diameter of 4 m. The in situ stress field is approximately hydrostatic, increasing linearly with depth with a magnitude equal to that due to the weight of overburden rock. The properties of the rock mass are given in Table 1.

The geometry of the blasted rock surface is characterised by a continuous series of asperities given by the surface sampling results in Table 2. Fibre reinforced shotcrete with the mechanical properties given in Table 3 and the cover dimensions given in Table 4 will be used to support the shaft.

5.1 Rock Mass Characteristic for the example excavation

A shaft problem has been chosen in order to show that the rock mass demand and thus the required capacity of the support system change with depth. The rock mass characteristic surface, or demand surface, is shown in radial displacement - radial stress - depth space in Figure 5. This surface comprises the radial displacement - radial stress characteristics for depths z, where $0 < z < 500$ m. For example, consider Figure 6 which shows the rock mass characteristic at $z = 370$ m where the magnitude of the hydrostatic stress field is conveniently about 10 MPa. The rock mass characteristic at this particular depth will be used to explore a number of features of the shotcrete/rock system behaviour when rock surface roughness is taken into account during shotcrete design.

Table 1. Rock mass material properties for the example excavation problem.

Rock Mass Property	Value
Uniaxial compressive strength (σ_{rc})	100 MPa
Material constant (m)	1.5
Material constant (s)	0.00025
Modulus of elasticity (E_r)	10 GPa
Poisson's ratio (ν_r)	0.25
Rock unit weight (γ_r)	27 kN/m^3
Fractured rock material constant (m_r)	0.25
Fractured rock material constant (s_r)	0.0001
Fractured rock unit weight (γ_f)	20 kN/m^3

Table 2. Rock surface geometry properties for the example excavation problem.

Rock Surface Geometry	Value
Proportion of asperities in surface	1
Proportion of flats in surface	0
Average asperity length or pitch (L_{av})	450 mm
Average asperity depth (d_{av})	200 mm
Radius - highest asperity tip (r_{dmin})	2.050 m
Radius - deepest asperity root (r_{dmax})	2.250 m

Table 3. Shotcrete material properties for the example excavation problem.

Shotcrete Property	Value
Uniaxial compressive strength (σ_{sc})	50 MPa
Shear strength (τ_{sc})	20 MPa
Shear bond interface strength (τ_{srb})	9 MPa
Modulus of elasticity (E_s)	20 GPa
Poisson's ratio (ν_s)	0.25

Table 4. The five shotcrete cover techniques used to line the example shaft.

Cover Technique	Average Cover Dimensions	Value
I	Coat thickness (t_c)	50 mm
II	Depth at asperity root (d_r) Coat thickness (t_c)	100 mm 25 mm
III	Depth at asperity root (d)	200 mm
IV	Cover on asperity tips (t_M)	25 mm
V	Cover on asperity tips (t_{SE})	50 mm

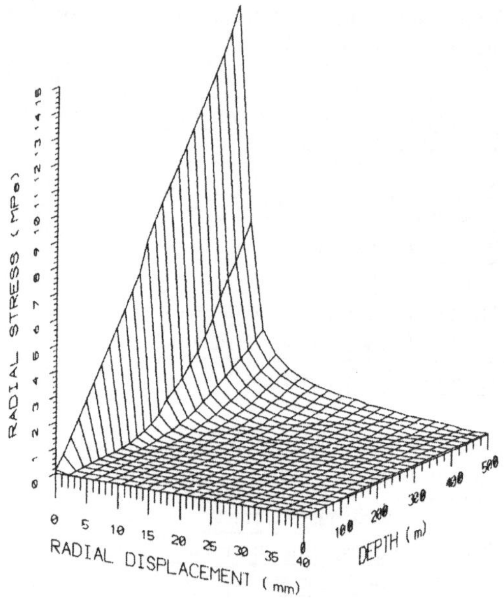

Figure 5. Rock mass characteristic surface.

5.2 Conventional shotcrete thickness design

Firstly, consider the shotcrete support response which results from the conventional assumptions of a perfectly prismatic, annular shotcrete lining applied to a perfectly smooth, prismatic circular shaft.

Figure 7 shows the rock mass characteristic at z = 370 m together with the radial stiffness, radial stress capacity and radial displacement capacity relationships obtained under the conventional assumptions for a series of shotcrete linings with thicknesses of 25 mm, 50 mm, 75 mm, 100 mm and 125 mm. In each case, the radius of the shotcrete extrados is set to 2.15 m which is the mean radius of the excavated shaft.

5.3 Shotcrete cover technique design

Now consider Figure 8 which shows the radial stiffness, radial stress capacity and radial displacement capacity that result when the roughness of the shaft is taken into account by implementing the equations given in Section 4 for the particular details set out in Tables 1 to 4. Figure 8 shows that at a depth of 370 m the dimensions specified for shotcrete cover techniques I and II do not produce a shotcrete response that intersects the rock mass characteristic whereas those for shotcrete cover techniques III, IV and V do. It also shows that the stiffness of the shotcrete/rock system increases as the notches in the rock surface are progressively filled with shotcrete. Furthermore, the radial stiffness, radial stress capacity and radial displacement capacity of the shotcrete/rock system continue to increase as additional shotcrete is applied over the filled surface. However, how does one dimension the shotcrete required to support the entire shaft? Clearly, as the rock mass characteristic changes the shotcrete demand changes.

The dimensions given above for shotcrete cover technique III could be specified over the complete shaft depth but would be found to be grossly conservative in the range $0 < z < 300$ m and inadequate in the range $400 \text{ m} < z < 500$ m. For example, Figure 9 shows that at a depth of 500 m cover technique V would be required.

Surprisingly, the simplest way to think of this optimisation problem is to consider it in three dimensions. The rock mass characteristic for the shaft was given previously as a surface in Figure 5. Thus the shotcrete response required to intersect this surface is also a surface. If a linear radial stiffness model is retained for the shotcrete response then the response surface is a plane. In this case the required response surface intersects the rock mass characteristic surface to form a line. We seek the equation to this line and the specification of shotcrete dimensions to achieve the line. However, there is an additional complication associated with the initial conditions involving the time taken to apply and cure the shotcrete.

To date, the application of shotcrete and the development of the peak mechanical capabilities of the shotcrete have been assumed to occur instantaneously and to coincide with the onset of stress redistribution and displacement of the rock mass. This does not occur in practice because a component of stress redistribution occurs immediately after the excavation is advanced one cut and it takes a given time for the shotcrete to be applied and then to cure, during which time further stress redistribution will have occurred. It is also possible that the face may have advanced before the shotcrete reaches its peak mechanical performance. In essence, these delays collectively result in the rock mass characteristic being intersected by the shotcrete response relation after a greater amount of stress redistribution and radial displacement has occurred. Figure 10 shows the effect of delaying shotcrete application and the development of peak mechanical performance until 1 mm of radial displacement of the rock mass.

Such considerations complicate design and lead directly to the requirement for an excavation and support strategy that manages the timing of

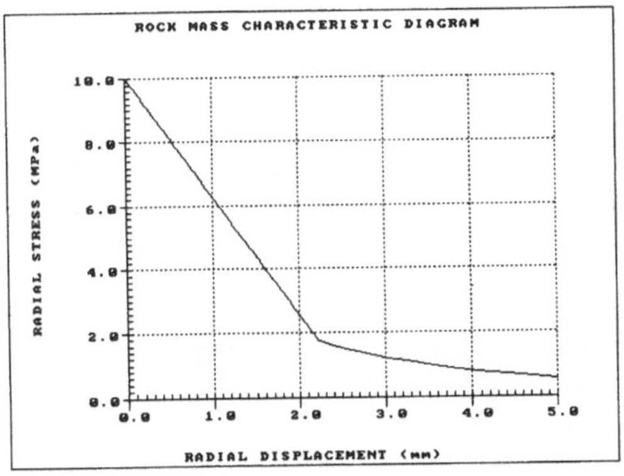

Figure 6. The rock mass characteristic for the circular shaft example at a depth of 370 m.

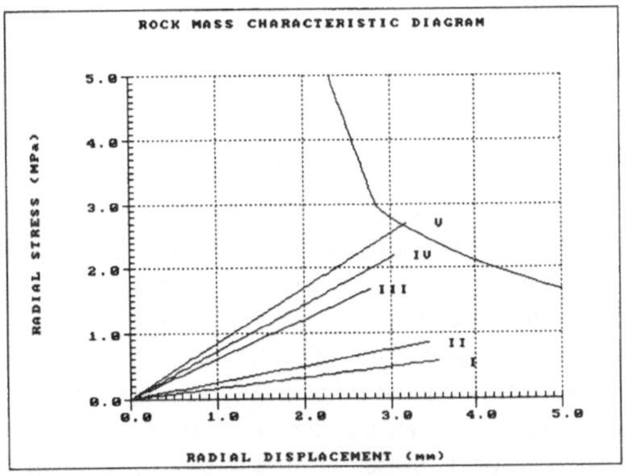

Figure 9. The rock mass characteristic line at a depth of 500 m and the radial response relations for the five shotcrete cover techniques.

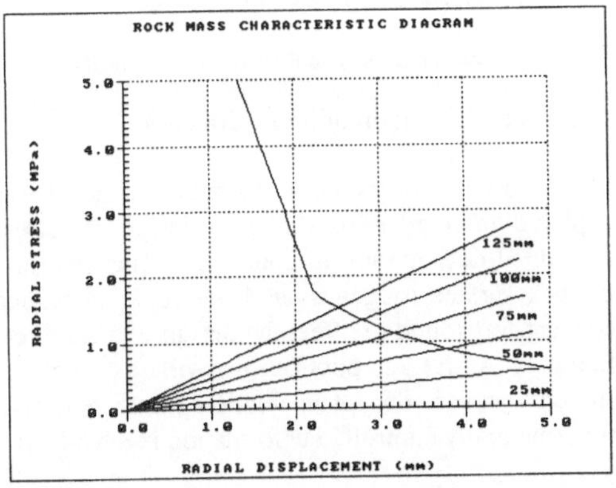

Figure 7. The rock mass characteristic (at z = 370 m) and the conventional radial response relations for prismatic, cylindrical, elastic shotcrete rings.

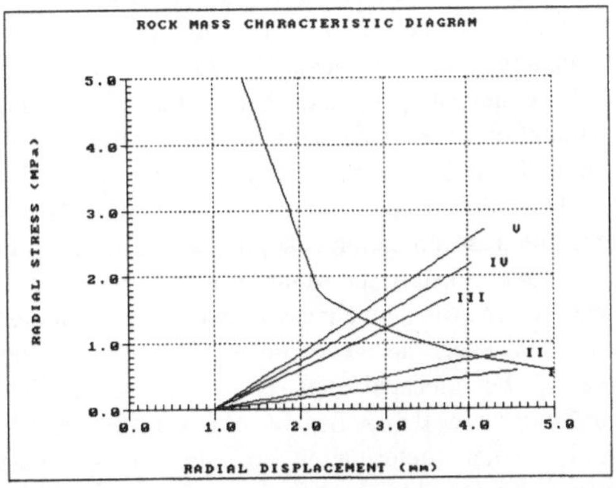

Figure 10. The rock mass characteristic (at z = 370 m) and the shotcrete cover response relations. Response is activated at 1 mm of radial displacement.

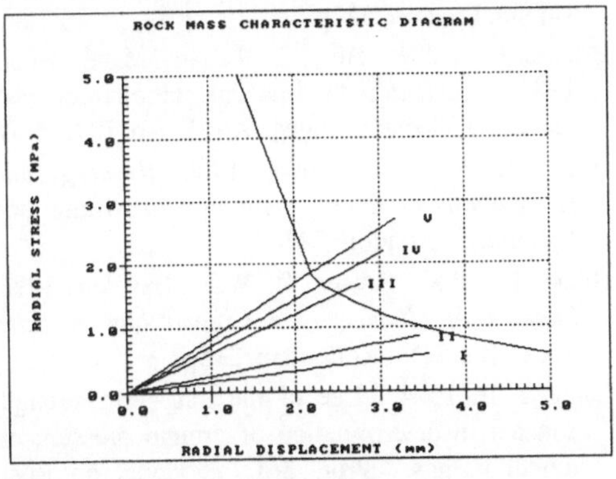

Figure 8. The rock mass characteristic (at z = 370 m) and the radial response relations for the five shotcrete cover techniques where 100% of the rock surface is rough.

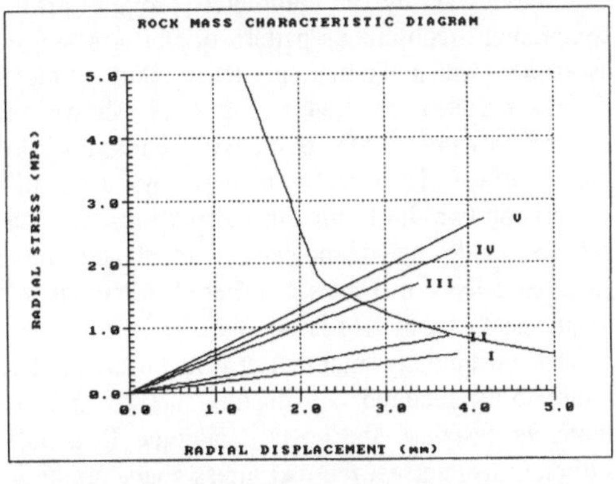

Figure 11. The rock mass characteristic (at z = 370 m) and the response relations for the five shotcrete cover techniques where the rock surface is 33% rough and 66% smooth.

excavation advance and the application of shotcrete support. This strategy would need to be based on in situ measurement of the radial deformation of the rock or the diametral convergence of the excavation. In simplistic terms, given a set of initial conditions and a limit on the allowable radial displacement at depth z, a 'required design line' can be drawn on the characteristic surface. For example, assume that the shotcrete (incorporating accelerating and curing agents to speed the development of material properties) is applied immediately after each cut and that 1 mm of radial displacement occurs before the activation of response. If the radial displacement is limited to 4 mm at z = 500 m, then the characteristic surface would be intercepted by shotcrete dimensions of cover technique II at z = 0 increasing uniformly to cover technique III at z = 500 m. The volume of in-place shotcrete required for the shaft is calculated to be 487.5 cubic metres using the procedure given by Windsor (1998).

It is interesting to note that if the conventional assumptions are made, then a lining thickness of about 70 mm to 75 mm would apparently limit radial displacement to about 4 mm at z = 500 m. If this 'thickness' specification is applied to the entire shaft then approximately the same volume is required as that for when the roughness is explicitly accounted for. However, the theory presented here indicates that if the allocated shotcrete volume is applied uniformly over the shaft, it would be insufficient to limit radial deformation of the rock mass surrounding the shaft in the interval 250 < z < 500.

5.4 *Effect of rock surface roughness*

In the previous example, the rock surface roughness was modelled by a completely rough profile comprising a continuous pattern of uniform shaped asperities which results in a roughness factor (Windsor, 1998) of 1.34. Figure 11 shows the shotcrete response for the five cover techniques if the rock surface is now modelled by a profile comprising two thirds smooth flat portions, with the balance made up of notches. If all the other parameters listed in Tables 1 to 4 are held constant, a roughness factor of 1.11 is calculated.

The substantially smoother profile means that less shotcrete is required to fill the surface and the shotcrete response for cover technique II is now sufficient to intersect the rock mass characteristic at z = 370 m. The shotcrete required in the metre strip 369.5 m < z < 370.5 m would be 0.43 cubic metres compared with 1.35 cubic metres for the rougher case. The implications for blasting practice are clear.

6 CONCLUSIONS

The conventional concept of a shotcrete 'thickness' and its meaning in relation to the rough excavation surfaces characteristic of excavations created by drill and blast methods have been questioned. For these excavations, thickness has been abandoned and replaced by a new concept of the shotcrete cover technique and associated cover dimensions. This concept was introduced in an attempt to improve the structural design of shotcrete linings.

The cover technique concept is thought to be of fundamental importance to:

1. The structural design of shotcrete linings.
2. The specification of shotcrete linings.
3. The application procedures for shotcrete.
4. The test measurement of in situ capacity.
5. The verification of lining dimensions.

In this discussion, the cover technique concept was explored in relation to shotcrete design for stress controlled deformation mechanisms. The inclusion of rock surface roughness and the cover technique concept was found to affect the design of a shotcrete lining. A future publication will explore the shotcrete cover concept in relation shotcrete design for structurally controlled deformation mechanisms.

REFERENCES

Brady, B.H.G & E.T. Brown 1985. *Rock Mechanics for Underground Mining*. George Allen and Unwin: London, 527p.

Brown, E.T., J.W. Bray, B. Ladanyi & E. Hoek. 1983. Characteristic line calculations for rock tunnels. J. Geotech. Engng., ASCE, No. 109, 15-39.

Hoek, E & E.T. Brown 1980. *Underground Excavations in Rock*. Institution of Mining and Metallurgy: London, 527p.

Hoek, E., P.K. Kaiser & W.F. Bawden 1995. *Support of Underground Excavations in Hard Rock*. Balkema: Rotterdam, 215p.

Ladanyi, B. 1974. Use of the long-term strength concept in determination of ground pressure on tunnel linings. Proc. 3rd Int. Cong. on Rock Mech., 1150-56. Nat. Acad. Sci: Washington.

Windsor, C.R. 1998. Structural design of shotcrete linings. Proc. 1998 Aust. Shotcrete Conf., IBC Conferences: Sydney. 1-34.

4 Support and reinforcement in metalliferous mines

Rock Support and Reinforcement Practice in Mining, Villaescusa, Windsor & Thompson (eds)
© 1999 Balkema, Rotterdam, ISBN 90 5809 045 0

Keynote lecture: The reinforcement process in underground mining

E.Villaescusa
Western Australian School of Mines, Kalgoorlie, W.A., Australia

ABSTRACT: This paper proposes a framework for reinforcement design in which the initial step is to determine the factors controlling stability of unsupported spans. Reinforcement schemes likely to match expected ground behaviour within an overall mining process are then described. Suggested practical measures, including documentation, which can be considered following a rock fall are discussed. Practical ways to optimize reinforcement schemes on an actual underground mining environment are considered. Instrumentation, laboratory tests and quality control issues required during an optimization program are also analyzed.

1 INTRODUCTION

The purpose of rock support and reinforcement is to ensure excavations remain safe and open for their intended life span. The effectiveness of a reinforcement strategy is important for two main reasons, these being safety to personnel and equipment and achievement of the most economical access to extract ore. Within this context the use of the most economical and efficient process for rock reinforcement is extremely important in order to achieve the best practices while maintaining safety and reducing operating cost.

2. GENERAL THEORY OF REINFORCEMENT

The type of support and reinforcement needed in a particular location is dependent on several factors including the available rock mass strength, the geometry of the excavation, the stresses present in the rock, the blasting practices and the weathering process. The reinforcement simply supplements and controls the load carried by the surrounding rock, a majority of which is carried within the rock mass itself. Two stabilization techniques are used to improve and maintain the load bearing capacity of a rock mass near the boundaries of an underground excavation:

- Rock Reinforcement - where the supporting members are an integral part of the reinforced rock mass, e.g. grouted bolts, split sets and cable bolts.

- Passive Support - where the supporting members are external to the rock and respond to inward movement of the rock surrounding the excavation, e.g. mesh, straps and shotcrete.

The reinforcing elements provide effective stabilization by helping a rock mass to support itself. This is achieved by preventing unraveling and enhancing the self-interlocking properties of a rock mass. A reinforcement pattern strengthens the exposed rock mass around an excavation by preventing the detachment of loose blocks, and by increasing the shear strength of the geological discontinuities intersected by the reinforcing elements. This results in a reinforced zone that helps to re-distribute stresses around the excavations and also minimizes dilation of the pre-existing geological discontinuities. Careful blasting and correct scaling reduce the amount of loose rock which has to be supported, thus enhancing the self-stabilizing behavior of a rock mass.

In most underground mines the primary form of excavation stabilization is provided by the reinforcement pattern installed within the rock mass. Passive support, such as that provided by mesh, is only needed to provide surface restraint at the excavation boundaries. The reinforcement controls the overall excavation stability through keying, arching or composite beam reinforcing actions, while the mesh supports the small loose pieces of rock detached within a bolting pattern. Adequate scaling will reduce the potential amount of loose rock, thus minimizing the work carried out by the mesh. In most cases, mesh effectiveness is a function of the reinforcement elements appropriately matching the observed rock mass behaviour throughout the life of the excavation. If this is achieved, the development of excessive loads in the mesh is likely to be minimized. Timing of the mesh installation is critical in order to minimize damage from blasting from a closely located advancing excavation face.

3 THE PROCESS OF SUPPORT AND REINFORCEMENT

A generalized approach to support and reinforcement is presented in the flow chart shown in Figure 1. Once an excavation has been created (or proposed), the first step in the process is to consider the stability of the unsupported spans. The general factors controlling unsupported behavior are the excavation geometry (shape, size and orientation), the rock mass strength, groundwater, induced stresses and blasting practices. Likely failure modes such as gravity fall, plane and wedge shear, buckling, spalling, bursting, toppling, swelling, etc. should also be determined at this stage. An assessment of unsupported stability must consider the potential size of the exposed in-situ blocks with respect to the exposed spans.

If unsupported instability is detected, the excavation can either be re-designed (e.g. to have smaller spans or different shape and orientation) or alternatively, the excavation must be reinforced. In the latter case, a reinforcement scheme likely to match the expected ground behavior determined during the initial unsupported assessment must be selected. The loading conditions from mining are critical and factors such as long term exposure (weathering), or whether deep reinforcement may be needed must be decided in order to choose the best support and reinforcement elements.

Following the selection of a reinforcement scheme, an assessment of stability is needed. Empirical methods or experience under similar rock mass conditions are used to provide an initial overall assessment of excavation stability under the reinforcement scheme chosen. In cases where the chosen reinforcement design achieves stability, optimization with respect to the overall excavation productivity can be achieved. Instrumentation of the rock mass and its reinforcing elements in relation to the excavation process is required to provide confidence in the optimization process (Greenelsh, 1985; Thompson and Windsor, 1993; Villaescusa and Schubert, 1998). Significant increases on development productivity and overall cost reductions (while increasing safety) can be achieved if the reinforcement patterns are optimized (Villaescusa et al, 1994).

In cases where failure through reinforcement is experienced, information such as volume, weight, mechanism and structural control of the failure geometry as well as the observed reinforcement performance can be used as an integral part of further detailed stability analysis. Blasting practices, stress re-distributions, and operating practices during reinforcement installation must be considered. Documentation of the reinforcement performance is very important for future references and potential optimization of the design and practices in areas of a rock mass having similar conditions.

4 ASSESSING UNREINFORCED STABILITY

When an excavation is made, the load that was originally supported by the excavated rock has to be transferred to the surrounding rock mass. This causes an increase in the rock stress near the surface of the opening. The magnitude of this increase depends upon the geometry of the opening, the magnitude and orientation of the in-situ stresses, and the rock mass strength. If the rock mass is weak, or the depth is too large, the redistribution of the load may raise the rock stress above the strength of the rock, thus causing the rock mass to fail. The failure can be sudden, such in a rock burst, or gradually by progressive loosening of slabs and rock blocks. Even when a rock mass has failed, immediate collapse may not necessarily occur, due to the self-interlocking effects of the broken rock pieces. The factors controlling unsupported behaviour in a rock mass are schematically presented in Table 1.

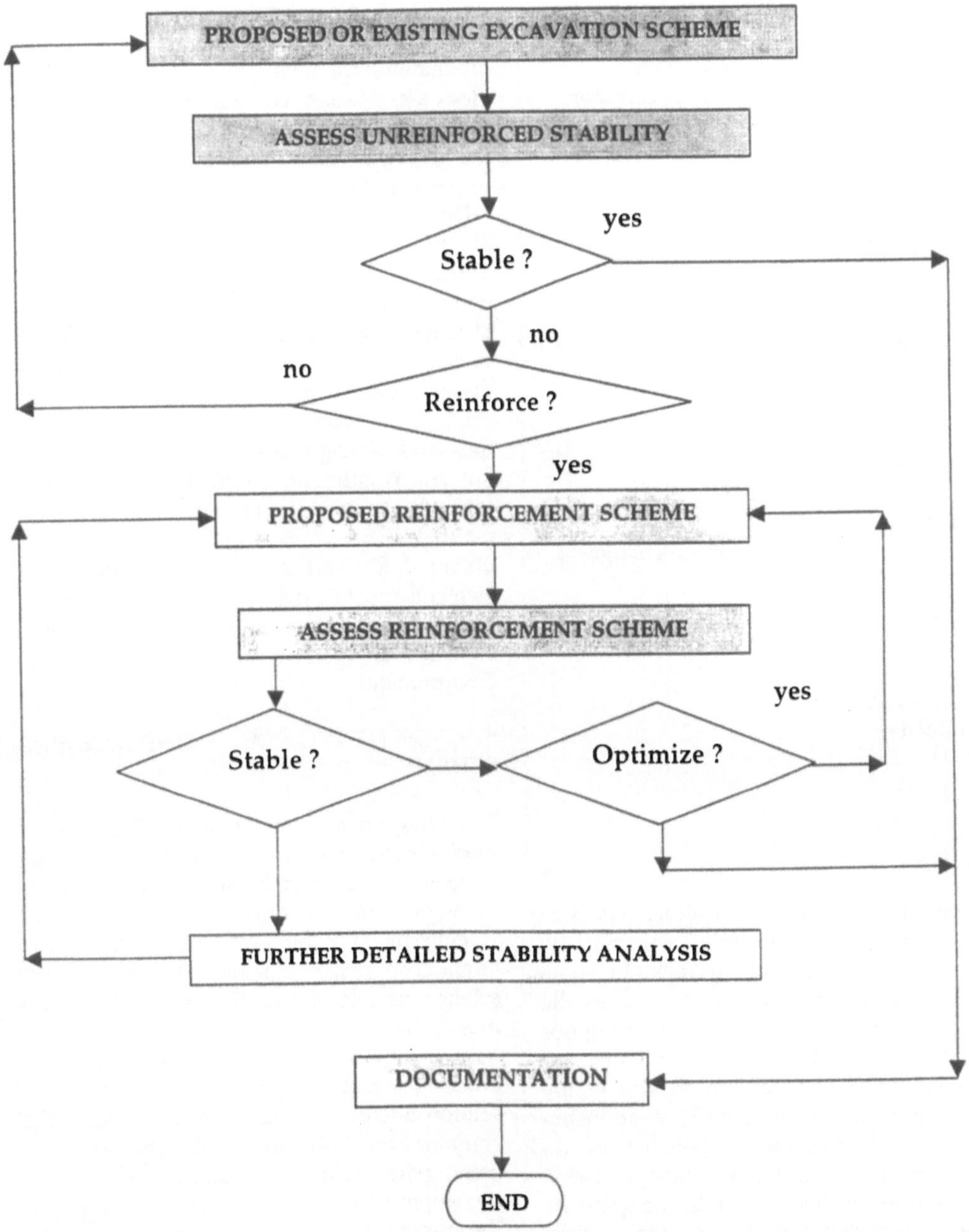

Figure 1. A generalized approach to rock reinforcement (modified from Windsor and Thompson, 1992).

Table 1. Factors controlling un-reinforced stability

ASSESS UNREINFORCED STABILITY
Excavation geometry • Shape, size, orientation, hydraulic radius **Rockmass strength** • Large structures (orientation, wedges) • Jointing (number, orientation, frequency, length, planarity, termination) • Intact rock (type, UCS, E, μ, cohesion, friction) **Failure mode** • Gravity fall • Toppling • Sliding • Buckling • Spalling, • Bursting • Swelling **Block size** • Blocky (larger than 1m) • Broken (0.3 to 1m) • Fragmented (smaller than 0.3m) **In-situ stress** • Magnitude (absolute, stress changes) • Orientation **Blasting** • Development • Stoping • Mass blasting **Groundwater** • Dry • Wet • Minor inflow • Flowing

Experience indicates that excavation shape, size and orientation significantly control the behavior of an excavation in rock. In some cases, the size of the equipment, such as the dimensions of the development mining jumbos and the longhole production drilling equipment, control the minimum size of the underground drives. A critical relationship exists between the size of the potential in-situ block sizes (potential falling/sliding blocks) and the exposed spans. The larger the span, the greater the probability of potential falling blocks. This is likely to occur in areas of a rock mass where continuous structures (such as faults, shears, etc.) are present. By enlarge, intact rock bridges are likely to limit the size of the potentially unstable blocks (Windsor, 1996).

Rock mass characterization techniques such as those described by Call et al, (1976); Windsor and Thompson, (1992); Villaescusa (1992), are required to determine the basic parameters of the rock mass discontinuities, including potential failure modes and estimates of joint orientation, frequency, size, termination and block size. Rock mass strength is a function of the geometric nature and surface strength of the geological discontinuities and the physical properties of the intact rock bridges.

The un-reinforced analysis also requires the identification of areas of the rock mass likely to undergo large stress re-distributions or blast damage during a subsequent stoping process. Damage from blasting or stress change can cause the opening of pre-existing discontinuities or cracking of intact rock bridges. This damage effectively decreases the rock mass strength, and in some cases leads to failures. Weathering due to water permeating through the fracture pattern in a rock mass may also affect un-reinforced stability. The potential for corrosion of the reinforcement elements must be determined by analyzing the water quality to determine if aggressive elements are present in a particular mining environment.

5 SELECTING A REINFORCEMENT SCHEME

Following an assessment of the likely failure modes predicted from the interaction of the excavation (geometry and purposes), the network of geological discontinuities, weathering and the loading conditions from stress and blast damage, a reinforcement scheme capable of matching the expected rock mass behaviour must be selected.

In most underground mines, the most efficient and economical method of reinforcement is based on a pattern that combines rockbolts and mesh. This scheme is compatible with a majority of current development mining strategies, and represents the minimum level of reinforcement expected for excavation stabilization.

Depending upon the potential size of the blocks exposed, meshing may not be necessary during the initial bolt installation. However, it is a good practice to install it, given that small scats or slabs may be broken from intact rock (between bolts) later in the life of the excavation. The option of spraying shotcrete is expensive and it can only be justified when the block size is very small, such as in the case of intersected shear zones or to support very weak rock masses.

In general, a stabilization scheme can not be selected without consideration of the drilling equipment and reinforcement installation available at a particular mining site. An optimal mechanized system must be able to install rockbolts and mesh in a single pass in order to increase productivity and reduce exposure of personnel during installation. To date, such equipment appears to be excessively large, and full mechanization of reinforcement installation (including meshing and grouting on a single pass), may not be possible for most mining situations where narrow orebody development is required to control dilution.

In some cases, excavation walls and backs should be stabilized using a different reinforcement scheme. This strategy allows the reinforcement to match different wall behaviour, and can also be used to minimize cost. However, the final choice is always controlled by the availability of equipment as discussed previously.

In situations where adverse geologic structures are present or the excavation size is increased significantly, the potential for deep failures usually require cablebolt reinforcement. Cablebolts provide deep anchorage and are used to stabilize large potentially unstable areas where normal rockbolts would prove inadequate. Cablebolts may also be needed in areas likely to undergo excessive loading from stress re-distributions or blasting.

Figure 3 presents a generalized approach to the selection of a reinforcement strategy that takes into account the purpose of the excavation (long term), the environmental effects (corrosion potential) and the geometry of failure (cablebolt requirements). The strategy is based on the assumption that the basic excavation support and reinforcement is provided by rockbolts and mesh.

6 ASSESSING A REINFORCEMENT SCHEME

The effectiveness of a particular reinforcement scheme can be assessed using empirical strategies, theoretical methods or geotechnical instrumentation (Windsor and Thompson, 1993; Villaescusa and Schubert, 1998). The first step is to determine the overall stability of an excavation followed by a stability assessment of typical individual blocks according to the identified failure mode and data from the rock mass characterization program.

Empirical evidence from other mines having similar rock mass strength and loading conditions, combined with empirical methods such as the Modified Stability Graph or other rock mass classification schemes, can be used to establish the overall initial reinforcement design (See Table 2). Instrumentation of typical reinforcement arrangements coupled with observations of rock mass behaviour can be used to improve the initial reinforcement designs. Detailed stability of typical or worst-case scenario wedges and blocks can be carried out using three-dimensional programs such as SAFEX 3D (Thompson and Mathews, 1989). Data from detailed rock mass characterization programs can be used to determine the joint set characteristics needed to determine the likely size of the largest rock block to be reinforced (Windsor, 1996).

The reinforcement scheme selected can be tested and optimized in-situ. In cases where empirical evidence suggests stability, the interaction between the reinforcement scheme and the rock mass can be instrumented in order to achieve an optimal match with respect to demand, capacity and cost. Figure 4 presents a process of assessment that includes practical safety measures in cases where failure through reinforcement is experienced in an operating underground mine. All relevant underground personnel must be informed and preventive measures such as "no entry" signs must be put in place.

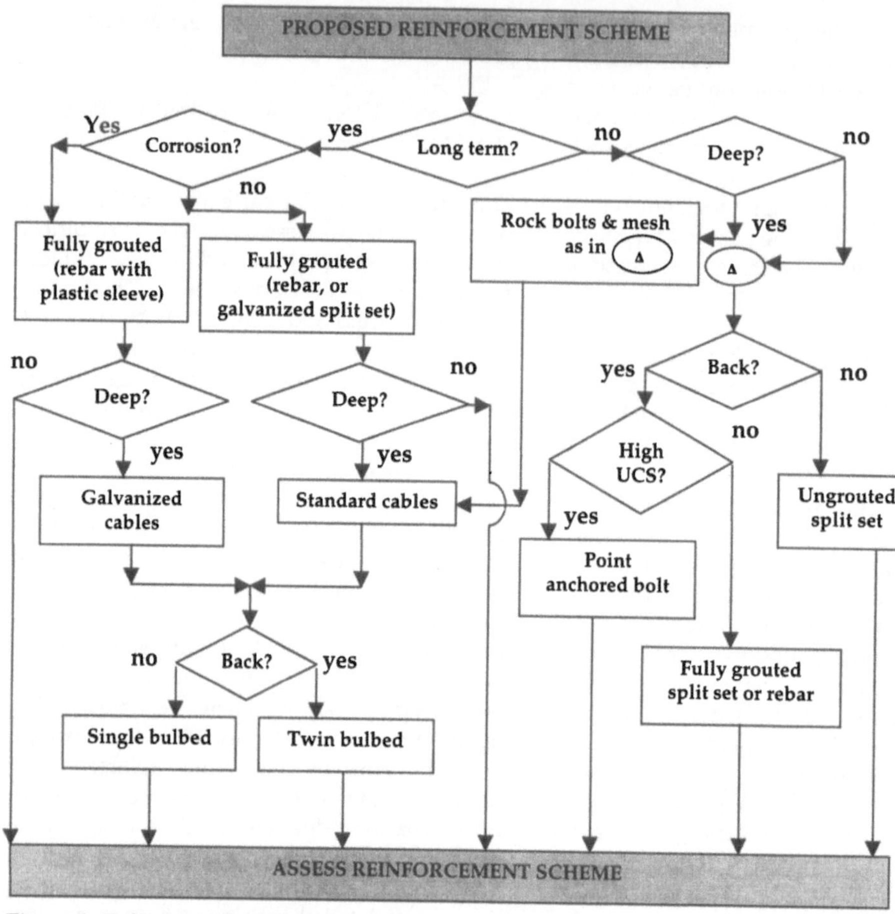

Figure 3. Different reinforcement schemes.

Inspection and documentation of an actual failure must be done by experienced underground rock mechanics personnel in order to gather data to facilitate an additional detailed stability analysis as shown in Table 3. Empirical evidence from failures can be used to improve the understanding of the interaction between the rock mass, the reinforcement schemes and the loading conditions from mining.

Detailed stability analysis may include calibrating a numerical model program to simulate the extraction sequences and assess the role of stress redistribution on the rock mass environment (Bywater et al, 1983). Information on the failure mechanism and the performance of the installed reinforcement are vital to determine the effectiveness of the support and reinforcement scheme. Evidence of corrosion or excessive damage to the bolts, cables, mesh or plates within the failure or within the immediate area will provide an indication on the loading conditions contributing to the failure.

Information on the shape and size of the failure indicates whether appropriate embedment lengths were selected during the initial design. Any evidence on the failure trigger mechanism, including rock mass damage caused by blasting, is critical if further failures of similar nature are to be prevented. In all cases, detailed documentation must be completed and in some, a presentation of the results should be made to the mine planning and operating personnel (See Table 3 and Figure 4). The final documentation effectively closes the design process and represents an important step into optimization of the design and practices. It also aids in the prevention of future failures.

Table 2. Methods to assess reinforcement effectiveness.

ASSESS REINFORCEMENT SCHEME
Overall assessment
- Empirical (back analysis)
- Instrumentation
- Observation
- Modified Stability Graph
- Rock mass classifications
Individual wedge/block analysis
- SAFEX (2D and 3D)
- Dead weight / analytical
- Worst case scenario controlled by:
- Size of excavation
- Maximum trace length
- Maximum joint spacing
- Empirical (back analysis)

Table 3. Typical information from an underground failure

DETAILED STABILITY ANALYSIS
Numerical modelling
- Absolute and stress change
Rock mass failure mechanism
- Structurally controlled
- Bursting of intact rock
- Swelling, squeezing
Volume and weight of failure geometry
Damage at multiple/single locations
Suspected trigger mechanism
- Blasting
- Scaling
- Drilling
- Reinforcement installation
- Water
- stress
Damage to support / reinforcement
- None
- Complete failure
- Plate pulled off
- Bolt sheared
- Cable pig-tailed
- Weakened by corrosion

7 OPTIMIZING A REINFORCEMENT SCHEME

The optimization of a reinforcement scheme is likely to improve safety and also increase productivity in most mining development situations. The optimization usually starts with an initial review of practices and the quantification of the potential economic benefits likely to be achieved if the practices can be improved.

Field observations can be used to indicate areas of a mine that are being consistently over or under reinforced, or where no technical reasons are being used to place a particular type of support or reinforcement. A list of items which can be considered during the initial stages of the optimization process are listed in Table 4.

A workshop given by the technical personnel to the mining crews can be used to present the results of the initial review of practices, including a description of long-term cost-control key performance indicators. Experience suggests that additional workshops must be held at approximately six month intervals in order to review the performance of the applied reinforcement schemes. The workshops provide a forum for feedback and identify areas where further research such as additional laboratory testing of new products or field instrumentation, is needed. In general, the workshops are most effective following a review of underground practices, or after the completion of significant reinforcement monitoring or laboratory testing work.

Commitment and feedback from the work force is vital to achieve optimization of the reinforcement process. Consequently, training of all the underground personnel involved in the rock support and reinforcement process is perhaps the single most important factor in order to achieve the best (safe and efficient) practices. Introduction to ground support should be a mandatory training module, and a requirement to other related modules such as installation of rockbolts, operation of grout pumps and installation and plating of cables, etc.

Feedback from the individual mining and support crews is invaluable if optimization is to be achieved. By training the workforce and

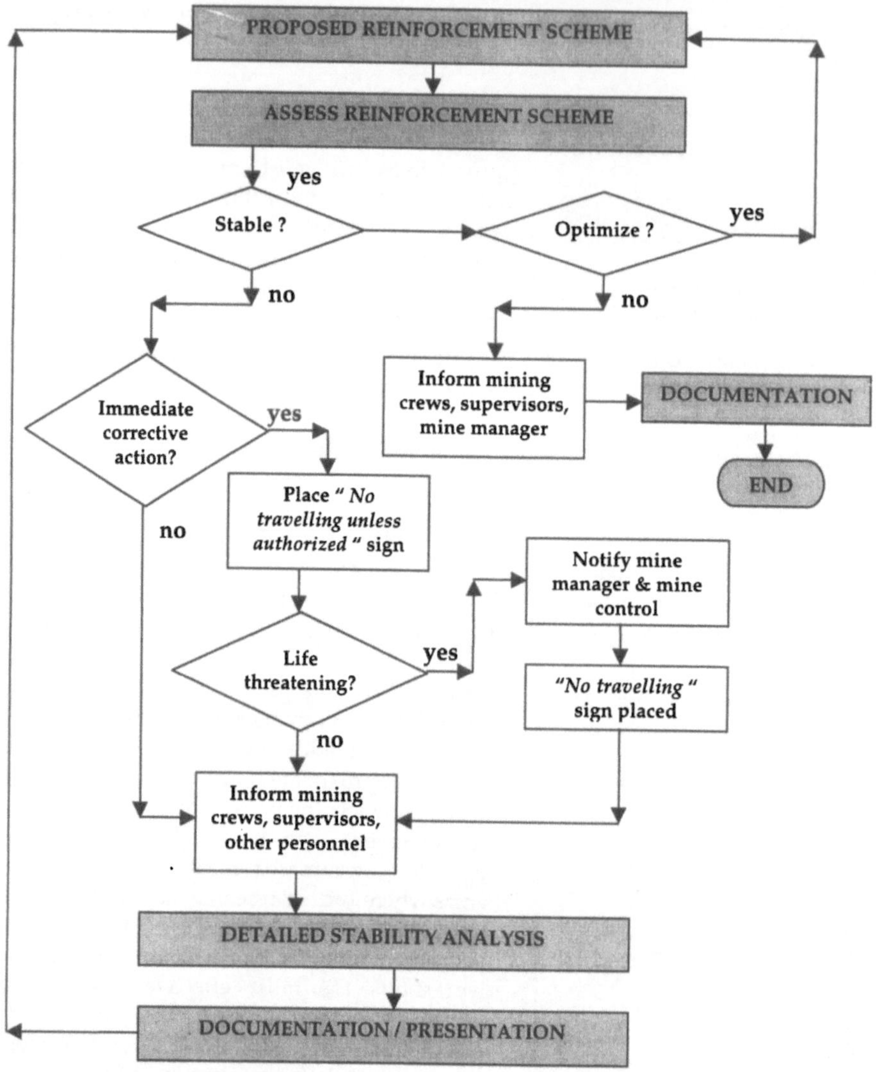

Figure 4. Assessing a reinforcement scheme.

using their feedback, it is possible to continuously improve the installation practices leading to significant savings and increased safety.

Figure 5 presents the results of long term cost reduction program in support and reinforcement undertaken by Mount Isa Mines for over a period of three years. (Villaescusa et al, 1994). Savings of over 2 Million dollars per year were directly related to review of practices, laboratory test work, in-situ instrumentation and systematic training to the work force (Villaescusa et al, 1994).

8 QUALITY CONTROL

The performance and ultimate capacity of a reinforcement scheme can be affected by sub-standard installation practices. Faulty installations are difficult to detect, given that the only visible part of an installed element is the plate, nut and a short length of the bolt indicating the orientation of installation with respect to the excavation wall.

A definite indication of poor installation is established in cases where the reinforcement fails well below its designed capacity. For

Table 4. Issues to be considered during an initial review of practices.

Support and reinforcement standards and rationale.

- **Bolt patterns (back and walls)**
- **Scaling methodology**
- **Typical unsupported spans and timing of reinforcement installation**
- **Ground conditions (number, size, spacing, orientation of joint sets, block size and shape, stress/blast damage, water)**
- **Failure mechanisms (back and walls)**
- **Excavation geometry (typical size, including turn-out)**

Equipment.

- **Size and drilling limitations**
- **Scaling**
- **Type of grout pumps (mono pump, piston based)**
- **Plate installation (impact wrench, specialized jack)**

Methodology of installation.

- **Mechanized or manual (installation, anchoring, meshing, plating)**
- **Drill, blast, muck cycles and their support and reinforcement**
- **Cablebolting**
- **Grouting**

Organizational role.

- **Crew organization**
- **Role of supervisor (quality control checks, communication with geologists and rock mechanics engineer)**
- **Role of underground manager**
- **Role of rock mechanics engineer (stability assessment, communication channels with operations and technical team)**

Training.

- **Induction modules**
- **On the job training**
- **Rock mechanics engineer (training to crews, techniques and hazard recognition, design and assessment, minimum reinforcement standards)**

Long term effectiveness.

- **Key performance indicators on cost**
- **Failures on supported ground**
- **Degree of re-habilitation needed**
- **Corrosion**

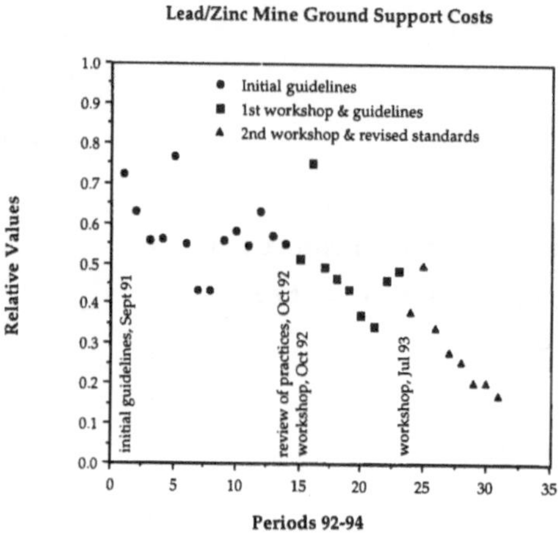

Figure 5. A reduction in support and reinforcement costs at Mount Isa Mines (Villaescusa et al, 1994).

example, in an installed fully grouted rebar, it is very difficult to determine the bonded length and the position of the grout along the axis of the bolt. Because the full bolt capacity is mobilized with very short embedment lengths, pull testing of a fully grouted element is almost meaningless, and very little information will be gained about how much of the bolt is actually grouted. It is therefore extremely important that crews are trained to use the proper grouting method, and that they are provided with the best grouting equipment and training.

A quality control and training program is needed to maintain a *minimum* level of standards, likely to provide the required support and reinforcement capacity throughout the life of an operation. *Systematic* in-situ pull testing schemes are recommended when a new reinforcement element is introduced into the system. *Random* pull testing of on-going reinforcement practices (not applicable to fully grouted systems, unless debonding is used) as well as UCS testing of grout samples will indicate the capacity of the reinforcement being installed as well as the actual installation practices.

8.1 *In-situ pull testing*

Pull tests in-situ are required to determine the degree of encapsulation and load transfer capabilities achieved by a particular reinforcement scheme. These tests are required when a new reinforcement element is introduced to a mine, and also to optimize on-going installation practices. All the pull testing programs must be based and designed on a proper understanding of the load transfer concept (Windsor and Thompson, 1992). These programs are undertaken to determine the degree of encapsulation and the critical embedment length values for grouted systems, anchor capacity and any subsequent loosening for point anchored systems, and load transfer capacity for frictional systems. All tests must be carried out following the International Society for Rock Mechanics suggested methods for rock bolting (Brown, 1978).

In the case of a fully grouted system, the use of short holes to determine critical load transfer distances can be misleading. Although the critical embedment length is established, very little information on the actual encapsulation capabilities of the system is determined this way. The best information is gained when full-length elements are pull tested. In order to mobilize short lengths at the toe end of the hole, a plastic sleeve must be placed to cover the portion of the element that is not being tested. In doing so, the reinforcement element is *de-bonded* by the plastic sleeve along its entire length, except for a very short length (usually

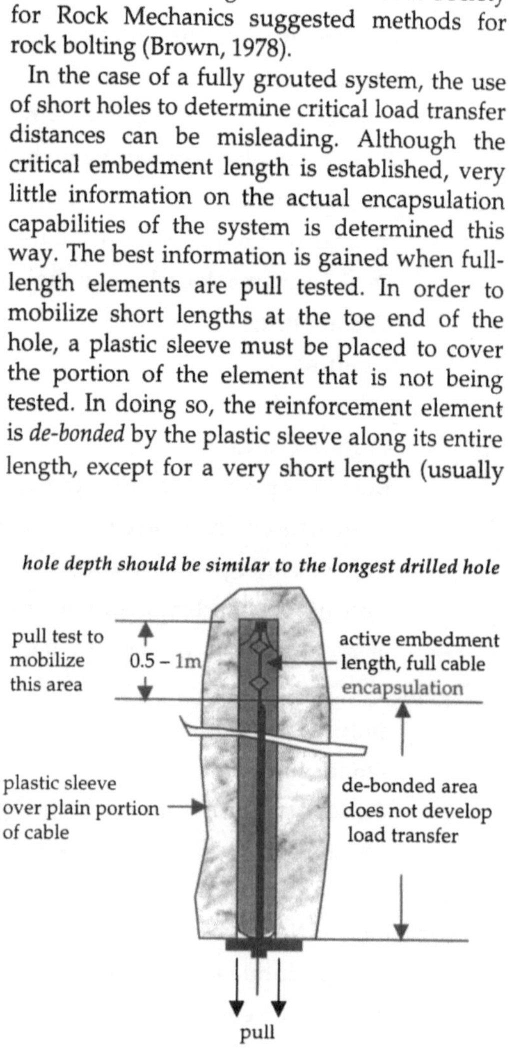

Figure 6. De-bonding techniques to pull fully grouted elements (not to scale).

0.5 or 1m) at the toe end of the hole (no plastic sleeve is placed in this portion), where all the load transfer is to be developed (See Figure 6). This procedure ensures that the actual encapsulation achieved by the grouting system, at the anchoring part of the reinforcement, is determined.

An example of such a test would be a 6m long cable, installed with a 5m plastic sleeve, which leaves the last metre exposed at the toe end of the hole. The cable could be manufactured in such way that bulbs are available in the exposed metre, where the load transfer and encapsulation will be tested, while the rest of the cable would be plain to facilitate the de-bonding. Pull testing the cable would determine if the grouting installations are effectively working, and what load capabilities are being mobilized at the anchoring end of the cable. Pulling a full-length grouted cable that it is not debonded is meaningless, because it is impossible to know which portion of the cable is developing load transfer. Pulling cable lengths installed in short holes do not provide information on the encapsulation capabilities of the system.

8.2 *UCS testing*

The strength of the grout significantly controls the load transfer capabilities of a fully bonded system. The typical failure modes for fully grouted systems are schematically presented in Figure 7. Failure through slippage at the *steel-grout* interface can occur when weak grouts and plain cables are mobilized. *Rupture* of the steel is achieved with moderately strong grouts and modified cable geometry, such as bulbed cables. Some slippage is always experienced before cable rupture is experienced. Failure of the rock mass around the cables has also been observed in very weak or severely damaged rock masses.

Failure at the *grout-rock* interface may be possible when the hole size is very small and not enough surface area (which controls the shear strength) is available at this interface. Observations of this type of failure are common in resin grouted reinforcement of highly broken rock masses. Failure of a rockmass around ι cement-grouted dowel has been observed in heavily broken rock masses, indicating that very small block sizes are present, rendering the rockbolt practically useless. In such cases, a different stabilization scheme such as shotcrete support should be used.

Uniaxial Compressive Strength of grout samples is the recommended procedure to quantify the strength and any variability in the actual water cement ratios used underground. Laboratory samples taken from piston-based grouting programs that continuously mix and pump from a single container show high variability. In order to obtain representative values, the sample must be random and taken at different stages throughout the grouting cycle. At least 3 samples, but preferable 5, are required from a single grouting job.

A review of the literature (Hutchinson and Diederichs, 1996) indicates that a high variability can be expected for the UCS values at low water cement ratios (between 0.3 and 0.35). This is mostly due to differences in the mixing time, the type of mixing and pumping machine used to collect the samples.

8.3 *Storage facilities*

Deterioration of reinforcing products must be prevented at the manufacturer facilities, during transportation to the mine site, during storage (surface and underground) and immediately before installation by the reinforcement crews. In most cases, the reinforcing products are manufactured in large industrial centres, far away from the remote mining sites. Damage during transportation to the reinforcing elements (mainly within the threaded lengths), should be avoided. The threaded portion of bolts must be oiled. Bundles of bolts are then to be covered with plastic bags at the threads, and the bars secured in such a way that movement between the individual bars is eliminated. The nuts should be oiled and packed in woven polypropylene lined plastic bags. The bags can then be transported and stored in sealed steel drums. The mechanical anchors, barrel and wedges, etc. should be supplied in sealed, stiff cardboard boxes. Plates must be securely strapped onto pallets, and clearly marked to differentiate between point anchored, rebar and cablebolting plates.

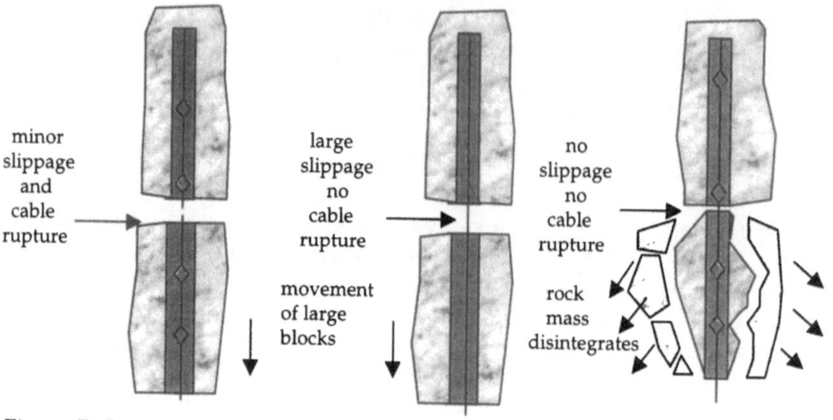

Figure 7. Common failure modes observed underground.

Fully covered surface facilities should be provided for all the materials likely to rust or corrode if left unprotected. Surface storage facilities that are linked by computer to the underground supply system are needed to indicate year to date and monthly average reinforcing usage. This will help to plan and maintain sufficient supplies for all consumables and also helping turning them over, such that they are not stored for long periods of time without being used. This will minimize the chances of deterioration.

Underground stores, where all the reinforcing consumables are kept, must be well organized with clearly marked sections for different consumables. They should have concreted floors, locked doors and be linked by computers to the local surface supply system. A store person should be available at all times. This person is responsible for storing materials as they come from the surface and also requesting additional consumables when the stocks are becoming low.

Underground crews have the responsibility of selecting an appropriate number of consumables according to their daily tasks. The reinforcing elements must be kept off the ground and out of water at all times. If not completely used, they must be returned to a clean place in order to avoid damage and to minimize wastage.

9 CONCLUSIONS

An integrated approach to reinforcement is required to optimize design, installation and practices. Rock mass characterization schemes define the likely failure modes under the different loading conditions induced by mining. A reinforcement strategy likely to match the expected long term behaviour must be chosen. Optimization to enhance safety and reduce cost can be achieved through the implementation of monitoring, training and quality control programs.

10 ACKNOWLEDGEMENTS

The author acknowledges the financial support from the Australian Centre for Geomechanics and Curtin University of Technology for his position as Professor of Mining Geomechanics at WASM. The Australian Centre for Geomechanics has received funding for this position from the Government of Western Australia, Centres of Excellence Program.

11 REFERENCES

Brown, E.T., (Ed), 1978. Rock characterization testing and monitoring. Commission on Testing Methods, International Society for Rock Mechanics, Pergamon Press, Oxford.

Bywater, S., R. Cowling and B. Black, 1983. Stress measurements and analysis for mine planning. In Procc. Fifth ISRM Congress, Melbourne Australia, pp D29-37.

Call, R.D., J.P. Savely, and D.E. Nicholas, 1976. Estimation of joint set characteristics from surface mapping data. Procc. Monograph on Rock Mechanics Applications in Mining. W. Hustrulid (Ed)., pp65-73, AIME, New York.

Greenelsh, R.W., 1985. The N663 stope experiment at Mount Isa Mines. International Journal of Mining Engineering., Vol 3, pp 183-194.

Hutchinson, D.J. and M. Diederichs, 1996. Cablebolting in underground mines. Bitech Publishers, 406p.

Thompson, A. and S.M. Mathews, 1989. Design packages for reinforced excavations in rock. . Procc. Conf. Computer Systems in the Australian Mining Industry., The University of Wollongong.

Thompson, A. and C. R. Windsor, 1993. Theory and strategy for monitoring the performance of rock reinforcement. Procc. Geotechnical Instrumentation and Monitoring in Open Pit and Underground Mining. T. Szwedzicki (Ed), Kalgoorlie, pp 473-482, Balkema, Rotterdam.

Villaescusa, E. 1992. A review and analysis of rock discontinuity mapping methods. Procc. Sixth Australia-New Zealand Conference on Geomechanics. Christchurch, pp274-279.

Villaescusa, E., L.B. Neindorf, and J. Cunningham, 1994. Bench stoping of lead zinc orebodies at Mount Isa Mines Limited., Procc. MMIJ/AusIMM Joint Symposium, New Horizons in Resource Handling and Geo-Engineering, Ube Japan, pp351-359.

Villaescusa, E. and C. Schubert, 1998. Monitoring the performance of rock reinforcement. Procc. Int. Conference on Geomechanics/Ground Control in Mining and Underground Construction, Wollongong.

Windsor, C.R. and A. Thompson, 1992. Reinforcement design for jointed rock masses, Procc. 33rd US Symposium on Rock Mechanics, Santa Fe, pp521-530, Balkema, Rotterdam.

Windsor, C.R. and A. Thompson, 1993. Rock reinforcement – technology, testing, design and evaluation. Comprehensive Rock Engineering. J. Hudson (Ed). Vol 4, Chp 16, pp 451-484, Pergamon Press, Oxford.

Windsor, C.R., 1997. Rock reinforcement systems. Int. J. Rock Mech. Min. Sci. Vol 34, No. 6, pp 919-951.

Rock Support and Reinforcement Practice in Mining, Villaescusa, Windsor & Thompson (eds)
© 1999 Balkema, Rotterdam, ISBN 90 5809 045 0

Integrated ground management – An essential component of our licence to operate

A.S. Logan
BHP Cannington, Townsville, Qld, Australia

ABSTRACT: The Broken Hill Proprietary Company Limited (BHP) Cannington Mine is a silver, lead and zinc mine 250km south east of Mt Isa in north west Queensland. At full production, it will be the biggest, single, silver mine in the world. The BHP Cannington vision is creating success through people with passion. To maintain our licence to operate, we value being simply safe in our approach. Systems are thus 'built in not bolted on'. Ground management is integrated with safety, operational and personnel management. This paper will discuss this integration of ground management strategies to both provide safe work environment and develop a productive and efficient mine.

1 INTRODUCTION

1.1 Location

The Cannington silver-lead-zinc deposit is located approximately 250 km south-east of Mount Isa in Queensland, Australia (Figure 1). Similar to Broken Hill in the 1880's, Cannington is located on semi arid grazing land at the head waters of one of Australia's major watersheds. Relief in the area is of the order of only a few metres and the area is subject to monsoonal flooding.

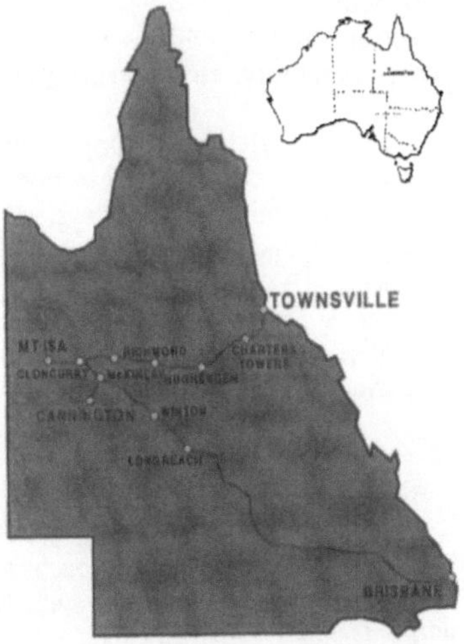

Figure 1. Location Map

1.2 History

The discovery of Cannington was the culmination of several years of exploration for Broken Hill style deposits in Australia. The deposit lies hidden under soil and sedimentary rocks. In 1990, a hole testing an anomaly detected by airborne magnetic survey, intersected 20m of mineralisation averaging 12.1% lead, 0.6% zinc and 870 g/t silver.

Drilling during 1990 and 1991 confirmed a significant silver-lead-zinc resource. Detailed feasibility work started in 1993 and centred around the construction of an exploration decline using a mining contractor.

Following the underground exploration , the resource was assessed to be 43.8 million tonnes at 11.6% lead, 4.4% zinc and 538 g/t silver.

Cannington mines and processes 1.5 million tonnes of ore per year to produce 24 million ounces of silver contained in 265,000 tonnes of lead concentrate and 110,000 tonnes of zinc concentrate.

Cannington is part of the recent trend to small-medium sized underground Australian base metal mines. Key constraints include:

- spatially constrained orebody;
- limited economies of scale;
- high project sensitivity to head grades, recoveries and metal prices;
- variable and poor ground conditions; and
- remote location - fly in fly out.

1.3 Underground Mining

The Cannington deposit is being developed as an underground mine utilising both sublevel open stope and bench mining methods. Orebody geometry and faulting systems complicate extraction strategies and stope shapes.

Production of ore from underground commenced in April 1997. Trucks have transported ore up the access decline to the surface for the first 18 months of stope production. The Fowler Hoisting Shaft was commissioned in May 1998 to a depth of 645 metres and will hoist ore for the remaining 20 year mine life.

Ground conditions at Cannington are highly variable, ranging from good to very poor rock mass classifications.

2 INDUSTRY TRENDS IN GROUND MANAGEMENT?

Ground management philosophies in the Australian mining industry are in a pivotal stage of evolution. Current Australian support strategies are being questioned and alternative approaches are evolving.

The issues facing the industry include:

1. People

- The ability of industry staff and operators to recognise hazards is decreasing as practical mining skills are diluted with the increasing number of underground operations;
- increasing skill level of personnel required to operate computerised equipment, achieve higher standards and work within increasingly self managed work teams; and
- boredom operating automated equipment and potential reactions to physical inactivity.

2. Technical

- An increasing level of uncertainty in the long term performance of grouted dowel bolts with the recognition that cracking of the grout annulus occurs and full encapsulation is difficult to consistently achieve in practice;
- a recognition of the unsatisfactory performance of friction bolts for blocky roof support;
- increased use of remotely applied fibrecrete; and
- a trend towards remote support installation equipment.

3. Management

- Management is becoming increasingly accountable to regulatory authorities for safe working conditions;
- ground support installation standards vary significantly across the Australian Mining Industry. Consider the relative testing effort that goes into assuring continued hoisting rope performance over time versus testing of ground support and the risk associated with each; and
- the increasing use of mining contractors with civil tunnelling backgrounds has seen the introduction of civil industry quality assurance systems.

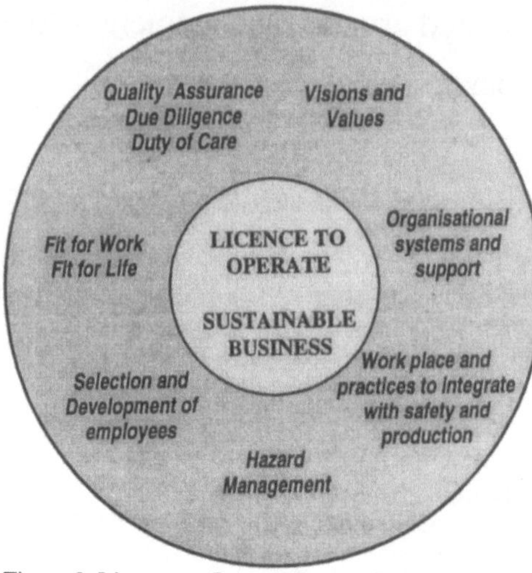

Figure 2. Licence to Operate Supporting Systems

3 LICENCE TO OPERATE

We consider that we not only need to maintain a licence to operate in a direct regulatory sense but also with the many people and groups that have a stake in Cannington's success. These stakeholders include BHP, our employees, government authorities, our customers and the community. Safety is a fundamental component of this licence to operate.

So, how do we operate in these highly variable ground conditions and maintain this licence to operate and a sustainable business with all our stakeholders?

We use integrated support systems (Fig. 2).

Ground management though, must also consider three important viewpoints:

- our people -selection, training, fitness for work, quality assurance, operability, empowerment and living our values.
- technical - how to cost effectively support the ground? work place design, equipment selection;
- management - organisational systems, accountability, due diligence, duty of care and statutory compliance.

The systems shown in Figure 2 act to integrate our operational people, technical and management viewpoints allowing them to better work together.

4 OUR VISION AND VALUES

Our vision and values underpin our approach at Cannington. Our vision is *"creating success through people with passion"*. Our values are

- Simply safe;

- Care & respect for our environment;
- Recognise & Value Our People;
- Take Pride in Our Success;
- Promote Effective Workplace & Community Communications; and
- Strive for Improving Quality

Both BHP and Cannington recognise that people are the means by which to maintain a licence to operate, deliver results and competitive performance. People are thus a major focus of our efforts.

Cannington's *Simply Safe* value equates to accepting the responsibility at all times for the well being of yourself, your team members, others in the workplace, your family and the equipment with which you work, thereby maintaining a productive and rewarding work environment and lifestyle.

The safety system framework at Cannington is provided by the National Occupational Safety Association (NOSA). This has been customised to meet the needs of the operation and renamed CanWorkSafe (Lennox et al 1998).

Cannington has also adopted the Positive Attitude Safety System (PASS) to facilitate moving from procedural to team cultural style where people are positively empowered with a CAN DO attitude to fix issues themselves or as teams.

5 BUILT IN WITH OUR PEOPLE

Ground management and systems are 'built in not bolted on' to the mining cycle, operations and our people.

5.1 Procedures

Our recruitment process was thorough giving us expertise, knowledge and experience in our work force. With one of our values being to *recognise and value our people*, we asked the operators to write their own operating procedures. This made the procedures a living document that would be easy read, had creditability, were useable and could be easily updated (Nobelius et al 1998). An example of an underground procedure is shown in Appendix A - Scaling.

The individual procedures were integrated together by considering their contribution to overall mining activities. Figure 3 shows the Cannington Development Mining Cycle. The cycle phases are shown on the inside and the procedures on the outside. The rock fall hazard is also shown for each cycle phase.

Development ground conditions are assessed into three classes and the appropriate entry, ground

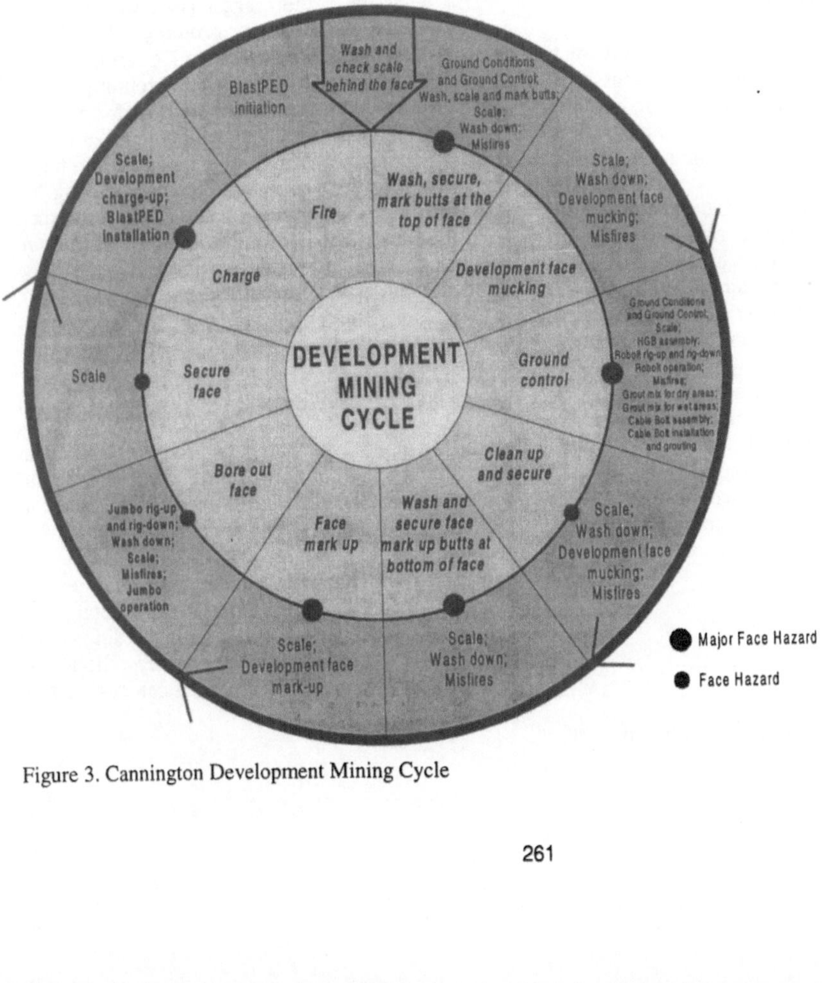

Figure 3. Cannington Development Mining Cycle

securing and support strategy applied. The three classes used are

- *good* (green) - hand scaling from the muckpile and remotely installed rock bolting;
- *poor*; scatty (yellow); hand scaling, remotely installed rock bolting; subsequent fibrecreting as required; and
- *very poor*; unravelling (red) - no personnel entry; initial 50mm layer of remotely applied fibrecrete; pattern bolting; subsequent fibrecreting as required.

Face locations on the shift change notice board are marked with green, yellow or red magnetic labels for effective communication of ground classes and ground management approaches.

The classification system was not seen as a diagnostic decision making system. Rather, it was used in conjunction with, and to calibrate assessments based on, operator experience and other logistical considerations. Application of this system is continuing to evolve from a contractual to coaching style as our work force has matured and their skills developed.

We believe this approach has built on the experience breadth of our mining personnel with the rigour, consistency and transparency of geotechnical systems.

5.2 Performance Management

Cannington adopted a salary rather than contract payment system as we believe it is consistent with being *simply safe* and *striving to improving quality*. We see improving ground support installation quality as one of the keys to best practice ground management.

This salary system is used at Cannington for all positions. A four level salary structure is in place for mine and mill operators. In each level there is a salary range based on competence. The salary structure for operators is underwritten by a Certified Agreement with the Australian Workers Union (Farcich et al 1998)

The objective of this system is to provide a mechanism to reward not only differences in skills and knowledge but also in performance. Personnel and their supervisor also conduct an annual performance review. Performance against agreed goals and behaviour in accordance with Cannington's values such as safety and quality, is discussed and agreed.

6 SUPPORT SYSTEMS

6.1 Selection Criteria

The technical selection criteria or 'wish list' of support systems at Cannington include:

- capability to be remotely installed reducing operator exposure to unsecured ground;
- simple installation thus high productivity and confidence in quality;
- performance in a wide range of ground conditions;
- flexibility in installation methods;
- long term performance (corrosion resistance);
- cost effective with respect to development advance (capacity and price); and
- product support and development.

6.2 Hollow Groutable Bolts

During the early stages of hard rock development, use of 3m hollow groutable bolts (HGBs) was implemented. This offered the advantages of:

- action as both immediate and long term support tendon;
- improved confidence of full encapsulation; and
- reduced installation time.

HGBs have proven extremely successful in blocky ground conditions at Cannington. Pull out tests indicate grouted bolt capacities of 15-17 tonne and ungrouted capacities of 7-8 tonne. No significant rock falls have occurred due to failure of fully grouted HGBs but several have occurred in areas of ungrouted HGBs.

This leads on to the biggest issue we face with continued use of the HGB- *grouting*. For safety reasons, continued production and development is contingent upon all necessary grouting being completed, as we see ungrouted HGB performance is not assured. Delays may allow loosening or corrosive groundwater time to eat away at the bolts. The bolts must first be installed remotely using the Robolter, and then grouted later by a three man grouting crew. This slows the mining cycle down, and ties up valuable resources. Production tonnage and development advance suffer.

6.3 Cable Support

Cables are installed utilising a toe to collar grouting method. A water to cement ratio of 0.35-0.45 is used. A mono type pump is used and assists in maintaining a high quality of thick grout. Barrel and wedges are installed on both strands and tensioned to nominally 2-4 tonnes approximately 6-8 hours after grouting.

Four to six, 6m twin strand, bulbed cables are installed in each development intersection. A geological assessment of the ground conditions and design is completed for each intersection and reviewed by a competent geotechnical engineer or geologist.

6.4 *Excavation*

6.4.1 Scaling

Protection of underground personnel is the key to Cannington's continued operation. Almost half of reported rockfalls at the mine are less than one cubic metre in volume, which means that continued scaling is vital. As discussed previously, we hand check scale from the muckpile in good to poor ground conditions. We believe this provides both an effective means to secure the ground and allows our operators to better know the conditions underneath which they will be working.

The underground scaling procedure at the time of writing is shown in Appendix A.

Ground support crews are also being formed at the mine dedicated to ensuring that development 12m back from the face is safely scaled at all times.

In new development rounds, unsupported ground is scaled from the muckpile by a mining crew. If the ground is assessed as hazardous, very poor, or cannot be scaled to solid, the heading is fibrecreted. Fibrecrete has proved to be a very effective form of support, providing protection for personnel in scatty ground, and for machinery where unstable ground must be bolted.

6.4.2 Arched Profile

Standard drives at Cannington are mined in a 5m x 5.2m arched profile. The arch geometry enables the rock mass to better support itself, and decreases its reliance on installed support.

6.4.3 Perimeter Blasting

Drive perimeters are drilled with an increased hole density, and fired using Trimex. This gives a smoother profile, and decreases blast damage to the perimeter.

6.5 *Alternative Systems*

At the time of writing, a trial of Atlas Copco Swellex Bolts has been ongoing at the mine since July 1998. The Swellex bolts are chromium alloy friction anchors which are resistant to corrosion and can be remotely installed as a single pass, potentially longer term support system. Initial results have been encouraging.

Chemical bolts are also being considered as a possible HGB replacement.

7 STOPE SUPPORT

7.1 *Support Before Extraction*

Stope failures pose a risk to both mine personnel and to the economic viability of the operation.

Stope support is assessed through the use of the Modified Matthews Stability Method. At this stage, only a limited number of stopes have been extracted at Cannington, therefore the standard stability curves are used.

The points for each surface plotted on this graph indicate which walls may be unstable when the stope is mined. This information is used to indicate the areas in which support may be required.

The stability of the stope surfaces is calculated at each stage of extraction. This information can be used in the risk assessment process, and aids ring sequencing.

Surface support is also applied to the accesses and draw points to successfully limit stoping induced damage.

Once a stope has been extracted, its performance is compared to the predictions made in the Matthews stability assessment. A comprehensive database of actual performance and predicted performance is being compiled. Site specific stability curves will be created using this information for the main rock types which form the walls of the stopes.

7.2 *Void Filling*

Timing filling of stoping voids is critical not only maintaining production but also limiting adverse ground subsidence and stress redistribution effects on excavations near these voids.

We adopted paste technology for filling these voids. A 79-82% solids paste is reticulated underground by pipeline through a series of boreholes. The paste fill consists of the complete portion of the tailings produced by the mill mixed with approximately 2-5% cement. Cannington's paste will generally have unconfined compressive strengths up to 0.7 to 0.8 MPa after 28 days curing.

Paste is currently being used to fill stopes using a primary, secondary sequence. Stopes are generally 50 metres tall with 20 to 30 metre square bases. Cement content varies with fill exposure area (height and width) and the time between fill and exposure. Cannington has yet to develop and correlate actual fill strengths to strength models. However, cement content (and fill costs) may be optimised by adjusting fill exposures and curing times.

The use of paste fill limits water drainage in comparison to conventional deslimed hydraulic tailings. With reduction of water in the rock mass surrounding stoping areas, improved ground behaviour is also envisaged.

8 TECHNICAL SERVICES

Two geotechnical geologists working alternate rosters to give full site coverage, work from within the Cannington Technical Services group. Their role is to:

- provide day to day operational and quality assurance support;
- stay in touch with industry best practice and investigate the application of improved systems;
- develop models to predict future ground performance issues; and
- assist mine planning to improve extraction designs.

8.1 Hazard Surveys

The early identification of hazards is an essential component of the mining philosophy at Cannington. Underground Geotechnical Hazard Surveys are carried out on a weekly basis by geotechnical and operating personnel. Every active stope and heading in the mine is assessed in these surveys, and a updated list of any remedial action required compiled.

Once the Survey has been completed, this list is delivered to the Mining Shift Superintendent with appropriate plans, and any action required noted. Any work required is overseen by the Mining Shift Superintendent in the following week.

The number of hazards recorded underground has significantly decreased since the adoption of this Hazard Survey procedure.

8.2 Quality Assurance Testing

Poorly installed ground support can be more dangerous than no support at all.

Pull testing is carried out at Cannington to determine the load bearing capacity of grouted and ungrouted HGBs.. Fibrecrete thickness tests are carried out to ensure the design thickness is being applied, and installation procedures discussed with operators regularly.

8.3 Instrumentation

Movement monitoring is essential for some of the large and highly travelled excavations at Cannington. Large scale movement is usually preceded by movement on a small scale, and this must be detected early to allow timely actions to be implemented.

Ground movement is monitored at Cannington using rod and resistance wire extensometers, glass slides and convergence measurements. Life of mine excavations such as the Multiplate Arch Portal, Crusher Chamber and the Decline are monitored on a regular basis.

Underground personnel are given ownership in monitoring projects in their areas. Ore handling operators keep track of extensometers and glass slides in their work area, with the guidance of geotechnical staff.

8.4 External Geotechnical Assistance

External advisors are commissioned regularly to audit procedures at Cannington. They provide a mentoring role for geotechnical staff, run training programs on the basics of ground behaviour for mining personnel, and assist in times of peak work load.

We maintain regular contact with a range of geotechnical advisors. We see that the regular contact ensures that we are in touch with practical industry developments and we had the ability to look from outside the 'fish bowl' at our performance and direction. Using a range of contacts, broadens our effective knowledge base and allows us to select the best person for each specific task.

9 CONCLUSIONS

Cannington is securing a strong licence to operate with continual improvement in its safety performance. An integrated ground management approach, built in with our people continues to be fundamental to our simply safe approach and strive for improving quality.

10 ACKNOWLEDGMENTS

Permission to publish this paper was given by BHP Cannington.

The author wishes to acknowledge all the people who have given assistance in the preparation of this paper.

11 REFERENCES

Lennox, A.W., McGuckin, P.H. and Seedwell, V., 1998, Learning about Safety, *AusIMM Underground Operators Conference, Townsville, 30 June-3 July, 1998.*

Nobelius, J., Milburn, A., Johns, M. and Hill, J., I Can Read the Writing but I Can't Write the Reading, *TRAM Conference, Sydney, August 1998.*

Farcich, A. Kemp, M.S, and Pratt, A.G.L, 1998 Recruiting for an Underground Workforce at Cannington, *AusIMM Underground Operators Conference, Townsville, 30 June-3 July..*

APPENDIX A
Example of Underground Operations Procedure
Scaling

Task	Step	Key Points
1. Prepare for scaling	1. Check shift board in muster room for indications of ground conditions. 2. Wear all correct **personal protective equipment**. 3. Check that ventilation is adequate. 4. If ventilation inadequate, contact service crew or supervisor. 5. Wash down area before scaling according to *Washing Down Procedures*.	• The classifications for ground conditions are: **Green - Good** **Red - Very Poor** WARNING If very poor ground conditions exist **do not scale**. Barricade area off and notify your supervisor.
2. Scaling down	1. Commence barring down ensuring you: • use a straight, sharp and correct length bar • maintain good footing • have a clear safe retreat • work from good ground to bad ground sounding ground as you go • keep scaling bar at 45% angle • pull up on the scaling bar, do not push down • watch for unexpected falls • drop the bar if a rock slides down it; and • **do not** scale close to another person 2. If loose rocks cannot be reached, use a manitou and a work platform.	WARNING **Do not** use a scaling bar that is too short as this will put you at risk of being hit by rocks being scaled down. WARNING When scaling, keep bar at your side, **do not** hold out in front of you as serious injuries may result if large rock/s fall and push the bar back toward you. WARNING If you think the ground conditions are unstable or scaling cannot be carried out safely, barricade area off using red and white "no entry" tape. **Notify supervisor.**
3. Clean up	1. Roll up hoses. 2. Replace scaling bar on L-pins or put back on vehicle.	• Make sure equipment is put away to prevent it from being run over or lost.

Rock Support and Reinforcement Practice in Mining, Villaescusa, Windsor & Thompson (eds)
© 1999 Balkema, Rotterdam, ISBN 90 5809 045 0

Rock bolting practice and the geotechnical management system at Darlot Gold Mine

Dianmin Chen & David Collopy
Homestake Gold of Australia, Darlot Gold Mine, Leonora, W.A., Australia

ABSTRACT: Ground support is critical for mine safety. A number of geotechnical factors should be considered in the determination of a ground support system. The ground support design must be adjusted to cope with different ground conditions. Due to limits of other rock bolt systems a resin bolt system has been adopted to improve the ground support at Darlot Gold Mine. Currently the Secura resin bolting system is being utilised as the primary ground support. This bolt system has a higher load transfer capability, is relatively simple to install and has a relatively low cost. Based on the geotechnical conditions the ground support regimes have been defined. The ground support requirements are determined based on geotechnical inputs such as a rock mass rating, purpose and life of excavation. Other geotechnical measures have also been taken to maintain a good ground support practice. These include continuous monitoring of the performance of rock bolting systems, strict quality assurance, promotion of a geotechnical-sensitive culture and training of all underground personnel.

1 INTRODUCTION

Rock fall has been the major cause of fatalities in underground metalliferous mining in Western Australia from 1980 to 1997. There have been more than 600 rock falls reported during this period and 32 people lost their lives due to rock falls according to the report from The Department of Mineral and Energy (DOME) (Anon (a), 1998). Since 1995 there have been 237 reported rock falls in WA underground mines. Of these nine have resulted in fatalities and 68 in injuries.

To reduce the risk of injuries due to rock falls, effective ground support systems are essential. To determine a suitable ground support system, a series of geotechnical assessments have to be carried out. This paper presents our experience in selecting an effective rock bolting system, implementing a comprehensive geotechnical management system and improving underground safety at Darlot Gold Mine. The work involved in this project includes:

- Evaluation of the past rock bolting practice.
- Assessment of rock bolting systems available.
- Trial of different bolting systems.
- Determination of suitable systems for our particular needs.
- Quality assurance and other procedures.
- Design of ground support parameters for different ground conditions.
- Timely geotechnical input from geotechnical data acquisition;
- Change of design parameters based on the updated geotechnical conditions;
- Ground awareness training course for all levels of underground personnel;
- A geotechnical report system for promotion of a geotechnical-sensitive culture.

2 DARLOT GOLD MINE

Darlot Gold Mine is fully owned by Homestake Gold of Australia. The mine site is located 122km north of Leonora. The mine is accessed by road via Leinster or Leonora and by air service from Perth.

The Darlot region has a long history of gold mining, with the first gold discovered in 1894 (Anon (b), 1994). This discovery started a gold mining boom, lasting until 1910, with the region proving to be one of the richest alluvial gold fields in WA. The area was extensively worked for gold in quartz lodes, deep leads and alluvium. Much of this activity was centred on the hills and creeks below the Mt Pickering breakaway. Several mines were involved in the mining activities, all of which are located within the current mining lease. By 1920 however, the majority of the historically successful mines had closed and no large scale exploration was carried out until 1986. In 1992 Darlot Gold Mine become part of Plutonic Resource Group and just recently Homestake has taken over all Plutonic operations. Open cut mining of Darlot orebody continued until last quarter of 1995, while underground mining of Darlot orebody commenced in October 1995.

As one of the most recent and exciting gold deposit discoveries in WA, the Centenary orebody lies 1.2km east of Darlot mine. The development of this orebody is currently under way and ore production from stoping commenced late in 1998. Several mining methods are used in the mine with the majority of the Centenary orebody to be mined by long hole stoping mining methods. The current ore production is 700,000 tonnes per annum with a potential to increase to more than 1.0Mt/y.

Centenary is a massive gold deposit, 300 – 500m below surface. Lamprophyre intruded into the central part of the orebody and these intrusions have made the rock mass condition variable. A large lamprophyre shear zone has also been identified within the central lamprophyre intrusion body and this sheared lamprophyre is highly fractured and very unstable as it is exposed. The mineralisation is hosted in magnetic dolerite and usually associated with quartz veining. The grade of gold is normally proportional to the density of quartz veining.

3 GEOTECHNICAL CONDITIONS

3.1 *Rock strength*

The rocks are strong to very strong and the uniaxial compressive strength test results are summarised in Table 1.

Table 1. Uniaxial compressive strength test results.

Rock Type	UCS (MPa)		
	Mean	Std Dev.	No. samples
Magnetic Quartz Dolerite (MMQD)	257	63	13
Magnetic Dolerite (MMD)	277	87	8
Lamprophyre (ULP)	143	31	4
Felsic	195	3	3

3.2 *Geological structures*

A major faulting system is well developed in Centenary. The dominant faults/fault zones are NW-SE striking. These faults extend hundreds of meters along the strike and plunge. The thickness of these structures varies from a few hundreds of millimetres to several meters. The infill materials of these faults can be significantly different, varying from quartz to sericite / chlorite. Some parts of a fault can be well healed and other parts open. This great difference in infill of faults makes the prediction of fault behaviour very difficult.

Other structures include two groups of quartz veins, one flat dipping and one moderately dipping to south. The flat dipping quartz veins have been considered to be the most critical structure for the stability of an opening.

The representative geological structures are summarised in Table 2 and Figure 2.

Table 2. A summary of major joint sets.

Set No.	Dip (°)	DipDir (°)	Type*
1	85	050	F. J.
2	10	280	V
3	55	180	V. J.
4	60	020	J

* F – Fault; J – Joint; V – Vein.

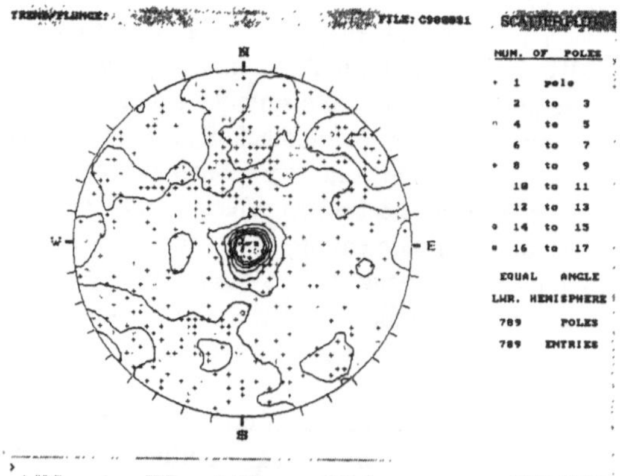

Figure 1. Joints mapped from Centenary.

Table 3. Stress field in Centenary.

Principal Stresses	Magnitudes (MPa)	Dip (Degrees)	Bearing (Degrees)
Maximum	33.7	8	088
Intermediate	21.1	6	178
Minor	10.3	80	304

3.3 *Stresses*

Virgin stresses have been measured using CSIRO HI cell overcoring techniques. The maximum principal stress is 33.7MPa at the depth of 330m, which is higher at that depth compared to some of the underground mines in WA. The results of the stress measurement are shown in Table 3.

A stress analysis using MAP3D has been carried out and it shows that a maximum mining induced stress could be up to a level of 75MPa in the areas of stress concentration, while other parts of the mine would be in a tension condition (AMC, 1998). Based on this modelling a mining strategy has been developed to minimise the impact of high stresses on the mine production.

3.4 *Ground water*

Some ground water problems have been encountered in Centenary. Water runs out from diamond drill holes and cable bolt holes through a shear zone that may connect with possible upper level aquifers. The water is salty with sulphides, a laboratory test of the underground water indicated that a corrosive rate of 0.048mm/y for galvanised steel and 0.63mm/y for BHP steel. This indicates that a normal galvanised bolt may last only one to two years if water is present.

4 PAST ROCK BOLTING PRACTICES

4.1 *Universal bolts*

During the early stage in the underground development the point anchor type Universal bolt was used primarily. The Universal bolt used at Darlot has a high tensile strength of 31 tonnes and it has the following advantages:

- Relatively easy to install;
- Able to take load immediately after being installed;
- Relatively low cost.

However, the Universal bolt has its disadvantages as follows:

- Loss of support capacity if the collar becomes loose;
- No resistance in shear movement until the bolt body contacts with rock;
- Need to retain the tightness of the nut by checking periodically.

Frequent blasting in the adjacent areas is usually responsible for loose rock in the collar due to the vibration. Once rock at the collar becomes loose the bolt could be dangling in the hole and provides no support to the ground as shown in Figure 2. If more bolts fail to support the ground rock falls may occur, which could be dangerous for equipment and personnel. Figure 3 shows that an area supported by Universal bolts failed.

Figure 2. A Universal bolt provides no support at T277.

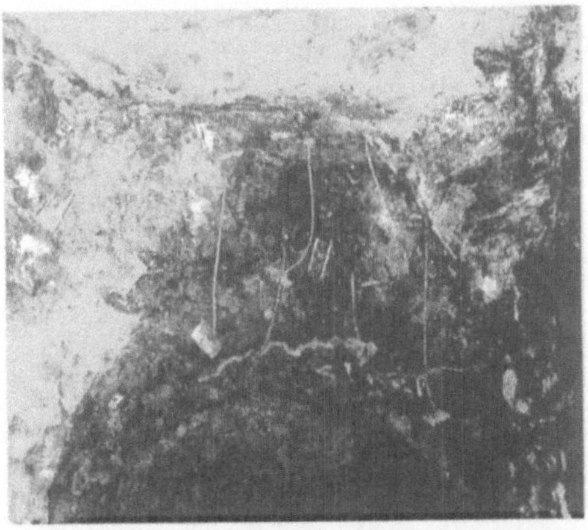

Figure 3. Wedge fall in an area supported by Universal bolts at T242.

This rock failure caused some damage to the equipment.

In areas with no nearby blasting and no water problems the Universal bolt was successful. However, it could become a problem if water causes corrosion. Therefore the Universal bolt cannot be used as a permanent support because the bolt is not grouted to protect from corrosion.

4.2 *Split Sets*

Split Sets have been widely used at Darlot Gold Mine in both development and stoping areas. The typical Split Sets used in Darlot are 46mm (in diameter) with variable lengths of 3.0m, 2.4m, 2.1m and 1.5m. The pull out tests on 3.0m long Split Sets indicated a pull-out force of 15.0 tonnes. This support has been used in Jumbo stoping areas and other short term excavations. Split Sets are easy to install and provide quick support. The major shortcomings of this support are:

- Limited support height due to a low load transfer capability, i.e. a longer anchor length to provide sufficient support.
- Low anti-corrosion capability if water present.
- Lower support capability.

4.3 *Grouted Split Sets*

Due to the shortcomings Split Sets are not used for a permanent support. To solve this problem grouted Split Sets have been introduced. The successful application of this system has been reported (Villaescusa & Wright, 1997). It has been noted that the grouted Split Set may still corrode where underground water present. The grouted Split Sets can then be reinforced by a single strand cable with a face plate. This grouted single strand cable can act as an anchor type support even if the Split Sets become corroded. However this system requires more than one step operation. If the operation is under pressure the grouting work will be delayed and it may cause some stability problems. Excessive load may cause the ungrouted Split Set to slip and rock failure may occur, as shown in Figure 4. The advantages of this system are:

- It turns the temporary support to permanent support.
- Higher shear resistance.
- Increased bonding strength.

The major disadvantages are:

- Three-step operation, i.e. installation of Split Sets, grouting cables and face plating.
- The cost per bolt is higher compared to other systems.

4.4 *CT bolts*

A trial of CT bolts was carried out in Darlot. This system provides an immediate support to rock masses but requires post grouting. This is a two pass installation system and more than one step operations are involved. In addition, the unit price per bolt is high. This bolt is not used at Darlot Gold Mine at the moment.

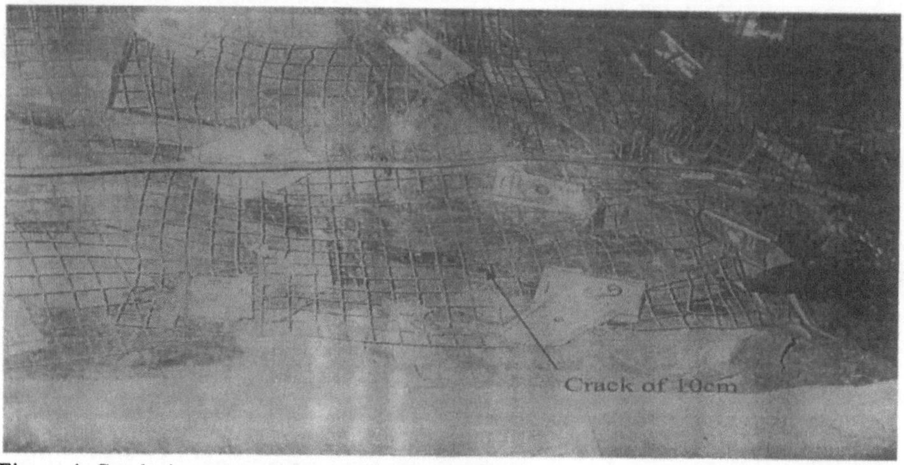

Figure 4. Cracks in an area supported by Split Sets.

4.5 *Cable bolts*

Cable bolts have been widely used in all stoping areas and intersections. Different cable bolts, mainly single or twin strand plain and bulbed cables, are used. In most cases the cables are fitted with face plate of 150 x 150 x10mm. Length of cables is varied depending on the ground conditions and locations. The practice in Darlot has revealed that cable bolts are effective in controlling rock failure and provide a safe access for production.

4.6 *Other supports*

Other supports used in Darlot include mesh, shotcrete/fibrecerete, W straps and steel sets. Mesh and shotcrete/fibrecrete are used in some of the permanent accesses and key locations such as the decline portal, stope drawpoints and other permanent facilities. W straps are usually used where more than one free face within a blocky rock mass condition is exposed.

5 RESIN BOLTING SYSTEMS

Due to the limitations of other bolts discussed above, a resin bolt system has been introduced since May 1998. It was expected that the resin bolting system would meet the following requirements:

- The support capability of more than 20 tonnes.
- One-off operation for installation.
- Low maintenance.
- Good anti-corrosion ability.
- Pretentioning and immediate support.
- Higher shear resistance.
- Reasonable cost per bolt.

5.1 *The trial of Posimix resin bolts*

The Posimix resin bolting system consists of a Posimix bolt, fast set resin cartridges, slow set resin cartridges, plate and nut. The procedure of installation for mesh and resin bolt is briefly described below:

- Drill hole to the required depth (2.95m in Darlot).
- Shoot the fast set resins (spin time 10 seconds).
- Shoot the slow set resins (spin time 20 seconds).
- Push a bolt (3.05m long) into the hole and spin while pushing to the bottom of the hole.
- Hold for the resin to set (approximately 30 seconds).
- Tighten the nut.
- Start the next cycle.

The equipment used for installation is Tamrock Robolter 320, as shown in Figure 5.

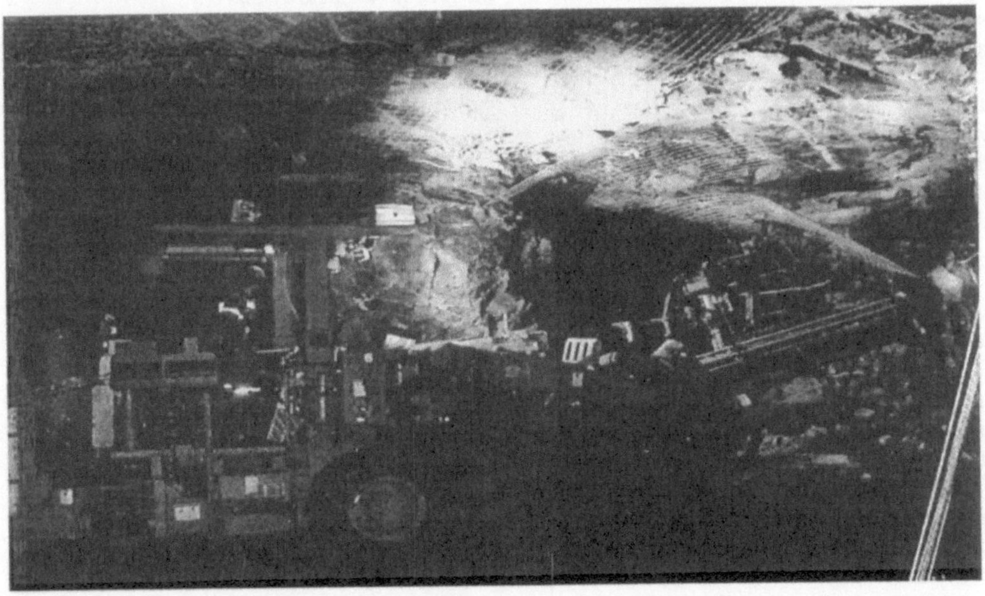

Figure 5. The Robolter used in Darlot Gold Mine.

5.2 *Test on Posimix bolts*

Tests have been carried out to check the performance of the Posimix bolts. A load transfer test on Posimix bolts indicated that a bonding length of 300mm could provide 9 tonnes support force for the bolt in a 35mm hole. It means that one metre anchor length may provide sufficient bonding strength for the bolt to achieve its ultimate tensile strength.

A number of tests were carried out to check if full encapsulation was achieved. Holes were drilled in walls and the length of the hole was measured. After the bolt was installed the grout condition was checked. If the bolt was not fully grouted more resin was required. For example if a 33mm knock-on bit and a 22.7mm (diameter) Posimix bolt was used, then four 880mm long resin cartridges were required to achieve a full encapsulation.

Pull-out tests were undertaken to understand the performance of the bolts installed. The typical pull-out load used for the test was 18 tonnes. The maximum load tested was 25 tonnes.

5.3 *Discussion on Posimix bolts*

The test of Posimix bolts indicated that the bolts provided adequate bonding (or friction) strength to achieve their tensile ultimate strength. However, the quality check of the grout indicated that some of the bolts were not fully encapsulated and that the maximum un-grouted bolt length was 1.0m. The main reason for the un-grouted part of bolts was that the hole diameter (33 - 35mm) was too big. A total of 3.2 - 3.8m long resin cartridges (25mm in diameter) would be required to achieve a full encapsulation. These resin cartridges had to be squeezed into the hole, which may have caused the resin to burst. The burst resin could have set prior to spinning and it would then be difficult to push the bolt into the hole. Some bolts had left a long tail outside the hole.

The installation operation was not as efficient as expected and the cost per bolt due to the amount of resin required was considered to be higher by the contractor. The partially grouted bolts also provide little resistance to shear movement on joints. As indicated previously, flat dipping joint sets dominate in Centenary and shear movement is likely to occur under high horizontal stress conditions. A full encapsulation would then be critical for an effective rock bolt system.

Due to the amount of resin required to achieve a full encapsulation it was difficult to put so much resin into the hole. A modification of the resin inserter was required so that a smaller hole could be drilled. Full encapsulation and a cost reduction per bolt could therefore be achieved. More importantly, the installation would be easier and utilisation of the Robolter could be improved.

5.4 *Trial of the Secura bolts*

A modification of the resin inserter was carried out by our contractor, which aimed at drilling a smaller hole whilst still meeting our ground support requirements.

The modified drill boom allowed the Robolter to drill a 29mm hole and shoot 25mm resin cartridges. A trial was carried out in the C1137 area. Three Secura bolts (a nominal diameter of 21.7mm) were installed under close supervision. One (1) 330mm fast set and two (2) 1000mm slow set resin cartridges (25mm in diameter) were shot into the holes. The bolts were pushed to the bottom of the hole while spinning and resin ran out from hole thus full encapsulation was achieved. These bolts were pulled to 22 (nut ripped off) – 25 tonnes. The trial was successful.

5.5 *Test of the Secura bolts*

After a trial of the Secura bolts, the system was further tested to find its load transfer capability. Three bonding lengths, 300mm, 600mm and 900mm, were tested underground. To ensure the designed encapsulation length, poly-tubes were used to cover the bolt body that should not be grouted. After all three shortened bolts were installed pull-out tests were performed. An initial load of 2.5 tonnes was applied to the bolts to obtain a base reading of the bolt displacement. The load and corresponding displacement were then recorded, as shown in Figure 6.

5.6 *Discussions on the Secura bolts*

It was found that the bonding strength of Secura bolt in a 29mm hole was higher compared to a Posimix bolt in a 35mm hole. With 600mm bonding length the Secura bolt reached close to its minimum ultimate tensile strength of 23 tonnes.

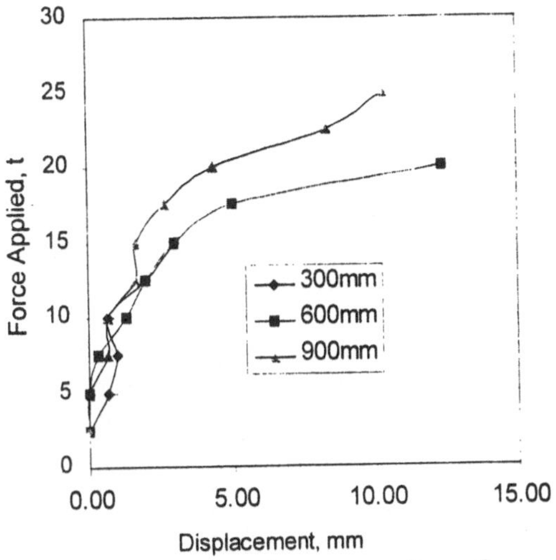

Figure 6. Load transfer test results – Secura Bolts.

As only a small amount of resin was used, the chance of resin burst in holes was reduced. Bolts could be easily pushed into holes and rebolting due to difficulties in spinning/mixing was reduced. Full encapsulation was achieved. Unit price per bolt was also reduced. Although there are still some problems in achieving a consistent quality of installation the Secura bolts are now preferred by the contractor and the mine.

6 GEOTECHNICAL MANAGEMENT SYSTEMS

After the primary rock bolting system has been selected the correct support parameters are to be determined. These parameters are related to the ground conditions and other factors. In the event of additional geotechnical information ascertained from underground excavations these parameters may change.

6.1 *Rock mass ratings and ground support regimes*

Four rock mass regimes have been identified using the Mining Rock Mass Rating (MRMR) system (Laubscher, 1990). Minimum support requirements for these regimes were designed based on empirical methods and engineering considerations. The details of these regimes and their corresponding *minimum* support requirements are shown in Table 4.

Table 4. Ground support regimes.

Regime	MRMR	Support Requirements
1	35 and less	Pattern resin bolting, mesh to half height of walls, Split Sets on walls
2	35-45	Pattern resin bolting, mesh in the backs, Split Sets on walls
3	45 - 55	Pattern resin bolting in the back, Split Sets on the walls
4	55 and above	Spot bolting

All development faces are geotechnically mapped after each round. The rock mass rating is then calculated according to the information recorded. This rating is used for the determination of ground support regimes.

6.2 *Geotechnical management system*

The Centenary area is still under development and a monitoring system has yet been established. Frequent observation has been carried out in all developed areas. If there are any signs of deterioration of rock masses and inadequate support, the additional support can be installed. However, it is difficult to have supervisors or geotechnical personnel underground all the time (particularly in a fly in-fly out roster). The involvement of the underground work force is essential to understand what is happening geotechnically in underground. A program for the promotion of a geotechnical-sensitive culture was proposed. This program included training all of the underground work force and implementing a geotechnical event report system. The training course was specifically designed for Darlot. It included basic geotechnical skills, ground support methods and instability identification.

The geotechnical event report system was designed for reporting any events that may have an impact on ground stability. The area concerned was then inspected by an experienced person such as a geotechnical engineer, a mining engineer, a supervisor or a foreman. Feedback was issued to the mine management, contract management and reportee. This feedback includes an analysis of the report and action plans where necessary.

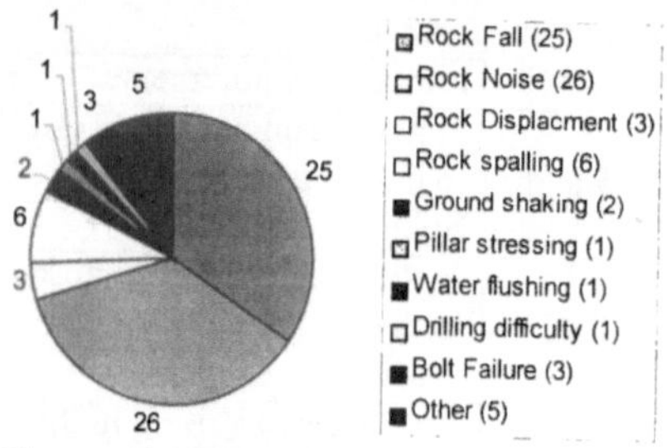

Figure 7. Geotechnical events reported (seven months period).

The reported geotechnical events are mostly rock falls and mine microseismic (rock noise) related activities. Others include ground water problems, drilling difficulty, stressed pillar conditions, rock displacement/deformation and fault/shear zones. A summary of these reports is shown in Figure 7.

The actions taken after these reports include:

- Additional ground support,
- Change of ground support regimes,
- Reviewing of mine design,
- Change of mine design, and
- Justification of geotechnical monitoring.

7 CONCLUSIONS

After trialing different resin bolting systems Darlot Gold Mine has committed to use the Secura resin bolts as the primary ground support in most of the Centenary development and stopping areas. It is considered that this ground support will meet our ground support requirements. No major failure in the area supported by Secura and Posimix bolt system has been reported. From the use of this bolting system and the implementation of geotechnical management systems the following conclusions can be derived:

- Hole size is the critical factor to achieve an effective resin bolting system.
- The modified Robolter drill boom can be used to drill a smaller hole (29mm in diameter). This reduces the use of resin by 38%, compared to a hole size of 35mm. Therefore cost per bolt is reduced.
- Less resin in the hole allows for easier installation of the bolt with the Robolter at Darlot.
- A full encapsulation can be easily achieved by using the 29mm hole.
- The number of resin cartridges is reduced, which can reduce the possibility of resin blockage or bursting.
- The resin annulus is reduced to approximately 3mm. The mixture of resin is relatively easy.
- The load transfer capability is improved by using 29mm hole and 22mm bolts. This could lead to a reduction of bolt length.
- The resin bolting system installed by a Robolter is a one-off operation and no work is done under unsupported ground.
- Quality assurance is critical in maintaining a consistent resin bolting practice.
- The resin inserter tube is small and it is difficult to quickly align the inserter with the hole.
- Resin may be lost in cracks and full encapsulation could be in doubt.
- Resin must be stored in the conditions specified by the supplier. Underground conditions at Darlot are suitable for resin storage.
- The geotechnical management system is essential to have a better coverage of geotechnical conditions.
- Training of all levels of the underground work force including management is important to promote a geotechnical-sensitive culture.
- Feedback to all geotechnical events reported is the key of the success to the geotechnical management system.

ACKNOWLEDGEMENTS

The authors would like to thank Darlot Gold Mine of Homestake Gold of Australia for the permission of publishing this paper. Special thanks are due to Strata Control System, ANI Support and Byrnecut Mining Company for the assistance in carrying out this project.

REFERENCES

AMC, 1998. Centenary Numerical Modelling. Internal Report for Darlot Gold Mine, Plutonic Resources.

Anon (a), 1998. Surface rock support for underground excavation, Code of practice (draft). Mines Occupational Safety and Health Advisory Board, WA.

Anon (b), 1994. Darlot – The Centennial Publication (1894 – 1994). Plutonic.

Laubscher D., 1990. A geomechanics classification system for the rating of rock mass in mine design. *Journal of South African Institute of Mining and Metallurgy*. Vol 90 No. 10, p257-273.

Villaescusa E. and J. Wright, 1997. Permanent excavation support using grouted Split Set bolts. *The AusIMM Proceedings*.

Rock Support and Reinforcement Practice in Mining, Villaescusa, Windsor & Thompson (eds)
© 1999 Balkema, Rotterdam, ISBN 90 5809 045 0

Developments in ground control at Outokumpu's Forrestania Nickel Mines, Western Australia

C.Carr & P.Lappalainen
Forrestania Nickel Mines, Varley, W.A., Australia

M.P.Sandy
Australian Mining Consultants, Perth, W.A., Australia

ABSTRACT: Underground development at the Cosmic Boy deposit, located 400km east of Perth, Western Australia, commenced in April 1992, with initial production in January 1994.

The main economic target is a 2m to 8m thick disseminated nickel sulphide orebody that forms the basal member of a serpentinised, ultramafic flow sequence. The orebody lies either directly on the contact with an underlying Banded Iron Formation (BIF).or is separated from it by a mafic dyke 0.5 to 2m thick.

In common with many ultramafic rocks, the orezone and immediate hangingwall at Cosmic Boy exhibit significant, time-dependent deterioration. In some areas where shearing is well developed, ground conditions are extremely poor and unraveling develops rapidly, even in small exposures. These problems are exacerbated under elevated stress conditions.

Early application of 'skin' support comprising mesh reinforced shotcrete is required in many areas to retain surface rock. Cables are also installed for deeper anchorage, longer term support, and as stope hangingwall support, in order to minimise waste rock dilution.

Developments in support and reinforcement practice at the operation are described.

The mining method has also been changed to better suit the difficult ground conditions, and to reduce hangingwall dilution. A sustained improvement in head grade has been achieved since the new method has been implemented. Extraction sequencing plays a key role in managing the effects of stress redistribution. A life-of-mine extraction sequence has been developed for the operations.

1 INTRODUCTION

1.1 *Location and Regional Setting*

The Cosmic Boy Mine is situated approximately 400km east of Perth, Western Australian in the central part of the Forrestania greenstone belt. The belt trends north-south and comprises two major sequences: a lower mafic-ultramafic sequence containing numerous banded iron-formation (BIF) horizons and an upper sequence of pelitic to psammitic schists. Enclosing the greenstone belt is a terrain comprising deformed and recrystallised granitoids and gneisses which have been intruded by younger, undeformed plutons of granite and adamellite. A series of east-west-trending Proterozoic dolerite dykes intrude the Archaean rocks.

A number of nickel sulphide and gold deposits have been discovered in the Forrestania greenstone belt.

All of the economic nickel sulphide deposits occur within the Eastern and Western ultramafic belts, mostly in the lowermost one or two ultramafic units (Frost *et al*, 1998).

1.2 *Mineralisation*

Cosmic Boy is the largest nickel sulphide resource at Forrestania, containing about 4.0 Mt at 2.4% nickel in two parallel ore zones. Production is currently from the larger and more easily mined basal orebody. The basal orebody is about 800m in strike length and extends from the weathering base to about 500m depth, at the base of a 40m to 60m thick olivine mesocumulate unit, overlying a prominent BIF unit.

The main economic target at Cosmic Boy is a 2m to 8m thick disseminated nickel sulphide orebody that forms the basal member of a serpentinised, ultramafic flow sequence. The orebody lies either directly on the contact with the underlying BIF or is separated from it by a mafic dyke 0.5m to 2m thick.

On average, the orebody dips between 40° and 50° to the west, although local variation from 20° dip to sub-vertical are possible. These are often associated with major, transverse sub-vertical fault zones.

1.3 *Mining Operations*

Underground development commenced at Cosmic Boy in 1992, with initial production in 1995. Annual production is around 400,000 tonnes at a head grade of 1.7% Nickel. Recent improvements to stope design and operating practice have increased the head grade to above 2.3% nickel. Primary access is via a footwall decline located in the main footwall BIF.

Most of the ore to date has been produced from long hole bench stoping and development. Some of the flatter-dipping areas have been mined using cut-and-fill. Bench stopes were mostly backfilled with waste rock in the early stages of mining, but the flat dip makes it difficult to provide effective hangingwall support with dry fill.

A number of deep stope hangingwall failures have occurred, and backfilling of these stopes has sometimes not been possible. Some of the larger unfilled stope voids have continued to fail or 'cave' through several levels.

The lower parts of the orebody are being extracted using small, 'blind' uphole stopes with yielding rib and crown pillars, retreated down dip.

2 GEOTECHNICAL ISSUES

2.1 *Ground Conditions*

Conditions in the footwall BIF are usually fair to good with Q values (Barton et al, 1974) in the range 4-15.

The RQD is typically above 90% and jointing is widely spaced. Based on a small number of tests intact rock strength in the massive BIF is 300-350 MPa.

At some locations in the lower levels of the mine, minor strain bursting has been experienced. This appears to reflect local variation in the stress field, and is associated with proximity to one of the major transverse faults (the 'R' fault). 'Coining' or 'discing' in drill core is common in the BIF in the vicinity of this fault, suggesting that the stress may be locally elevated or more deviatoric than elsewhere in the mine. Experience in other Western Australian underground mines shows that lateral as well as vertical variation in the stress field is not uncommon, particularly where major structural features are present. In common with other mines in the Forrestania greenstone belt, the pre-mining stresses appear to be elevated relative to the depth; the maximum horizontal stress may be 3-5 times the 'overburden' stress.

Ground conditions in the ore zone and hangingwall are quite variable and range from massive, poorly jointed ultramafic, with Q values >15 to (more typically) blocky conditions with several joint sets (Jn = 12-15) and thin, low friction (talc-chlorite) joint infill (Q = 0.9-2.5). In some areas the ore zone and hangingwall are intensely sheared, with abundant low friction minerals present (Q = 0.3-0.9).

A series of major sub-vertical northeast - southwest trending fault zones transect the deposit, spaced to 50m to 100m apart. Ground conditions are often poor within the fault zones and thick infills of talc, chlorite or other fibrous serpentinite minerals may be present where the faults intersect the ultramafic. The faults have less effect on conditions in the underlying BIF. Some of the major faults may be represented by rehealed, siliceous breccias, while others are difficult to trace. The faults can offset the ore contacts by up to 15m; a number of fault windows are evident on the orebody long projection.

2.2 *Observed Ground Behaviour*

In the BIF, apart from minor strainbursting in isolated sections of the lower levels there is little obvious response to excavation and the stress changes associated with orebody extraction.

The stress field has not been measured at Cosmic Boy. Evidence discussed above suggests that the pre-mining stress field is likely to be complex, variable and affected by major structures. A large number of measurements would probably be required to develop an adequate understanding of this variability and the factors controlling it.

At this stage there are no plans to extend mining to greater depths than the current lower levels. If deeper, economic resources are identified it may be necessary to undertake a stress measurement programme to define the local stress-depth gradient.

Rock mass strength within the ore zone and hangingwall ultramafics is also highly variable. In the more massive ultramafic UCS_{50} values of 50-70 MPa would be expected. The more sheared material is not amenable to laboratory strength testing, but the UCS_{50} would be predicted to be less than 10 MPa. The combination of variable strength and (possibly) variable stresses leads to a wide range of ground behaviour.

The ore drives have generally been driven along the footwall of the ore zone, with the contact shear between the BIF and the overlying mafic or ultramafic exposed in the eastern sidewall (footwall) of the drive.

On initial development before production commenced, most ore drive exposures required only light pattern bolting or spot bolting to maintain stability by securing isolated, structurally-defined blocks and wedges in the backs and walls.

In areas where the conditions are severely affected by shears and in particular the 'R' fault, rapid unravelling has occurred (Figure 1), in some cases causing the complete collapse and abandonment of the development.

However, in the areas not affected by shears/faults deterioration occurs more slowly, and is dominated by shear and dilution on discrete structures, consistent with a maximum principal stress approximately orthogonal to the stratigraphy.

Where unfavourable combinations of structures are present in the walls (particularly the hangingwall, but also in the footwall in conjunction with the footwall shear) blocks or wedges can be 'squeezed' out.

Deterioration usually first occurs as movement on the footwall contact shear, often accompanied by seepage indicating that the shear has locally dilated and connected through to development or stoping up dip.

In areas where the stresses normal to the orebody are elevated (for example in abutments or pillars), the shearing may be in response to bulking of the rock mass caused by induced fracturing and shear on structures.

As conditions deteriorate further, shearing may also develop sub-parallel to the footwall contact towards the drive hangingwall. This may cause loosening of the backs, assist in squeezing blocks out of the hangingwall, or result in local floor heave.

Shearing on the footwall contact shear becomes more accentuated, sometimes causing localised rockfalls where the footwall is unsupported, and allowing more pronounced floor heave. Where the footwall is meshed and bolted, deformations can be large enough to tear the mesh (Figure 1) and 'guillotine' rock bolts that are installed across the shear.

This style of behaviour is summarised in Figure 3. This first became apparent at access crosscut-ore drive intersections, particularly in pillars in areas of relatively advanced extraction.

Figure 1. Unravelling in sheared ultramafic. Unreinforced shotcrete has been ineffective in controlling the failure.

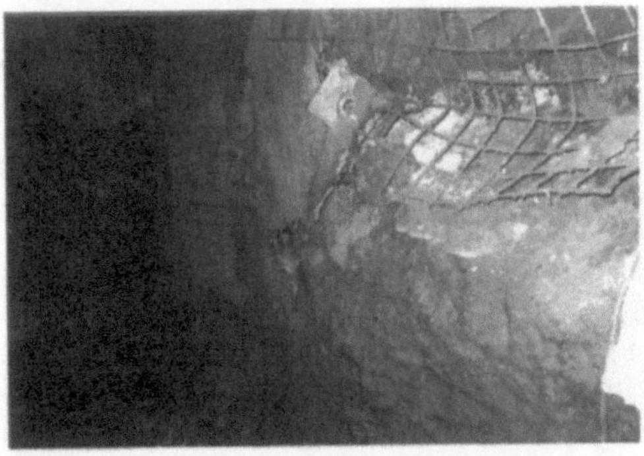

Figure 2. Onset of shearing on the footwall contact shear.

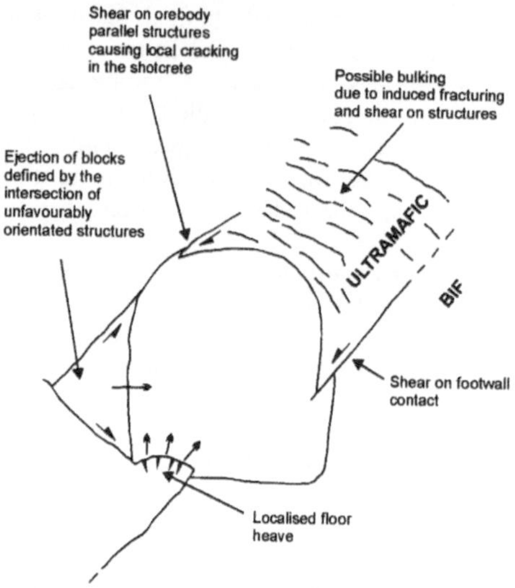

Figure 3. Interpreted ground behaviour in Cosmic Boy ore drives, (looking north).

Early deterioration of the walls was also observed soon after development in ore drives in the lower (down dip) regional abutment to stoping. The implication was that redistributed stresses in the regional pillars and abutments were reaching levels sufficient to cause severe drive deterioration. Another factor was the absence of cable support in the intersections. Initially, cable support had only been installed as part of stope hangingwall reinforcement, hence the oredrive backs were not usually supported, particularly in access intersections where the final pillar would be recovered in one blast.

3 GROUND SUPPORT

3.1 *Initial Approach*

In the early stages of ore drive development ground support comprised pattern friction stabilisers with meshing where required.

Unreinforced shotcrete was trialled in a few areas to control the loose scats that began to develop in the backs and upper sidewalls. The main objective was to speed up the development cycle.

It quickly become clear that ongoing deformation in the drive walls could cause the detachment of the shotcrete, presenting an additional hazard (Figure 4). Some areas had to be rebolted and meshed to control the deteriorating conditions including the detached shotcrete.

Figure 4. Detachment of unreinforced shotcrete, due to ongoing deformation.

3.2 *Recent Developments - Current Support Practice*

Following the re-bolting and meshing of some of the areas that had experienced ongoing deterioration, an application of plain shotcrete was placed over the mesh, from grade line to grade line.

The effectiveness of this type of support was immediately apparent (Figure 5). Even in areas where the rock mass had severely dilated or was affected by proximity to caved stopes, the rate of deformation was slowed substantially.

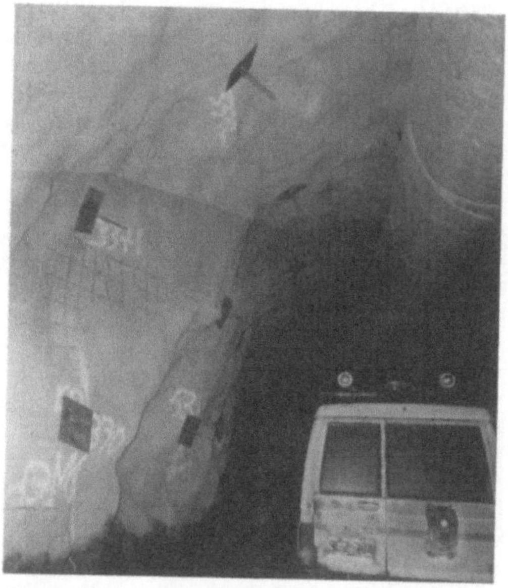

Figure 5. Rehabilitated ore drive support comprising mesh, friction stabilisers and nominal 50mm of shotcrete. Twin strand plain cables are installed on a 2.5m x 2.5m pattern.

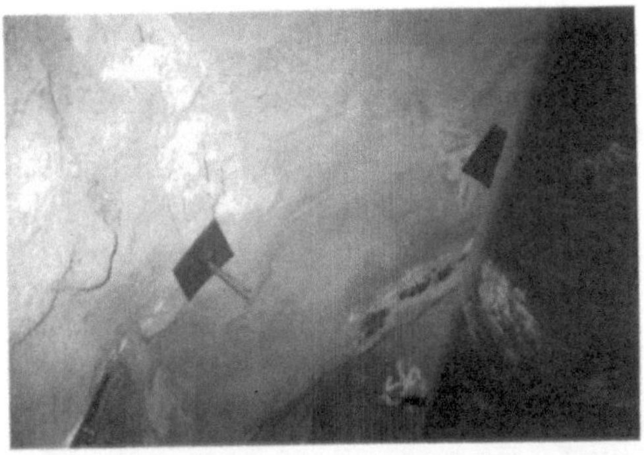

Figures 6 (above) and 7 (below). Ongoing deformation causing localised spalling of plain shotcrete off the mesh.

In the most severe conditions, further deterioration of the reinforced shotcrete can still occur (Figures 6 and 7).

Re-application of shotcrete can provide a temporary solution, but in many cases the movements on the major structures continue, causing further damage to the new shotcrete.

In some cases, an external application of mesh, secured with split sets, may be more effective. Alternatively, the use of fibrecrete would reduce the potential for failed shotcrete to detach and fall. In any event, some rehabilitation is inevitable given the ongoing deformation associated with the ground conditions, and the operational life of the ore drives.

Current support practice in 'typical' (massive) ore zone ultramafic conditions is to install bolts and F51 mesh from grade line to grade line, with the application of shotcrete no further than 5 cuts from the face.

In the sheared ultramafic, an initial application of 75mm of fibrecrete is placed within 4-6 hours of firing, followed (after 24 hours) by bolting and meshing. A second application of (plain) shotcrete is placed over the mesh. Twin strand cables are then

Figure 8. Cables installed through the footwall contact shear that have undergone more than a metre of shear (view looking up dip).

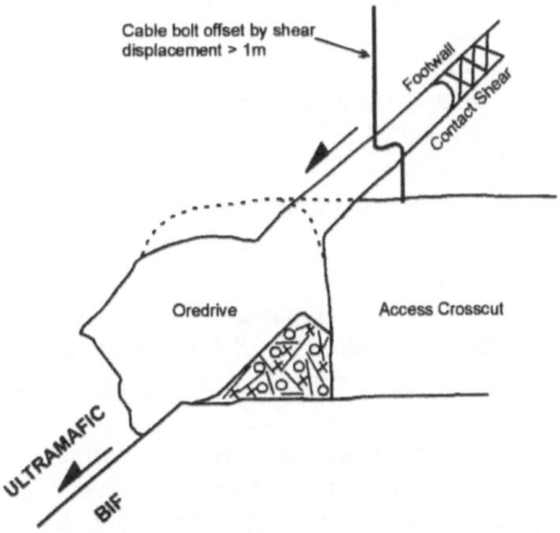

Figure 9. Cross section showing interpretation of ground behaviour in Figure 8. The hangingwall has sheared down dip relative to the footwall.

installed both for stope hangingwall support, and to provide deep anchorage reinforcement for the 'skin' (surface) support.

3.3 *Future Investigations*

Ongoing damage is observed, even in the more ductile reinforced shotcrete 'skin' support. Large scale shear movements have occurred on the footwall contact shear, and other major structures. Figure 8 shows an extreme example, where cables installed from an access crosscut have been subjected to a shear displacement of more than 1m. This is shown schematically in Figure 9.

Most of the cables would be expected to experience lower levels of (predominantly axial) deformation, however, the ability of even plain strand cables to accommodate these deformations may be a concern. Conversely, it could be argued that stiffness measured in the laboratory could be misleading, as the tests are conducted in a very short time frame compared to rates of loading/deformation in the field. Studies are in progress at Cosmic Boy of the performance of cables deliberately designed to yield.

4 EXTRACTION SEQUENCE

4.1 *Early Production*

The mining methods employed at the start of production at Cosmic Boy were uphole benching and cut-and-fill. The lower levels of a production area or panel, would be developed and then retreated back to the access crosscut. Backfilling using dry rockfill would then follow before commencing production from the next sub-level bench stope or cut-and-fill lift up dip.

Where conditions allowed, a stope would generally be left open until the crosscut was reached. The hangingwalls were reinforced with cables to improve stability.

A number of stope hangingwall failures occurred and stope sizes were reduced by the use local pillars, and by decreasing the strike length opened before introducing backfill.

With a single, central decline, the extraction sequence options generally involved closure on central pillars over the access crosscuts. To obtain early production, some stoping panels were commenced at intermediate levels, rather than on the lowest level. This resulted in the formation of a number of regional sill pillars that will have to be recovered as 'reducing' pillars.

4.2 *Sequence for Remaining Resources*

The proposed mining method for the lower production levels comprises a series of small, blind uphole stopes (each typically about 2500t) retreated down dip, with yielding rib and crown/sill pillars. This method is expected to eventually result in localised caving of the hangingwall.

The time-dependent behaviour associated with ultramafics results in ongoing deterioration of pillars and abutment areas. The general extraction strategy adopted for the lower levels at Cosmic Boy is to extract areas continuously and avoid extended periods when sequences are essentially 'stationary'.

The main objectives are, as far as possible, to retreat the stope 'front' continuously into one or both abutments, and to avoid the formation of isolated pillars, that need to be recovered as part of the mine plan.

5 CONCLUSIONS

- The style of ground behaviour commonly observed at Cosmic Boy results in large differential movements on the footwall contact shear and other structures in the ultramafic.
- Since the early stages of mining, significant improvements in ground control have been made in the Cosmic Boy operation.
- In particular, the introduction of mesh-reinforced shotcrete in conjunction with cables as standard support in ore drives has allowed access to be maintained in pillar and abutment areas, despite the increasing effects of redistributed stress.
- The support system integrates effective surface 'skin' support (the mesh-reinforced shotcrete) with deep anchorage reinforcement (the cable bolts); neither component would be capable of adequately controlling the ground behaviour experienced at Cosmic Boy on its own.
- The rehabilitation of a number of areas has also been successfully achieved through application or re-application of shotcrete over existing support.
- Even though mesh-reinforced shotcrete can tolerate much larger deformation than fibre reinforced shotcrete, some damage is inevitable. This can cause the shotcrete to crack and fall off the mesh. Re-application of shotcrete can provide a temporary solution but in many cases the movements on the structures continue, causing further damage to the new shotcrete.

In some cases, an external application of mesh, secured with split sets, may be more effective. Alternatively, the use of fibrecrete would reduce the potential for failed shotcrete to detach and fall. In any event, some rehabilitation is inevitable given the ongoing deformation associated with the ground conditions, and the operational life of the ore drives.

- 'Stress Management', through the use of extraction sequences that generally retreat continuously towards abutments and avoid the formation of isolated pillars, can significantly reduce the ground control problems associated with time dependent behaviour in weak rocks.
- The operation currently uses around 40 Km of twin strand cable, 4000m^3 of shotcrete and 50 000 friction stabilisers annually.

ACKNOWLEDGEMENTS

The authors wish to thank Outokumpu Mining Australia for permission to publish this paper. The assistance of their colleagues in the preparation of this paper is gratefully acknowledged.

REFERENCES

Barton, N., R. Lien & J. Lunde 1974. Engineering classification of rock masses for the design of tunnel support. *Rock Mechanics.* 6:189-236.

Frost, K. M., M. Woodhouse, and J.T. Pitkajarvi, 1998. Forrestania nickel deposits, in *Geology of Australian and Papua New Guinean Mineral Deposits* (Eds: D A Berkman and D H Mackenzie), pp 365-370 (The Australasian Institute of Mining and Metallurgy: Melbourne).

Rock Support and Reinforcement Practice in Mining, Villaescusa, Windsor & Thompson (eds)
© 1999 Balkema, Rotterdam, ISBN 90 5809 045 0

Choices in ground reinforcement at Mt Charlotte Mine

P.A. Mikula
Kalgoorlie Consolidated Gold Mines, W.A., Australia

ABSTRACT: Several types of ground reinforcement have been used at Mt Charlotte mine over the last few decades. A 100 kN shell anchor bolt was used in development drives for many years, but was replaced with a resin anchor rebar bolt, at a lower installed density, with excellent results. Secondary support comprises a fully grouted rebar for tight conditions, and a yielding element, the cone bolt, where dynamic loading or major movements are expected. The cone bolt has performed well under seismic conditions. For control of overbreak in stopes, plain strand cable has given way to purpose-designed modified strand cable. Despite the wide range of ground reinforcement hardware now available, suppliers generally do not provide enough information to ensure quality designed and installed ground reinforcement in mines.

1 INTRODUCTION

Many years have passed since bulk mechanical mining commenced in 1962 at Mt Charlotte, which is located adjacent to Kalgoorlie in Western Australia. This paper is a study of the changes in the types of ground control reinforcement used, reasons for the changes, resulting benefits, and problems with current reinforcement.

2 MT CHARLOTTE MINING ENVIRONMENT

At Mt Charlotte, gold ore occurs as large zones of quartz vein stockwork in a dolerite rockmass. Two major fault sets cut the rockmass, one steeply dipping to the W, and one shallow dipping to the SW. There are four rough wide-spaced joint sets, however only three sets are developed at any one location underground (Trueman & Harries 1997). Joints are rough/irregular undulating (J_r = 3), with unaltered walls (J_a = 1). RQD is close to 100%. Overall the rock mass comprises well-interlocked blocks, and classifies as very good quality (RMR 82, Q' 33). Intact rock strength is about 175 MPa, while rock mass strength is about 100 MPa. The mine has some very saline groundwater seepage, but is generally regarded as dry.

The stress field is field is high and deviatoric: $\sigma 1$ is subhorizontal, oriented NW-SE; $\sigma 2$ dips shallowly to the NE; $\sigma 3$ is subvertical. Typical magnitudes at 1 km depth are $\sigma 1$ = 75, $\sigma 2$ = 40, and $\sigma 3$ = 25 MPa.

The mining environment is described by Mikula (1993). The main mining method at Mt Charlotte is open stoping. Typically, large stable open stopes are mined, and then rib/crown pillars are mass blasted, with waste rock fill from above cascading onto the broken ore. Extraction is from drawpoints below, until waste rock dilution becomes excessive. Waste rock fill is continually added from the surface.

3 DEVELOPMENT REINFORCEMENT

The main failure mechanism in development is formation of wedges in the backs. However, wedges that can move are rare because of the well-interlocked blocky rock mass.

3.1 Cut and fill mining 1962 – 1968

Cut and fill extraction down to 140 m (5 level) from 1962 to 1968 was use to selectively mine high-grade portions of orebody in oxidised and partly oxidised rock. Stresses were low, drives were small and reinforcement of any sort was infrequently used. Some large spans were lightly bolted with a 16 mm integral head shell anchor bolt of 100 kN ultimate load capacity.

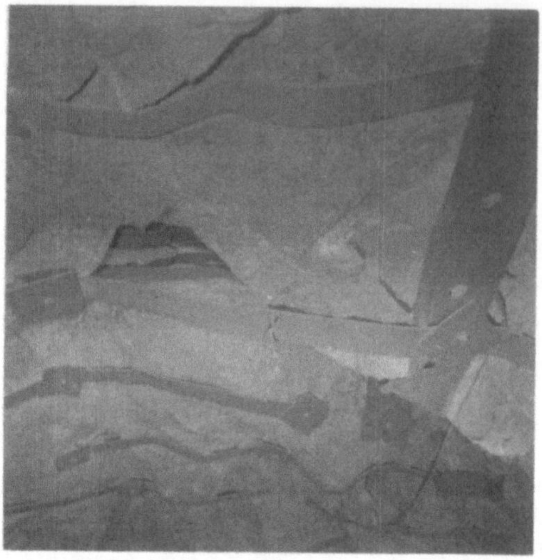

Figure 1. The integral head shell bolt, with both the flat strap requiring 0.9m bolt spacing, and the more flexible W-strap.

Figure 2. Several episodes of bolting and surface support were required in a drive at the crusher chamber.

3.2 Open stope mining 1968-1991

Long hole open stoping was used below 140 m. Initially drive profiles were arched and bolt usage in backs was infrequent. Some unbolted drives have lasted over 30 years in excellent condition. However over time and with increasing depth (to 920 m) larger and higher profile diesel equipment was acquired, and the size of development increased accordingly (to 6m wide by 5m high). More severe blasting inflicted more damage on the rock. Half-barrels were rare, and blocky irregular backs were common. Also there was a change to mining square cut rather than arched backs.

The shell bolt was used initially with a small bearing plate. For surface rock support, a flat steel strap was introduced. This strap had holes at 0.9 m spacing, so that became the installed bolt spacing (Fig. 1). The W-strap gradually replaced the flat strap from 1986. Split sets were not used in this period, except for a few installed on plats by contractors working on the new Cassidy Shaft from 1984 to 1986.

3.3 Introduction of rebar

- A report by Golder Associates (1979) noted the shell bolts were susceptible to vibrational loosening and inability to contain stressed ground. A trial of cement grouted cable strand dowels was suggested. Expected benefits were no vibrational load loss, added strength, surface spalling control and maximum restraint. However the trial did not proceed. The shells continued to be used till as late as 1991, when a change to a different type of bolt was recommended (Morley 1990) because:
- Corrosion of shell bolts could lead to bar failure either at the integral head, or at the threads for the shell. A conservative life estimate was 5 years in dry conditions and only 2 years in wet. Rebolting was often required (Fig. 2).
- Vibrational and creep loosening could occur. The bolts required retorquing, but this was rarely done. Many bolts lost all tension and hung in the hole until corrosion allowed them to fall out.
- The bolts had limited thread length, needing correct hole depth.
- There was a tendency to accept these bolts as providing adequate ground support, despite the issues noted above.

The 2.4 m long, 20 mm diameter dywidag bolt, and later the gewibar and the threadbar (all these are resin-anchored rebar bolts) were first trialled in 1986 and after a changeover period replaced the shell bolt (Fig. 3). The advantages of the rebar were:

- Higher strength (170 kN yield, twice that of the shell bolt).
- Ability to install any desired length.
- Longer lasting in corrosive conditions (greater diameter, and resin encapsulation protection).
- No ongoing maintenance.
- Components (bar, anchor, nut) were matched in strength.

The rebar, however, came with special requirements. Hole diameter was critical. The ideal range was 26 to 28 mm, with up to 30 mm being tolerated (Fig. 4). Perhaps because of the problems seen with the shell bolts, it was considered worthwhile to engineer an appropriate installation system. A purpose-designed

Figure 3. A typical 6m wide by 5m high drive with rebar reinforcement at 0.6m to 0.9m bolt spacing, plus W-strapping.

rock bolting unit was constructed on site to drill 28 mm diameter holes. This has resulted in Mt Charlotte enjoying excellent development bolting.

Chemical resin storage is also critical. Resins deteriorate at high temperatures. Efforts are made to store resin underground (constant 25 °C), and to cycle stock rapidly. The product distributors now also take measures to control stock temperature exposures, and the delivery point is underground via the decline.

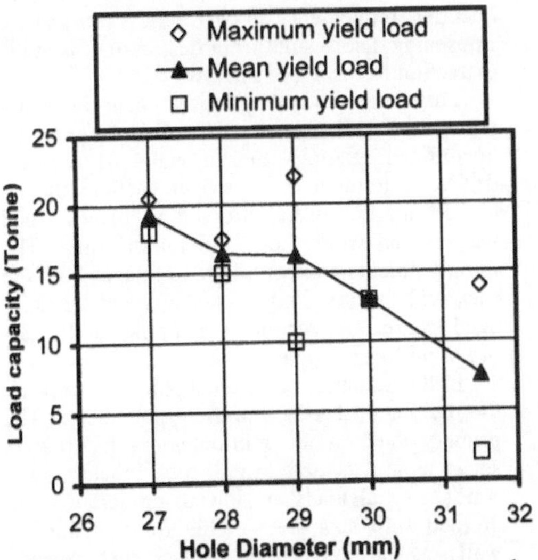

Figure 4. Summary of 29 underground bolt load tests on rebar installed with one 25 x 500 mm resin cartridge in boreholes of various diameters. Note, only one test at 30 mm diameter.

3.4 Changing the support density 1991 – 1998

Because of prior experience with the shell bolts, plus the heavy blasting and increasing stresses with depth, the rebar bolts were typically set at 0.6 m or 0.9 m spacing. This meant that in a 6 m x 3.3 m jumbo development cut, it was usual to find 30 to 35 bolts installed, complete with W-straps.

Experience and calculations using SAFEX software indicated the ground was over-bolted and a bolt spacing of 1.2 m to 1.5 m would be generally adequate. However, it was not easy to change the accepted practice. Mt Charlotte used several strategies over a two year period to achieve a reduction of 45 to 64% in bolts required:

a) Operator training. A ground control training course for all rock bolters and cable bolters was well received and gave the operators confidence in the variations to the bolting standard.
b) Arched profile backs. The flat drive profile was changed to a low arch, which was better suited to the stress field and rock behaviour. With the arch, any rock wedges in the backs would be smaller and each bolt would be more likely to have a secure anchorage. All development except drawpoints is now arched.
c) Changed blasting pattern. The number of perimeter holes was increased and a lower density explosive product was adopted in order to reduce blast damage to the rock. More half-barrels are now visible in each cut.
d) Trials without W-straps. Sections of development were trialled using butterfly plates instead of W-straps for local surface support. Operators were encouraged to open out the bolting pattern to suit the ground. It was observed that without the strap to suggest the 0.6 m or 0.9 m spacing, the typical spacing became 1.0 m to 1.2 m and only 15 to 20 bolts were required in a cut. Current practice is to use butterfly plates unless the ground requires W-straps.
e) Scaling. Check scaling of backs increased in importance as the changes took effect. A record of the scaling history of each drive is now kept.
f) Proof load testing. A contractor undertakes monthly random proof load testing to 12.5 kN of bolts installed during the month. Currently 98.5% of bolts pass the test.

3.5 The Universal bolt

A decline access was developed in 1997 by a Contractor, and the mechanical anchor universal bolt was used for primary support. While this bolt is easy and cheap to install, it has drawbacks in that the shell anchor can be damaged during installation, the bolt can loosen over time due to vibration, and it is susceptible to corrosion. The bolts were accepted because the decline (after completion) was in a low vibration zone, the area was known to be dry, the decline lifetime was short, and the Contractor was aware of the potential installation problems. A program of nut torque checking was accepted as the cost of using this bolt.

4 SECONDARY REINFORCEMENT

Additional reinforcement is installed depending on the use of the drive and the geometry, geological structure and stress environment of the area. The main ground control issues are:

- Movement on major 45° dipping joints;
- Wall slabbing or spalling under stress;
- Destressing causing relaxation and opening of joints in walls
- Seismic damage, mainly to drive shoulders.

4.1 Stiff reinforcement

A stiff reinforcement is installed when the objective is to limit minor movement on geological structures within the backs. These situations include wide spans at intersections, drawpoint backs and brows, and accesses close to stopes.

Usually, 4m long cement grouted rebar is used. On occasion, 8 m cable has been installed where high anchorage is necessary. Full encapsulation is used, so that the support develops early resistance to movement where needed, and it continues to act even if the collar fixture is unserviceable.

This support was first installed in about 1990 and for the first 3 years a high bolt density (1.1 bolts/m^2) was used. However, experience indicated that this was conservative, as most short-life drives remained totally undamaged despite major stoping adjacent (Fig. 5). The density now used is 0.4 bolts/m^2, except at drawpoint brows where 0.9 bolts/m^2 are installed. Bolt duty at drawpoint brows is particularly onerous due to impact, abrasion and vibration.

4.2 The need for yielding reinforcement

Yielding support is required in two situations, dynamic loading and major movement.

a) Dynamic loading, which momentarily can significantly exceed gravity, imparts kinetic energy to the rock on the edge of an excavation. If the rock fractures in tension, or separates on geological structure, rock blocks can be projected into the opening. A 1 m^3 rock block may have a kinetic energy of 50 to 100 kJ.

   The rebar, especially if fully encapsulated, has far too low a strain capacity. It is able to absorb only 4 to 5 kJ before breaking and so is unsuitable in seismic conditions. The damage from seismic events at Mt Charlotte often includes broken rebar. Cone bolts are a suitable reinforcement, as the yielding anchorage provides high displacement. Split sets may also be also suitable, as they allow high displacement, but at less load capacity. However grouted split sets have much less displacement capacity, and are not suitable.

Figure 5. This drive was conservatively bolted, first with 2.4 m resin-anchor rebar in the backs, and then with 4 m cement grouted rebar in backs and walls. Although being adjacent to a large stope, the drive was undamaged.

b) Major movement. Loading of the rock mass, seismic or not, often forces significant shear or opening movement on geological structures. Loosening also results from destressing as stope extraction forms stress shadows.

   Large movements typically occur in drawpoint pillar walls, which can crack through under unconfined loading, and in walls of adjacent drives. If joints and faults open sufficiently (up to 25 mm has been observed in places) rock wedges can work loose and tumble out. The mining process can contribute to this, as both the rock and the installed bolts/straps can be damaged by loaders, secondary popping, and clearing of millhole hangups.

   In this situation, reinforcement must restrain the rock against large imposed movement. The primary load on the reinforcement is often in shear, as gravity acts to pull rock blocks downwards. Axial loads are low to moderate. Ungrouted split sets are suitable for this duty in walls, as they have high displacement capacity and sufficient load capacity. However, problems with split sets need to be considered:

- Air/water corrosion. Split sets have a short life in wet areas.
- Diesel fumes corrosion. Fumes from equipment exacerbate corrosion, even in dry areas. Split sets used to pin mesh in the 14 Level diesel workshop lasted less than 5 years.

4.3 Kinds of yielding reinforcement

Mt Charlotte has tried two special yielding elements in development drives since 1993:

a) Debonded birdcage strand. These were labour intensive to manufacture, but easy to install. Both ends of a 4 m strand were birdcaged, and the rest of the strand was debonded in a plastic sleeve. The whole element was grouted into the borehole. The cage element at the far end of the hole was made short, so that it would be unable to develop enough anchorage to break the strand before it started slipping. The birdcaging near the collar was intended to lock into the rock and effectively take the place of the surface fixture. The barrel and wedge anchor and bearing plates attached at the collars were quickly damaged by loader impact. About 500 of these strands were installed in I1 stope drawpoints on 32 Level in 1994 and performed very well. The drawpoints did not require any remedial bolting for the duration of mining. The worst seismicity these drawpoints experienced was a Richter magnitude 2.9 event 50 m away.

b) Cone bolt. A dozen cone bolts were imported from South Africa and trialled in 1993 in the I1 drawpoints. They were easy to install, but like the birdcage strand, the surface fixtures were susceptible to loader damage. The cone bolt is designed with low stiffness, to yield under high dynamic loading conditions. It was not practicable to perform dynamic load testing at Mt Charlotte. However, static tests were done, with good results (Fig. 6). Combined with South African experience, it was concluded that the cone bolts were likely to live up to expectations.

Presently, about 2000 cone bolts have been installed at Mt Charlotte, mainly in drawpoint pillar walls. The worst seismicity experienced by cone bolts was a Richter magnitude 3.5 event about 30m away from the S2 stope drawpoints. After the event, all the drawpoints remained in serviceable condition, with only scaling and minor rebolting required (Fig. 7). No broken cone bolts were observed. Cracking and slabbing locally in drawpoint walls resulted in loss of ground around many cone bolt collars. Given the magnitude of the event, it is thought that the drawpoints would have required major rehabilitation in the absence of cone bolts. The main cautions with cone bolts are:

- Occasional collar end damage (bent, broken) by loaders and trucks.
- Shearing movements in the rock mass can pin the bolts and prevent yielding operation.

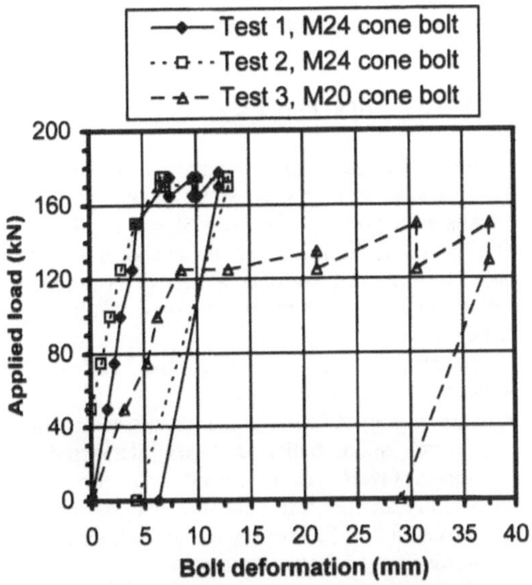

Figure 6. Results of static load tests performed underground on 3 m long by 24 mm and 20 mm diameter cement grouted cone bolts. The grout water/cement ratio was 0.4, with no additives, and the testing occurred 3 days after grouting.

Figure 7. Cone bolts with W-strapping were installed in drawpoint walls in the S2 stope at 800 m depth. Here the wall remained in place despite a close seismic event. Elsewhere there was loss of ground around bolt collars.

At Mt Charlotte this does not seem to be an issue, because in drawpoints, fractures open at least as much as they shear.

- The wax coating on the bolts deteriorates under sunlight, high temperatures and wet conditions. Therefore all cone bolts are stored in a dry place underground.

5 STOPE REINFORCEMENT

An important requirement for most of the history of Mt Charlotte was to keep stope backs flat. Stope crowns were prepared for mass blasting by horidiam drilling, i.e. drilling horizontal blastholes across the backs from a rise positioned at one side of the stope. Holes were lost if overbreak occurred and backs became arched.

Several adverse types of ground behaviour are sometimes observed in stopes:

- Shearing on 45° structures in backs, walls and rib pillars. Major faults or shears intersected most stopes. Overbreak from backs was common, especially as structure exposure lengths increased above 25 m. Failure tended to arch steeply upwards under structural control.
- Unravelling on joint structures, following destressing and loosening in walls, backs or rib pillars.
- Wedge failures by sliding from walls.
- Some failures are seismic in nature.

The major issue is the large size of rock blocks that can loosen on the 45° structures. It is often difficult to secure such blocks in a practical manner.

5.1 *Cable bolt usage*

Cable bolts were first trialled in 1979 in the crown pillar above E3 stope, in an attempt to control arching and overbreak. About 40 cables of 10 to 15 m length were installed, fanned upwards from drives at the base of the crown pillar. However, as mining proceeded, overbreak developed regardless of the cables. Many cables were left dangling with little evidence (no pigtailing or coiling) of having taken load. Apparently, these cables were installed as pure dowels, with no tensioning or collar fixtures (Lee, 1998). It was considered at the time (Martin et al 1984) that the cables were ineffective because:

- They were too short and too few.
- Dirt on cables caused a poor cable-grout bond.
- The flatter oriented cables were of no benefit.

With hindsight 15 years later, two more reasons can be added:

- Grout shrinkage would have lowered bond strength.
- Plain strand would have allowed cable/grout slip from shrinkage or destressing.

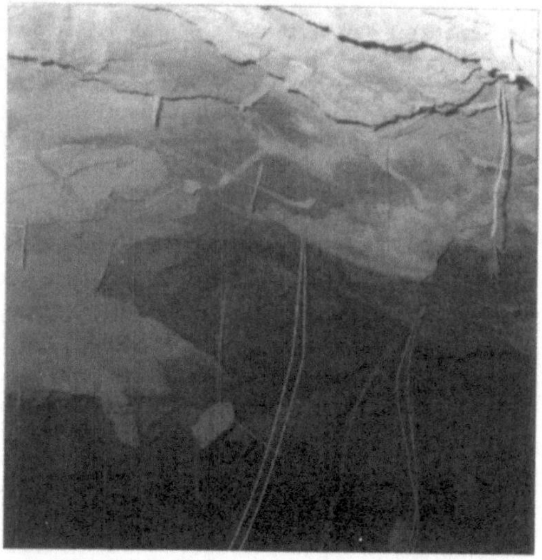

Figure 8. A cable bolt trial in F3 stope in 1982. Overbreak height was limited to 5 m.

More cable bolts were installed in the crowns of F1 and F2 stopes, this time with collar fixtures. However, these also made little difference to the development of overbreak, for similar reasons. Mt Charlotte then joined an AMIRA Project on Cable Bolting (Askew et al, 1983). Improvements were:

- Increased density of cable bolts.
- Installation procedures to keep cables clean.
- Installation from specially located cable bolt drives, to form "reinforcement ribs" running across the stope backs. This was first done in F3 stope in 1982. The result (Fig. 8) was overbreak limited to about 5 m height in the unsupported stope backs between the cable bolt drives. The flat orientated cable bolts that had been installed into these areas proved useless, as rock peeled off these cables. The cables performed best where they were oriented in tension.

  Later in G Block, crosscuts at 13 m centres were driven across the stope block, and all the backs were cabled, resulting in control of the amount of arching between the ribs. W-strap linking of the collars was an essential part of the technique.

After this, cable bolting of crown pillars became the norm at Mt Charlotte. The conventional grouting procedure (short grout tube and long air breather/grout return tube) was used, with a Spedel pump system, and the water/cement ratio was generally 0.40 to 0.42. However, failures of cable bolted backs still occurred from time to time. For example, a massive failure of the I1 stope backs occurred overnight in 1994, despite extensive cabling. It appeared that once the collar fixtures gave way, the rock peeled off the cables.

5.2 *Modified strand cable*

When birdcage and then bulbed cable first became available, they were thought to be too stiff for application at Mt Charlotte. It was then considered that only plain strand had the large strain capacity to accommodate the inevitable movements on 45° structures in stope backs. The stiffness of the cables had to be matched to the stiffness of the local rock mass. The birdcage was also considered to be too awkward to bend and manhandle into cable bolt holes. Birdcage was only used in 1992 to secure a wedge in the east wall of ROB5 stope.

However, in a historic trial in 1994, the brow of a major pass into I1 stope was successfully secured with a purpose-designed cable, namely a twin plain strand with just the remote end being birdcaged. This was the precursor of the bulb design now regularly used at Mt Charlotte. A review indicated that a purpose-designed twin strand bulb cable could satisfy both the secure anchorage and the high strain capability requirements. The change to the new cables commenced in I2 stope in August 1994. The new cables (shown in Fig. 9) featured:

- 0.5 m bulb spacing for the end 2 m, to provide a guaranteed anchorage, even in substandard grout and under deserting.
- 5 m spacing between bulbs for the rest of the strand, to provide an adequate length to accommodate movement on one or several planes.
- Offset bulbs on the strands, so that effective bulb spacing in the rock mass was 2.5m. If collar fixtures failed, no more than 2.5m rock thickness could peel off before meeting a bulb restraint.
- Collar fixtures on both strands, for maximum initial restraint, and continued restraint in case of failure of one fixture or strand.
- Dome barrel and wedge anchors with a matched twin-slot plate. This reduced the bending of individual strands as the fixture was tightened. Domes are used for rebar, and should also be used for cable.
- Twin fishhook prongs at the top of the cable bolt, for support in upholes before grouting. The practice of kinking cable strand (or any reinforcement element) to make it stay up the hole is actively discouraged at Mt Charlotte. A bent strand is weakened and contributes to premature failure.

The results with the modified strand cable in I2 stope were instructive. Cables were installed in a dense rib across the stope to support a 35 m span. As mining progressed, the backs took load and it became apparent that a large wedge defined by two major structures was loosening. The stope back profile was measured on several occasions using the Cavity Monitoring System. This indicated that initially the cable bolted rib carried the load without any failure, while the unbolted backs beside the rib collapsed to form a shallow dome. The lower portion of the rib eventually collapsed, forming an arch profile up to 5 m high. After some months, more of the rib collapsed, forming a new arch up to about 9 m high. This was the final profile at the time of the mass blast to remove the crown.

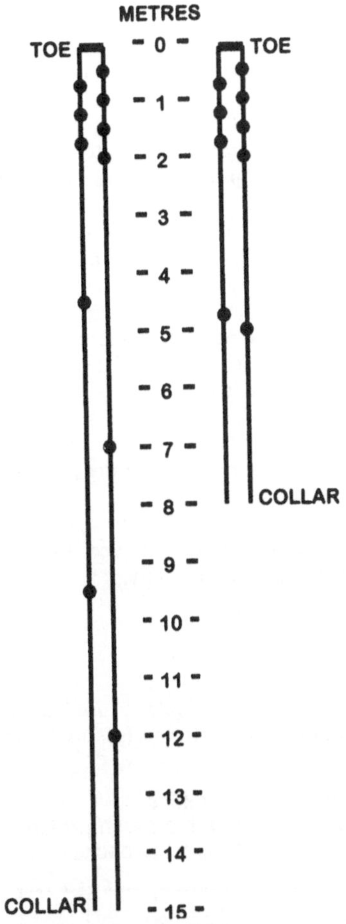

Figure 9. The design of the 15 m and 8 m twin strand Garford bulb cable bolts used at Mt Charlotte. The manufacturing process allows bulbs to be located to suit the reinforcement duty.

It was concluded that the bulbs were indeed locking into the rock blocks and actively restraining overbreak. Visual inspection showed that the cable bolts were breaking under load. There was no indication of rock peeling off the cables.

5.3 *Arched stope back profile*

Where a flat stope back is not required for horidiam drilling purposes, an arched stope back profile can be used. In 1997 in S2 stope, an arched back design, combined with judicious placement of cable bolt support was trialled as a cost-effective means of controlling back overbreak and helping the rock to support itself.

The principle was to blast a stope back profile that approximated the natural unsupported equilibrium shape. The correct profile was easy to establish empirically, as several dozen CMS surveys of previous stope profiles were available (Fig. 9). The main controls on the profiles were the minor span and the degree of end wall support.

With this strategy, the stope drill drives were also the cable bolt drives, and the cable bolts only had local rock support duty. No special cable bolt drives were needed. This strategy suited Mt Charlotte because the ore grade boundary is gradual, so that mining to a profile has little effect on overall tonnes or grade. Also the risk of overbreak and the cost of support are reduced.

The strategy worked well during the open stope period, and the cable bolts partly survived the mass blast despite associated high seismicity.

6 MESH AND SHOTCRETE

Mesh and shotcrete are not regularly used at Mt Charlotte. Chainlink meshing for surface rock support is typically installed in areas such as cribrooms, plats, and workshops, and in some drives where the rock mass quality is poorer within a few metres of the major faults.

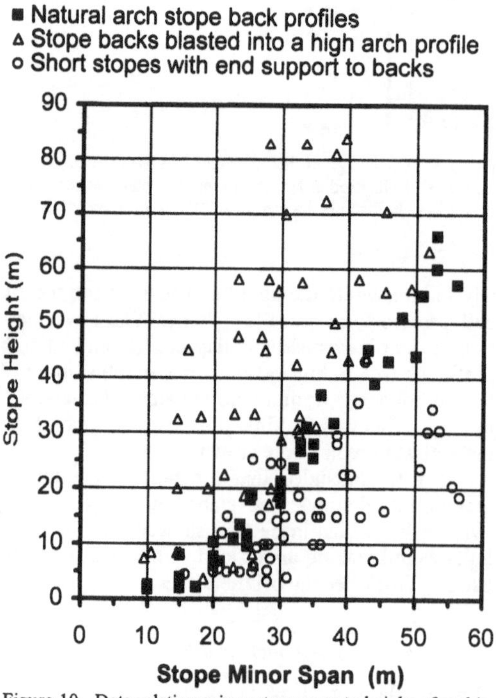

Figure 10. Data relating minor stope span to height of arching of stope backs. Unsupported backs of sufficiently long stopes tend to overbreak to a height indicated by the black squares.

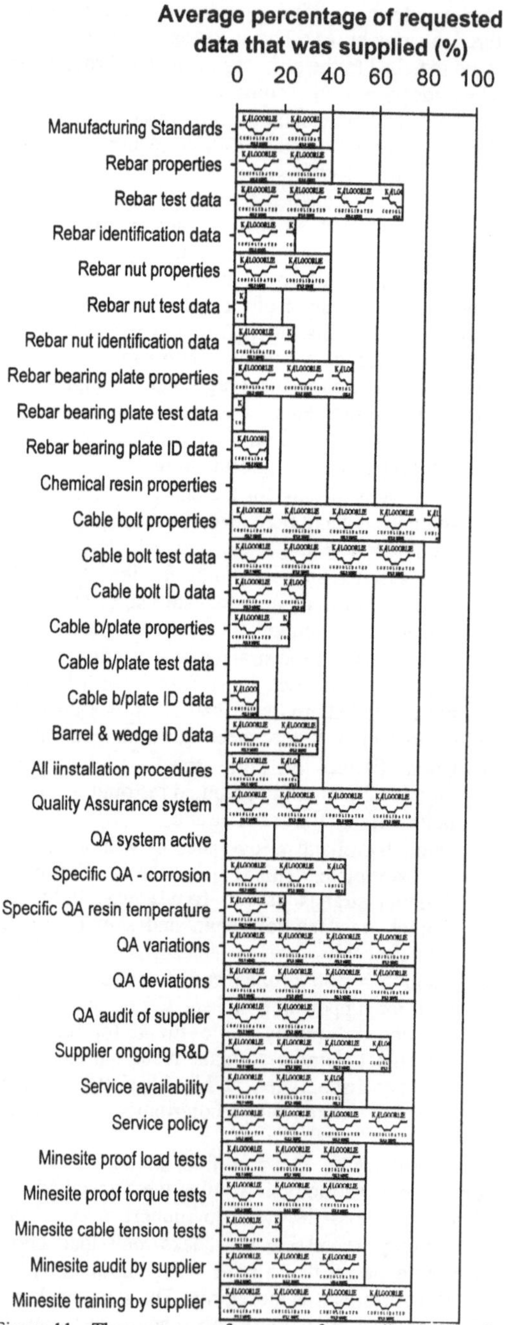

Figure 11. The average performance of Australian bolt hardware suppliers in providing information requested on various topics for a Tender evaluation process.

In 1997, 75 mm to 100 mm thick fibrecrete was placed experimentally in some drawpoints, but was damaged where ground destressing and relaxation occurred. In 1998 decline access from the surface became available. While this made shotcrete placement much more attractive, it has not been considered necessary except in rare instances where containment by mesh was impractical.

7 PASSES

Reinforcement of ore passes at depth has been a failure. Passes were lined with resin-encapsulated rebar, using a mix of medium and slow set resin cartridges. However, pass wall failure soon developed, due to impact from free-falling ore, and fracturing in the high stress field.

Keeping passes full of ore would alleviate this problem. This is theoretically possible at Mt Charlotte, as the ore is not reactive. One solution would be to bell out the top of the ore pass to form a surge bin and install a monitor to ensure the bin is never fully emptied.

8 SUPPLIER ISSUES

It is disappointing that at present there is no Australian Standard covering rock support and reinforcement elements. It is also disappointing to note that in a tender in 1997 for the supply of rock bolt hardware at Mt Charlotte, not one supplier could provide all the specification data requested, and some of the data supplied was poorly presented. Figure 11 summarises the performance of Australian suppliers.

The Department of Minerals & Energy WA, in their guidelines under Regulation 10.28 (DME 1997), indicate that suppliers of rock support and reinforcement elements should provide detailed instructions for correct installation and testing of elements, and information on factors that determine the quality of the installation. This will greatly assist mines in ensuring quality ground support choices.

9 CONCLUSIONS

Mt Charlotte has come a long way over the years. Ground support have changed in response to problems experienced underground, and the explosion in new hardware that has become available in recent times. Future developments that would be well received include:

- Standard ability to drill 26 to 28 mm diameter holes. The drilling of 30 mm and larger holes for rebar/resin installations is not good practice.
- Bolt integrity tester. Operators and supervisors should be able to easily check the integrity and load of every bolt at the time it is installed.
- Standards and Specifications. All bolt hardware should be manufactured under appropriate Standards, and proper specifications must be provided to end users.
- Collar damage. Bolts should be designed and/or installed to be immune to collar damage.

In summary, Mt Charlotte has a quality ground reinforcement strategy. The range of new hardware products means that the mine has been able to obtain the bolts and cables that are well matched to the mine environment.

ACKNOWLEDGEMENTS

Permission to publish this paper was provided by KCGM. Special thanks are due to all those who assisted in the preparation and reviewing of the paper.

REFERENCES

Askew, J.E, McAuley, T.J, Miller, D.R. & Reed, M.R. 1983. Cable Support of Open Stopes: Case Study – F3 Stope, Mt Charlotte Mine. AMIRA Report, October.

Dept of Minerals and Energy (DME) 1997. Guidelines: Geotechnical Considerations in Mines. Perth, WA, June.

Golder Associates 1979. Rock Mechanics Review, Mt Charlotte Mine, December.

Lee, M. 1998. Pers. comm.

Martin, R, Reed, R & Askew, J.E. 1984. Innovative aspects of mining at Mount Charlotte. In Regional Conference on Gold: Mining, Metallurgy and Geology, AusIMM, October.

Mikula, P.A. 1993. The influence of geotechnical monitoring on mine design at Mt Charlotte. In Geotechnical Instrumentation and Monitoring in Open Pit and Underground Mining, Szwedzicki (Ed), Kalgoorlie.

Morley, C 1990. Expansion shell rock bolts. KCGM Internal Memorandum.

Trueman, R & Harries, N. 1997. Geotechnical Environment of Mount Charlotte. JKMRC Report to KCGM, August.

Rock Support and Reinforcement Practice in Mining, Villaescusa, Windsor & Thompson (eds)
© 1999 Balkema, Rotterdam, ISBN 90 5809 045 0

Productivity gains through improved rock reinforcement at McArthur River Mining

C.J. Schubert & C.M. Stewart
McArthur River Mining, Darwin, N.T., Australia

ABSTRACT: Recent rock mechanics investigations have initiated improvements in rock reinforcement that have improved worker safety and the economical extraction of the shallow dipping orebodies at McArthur River Mine. This paper outlines the changes to mining and rock reinforcement practices and shows the productivity gains achieved in the areas of improved ground conditions, reduced dilution, a reduction of secondary ground support installation, more efficient mine production and improved mining recoveries.

1 INTRODUCTION

The McArthur River Mine site is located approximately 720 kilometres south-east of Darwin and 100 kilometres inland from the Gulf of Carpentaria. Mineralisation within the mine area extends approximately 1.5 kilometres north-south and 1.0 kilometre east-west. The central area targeted for selective mining dips between 15 and 20 degrees towards the east at a depth of 100 to 350 metres. The average thickness of the currently mined Two Orebody is 3.2 m. The immediate hangingwall to the room and pillar mine excavations consists of a 1.5 m thick bed of unmineralised siltstone overlayed by a minimal strength, 5 mm thick, clay tuff band. This clay band is effectively the first continuous parallel discontinuity to the hangingwall of the mining excavation.

The mine commenced production in March 1995. Current total ore reserves for the 2, 3/4 and 4 orebodies is 23.2 million tonnes @ 13.5% Zn, 6.3% Pb and 65 g/t Ag (Nihill et al. 1998).

2 GEOTECHNICAL SETTING

The mineralisation comprises eight stratiform orebodies, stacked one on top of the other; each separated by dolomitic siltstone and sedimentary breccias. The 55 m thick sedimentary sequence that hosts the orebodies is divided up into a series of geotechnically continuous layers of rock separated by thin clay filled tuff bands. These 0.5 to 1.6 m thick beds have a frequency of bedding plane breaks ranging from 3 to 10 breaks/ m. Underground exposures show that bedding plane breaks extend up to 3 m in length.

The orebody sequence is irregularly offset by four major vertical fault sets. Local vertical displacements on faults range from zero to fifteen metres. There is generally no alteration or disruption (drag folds etc.) of the rock conditions adjacent to the vertical faults. The faults are usually tightly healed and infill is commonly either dolomite or chlorite. Three prominent joint set orientations have been identified at McArthur River, which are all steeply dipping. The joints are frequently short and often terminate against clay tuff bands. Joint spacings are generally large (>2 m) and the majority of joint surfaces are unaltered. Intact rock strength values of approximately 157 MPa (average) have been tested.

The maximum principal stress (σ_1= 8.6 MPa) dips approximately 20° towards 140°. The measured vertical stress (σ_z= 6.4 MPa at 239 m depth) is almost equal to that expected from the weight of the overburden.

3 MINING METHOD

Room and pillar mining is the primary extraction technique. Three closely spaced tabular orebodies have been targeted for extraction over the life of the mine and due to economics the lowermost number

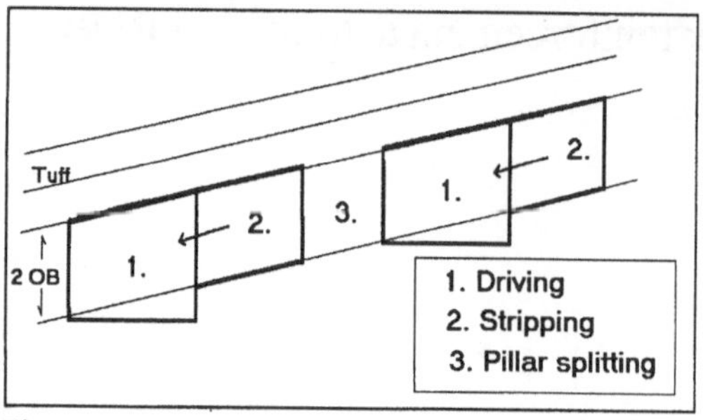

Figure 1. Schematic cross section of room and pillar mining sequences at McArthur River Mine.

Two Orebody is being mined first. Five metre wide pilot drives are driven on an apparent dip to access ore blocks. The roof of these drives is positioned on the hangingwall of the Two Orebody. The walls of the pilot drives are stripped leaving rib pillars between rooms. The rib pillars are split to form a series of shorter island pillars (Fig.1).

4 RECENT INPROVEMENTS TO MINING AND REINFORCEMENT

4.1 *Ground conditions*

Detailed geotechnical investigations of early rockfalls identified the presence of bedding parallel clay filled volcanic tuff layers as the structures most likely to control excavation stability. Although 4 sets of fault orientations plus associated joints have been identified, these are usually close to vertical and rarely effect roof stability because the potential to form wedge and key block failures is low.

The poorest ground conditions during initial mining occurred when the mining profile significantly overbroke the Two Orebody hangingwall into the weaker zone of rock containing clay filled tuff beds. The failure mode for this rock consists of buckling of the thin shale bedding planes unravelling until a stable span is created at the apex. The first of the continuous clay bands is positioned 1.5 m above the Two Orebody hangingwall.

Better control of mining profiles has resulted in a reduction of roof overbreak into structurally weaker strata. Maintaining stronger rock in the roof and mining to a consistent hangingwall bedding plane has resulted in the development a continuous roof beam, more stable ground conditions and reduced hangingwall dilution.

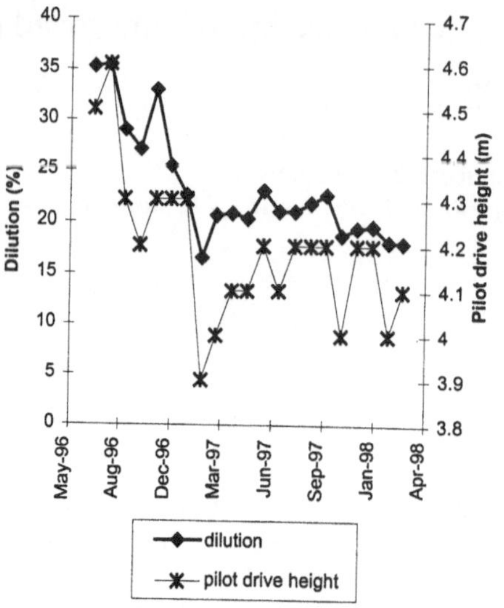

Figure 2. Monthly average pilot drive height and average pilot drive dilution.

4.2 *Reduced dilution.*

Minimum drive heights have been generally dictated by the height required to install the friction coupled roof bolts with a mechanised bolting rig. In order to lower drive heights and hence dilution, primary bolt lengths have been gradually reduced from 2.4m to 2.1m and currently to 1.8m. As a result of this, the average heights of production pilot drives have been reduced from 4.6 m to 4.1 m (Fig. 2). This has meant the levels of dilution have been reduced from 35.5 % to 18.1 % over the last 2 years (Figs 2,3).

Insitu monitoring was conducted to investigate the reinforcing effect of the friction coupled bolts on the stability of the Two Orebody hangingwall (Schubert et al. 1997). The results of the monitoring indicated the installation of the friction coupled bolts assist in reinforcing the 1.5 m thick rock beam but do not have enough load transfer capacity to prevent the beam from separating at the overlying clay filled tuff band (Fig. 3)

Calculations based on a voussoir beam indicate the 1.5 m thick intact rock beam will be self supporting after a minimal initial beam deflection and the formation of a flat arched confining stress profile within the beam. Insitu monitoring supported this prediction. It was determined a reduction of the bolt length from 2.4 to 1.8 m would not effect the self supporting mechanics of the reinforced beam. Mining to date has indicated this to be the case.

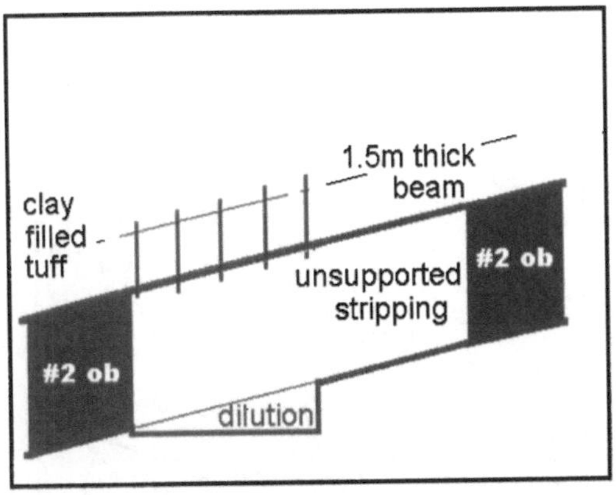

Figure 3. Diagram showing Two Orebody mining geometry, footwall dilution, and hangingwall rock beam plus bolting pattern.

4.3 *Reduction in ground support installation*

Although theoretical estimates can be used to predict the stability of rock beams, very little was known about the actual hangingwall rock mass response to room and pillar mining at McArthur River Mine. As precaution early rock reinforcement at the mine consisted of installing a second pass of deep anchored support prior to stripping the pilot drives from 5 to 8 m wide. Single rows of 6 m long fully grouted bulbed cablebolts were installed at 2m spacing along the edge of the drive prior to stripping.

An investigation was conducted to determine whether the deep anchored secondary support was required to maintain stability of the expanded roof spans. A series of cablebolts were instrumented with strain gauges to record the load carried by the bolts. Each strain gauge was constructed from resistance wire and measured deformation over a cable length of approximately 450 mm. The strain gauges were positioned to straddle sections of expected maximum strain for the cablebolts. Resistance wire extensometers were also installed in boreholes next to the instrumented cablebolts to measure the associated rock movement.

The maximum recorded load for the cablebolts during and after wall stripping was 16 kN. This is compared to their ultimate strength capacity of 250 kN. The monitoring indicated the roof deformation reached equilibrium at approximately 3 mm of vertical movement. The 16 kN load after 3 mm of movement is consistent with early load/ displacement behaviour of similar laboratory tested cablebolts (Villaescusa et al. 1992).

A second drive with similar dimensions to the above mentioned area was monitored with three sets of extensometers but was not supported with cablebolts. The wall of the 5 m wide drive was stripped in three firings to create a 9 m wide roof span over a strike length of 25 metres. The extensometers show there was no marked reduction in roof deformation for a 9 m span that is supported by cablebolts over one that has no cable bolt support. There was approximately 3 mm movement recorded for each of the tested rooms. Neither test area showed any tendency towards failure.

The cablebolt monitoring trial indicated the routine installation of 6 m long cablebolts was not contributing to the stability of the rock beam and roof spans. The pattern installation of deep reinforcement has been ceased resulting in a significant reduction in ground support costs and required resources.

Specific secondary ground reinforcement is not often required, however where the roof beam has been broken, or significant spans are to be exposed, deep support is installed (Fig. 4). Fully grouted 6m cable bolts were originally used for this function, however the insitu monitoring has indicated the laminated ground rarely requires support to this depth. Galvanised three metre long mechanically point anchored bolts are now used in most cases to support broken hangingwall beams. The point anchored bolts do not require grouting and can be installed with a conventional drilling Jumbo, allowing rapid and more efficient deep support (Fig. 4).

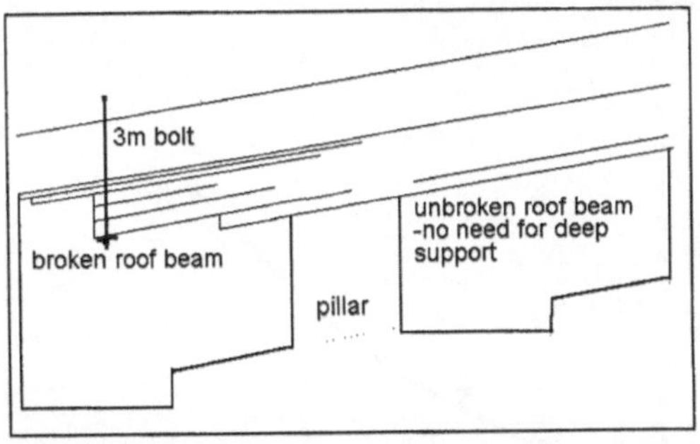

Figure 4. Schematic cross sections showing broken and unbroken beam geometry.

4.4 *Increased roof spans*

The nature of mining at McArthur River is such that a sequential increase in the roof span is achieved as the sidewalls of drives are progressively stripped. Better controlled mining along geologically defined bedding planes and the development of unbroken roof beams in stronger roof strata have allowed larger roof spans to be created with no extra support. The increased roof spans have meant improved recoveries and more efficient production.

While stable roof spans up to 22 m wide have been trialed and monitored (Fig. 5) current planning guidelines stipulate that back spans should not exceed a diametrical distance of 12 m without additional deep ground support (Fig. 6). These span dimensions combined with the regular placement of stable pillars have proven to be a reliable design guideline without the requirement for deeper secondary roof reinforcement.

4.5 *Increased orebody recovery*

Once the ranges of stable roof spans were determined for the local hangingwall conditions the design of stable mine pillars was critical. Individual pillar dimensions and the pillar placement is based on the local pillar strength verses pillar load requirements (Schubert & Villaescusa. 1998). As backfill is not currently implemented, long term sound pillar performance is required to ensure the global stability of the underground openings. Where conservative pillar designs are likely to have a detrimental economical impact, aggressive pillar designs could lead to catastrophic failures, worker safety concerns and sterilisation of overlying ore reserves.

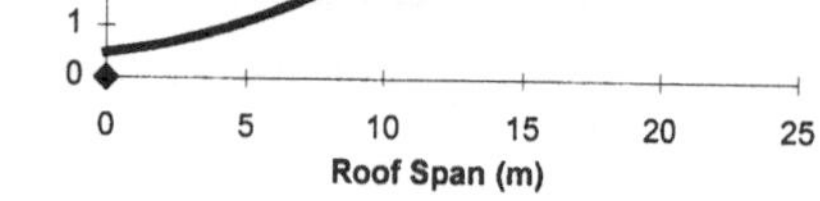

Figure 5. Measured deformation as a function of increased roof span at McArthur River Mine.

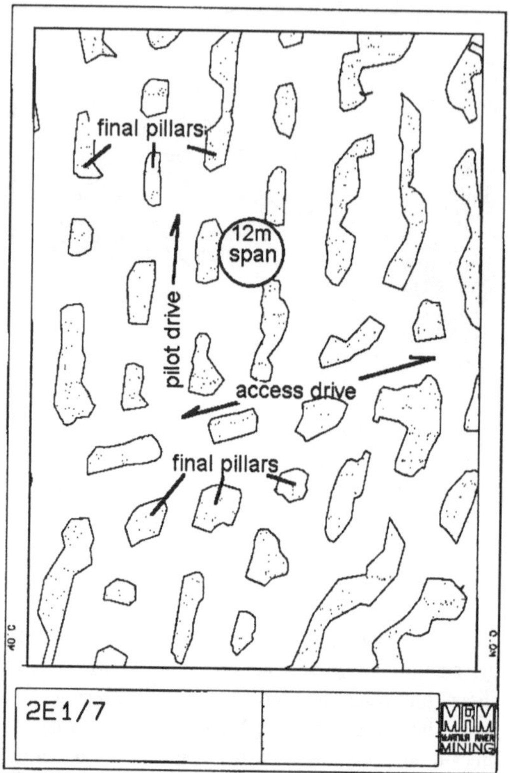

Figure 6. The plan surveys of a recently completed mining block. The extraction ratio for this area is approximately 76%. The maximum roof span has a diametrical distance of 12 m.

An improved understanding of factors that influence pillar strength has allowed a refinement of designs and an increase in orebody recoveries while maintaining stable conditions (Fig. 6). Local pillar geometries are designed to support a range of roof spans taking into account the local field stress and rock strength conditions. The improved design and positioning of access drives, pilot drives and final mine pillars has achieved an increase in regional mining recoveries from an average of 53 % to 67 % over the last 2 years.

5 CONCLUSION

Opportunities for rock reinforcement improvements were identified through an observation of local rock and support behaviour. Information on typical failure modes were gained through geotechnical investigations of early roof collapses and detailed geological mapping. Rock and reinforcement responses to large span mining were monitored with installed instrumentation. Recent improvements in mine design, mining and reinforcement practices have enhanced the self supporting performance of the hangingwall rock mass by maintaining (where possible) an intact rock beam. Productivity gains have been achieved due to improved ground conditions, reduced dilution, a reduction of ground support installation and enhanced mining recoveries

ACKNOWLEDGEMENTS

The authors wish to thank the management of McArthur River Mining Limited for the permission to publish this paper.

REFERENCES

Nihill, D., S. Pevely & T. Andreatidis 1998. McArthur River Mine -1998 Ore Reserves Statement. Unpublished. Internal Company Report.

Schubert, C.J., E. Villaescusa & C.M. Stewart 1997. Ground support design and monitoring at the McArthur River Mine Australia. Proceedings Int. Symposium on Rock Support. Lillehammer, Norway. 329-339.

Schubert, C.J, & E. Villaescusa 1998. An approach to hard rock pillar design at the McArthur River Mine, The AusIMM Annual Conference, Mount Isa (In Press).

Villaescusa, E., M.P. Sandy & S. Bywater 1992. Ground support investigation and practices at Mt Isa. In Proceedings Rock support in mining and underground construction, Sudbury, Canada, pp.185-193.

Rock Support and Reinforcement Practice in Mining, Villaescusa, Windsor & Thompson (eds)
© 1999 Balkema, Rotterdam, ISBN 90 5809 045 0

Reinforcement design investigations at Big Bell

M.P.Sandy
Australian Mining Consultants, Perth, W.A., Australia

J.R.Player
Wirralie Gold Mines, Cue, W.A., Australia

ABSTRACT: Underground mining at the Big Bell Gold Mine in Western Australia, dates back to the 1930's. The current operation began in October, 1993 and commenced with the recovery of remnant pillars under cave in old blocks that were previously mined in the 1940's-1950's. It was planned to extract virgin blocks using a 'core-and-shell' method allowing the core tonnage to be recovered cleanly in an open stope, followed by pillar wrecking (mass blasting) under cave.

Although the early stopes were generally stable, conditions in pillars were poor with blasthole closure, overloading of ground support and in some cases complete collapse over intersections. A decision was made in 1997 to introduce Sub-Level caving with the objective of simplifying the operation, avoiding the problems associated with the recovery of late stage pillars and to achieve the required economic production rate of 1.8 million tonnes per annum.

At Big Bell the mine geometry, pre-mining stresses, rock strength and tendency for brittle failure combine to produce intense fracturing over the oredrive backs. The fracturing is accompanied by dilation in the order of several centimetres. A well-defined crack parallel to the foliation usually appears at the back/hangingwall intersection. This style of behaviour was first identified above the backs of old (1950's) development, when new cross-cuts were being developed on 320 and 380 levels. Observation-based ground behaviour monitoring in new development indicated that the damage was widespread and (mostly) developed in areas where the effects of stress redistribution around extracted areas were most intense.

To develop a rational basis for reinforcement design, information was required on the timing, extent and degree of deformation associated with development and the subsequent stress changes due to the extraction of the orezone. A programme of deformation monitoring has been undertaken at selected underground sites since mid 1997. The results are presented and discussed in terms of the interpreted ground behaviour and implications for access development and oredrive support.

1 INTRODUCTION

1.1 *Location*

The Big Bell gold deposit is located near Cue in the Murchison Province of the Yilgarn Craton, Western Australia, about 540 km NNE of Perth.

The mine employs approximately 260 people, including contract workers. The majority of employees work on a fly-in-fly-out basis from Perth and Geraldton.

Current mine production is around 1.8 million tonnes annually from which 160,000 ounces of gold are recovered.

1.2 *Regional Setting, Local Geology and Mineralisation*

The Big Bell deposit is hosted by a greenstone sequence within the Murchison Province of the Yilgarn block. The greenstone sequence comprises the western limb of a regional anticlinal structure and in the vicinity of the mine is about 1500 metres thick.

The lithological contacts strike at 030° and dip to the east at about 72°. The deposit is pervasively foliated. Jointing is generally widely spaced. The orebody is transected by intrusives of variable dip and width and large shears, both flat and steeply dipping.

The mineralised zone at Big Bell is about 600m in length, up to 50m wide and has been explored to a depth of 1430m below surface (Player, 1998). Mineralisation is lenticular in plan and forms a tabular body plunging near vertically from surface. Gold has been recorded in all rock types apart from the intrusives. Economic mineralisation is generally restricted to the 'K-Feldspar Schist' (KPSH), 'Altered Schist' (ALSH) and 'Biotite Schist' (BISH) with the KPSH and ALSH dominating.

2. THE PROBLEM

The provision of a safe workplace and safe access to the working environment are major considerations for the operator of any underground mine. The current regulatory requirements in Western Australia have re-emphasised the responsibility of the mine owner for ensuring that the underground environment is, as far as practicable, free from hazard. In the area of ground support, this is achieved by requiring the mine's owners to develop a thorough understanding of the ground conditions that are likely to be encountered and the ground behaviour to be expected during development and stoping. Appropriate ground support then needs to be selected and installed.

During 1996 early indications of an unfavourable stress environment, with associated large-order deformations included the failure of Hollow Groutable Bolts ('HGBs') in the back of the decline and large deformations in the oredrives. The HGBs were rupturing at depths of about 15-30cm into the back. Damage in oredrive development included the crushing of narrow pillars developed to access the bottom of the old open stopes, hole closure and drive deterioration, particularly in rib pillars.

Due to the problems experienced in development, and blasthole stability, and required production rate a change of mining method was proposed from 'core-and-shell' stoping (primary open stopes with rib and crown pillar wrecking) to longitudinal Sub-Level Caving (SLC). The recovery of highly-stressed, late stage pillars could thus be avoided. Nevertheless, it was recognised that the SLC retreat faces would be subjected to high stresses and the associated ground control and blasthole damage problems.

In SLC, reliable, safe access to the brow is essential as all blasthole preparation, (cleanout and charging), activity takes place in this area, within metres of the cave front. In addition to the operator exposure, any delays caused by excessive hole cleanout, redrilling and support rehabilitation can severely impact on drawpoint availability and hence the overall production rate. Adequate control of the brows is one of the critical success factors in Sub-Level Caving.

2.1 *Geotechnical Issues*

The broad challenge at Big Bell is to successfully operate longitudinal SLC mining in an elevated and deviatoric stress environment. The maximum *in situ* principal stress direction is sub-horizontal and strikes at about 45° to the strike of the orebody. Mining widths vary from 9m to 35m over a strike length of 500m.

The orebody is schistose in nature with smooth, persistent lithological contacts and foliation planes. Up to seven joint sets have been identified but it is rare for more than two sets plus foliation to be present at a particular location. A continuous graphitic shear lies to the footwall of the orebody 10m to 15m from the mineralisation.

The orebody intact rock strength (normal to foliation) averages 150 MPa (UCS_{50}) with a Young's Modulus of 50 GPa. The average Q-value (Barton *et al*, 1974) for the ore zone is 2.9 but can range from 0.4 to 12.5. RQD is typically 90%-100%.

In situ stresses were measured using the CSIRO Hollow Inclusion Cell overcoring technique at two sites, 350m and 380m below surface (Sandy and Lee 1997). The pre-mining stress field at 365m below surface can be described as follows:

Table 1: Pre-mining Principal Stresses at Big Bell

Principal Stresses	Magnitude (MPa)	Dip	Azimuth *
Major	63	06°	224°
Intermediate	35	13°	315°
Minor	19	75°	109°

* Azimuth relative to mine grid north which is 30.8° east of AMG.

The primary failure mechanism that the reinforcement and support design is required to address is the intense fracturing above the backs of the oredrives caused by brittle, stress-induced intact failure of the rock. This was initially identified from exposure of the backs of old workings (Figure 1)

The same style of failure is observed in recent development (Figure 2).

Figure 1: Stress induced fracturing above the backs of an old haulage drive, 320 Level.

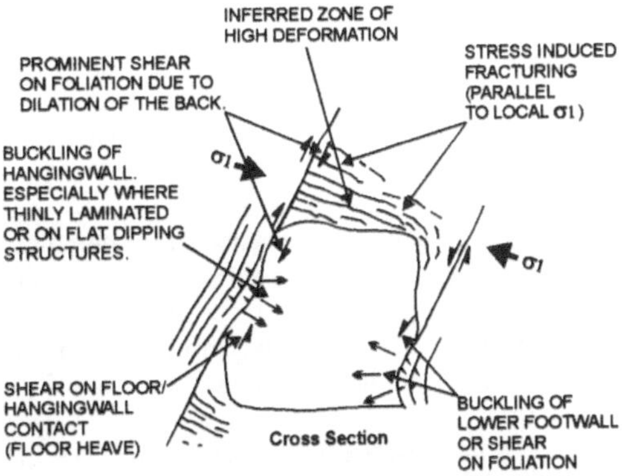

Figure 3: Observed deformation and interpreted ground behaviour.

Figure 2: Stress induced fracturing and dilated joints above the backs of a recent oredrive, 410 Level.

Figure 4: Severe damage caused to cable bolt surface fixtures by large deformations.

The fracturing above the backs of the oredrives is accompanied by dilation, in turn promoting shearing of footwall foliation planes particularly where these have been undercut by the drive profile. Dilation in the back also produces shear on foliation at the back/hanging wall intersection. An idealisation of these styles of ground behaviour is presented in Figure 3.

Near the cave front, a well-defined crack usually progresses along the hangingwall corner of the drive with the advance of caving. Figure 4 shows a hangingwall crack, with intense deformation around a cable bolt. The 'w-strap' that the cable bolt was holding has broken off. Hangingwall corner crack growth is encouraged when either a weak foliation plane or lithological contact is in the hangingwall corner of the drive and acts as a release plane for the sub-horizontal stress induced fractures in the drive back.

The availability of a weak structure may promote further dilation, which significantly increases the probability of back failure during stoping operations. It has also been observed that a relationship exists between the amount of local seismicity in a drive during development and the presence of large shallow dipping structures above the backs.

2.2 *Mining Problems*

Two main operational problems were experienced following the introduction of SLC:

- Loss of control of the brows, preventing access to the collars of the next ring of blastholes
- Blasthole closure due to movement on structures and stress-related damage

Another factor was the large development size necessary to accommodate large capacity production equipment. The depth of the failed zone in the backs and walls is scale dependent.

These contributed to significant disruption of the production cycle, in turn affecting the overall production rate from SLC.

Loss of control at the brow resulted in back failures ranging from 0.2m to 8m deep with about 1m being typical. The implications arising from brow failure include:

- Difficulty in gaining secure access to the holes for checking and when required, cleaning out by the drill rig. On most occasions, the rill would often prevent the drill rig from reaccessing the hole collars, thus requiring redrills.
- A regular requirement for redrilling of inaccessible stope rings, involved either 'dumping' multiple rings forward, drilling from the footwall access drive, or drilling from one oredrive to the next. These options usually resulted in less favourable drilling patterns, increasing the risk of a blast malfunction.

Current oredrives are 4.7m high by 5.0m wide by design, but are generally larger in areas where poor control was maintained and square or 'shanty back' profiles were used. The change to an arch profile and tighter control on the drilling of the development round significantly improved ground conditions.

The main geotechnical issue in the areas that are already developed, or that are being developed to the longitudinal SLC layout, remains the installation of support and reinforcement systems with appropriate characteristics to deal with the expected adverse high deformation ground behaviour.

3. INVESTIGATION STRATEGY

The recognition of the significant problems with the drive support and reinforcement led to the establishment of an investigation programme to determine:

- The deformation profile that was occurring in the backs and walls of the oredrives,
- The timing of ground deformation relative to both the development and caving cycles,
- Requirements for changes in development and support practice to maintain long term stability of the oredrives despite stress change, blast damage and time dependent effects,
- The suitability of the equipment deployed to install reinforcing and supporting elements,
- The most appropriate drive profiles, balancing drive stability and support costs against maximising draw width for the SLC,

The investigation was expected to provide the key information necessary for selection of the best support/reinforcement systems for the oredrives of Big Bell. This would be in terms of support stiffness, capacity, cost and ease of integration into the development and mining cycle, using as far as possible existing equipment.

3.1 *Bolt Selection Criteria*

In considering alternative reinforcement elements, compatability with the existing equipment and established ground support procedures is an important aspect.

The equipment available on site (Atlas Copco Boltec rig) was suitable for the installation of mechanical point anchored bolts up to 2.4m long. The bolts would be used in conjunction with RF81 (7.6mm wire in 100mm x 100mm squares) weldmesh in 6m x 2.4m sheets. W-straps for wall and brow support would be initially secured using friction stabilisers by a short boom jumbo.

Fully encapsulated bolts were preferred for brow support in the oredrives. They would be expected to provide increased control over the dilation of the ground associated with stress changes from the cave approaching, and possibly damage due to production blasting. Point anchor bolts could also be susceptible to anchor slippage, and if the plate loses contact with the rock they no longer offer any reinforcement capacity.

However it was expected that fully encapsulated bolts would be too stiff for the deformation following development and the deformation from caving. Installation initially only as a point anchor provides the bolt with a flatter (less stiff) load-response curve than a fully encapsulated bolt.

The tension applied to or developing in the bolt would be expected to increase the friction on sub-horizontal stress fractures in the backs, increasing the resistance to shearing and dilation along these surfaces, provided the bolt is not overloaded. Friction stabilisers would be expected to slip or yield rather than develop significant tensile stresses.

Although grouting at an early stage (after deformation associated with development has finished) should reduce the amount of shear that occurs in the back of the drive, late stage grouting (eg after the cave front passes over on the level above) reduces the potential for bolt overloading.

Early experiences with the HGB suggested that the bolts were easily ruptured in high deformation environments. The thin wall construction, coupled with very efficient load transfer once the bolts are grouted internally and externally, could lead to very high, localised stress levels in the bolts. However, a post-groutable bolting system was still attractive for operational reasons because it would separate the grouting component of bolt installation from the development cycle. The requirement for immediate support at the development face favoured the use of a point anchored bolt. Post-grouting also offers the possibility of modifying the stiffness of the bolt at an appropriate point in the development/stoping cycle. This has to be based on sound understanding of the timing and degree of deformation associated with various stages in the mining cycle.

The two bolts examined were the Tubular Groutable Bolt (TGB) and the Combi Tube (CT) bolt. The TGB is made from cold drawn steel and uses a mechanical anchor shell. The bolt is a smooth tube of narrow diameter, (hence requires higher grout embedment length to prevent pull out) with an external diameter of 17.4mm and internal diameter of 8mm. The bolt starts to yield at 132 kN and breaks at 157 kN.

The CT bolt is a solid bar point anchor style bolt with a plastic sheath for grout flow for encapsulation. The bar is ribbed and only requires a short embedment to break the bolt. The bar diameter is 20mm; minimum yield occurs at 144kN and minimum failure strength is 223kN. Shell slip for the bolt in Big Bell conditions occurs at 16t to 18t. There were some concerns over the suitability of the CT bolt for brow support because of the potential for the plastic sheath to decouple the bar from the grout and the rock, and the long term design life of the bolt for a short term application.

Table 2: History Of Extensometer Sites

Site	Date	Major Events
Site One 410 F79N	16/12/97	Back extensometer Installed into development 10 months old. No wall monitoring or load cells. Figure 5 shows a long section of the site and Figure 6 a cross section.
		Support in the drive combination of TGBs (predominantly in the vicinity of the test site) and SPA's, both tensioned to 10t with only a 39mm SplitSet plate (wrong selection). Minimal bolting used for wall support. Weld Mesh RF81. Mesh sheet size controls the bolt row spacing at 1.0m to 1.1m
	18/1/98	First effects of the cave going over on 380 level.
	6/5/98 and 15/5/98	Cable bolts installed on the originally designed burdens (twin strand, 4 bolts per row, rows 3m apart, 4.0m embedment outside pair, 6.0m embedment inside pair).
	13/6/98	Estimated last effects from caving on 380 level (40m away)
	24/6/98	Influence of cave on 410 level, extensometer responds to blasting 40m away. Monitor advancing production blasts.
	12/7/98	5 'gewi' bars (2.8m of embedment, grouted 20mm diameter threaded bar, 150 x 150 x 6mm dome plate) installed on each row of cable bolts.
	25/8/98	Change of ring burden and rows of 4 Hardi bolts and 5 gewi bars installed through w-straps 1.0m behind new designed ring positions and requirement for all brows to be w-strapped. The two sets of brow support are within 0.5m of each other. Figure 7 shows the multiple layers of reinforcing and support.
	11/10/98	Production blast takes out the extensometer.
Site Two 435 F55N	8/10/97	Extensometers installed in the back and the hangingwall, in cycle with the development. Figure 8. Square drive profile.
	13/10/98 to 19/10/97	Heading developed, 10m either side of the site reinforced with TGBs installed to 5t tension, RF81 weld mesh, installed to the face. Plates 150 x 150 x 6mm dome plates.
	18&19& 25/10/97	Wall w-strapping and spot bolting with SplitSets, for the full length of the oredrive.
	23/11/97	TGBs grouted – backs of the drive continued to creep, see Figure 13.
Site Three 435 F80S	13/8/98	Extensometers installed in the back and the wall, in cycle with the development. (Figures 5 and 7). Two load cells also installed. Figure 9 shows the monitoring site. Drive developed with an arched profile.
	14/8 to 18/8/98	Development continues. Nominal 1400J* fibrecrete sprayed direct and bolted through with TGBs on 1.4m spacing between rows. Brow support w-straps installed during development stage. Figure 10 shows the nominal bolting pattern and fibrecrete used for this site.
	15/8/98	Caving above influences site.
	21/8 to 30/8/98	Development continues along the drive, but does not influence extensometers; Support is with TGBs and mesh. The arched profile is not maintained as well in the bolts and mesh section as in the fibrecrete area.
	6/9/98	TGBs grouted.

* 1400J fibrecrete specification based on 55kg/m^3 RC 65/35N DRAMIX fibre.

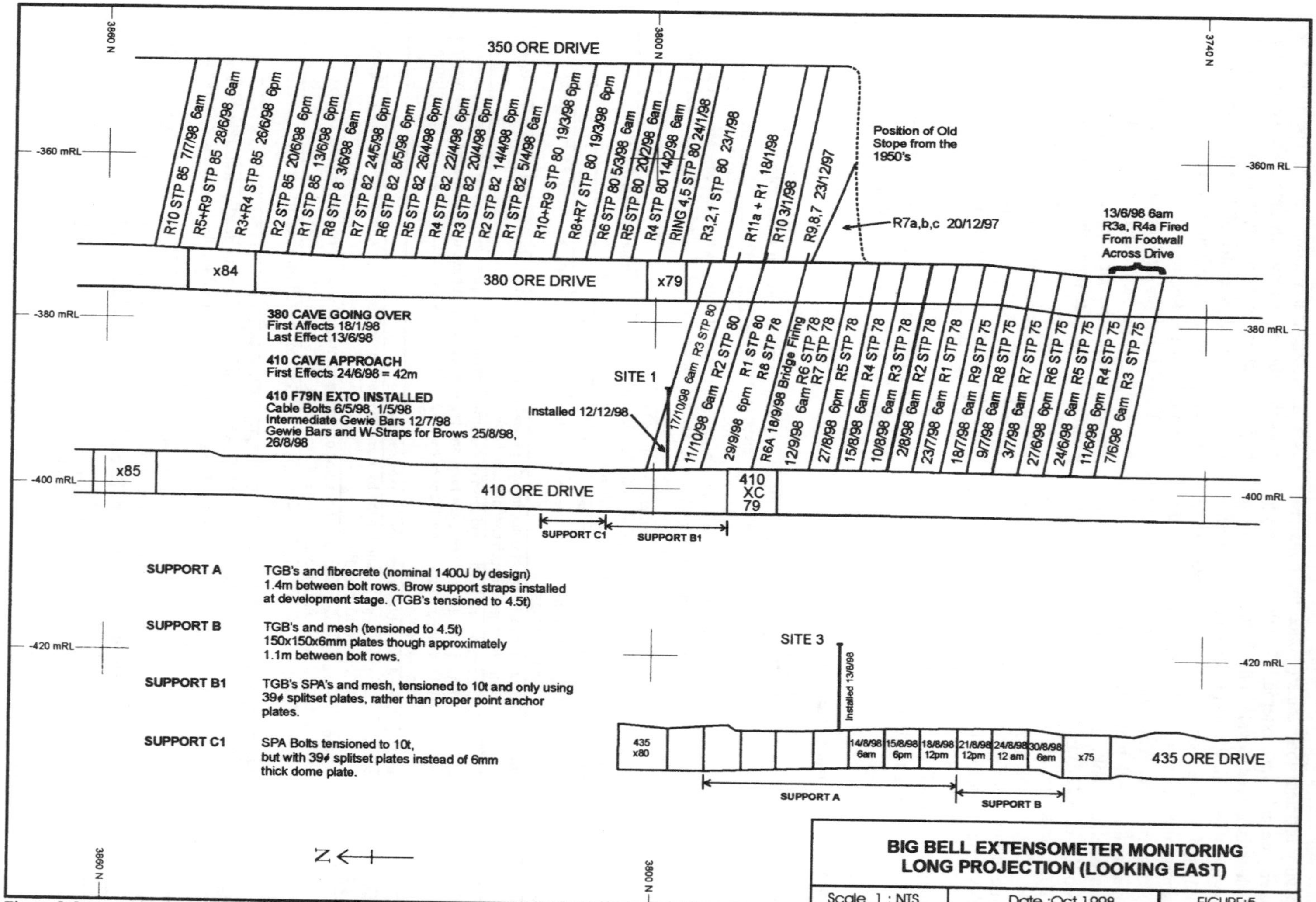

Figure 5: Long projection at extensometer sites 1 and 3 looking to the east.

3.2 *Drive Deformation Monitoring*

Understanding the development of deformation around the oredrives was required in order to select appropriate reinforcement and support systems. While observation was, and continues to be an important source of information about ground behaviour, it was clear that large deformations were occurring. Quantification was required. Timing of the deformation in the backs and walls of the drives was unknown, but time was not available to observe the whole cycle. Therefore some extensometers were installed to monitor the influence of caving, and others to gather the deformation history associated with initial development.

Manual rod extensometers consisting of 6 anchor points were selected. They were located typically in the centre or towards the hangingwall side of the back of the drive and drilled vertically (ie similar to the support elements) rather then parallel to foliation. Anchors were initially located at 0.5m, 1.0m, 2.0m, 3.0m, 6.0m and 10.0m. This changed to 1.0m, 2.0m. 3.0m, 4.0m, 6.0m and 10.0m following early results from Site 1 where significant deformation occurred between 2m and 6m into the back.

At some locations shorter extensometers were also installed in the footwall or hangingwall of the drive. Anchor locations were selected to provide detail on the deformation profile that was occurring above the back of the drive, both in the region of the reinforcement and beyond.

Table 2 presents the history of the monitoring sites that have been used for the compilation of data. The reinforcement used at these sites is also summarised.

Figures 5 and 8 show the advancement of the development heading at Sites 2 and 3 and the progression of the cave front in relation to the extensometers at Sites 1 and 3. Figure 6 is a cross section showing the relative location of Sites 1 and 3 extensometers on a common plane. Three extensometers have been installed that are yet to be affected by caving.

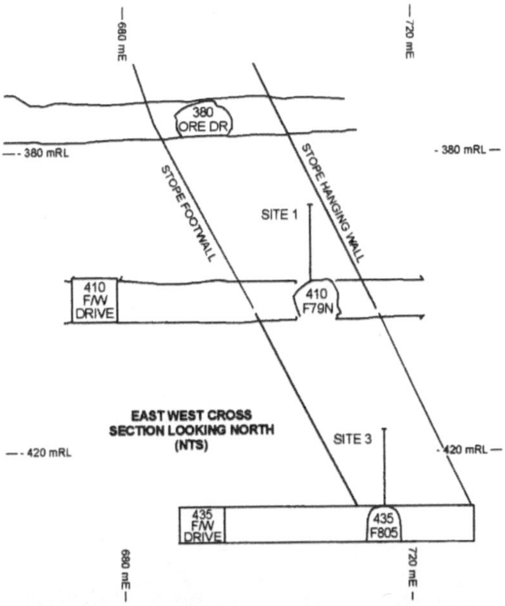

Figure 6: Cross section at the nothern end of the mine showing Sites 1 and 3

Figure 7: Multiple support phases at Site 1, cable bolts, gewi bolts and straps

3.3 *Obtaining Data*

An important issue in any monitoring programme is the frequency of readings. It was elected to use manual rather than remote reading heads for the following reasons:

- They are cheaper without the costs associated with a remote head, cable and the read-out unit.
- Potential damage to the cable or the remote head from passing equipment, or fly rock from nearby blasting, is avoided.

The disadvantages included the requirement to gain repeated access to the backs of the drive and ensuring operators would take the readings with sufficient frequency:

- The underground samplers were trained for this job, but an Integrated Tool Carrier (IT) was needed to lift the samplers to take the readings.
- The samplers had to have free time to take the readings.
- Access to the sites was sometimes prevented by drill rigs or loaders.

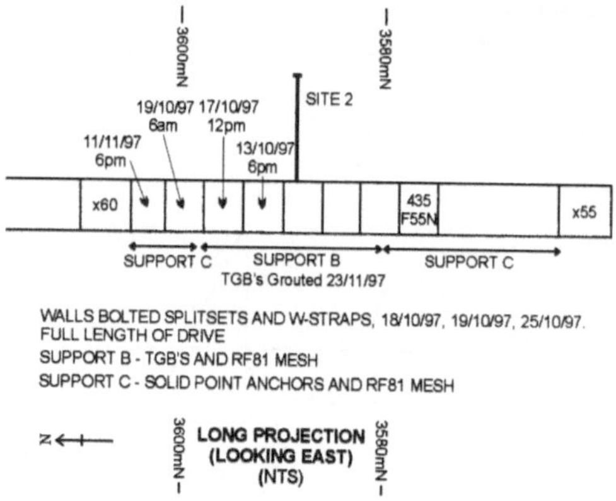

Figure 8: Long projection of development at Site 2 with the associated ground support.

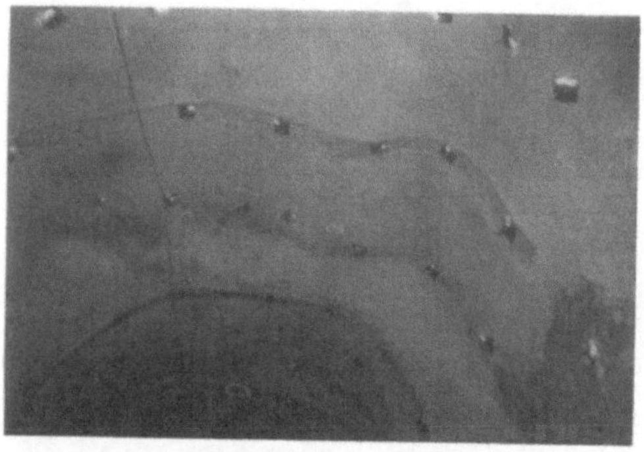

Figure 9: Sprayed fibrecrete and bolting at Site 3. W-straps installed at development stage.

- Consistent coordination is more difficult to maintain in fly-in/fly-out operations.

The above factors meant it was difficult to obtain readings more frequently than once per 12 hour shift, and readings could be as far apart as two weeks. The desired frequency of readings was dependent on the mining activities occurring in the vicinity and the influence they were expected to have on the stress field. In some cases the exact timing of an instrument response is missing because of the spacing between consecutive readings. Nevertheless, the results obtained are sufficiently detailed to determine:

- The movement associated with initial development,
- The creep or time-dependent response of the ground,
- The effect of stress redistribution from the cave above,
- The effects of production blasting and stress redistribution on the same level.

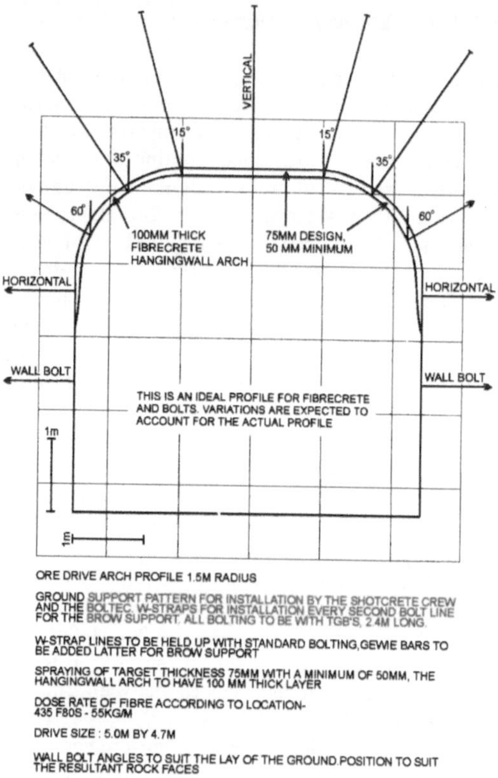

Figure 10: Nominal fibrecrete application and bolting pattern at Site 3.

Each data point was based on two readings of each anchor to a tolerance of 0.2mm. Any obviously incorrect results were removed to improve the consistency of the plots.

3.4 *Drive Deformation during Development*

At Sites 2 and 3 the extensometers were grouted into 76mm holes located 1.0m back from the face of the drives. Besides the extensometer an observation hole was drilled to observe changes in the back of the drive. The observation hole provided an opportunity to assess the sense of movement in the back of the drive (ie if the movement was dilation, shear or some combination). The exact position of open structures could also be determined.

Three trial sites to monitor the performance of fibrecrete and TGBs compared to the performance of TGBs and RF81 mesh have been established, of which only two are discussed in this paper. Additional test work is being undertaken to confirm the initial results.

Wall extensometers were installed to measure dilation and shear on footwall foliation planes both during development and in response to the creation of the pillar between oredrives during mining of the hangingwall drive. No convergance monitoring stations were established to assess complete drive closure.

Load cells have been recently included between the nut and the plate on some bolts. Insufficient data is available at this stage apart from showing that load increases on the bolt as the deformation increases in the back of the drive. There is also an indication that there may be some 'stick-slip' type of loading occurring in the point anchored ground reinforcement.

3.5 *Drive Deformation from Caving Above*

Sites 1 and 3 have recorded the influence of caving, as they were installed relatively close to the production faces, as shown in Figure 5. The two sites have substantially different installed support and drive life prior to the influence of the cave as described in Table 2. Of the other three extensometers installed, two have had the cave pass over two levels (45m) above without causing any change in readings.

Site 1 on the 410 level was established in a drive that had been developed 10 months earlier. Based on current understanding, the installed development support was not particularly well suited to the requirements. Significant improvements have been made since that period. Three phases of additional support were added for the brow and a total of 58mm of displacement was recorded. The 10m anchor must have been damaged during installation, as it never responded to any movement. The 0.5m and 1.0m anchors also had minimal response, but it is considered that there was no differential movement in this lower portion of the back. This may have been due to separation occurring just over 1.0m into the back, and the ground between this point and drive essentially remaining intact with the grouted TGBs.

Site 3 on the 435 level was established just in front of the cave as shown in Figure 5. The effect of the cave at this extensometer site is interpreted to be significantly reduced because of the improved support utilised, and 'younger' drive life. The site was used as it gave the best opportunity to assess the deformation associated with the drive's development. This site provided information on 'two stage' ground support and reinforcement. The 'upgrade' to production level support may only require gewi bars infilling in the w-straps that were installed during development. It is not expected that there will be any requirement for rehabilitation bolting or additional wall bolting or strapping in this drive.

3.6 *Test Programme Reinforcement and Support Elements*

When the initial problems where identified with the failure of HGBs under low nominal axial loads, the role of shearing in bolt failures was required to be investigated. Allowance for shearing on structures/fractures in the backs would hence influence the selection of a suitable bolt. Some failures also indicated that corrosion may have contributed to the failure despite the HGB being nominally a fully encapsulated bolt. Small ground movements may have cracked the grout, allowing saline ground water to attack the bolt at a localised point.

It was recognised that the ideal bolt would have adequate capacity to accommodate both shear and axial deformation. Both the TGB and CT bolts tested were considered to meet these criteria. The HGB was expected to be less able to accommodate shear because of the thin wall (3.2mm) construction and large diameter (19mm) internal cylinder of grout. The internal grout cylinder is relatively large in comparison to the wall thickness of the bolt. The surface area (and hence grout-steel bond) per metre of bolt length is significantly higher than for the TGB and CT bolt.

The TGB has a much smaller internal diameter (8mm) and thicker walls (4.7mm). It is suggested that the internal grout column is too small to significantly affect the deformation response of the bolt, when loading either in shear or tension. The smooth finish and small diameter of the bolt, reduces the bond strength per unit of bolt length, requiring an embedment length of greater then 1.0m to break the bolt in tension. The main purpose of the grouting of the TGBs is to reduce the shear and ground movement and maintain some reinforcing effect on the drive if the bolt is broken on the thread.

The manufacturers' performance data for the HGB, TGB and CT are presented in Table 3.

A series of pull tests was performed on all the bolts being investigated including post groutable point anchor and friction stabilisers. The tests assessed different embedment lengths, effect of bit diameter, and load-deformation or stiffness curves.

The load-deformation curves were based on

Table 3: Rock bolt manufacturers' performance data:

Specification /Type	Hollow Groutable Bolt	Tubular Groutable Bolt	CT Bolt
Internal Diameter	19mm	8mm	Solid
External Diameter	25.4mm	17.4mm 19mm on thread	20mm
Wall thickness	3.2mm	4.7mm	20mm
Ratio External to twice wall thickness	4.0	1.9	1.0
Ultimate Tensile Strength	11t – tests generally 12t and grouted 15t-16t	15.7t	22.3t – tests generally 25t
Yield Tensile Strength	9t – 10t	13.2t	14.4t – test generally 16t – 18t
Thread	Cold Rolled Rope thread	Cold rolled fine thread and cold drawn tube.	Machined fine thread.

manually operated pull test rigs. In these tests the rate of loading is generally much faster than the bolts would normally experience under natural (non-seismic) loading. As a result, the capacities may be underestimated and the stiffness over-estimated compared with actual field performance.

The test results were compared with available data from other mine sites, and published test work on other bolts to allow comparisons to be made for the various types of bolts. This aided in developing an understanding of a relative ranking of all bolts for stiffness, in relation to when they were installed, and whether they were grouted or ungrouted. This paper only briefly presents some of the results obtained from the test programme, as it is intended to present this work fully at a later date. Figures 11 and 12 summarise the pull tests results.

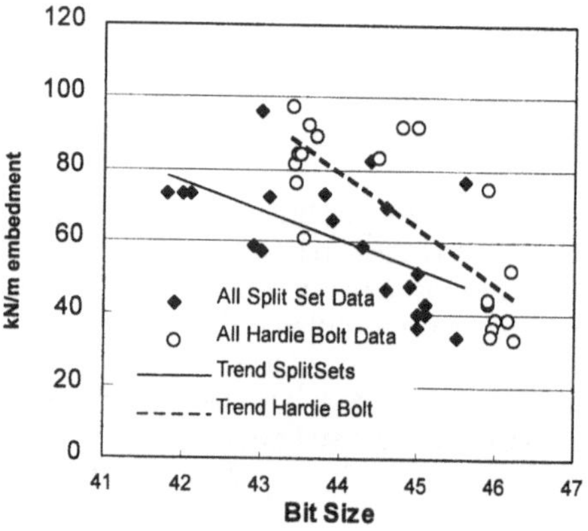

Figure 11: Trend Lines SplitSets and Hardie Bolts

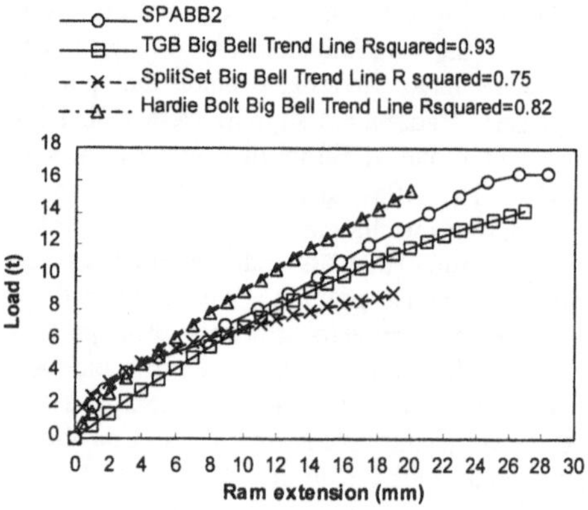

Figure 12: Trend Line Comparison of SplitSets, Hardie Bolts, TGB's and SPA.

3.7 *Stress Modelling and Monitoring*

The monitoring of the stress redistribution effects underground was largely by observation of the response of the backs and walls of development. These could be related to changes in the local excavation geometry, such as a cave front passing above the level, or approaching on the level.

Modelling undertaken for the central region of the orebody showed that stress concentrations would be expected to be highest at the corners of the cave opposite to the NE-SW principal stress direction.

This was observed to be the case based on ground behaviour around the old open stopes. However observations in the caving layouts showed that the footwall oredrive generally had more intense stress cracking above the backs of the drive and increased brow damage irrespective of the retreat direction of the cave. The localisation of the damage may be more a reflection of rock strength rather than different stress conditions.

The 2D modelling used for the assessment of the drives underneath the cave did not have the ability to include both the cave and the lower drives on the one run. Comprehensive 3D modelling of the development and caving may be undertaken following overcoring stress measurements on the 460 and 485 levels. A stress change cell will also be installed between the 410 and 435 level to monitor the effects of the cave passing over and the cave face approaching on the level.

The relationship between width of orebody, rockmass quality (Q), stress change and the influence of the hangingwall drive on the footwall drive requires further investigation. Observations underground show that the central part of the orebody typically has worse drive conditions (during development) and brow conditions (during production) than at the limits of the orebody.

4. RESULTS FROM TEST PROGRAMME

Observation plays a key role in any ground behaviour monitoring programme as instruments usually only provide information on what is happening at discrete locations. The observer needs to tie this to what is happening in the general vicinity of the monitoring. Observations of the general style of ground behaviour at Big Bell are summarised in Figure 3.

Other important observations include:

- A change of ground conditions back from the face when an oredrive was mined in particular, increased local seismicity.
- Changes in the oredrive conditions when the cave front passes overhead or approaches on the same level, including further crack growth in the hangingwall corner of the drive, sliding of blocks on the footwall foliation planes and increased fracturing of the ground on the back of the drive, with bagging in the mesh at the shoulders in some areas.
- Crushing and shearing of blast holes.
- Evidence of increased loading of support/reinforcement elements such as plate buckling, damage to collar fixtures or ruptured bolts.

4.1 *Results from Development Deformation Monitoring*

The deformation of the oredrive backs associated with initial development and then creep with time are shown in Figures 13 and 14. These correspond to Site 2 (mesh and bolts) and Site 3 (fibrecrete and bolts).

The monitoring shows that for Site 2 equilibrium is reached after three cuts, whereas at Site 3 no further changes occur after the second cut. A comparison of the deformation response to further development between drives supported with fibrecrete against mesh is shown in Figures 15 and 16. These show a 35% reduction in the back deformation, an 80% reduction in dilation of the hangingwall, where fibrecrete has been used.

The reduction in dilation in the back has occurred in the surface to 1m deep zone. This appears to correspond to the stiffer support being offered by the fibrecrete, and the better design utilising an arched profile. Additional fibrecrete in the hangingwall corner will also have assisted in controlling the deformation in this area.

The total recorded deformation of 26mm for Site 2 is close to inferred capacity of the TGBs with no pre-tension on installation. Allowing for 5t of tension applied to the bolts on installation it appears that there must be some loss of tension occurring during the life of the bolts, otherwise they would have failed. This is supported by the preliminary load cell results from Site 3 which show an increase before, followed by a drop of the load in the bolt just after the second, third and fifth development firings.

Caving has influenced Site 3 before any creep has time to occur, but it is expected that it would have been low for the fibrecreted site. Total creep of 5mm over 12months was recorded for Site 2. Virtually all creep occurred after grouting of the TGBs, which appear to be able to yield due to their design.

Inspection of Figure 13 shows that the 1-2m zone at Site 2 seems to be affected by four changes in the rate of deformation once creep has commenced, whereas the other zones remain constant. The first and third periods are at the same rate and the second and forth periods are at the same rate. No explanation is offered for this occurrence, it may be due to a general change in the stress field or a particular change to reinforcing elements.

First 16/11/97 to the 1/1/98 – 1mm per 44 days
Second 1/1/98 to the 5/4/98 – 1mm per 215 days
Third 5/4/98 to the 5/5/98 – 1mm per 44 days
Forth 5/5/98 to the 2/10/98 – 1mm per 215 days

4.2 *Deformation Associated with Advancing Cave Fronts*

The deformation profiles for the backs of the oredrives affected by caving are shown in Figures 17 and 14. These represent the deformation profile for Sites 1 and 3. Site 1 has had the complete 'caving history' with both the effects of a cave

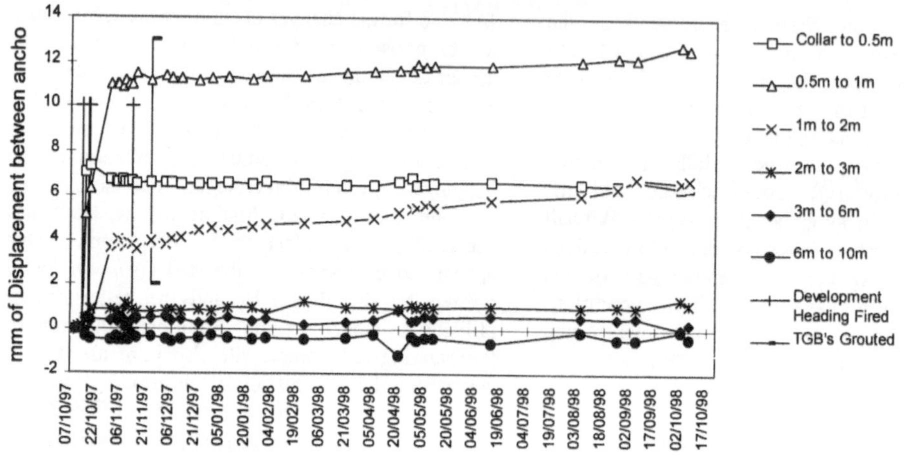

Figure 13: Site Two - 435 F55N
Deformation Profile in the Back, supported by TGB's and RF81 mesh. Bolt pattern 1.1m rows by 1.1 spacing

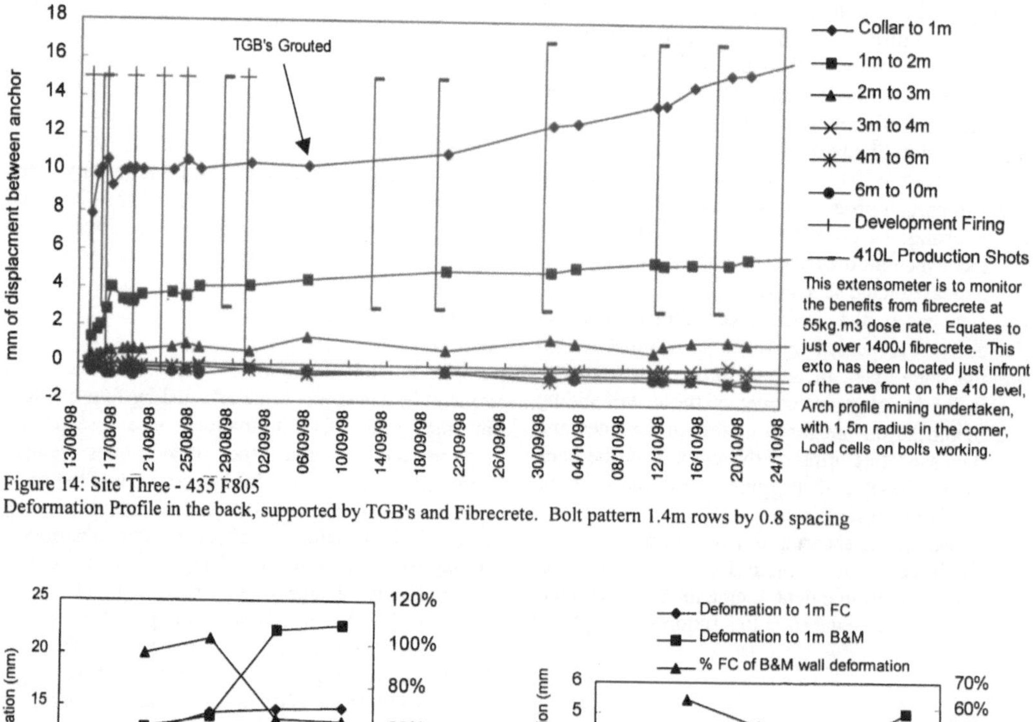

Figure 14: Site Three - 435 F805
Deformation Profile in the back, supported by TGB's and Fibrecrete. Bolt pattern 1.4m rows by 0.8 spacing

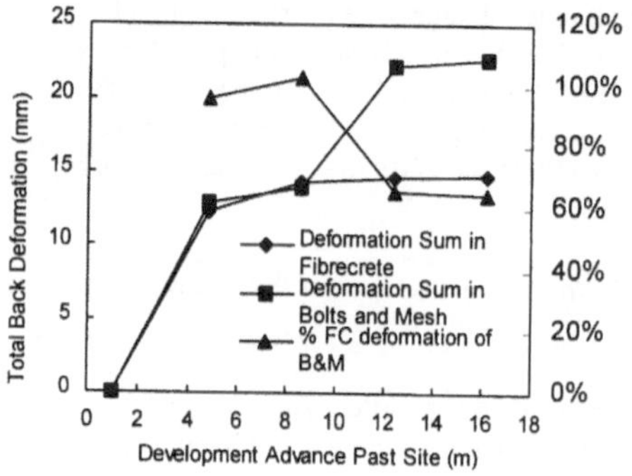

Figure 15: Comparison of Back Deformation from Site 2 and Site 3 - Fibrecrete and TGBs vs Mesh and TGBs.

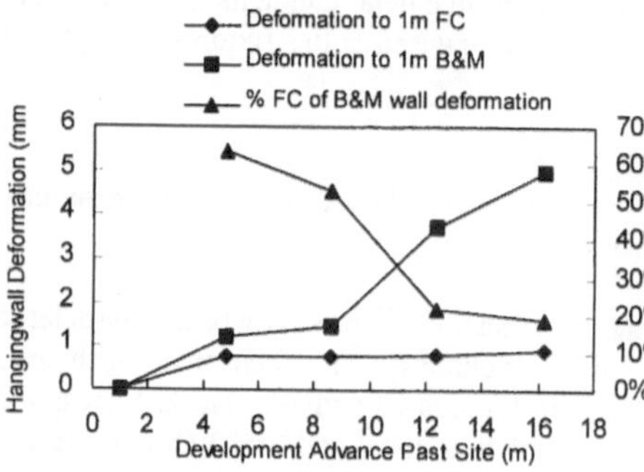

Figure 16: Hangingwall Dilation Comparison of Site 2 and Site 3 - Fibrecrete and TGBs vs Mesh and TGBs.

passing overhead and approaching on the same level being recorded. Site 1 will be the base case against which future monitoring results are compared for improvement in support and reinforcement practices. As discussed above, the frequency of readings has not been ideal as it has given an incomplete record of the deformation history.

A detailed discussion of the 1 to 2m zone at Site 1 is presented below, based on the assumption that known deformation rates can be used to interpolate between measurements.

A significant acceleration in dilation occurred some time after the opening of the slot at X79 above the extensometer, however there was minimal effect on the results with the widening of the slot to the footwall and the hangingwall sides.

Deformation at a slow rate (1mm per 14 days) started again once the cave continued moving to the north on 380 level. This rate would be expected to have continued until just before the firing of Rings 7 and 8 in STP80.

Projecting the deformation rates recorded following the firing of Ring 9 and 10 (STP 80) shows that these two shots (each of two rings) had a substantial effect on the level below, the average rate increased to 1mm every two days ie by a factor of seven.

This high rate slowed to 1mm every 4.5 days but did not change again until the cable bolts were installed, which resulted in a further reduction of the rate to 1mm per 14days.

Dilation then seemed to stop, presumably indicating that further advance of the caving on 380 level was no longer exerting an influence. This was after the firing of Ring 8 (STP 82) – 40m north of the extensometer. The caving on 410 level did not exert an influence until Ring 5 (STP 75), at 40m south of the extensometer. The deformation rate associated with the influence of blasting on 410 level is 1mm per 9 days. When the face is 18m from the extensometer and closer there is an increase in deformation after each blast, but insufficient detail to accurately define the rate changes associated with the firing of the remaining shots. This includes the effects of the installation of the second set of gewi bars.

The '3m to 6m' zone also responded when the face was at 18m. This may be related to movement on a particular continuous smooth foliation plane mapped in this area. Brow failure to the south has been associated with this structure.

The first set of gewi bars were exposed to less than 15mm of shear and dilation and the cable bolts only 20mm of shear and dilation. It is unlikely that this would have affected the capacity of either bolt. More importantly additional control was maintained over the backs during the dilation and shearing that occurred as a result of the cave passing overhead. This indicates that the gewi bars should be installed earlier to maintain brow integrity.

Figure 18 shows the brow after the firing of the last ring in front of the extensometer. The brow was stable and all bolts appeared to be intact. The brow was 'double reinforced', with both the cable bolt and gewi bar ring and the w-strap and gewi bar line.

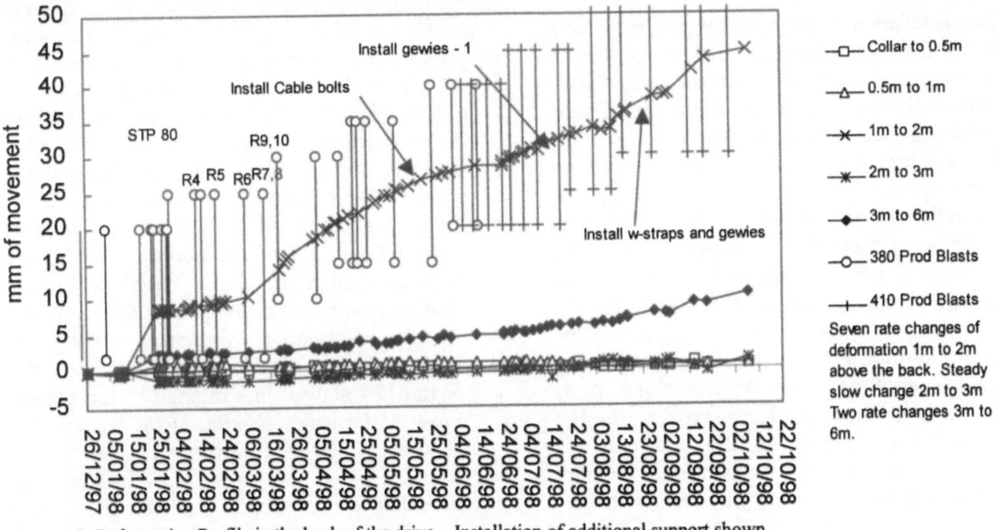

Figure 17: Deformation Profile in the back of the drive. Installation of additional support shown.

Figure 18: Site 1 brow conditions before charging.

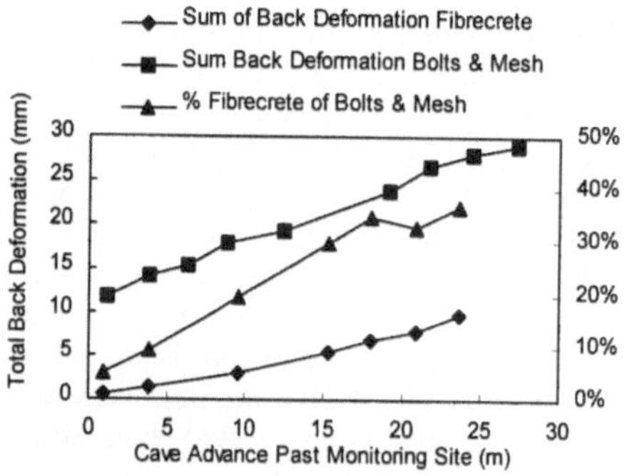

Figure 19: Comparison of total back deformation due to the influence of caving at Sites 1 and 3.

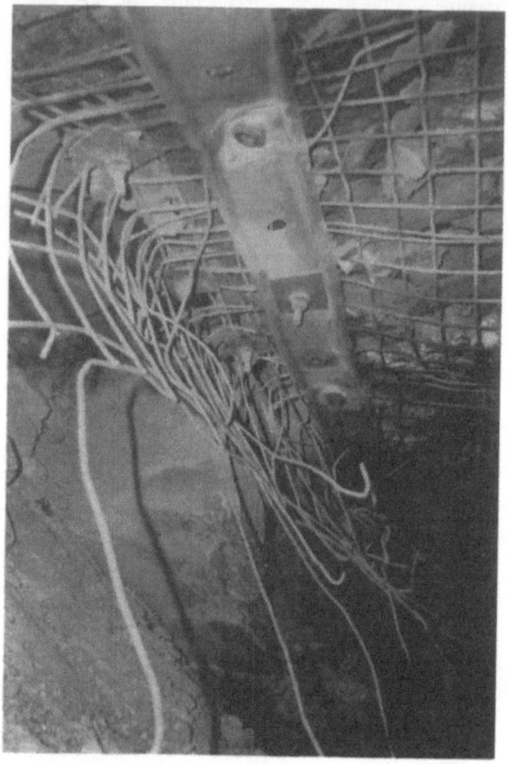

Figure 20: Current brow performance.

Site 3 (with fibrecrete and bolts) only showed 33% of the deformation in the backs compared with Site 1 (see Figure 19). Explanations for this may include:

- The installation of the extensometer may have post-dated the initial stress redistribution associated with the caving on the level above. This is difficult to assess until Site 2 is affected by caving from the level above.
- The site may be outside the influence of high stress changes as a result of the extra 5m separation between the sites. This is 25% extra separation, although it is situated closer to the footwall side of the ore body which should increase the stress concentration effects.
- The fibrecrete and bolts in conjunction with the arched profile are very effective in reducing and controlling the effects of stress redistribution from the level above. This last possibility is supported by the location of the deformation. Currently the significant majority is between the surface and 1m deep, rather than 1 to 2m deep as observed at Site 1.

5. CONCLUSIONS

Significant progress has been achieved in the past 12 months, and further work is ongoing. The monitoring programme achieved the original objectives of identifying the timing, location and magnitude of deformation in the oredrives. The conclusions to date are summarised as follows:

- The performance of TGBs in oredrive support is satisfactory provided they are used with suitably sized washer and 6mm thick dome plate. Bolts must be installed as close to orthogonal as possible. Further test work is required to confirm the optimum timing for post-installation grouting. Based on current understanding this appears to be following the initial deformation associated with the development.
- Figure 20 shows an example of the current brow performance looking along a strap line. Figure 21 is looking up the hangingwall foliation plane and shows brow loss past the next line to be fired. This was a common situation prior to the recent changes in brow

Figure 21: Previous brow performance.

Figure 22: Brow conditions prior to the introduction of W-strapping.

support practices as indicated by Figure 22.

- Higher strength w-straps will be assessed for their potential to overcome the effects of wear from blasted rock and blast damage.
- Gewi bars and w-straps should be installed at sufficient distance in front of the cave so that the grout has cured when the cave is 18m in front of the brow position. Gewi bars should be targeted to be installed 40m in front of the cave, as long as this does not expose them to the influence of the caving above.
- The firing of two rings together should be avoided because of the detrimental effect it has on the level below. This is difficult to achieve when blasting through intersections. The intersection below should be cable bolted where the cross-cuts cannot be staggered to reduce the influence from high stress change and blasting.
- More information is required on the performance of fibrecrete, particularly in the critical brow area. This will become available during production at Site 3. Results to date have shown substantial reductions in the levels of deformation in the immediate excavation back and where fibrecrete has been used. This has potential to reduce loading on reinforcement elements, and this could lead to a reduction in bolting pattern density.
- The use of remote heads for monitoring the effects of caving passing overhead on the level above and the approach of caving on the same level would allow the timing of deformation to be more accurately recorded.
- 3D modelling of stress redistribution will be undertaken following additional stress measurements to investigate the relationship between local stress changes and deformation.

ACKNOWLEDGEMENT

The authors thank the owners and operators of the Big Bell Mine for permission to publish this paper, and their colleagues and the operators that provided the data for this research.

REFERENCES

Barton, N., R. Lien & J. Lunde 1974. Engineering classification of rock masses for the design of tunnel support. *Rock mechanics*. 6:189-236.

Player, J.R. 1998. Big Bell coming back for seconds, thirds and fourths. *AusIMM underground operators' conference*, June/July.

Sandy, M.P. & M.F. Lee 1997. Big Bell mine overcoring stress measurement. *Unpublished Australian Mining Consultants report*, February.

Rock Support and Reinforcement Practice in Mining, Villaescusa, Windsor & Thompson (eds)
© 1999 Balkema, Rotterdam, ISBN 90 5809 045 0

The development of a ground support system at the Pasminco Broken Hill Mine

N.S.Rauert, P.W.Butcher, P.M.Ormsby & K.P.Tully
Pasminco Broken Hill Mine, N.S.W., Australia

ABSTRACT: The Pasminco Broken Hill Mine has a history dating back some 90 years. In that time ground support has played a major role in the variety of mining methods utilised at the operation. Over the years Broken Hill has played a major role of ground support and reinforcement development, and is still one of the largest consumers of rockbolts, cable dowels and shotcrete in the Australian mining industry.

In 1997, management considered it appropriate to review the ground support and reinforcement systems, procedures and quality of all previously installed ground support and reinforcement within the mine. Consequently, a ground support audit was carried out over the 90 kilometres of accessible openings within the mine. The audit was one of the largest undertakings of its type carried out and considered aspects such as:

1. Ground condition
2. Water presence
3. Installed ground support and reinforcement condition, coverage and age
4. Remedial ground support and reinforcement quantities and priorities.

The information was then used as a basis for establishing a ground support system utilising mine design, geotechnical and operational considerations to systematically allocate rockbolt, cable dowel and shotcrete resources.

1 INTRODUCTION

Broken Hill is situated in far west New South Wales some 1,165km west of Sydney and 508km north-east of Adelaide.

The Pasminco Broken Hill Mine is currently mining lead and zinc ore at a rate of 2.8 million tonnes per annum, primarily from underground. Presently operations extend to a depth of 1200m and are anticipated to reach 1300m. Longhole Open Stoping (LHOS) is now the dominant mining method.

Mining has been continuous at this site for over 90 years. The underground workings at the Pasminco Broken Hill Mine are some of the most extensive of any underground mine in the country, consisting of over 90 kilometres of accessible drives. The ground conditions within the mine show quite large variability and the installed ground support/reinforcement in these accessible areas is quite varied in type, age and condition.

Figure 1 shows an isometric view of the operation, highlighting the complexity and extent of the accessible underground openings.

During 1997, visual inspections indicated that extensive remedial work was required.

The inadequacies in the ground support was due to:

(i) The wrong type of ground support/reinforcement for the ground condition.
(ii) The deterioration of ground support due to ageing and the effects of water.
(iii) Incorrect installation.

In order to address this possible safety risk, a decision was made to carry out a Ground Support Audit at the operation (Rauert 1997). This initial audit took some three months to complete and was possibly the most detailed investigation of its type ever carried out in an operating mine.

The Ground Support Audit process involved site investigation, ground support/reinforcement requirement evaluation and recommendation of trials of alternate ground support/reinforcement technology.

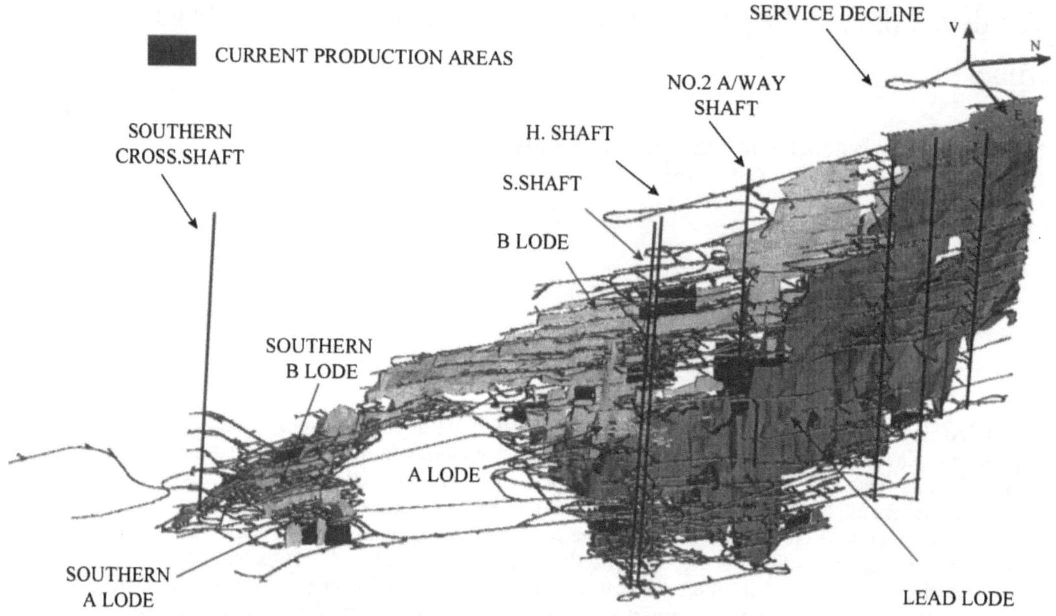

Figure 1. An isometric view of the Pasminco Broken Hill Operation showing the extent of development and mined stope openings. Obtained using the "Vulcan" geological and mine design modelling software.

The Ground Support Audit then became the basis of a ground support and reinforcement management system currently in use at the operation for managing the extensive ground support/reinforcement program at the Pasminco Broken Hill Mine.

2 GEOTECHNICAL CONDITIONS

Ground conditions are quite variable within the mine and range from very good in the upper portions of the mine to the highly stressed and structured in the Southern 'A' Lode area, and the very blocky low stressed problematic ground in the Lead Lode areas. Additionally the man-made geotechnical challenges of mining through backfilled areas in the Lead Lode Pillar mining areas.

Much of the orebody is very structurally controlled, with features such as the "main shear" forming the contact zone for most of the Lead Lode orebody. Several cross-cutting discontinuity features such as the "Flat Fault" and the "Zinc crack" zones makes stoping a ground support/reinforcement challenge.

The southern areas of the mine and an area in the central 14 to 15 level area are prone to microseismic activity, consequently the operation has a microseismic monitoring system for managing the seismic risk in these areas.

3 GROUND SUPPORT AND REINFORCEMENT

The Broken Hill operation has a long historical involvement with developments in ground support/ reinforcement technology. The operation played a key role in development of the Cable Dowel during the 1970s and is still one of the world's largest users of cable dowels.

In recent times the operation has pioneered the use of mechanised shotcrete in the Australian mining industry and currently uses an underground batching plant servicing two shotcrete rigs (owned and operated by Pasminco).

Another development in recent times was the development of the "Rib-Roc" cable dowel system for longhole stope wall and crown reinforcement (Rauert 1995).

Currently the operation utilises a comprehensive "toolkit" of ground support and reinforcement, consisting of rockbolting, cable dowelling and shotcrete.

3.1 *Rockbolting*

Friction anchor rockbolt reinforcement is used for immediate reinforcement in development less than three years in life and in areas where other rockbolt types are difficult to install due to hole closure.

The Ingersoll Rand CT Bolt is used for permanent reinforcement in development headings greater than three years in life or having particularly aggressive mine water inflows.

3.2 *Cable Dowelling*

Single cables are installed in holes seven to fifteen metres in length for permanent development reinforcement in poor ground condition areas and for stope drawpoint support. All cables being grouted and a plate installed.

Two cables are installed in holes seven to fifteen metres in length for stope crown reinforcement including "Rib-Roc" stope crown reinforcement. The grouted dowel is fitted with a plate and post-tensioned.

In the case of "Rib-Roc" stope wall reinforcement downholes up to one hundred millimetres in diameter are drilled from an adjacent access drive into the stope wall and have up to six cables installed and grouted.

3.3 *Shotcrete*

Shotcrete (no steel fibres) is applied as a permanent development support in poor ground condition and in pillar mining development areas.

Fibrecrete (with steel fibres) is applied as permanent development support in areas prone to high loadings.

4 GROUND SUPPORT AUDIT

4.1 *Ground Support Audit Scope*

The scope of the Ground Support Audit (in all accessible openings) was as follows:

1. To document type, age and condition (as much as practical) of current installed ground support/reinforcement.
2. To document a reasonable estimate of ground conditions.
3. Determine a reasonable estimate of ground support/reinforcement.
4. Together with planned future ground support requirements, estimate all ground support resources currently required for the Pasminco Broken Hill Mine. These resources include rockbolting, cable dowelling, shotcreting and backfilling of unnecessary openings.
5. Assess the applicability of different ground support/reinforcement technology; such as hollow groutable expansion shell bolts and grouting friction anchor (split set) rockbolts will be assessed as part of the project.

The project team for the audit consisted of both technical (Rock Mechanics) and operational (ground support supervisory) Pasminco personnel.

4.2 *Site Investigation*

The site investigation phase of the project involved dividing the mine into workable areas approximately 100 metres in length. All information collected pertained to each area. A total of 850 areas were inspected totalling nearly 90 kilometres of accessible openings within the mine.

The process was formulated to be expressed in practical terms, given the audit involved non-technical operational as well as technical (Rock Mechanics) personnel.

As part of the project a complete graphical record of installed major ground support and ground condition was assembled providing a useful model of geotechnical condition of the operation.

4.3 *Procedure*

The basic procedure involved inspecting each area and filling in a proforma sheet shown in Figure 2. Each area having a designated identifying code for a location, date, etc. Ground condition information, presence of water and installed ground support description were then recorded. Comments in terms of apparent ground support needs were also made.

4.4 *Ground Condition Estimates*

The methodology involved in determining ground condition used a very practical approach based on assigning a ground condition to the type of ground support historically used at the operation assuming the area was used only for access. More elaborate geotechnical methods of ground condition classification, such as rock mass classification systems, were considered far too complicated and time-consuming for the purposes of this exercise. Detailed descriptions of the four ground condition types included diagrams and photographic examples of each type.

The four types of ground condition are as follows:

1. Spot Bolt Category (Type A). Continuous rock mass with no significant discontinuities.
2. Pattern Bolt Category (Type B). Continuous rock mass with some discontinuities. Discontinuities being tight and high strength.
3. Cable Dowel Category (Type C). Discontinuous rock mass with open, low strength jointing.
4. Shotcrete Category (Type D). Discontinuous rock mass with open, low strength jointing and observable movement.

PASMINCO BROKEN HILL MINE GROUND SUPPORT AUDIT SITE INSPECTION

DATE:

INSPECTED BY:

ID CODE:

LOCATION:

GROUND CONDITION

CODE	A	B	C	D
DESCRIPTION *	SPOT BOLT	PATTERN BOLT	CABLE DOWEL	SHOTCRETE +C/DOWEL

* NOTE - GROUND CONDITION DETERMINED AS RELATIVE TO WHAT TYPE OF SUPPORT IS REQUIRED
AS PER REFERENCE SHEETS FOR A DRIVE TO BE USED AS ACCESS DRIVE ONLY.

PRESENCE OF WATER

IS WATER PRESENT? Y or N

COMMENT IF WATER IS PRESENT?

INSTALLED GROUND SUPPORT

ROCKBOLTS	COVERAGE (LOW,MED,HIGH)	PLATE LOAD(Y/N)	DEGREE OF RUST(L,M,H)	ACCEPTIBLE INSTALLATION(%)	INSTALLATION DATE	COMMENTS
SPLIT SET						
RESIN BOLT						
EXPANSION SHELL						

CABLE DOWEL	COVERAGE* (%OF AREA)	PLATE LOAD(Y/N)	DEGREE OF RUST(L,M,H)	DESIGN DWG NOS.	INSTALLATION DATE	COMMENTS
SINGLE CABLE						
MULTIPLE CABLES						

SHOTCRETE	COVERAGE* (%OF AREA)	CRACKING (Y/N)	SPALLING (Y/N)	THICKNESS (IN mm)	APPLICATION DATE	COMMENTS

* APPROXIMATE CABLE DOWEL AND SHOTCRETE COVERAGE TO BE MARKED ON 1:3000 SCALE PLANS

SETS	PADDOCKING (N/Y?)	ON SURVEY PLAN(N/Y?)	COMMENTS
STEEL			
TIMBER			
ARMCO			
OTHER			

MESH	COVERAGE (%OF AREA)	TYPE (SMALL/LARGE)	HOLDING LOOSE MATERIAL(Y/N?)	DEGREE OF RUST(L OR H)	DAMAGE (L/M/H?)	COMMENTS

GENERAL COMMENTS

Figure 2. Proforma sheet used for site investigation during the Ground Support Audit.

The results were also colour-coded onto a set of 1:3000 scale plans of the operation, together with any obvious major geotechnical information, such as major discontinuities, hence giving a geotechnical model of the mine and a useful tool for mine design.

4.5 *Presence of Water*

The presence of water is possibly the most important performance inhibiting factor for all types of ground support/reinforcement. Experience on site has demonstrated life of ground support/reinforcement is severely shortened when water is present. To illustrate, the life of friction stabiliser rockbolts can be as short as a few months.

4.6 *Installed Ground Support and Reinforcement Evaluation*

The type, coverage, condition and age of all installed ground support/reinforcement was recorded on the sheet in Figure 2, together with comments relating to the installed ground support, as well as visual estimates of ground support requirements for each area. Appropriate definitions were formulated in describing the installed ground support/reinforcement. All major ground support and reinforcement (shotcrete, cable dowels and non-surveyed sets) were also recorded graphically on a separate set of 1:3000 scale plans. Hence, a ground support model of the operation is created, forming a useful tool for mine design.

Estimates of the age of the installed ground support were made partly from survey records which would indicate when initial rockbolt support was done and partly from conversation with personnel.

4.7 *Data Processing of Records*

Relevant information from each sheet was entered into a Microsoft Excel spreadsheet. Later estimates of necessary ground support are added.

Another set of 1:3000 scale plans was also maintained showing the mapping domains or area locations.

A daily diary was also kept listing the day's activity and highlighting any immediate priorities.

5. GROUND SUPPORT REQUIREMENT EVALUATION

Estimates of ground support requirements were tabulated in the same spreadsheet as the site inspection results. Estimates for planned development ground support over the next twelve months were also calculated.

5.1 *Estimation Techniques*

Estimates for resecuring of all areas inspected were based on historical estimates per linear metre of drive.

Estimates for additional or planned development over the next twelve months were based on current design information and tabulated with the audit domain.

Estimates of life and the use of areas inspected were made and tabulated. This allowed a true estimate of ground support requirement. For example, a stope sill in Category B ground condition usually requires cable dowelling whereas an access drive in similar ground condition would not.

Conversely, areas requiring ground support/reinforcement but having no further use were assessed for backfilling.

Figure 3 shows the total ground support and reinforcement requirements for the twelve months following the audit compared to the previous twelve months.

5.2 *Risk Analysis for Scheduling Ground Support and Reinforcement Work*

In order to schedule the ground support work requirements a risk assessment matrix technique was used to set priorities for all the work as shown in Figure 3.

Figure 4 shows the risk matrix used for the risk calculation. An individual risk number was calculated for every area inspected, based on assumptions outlined in Tables 1 and 2.

The occurrence was directly related to ground condition and the consequence to the use of the area to which both an economic and safety rating was applied.

Table 1. Rating for Occurrence of a Rock fall based on over riding ground condition for each area.

	RATING	GROUND CONDITION
4	PROBABLE	D
3	POSSIBLE	C
2	REMOTE	B
1	IMPROBABLE	A

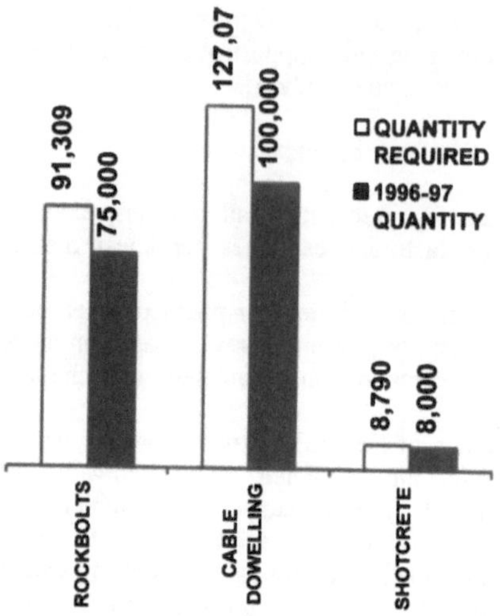

Figure 3. Graph of total ground support and reinforcement requirements estimated from the Ground Support Audit for twelve months as well as the previous twelve month period

Table 2. Ratings for Consequence for each area based largely on accessible drive usage.

	RATING	INJURY RISK	MATERIAL DAMAGE	BUSINESS INTER.
4	CATO-STROPHIC	VERY HIGH	>$1.5M	>1 MONTH
3	SERIOUS	HIGH	$0.1M-1.5M	1WK-1MON
2	MOD.	MED.	$0.01M - 0.1M	1DAY-1WK
1	MINOR	LOW	< $0.01M	<1 WEEK

6 RESULTS OF THE AUDIT

Results indicated some two thirds of the areas inspected required some remedial ground support/reinforcement work. The ground support work required over the following twelve months was approximately 120% of the total work achieved in the previous twelve months. Results also indicated two thirds of the operation's accessways were in reasonably good ground conditions.

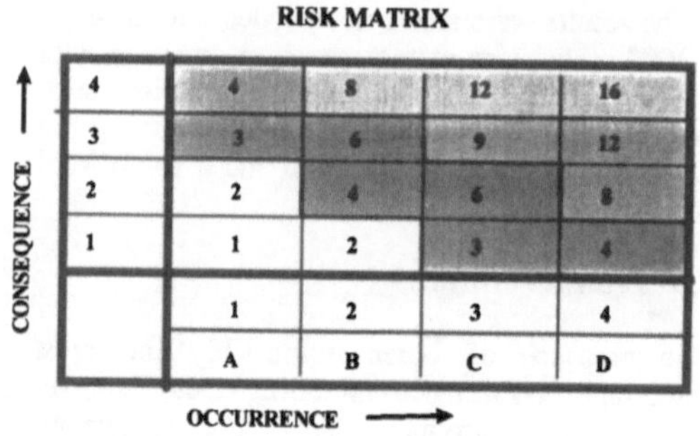

Figure 4. Risk analysis and matrix used in each area.

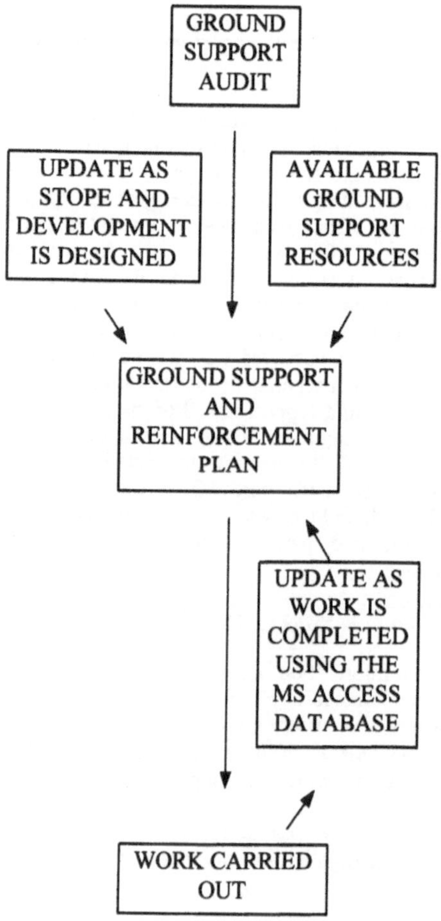

Figure 5. A flowchart of the developed ground support and reinforcement management system.

7 GROUND SUPPORT AND REINFORCEMENT MANAGEMENT SYSTEM

Operational resources in terms of manpower and equipment were immediately transferred from production and development duty and engaged in a mine resecure project.

An ongoing system of ground support management based on the ground support audit was developed at the operation to balance rehabilitative work requirements with ongoing mine development requirements for continued production. Figure 5 presents a flow diagram of the system.

The ground support audit data was transferred into Microsoft Access, in the form of a database linked to another Microsoft Access database, into which fortnightly achieved ground support/reinforcement quantities were entered. The data was linked by an area ID code, hence allowing comparisons to be made between actual and planned ground support/reinforcement requirements. Figure 6 shows a total comparison to plan of all quantities as at October 1998.

The database comparison also allowed data to be grouped by risk number which enables high risk areas to have ground support and reinforcement work done at higher priority. Figures 7a-f show a comparison of achieved to plan quantities by risk number; for rockbolting, cable dowelling and shotcrete categories as at October 1998.

Results in Figures 6 and 7a-7f indicate resecuring totals have been difficult to achieve. This is due, in part, to underestimation made in current development areas, as well as the majority of the resecuring work being undertaken in the highest risk areas (poor ground, significant deterioration and major infrastructure areas). the operation was essential in managing any future fall of ground risk.

8 CONCLUSIONS

The audit, by using simple clearly defined logic, has given the Pasminco Broken Hill Operation a greater understanding of the effectiveness of its installed ground reinforcement and support than ever before.

Areas with potential to develop into the unacceptable fall of ground scenarios can now be identified with more certainty.

The graphical information recorded on the 1:3000 scale plans for ground condition and installed major ground support will have a major use in mine design and future planning of the operation. The degree of documentation of the condition of accessible areas has never been so comprehensive.

The ground condition model selected based on site experience of different support and reinforcement types required for a particular ground type was very successful and could have a wide use in the industry.

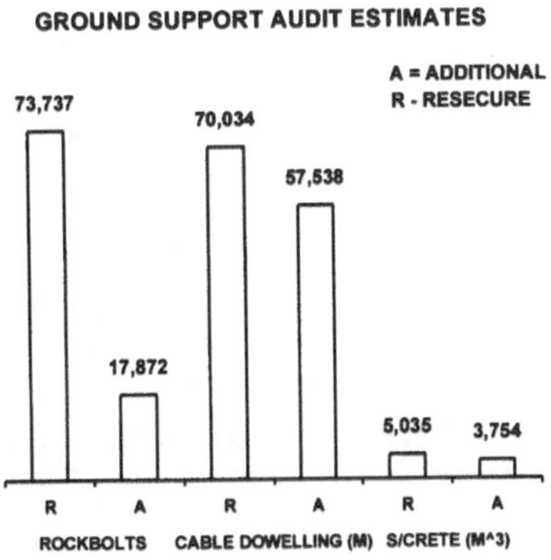

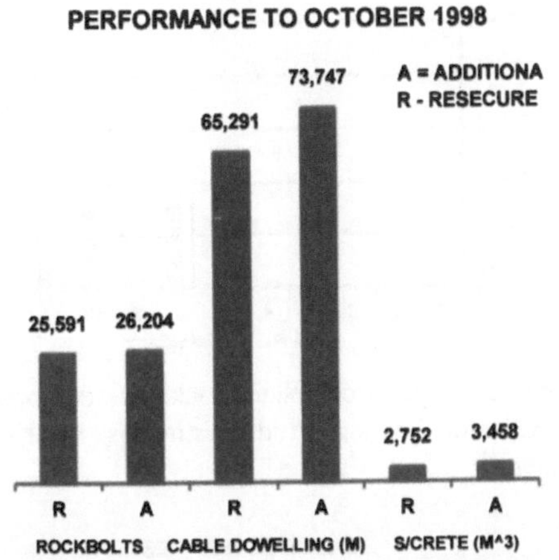

Figure 6. Total comparison of ground support and reinforcement to ground support audit estimates showing resecure and additional components.

9 ACKNOWLEDGEMENTS

The support of management and all other Pasminco personnel involved in the ground support audit and setting up the ground support

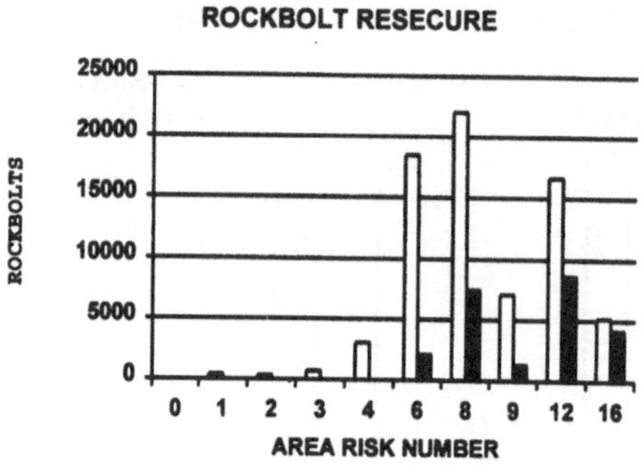

Figure 7a. Graph of completed rockbolt resecure and Ground Support Audit estimates by risk number.

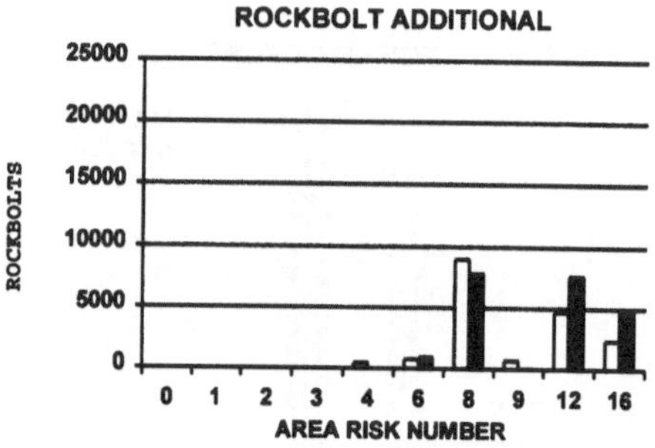

Figure 7b. Graph of completed rockbolt additional and Ground Support Audit estimates by risk number.

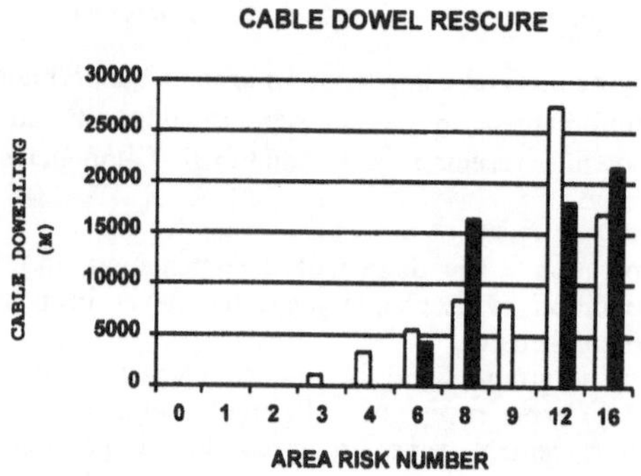

Figure 7c. Graph of completed cable dowel resecure and Ground Support Audit estimates by risk number.

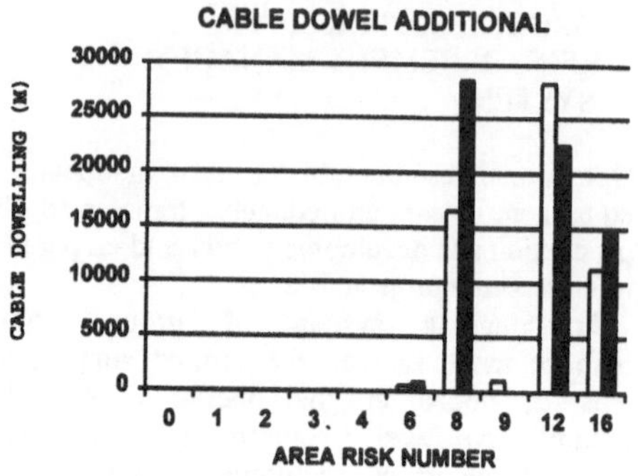

Figure 7d. Graph of completed cable dowel additional and Ground Support Audit estimates by risk number.

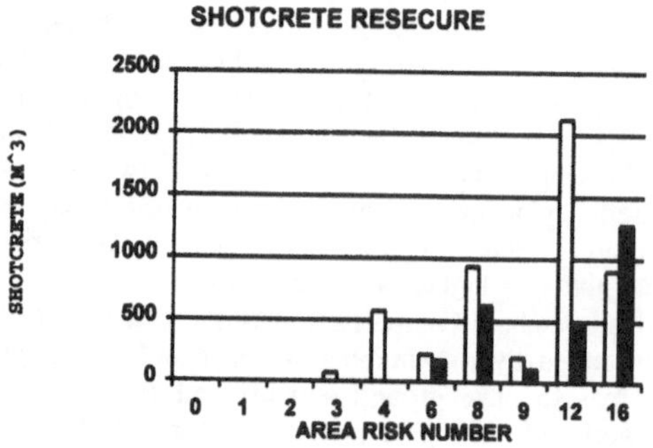

Figure 7e. Graph of completed shotcrete resecure and Ground Support Audit estimates by risk number.

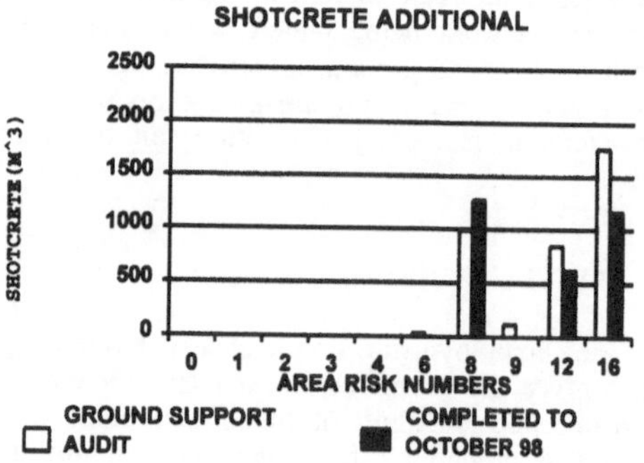

Figure 7f. Graph of completed shotcrete additional and Ground Support Audit estimates by risk number.

reinforcement system is acknowledged. Advice given by Alan Thompson and Chris Windsor (Rock Technology Pty Ltd) as part of the Australian Minerals Industries Research Association Limited P489 "Shotcrete Bolts and Mesh Project" is also acknowledged.

REFERENCES

Rauert N.S. 1995 *The Development of the Rib-Roc Cable Bolt Reinforcement System at the Pasminco Mining Broken Hill Operations*. AusIMM Underground Operators' Conference, Kalgoorlie.

Rauert N.S. 1997 *Pasminco Broken Hill Mine Ground Support Audit Report*. Internal report.

5 Support in burst-prone ground

Rock Support and Reinforcement Practice in Mining, Villaescusa, Windsor & Thompson (eds)
© 1999 Balkema, Rotterdam, ISBN 90 5809 045 0

Retainment support for dynamic events in mines

T.R.Stacey & W.D.Ortlepp
SRK Consulting, Johannesburg, South Africa

ABSTRACT: The paper deals with the testing of retainment support under dynamic conditions typical of the loading experienced during rockbursts. The testing was carried out for the following support types:

- rebar type rockbolts;
- yielding and smooth bar rockbolts;
- splitset and Swellex rock bolts;
- destranded hoist rope.

The dynamic testing method made use of test samples which attempted to simulate realistic in situ conditions, and the range of loading imposed using a drop weight system was equivalent to small to severe rockburst conditions.

1 INTRODUCTION

Tunnels under high stresses and deep mining conditions are often subjected to large static and dynamic deformations. Static deformations are in the form of squeezing out of failed rock, whereas the manifestation of seismic loading is the violent ejection of a volume of rock. In tunnels in deep level mines it has been observed that commonly approximately a metre thickness of rock may be ejected from the surface at velocities of up to 10 m/s or higher (Ortlepp and Stacey, 1994). It is usually not practically possible to contain the energy involved in this type of failure by means of stronger support, as illustrated in Figure 1, in which rockbolts and high capacity cables have failed in a brittle fashion. Instead, the support must yield and, in yielding, absorb energy. With yielding support of suitable deformation capacity, it should be possible to contain very severe static and dynamic deformations. Support systems for severe loading conditions should therefore be designed on an energy basis rather than on a conventional factor of safety approach.

In this paper the results of testing of retainment support under dynamic conditions typical of the loading experienced during rockbursts are described. The testing was carried out for the following support types:

- rebar type rockbolts;
- yielding and smooth bar rockbolts;
- splitset and Swellex rock bolts;
- destranded hoist rope.

Figure 1. Damage in a severe rockburst

The dynamic testing method made use of test samples which attempted to simulate realistic in situ conditions, and the range of loading imposed was equivalent to small to severe rockburst conditions.

2 PREVIOUS DYNAMIC TESTING OF ROCKBOLTS

The results of dynamic tests on specially developed yielding rockbolts have been described by Jager et al (1990). In these tests, yield, without failure, was shown to be satisfactory at velocities of up to 3 m/s. Impulse loading testing of these rockbolts was carried out by Ortlepp (1994), who showed that they were capable of yielding at velocities of up to at least 12 m/s. He also showed that these bolts could absorb more than 50 kJ each in yielding some 0.9 m, without showing any damage. In this testing programme, the behaviour of conventional rebar and smooth bar bolts was also evaluated. The rebar bolts performed very poorly and were unable to absorb the explosive input energy, as is illustrated by the spectacular failure of the support system under the explosive impulse in Figure 2. However, the performance of the smooth bars was promising, and they absorbed all the input energy in yielding by 0.4m.

These tests probably provided the first set of quantitative data on rockbolt performance under dynamic loading. Information on the dynamic behaviour of the more commonly used rockbolts was still not available, however. To rectify this situation, research was commissioned in 1997 to carry out further testing of rockbolts.

Figure 2. Failure of support in an impulse load test

3 RECENT DYNAMIC TESTING OF ROCKBOLTS

The test programme involved the dynamic testing of a range of rockbolts commonly used in the South African mining industry, using a drop weight impact system of loading. Tests were carried out on the following:

- 16mm diameter rebar (fully grouted, rigid);
- 16mm diameter smooth bar (partially bonded, semi-yielding);
- destranded hoist rope strand (lubricated, helix);
- 18mm compact strand cable (fully grouted, very stiff, prestressed);
- 39mm diameter splitsets (percussively driven, friction contact);
- standard diameter Swellex bolts (inflated, friction contact).

These are all referred to generically as tendons in the following, except when specific tendon types are dealt with. Although these tendons were chosen because they are common in South African mines, it is expected that the results will be of value in other countries as well, and will contribute to safer working conditions in the future.

3.1 *Test method*

The tendons were grouted into steel tubes, or installed in boreholes formed in simulated rock in steel tubes. The dimensions of the tubes were chosen so that the tubes provided confinement of the same order as that provided by the rock mass. Each "test specimen" consisted of two lengths of tube, butted together. These tubes were spot welded together at the butt for the preparation of the specimens. The welds were subsequently cut, and the butt provided the "joint" separation surface at which failure of the tendon could take place. Various lengths of tube were used to cater for different tendon embedment lengths.

The dynamic loading was imposed by means of a drop weight impacting on a swing beam, as illustrated in Figure 3. The test facility was a modification of the equipment, described by Ortlepp and Stacey (1996), which was used for dynamic testing of containment support. Impact loading velocities representative of severe rockburst loading were achieved - four values were over 20 m/s, five between 15 m/s and 20 m/s, seven between 10 m/s and 15 m/s, and the rest up to 10 m/s. Large input energies, sufficient to fail all of the tendon types, were easily achievable with the drop weight system.

3.2 *Test results*

In all of the tests the behaviour of the tendons was characterised by one, or a combination of two or all, of the following mechanisms:

- if the tendon was strongly bonded, or prevented from slipping, "necking" of the tendon commenced at the separation surface. Because of the small

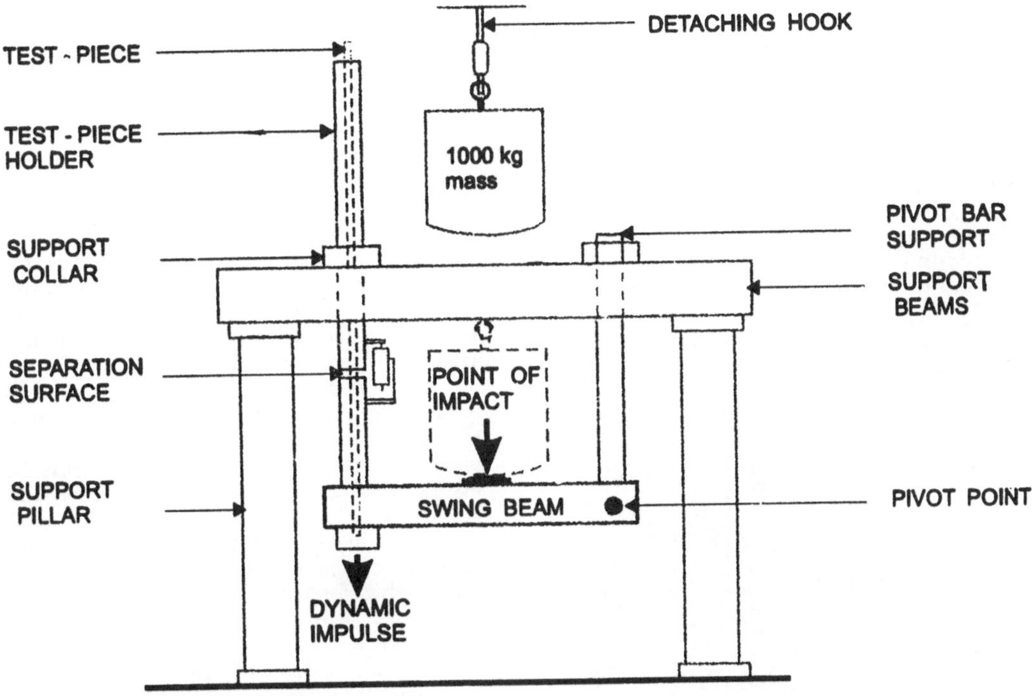

Figure 3. Dynamic testing facility

volume of steel involved in the yield process, the total energy consumed before failure takes place is small. This energy component is denoted as rupture energy;

- if failure of the bond at the grout/steel interface takes place, then yield of the the steel, with concurrent work hardening, also occurs. This is in the form of "necking" which, as a result of the work hardening, advances away from the initiation point. Theoretically, this advance, in both directions, could involve the whole length of the tendon. Slip of the tendon anchorage does not occur, due, for example, to the ribs on the rebar surface. The energy component developed in this progressive elongation of the tendon is termed elongation energy;
- a small decrease in the diameter of the tendon accompanies the progressive elongation of the steel. This causes progressive debonding to occur and, for smooth tendons, allows sliding to take place. In the case of friction rockbolts, the hollow steel tube makes it relatively easy for the pressure normal to the interface to relax sufficiently to allow sliding to take place. The energy component involved in this mechanism is termed sliding energy.

In initial tests, in which input energies were small, failure of the rebar tendons could not be achieved. It was found that the softness of the testing system enhanced the occurrence of the second mechanism described above - progressive debonding of the grouted tendon occurred, which, in turn, provided a longer length of "exposed" tendon along which elongation of the steel could occur, with consequent energy absorption. "Softness" in this context is relative, since the system was initially considered to be stiff. However, it became evident that this phenomenon was occurring with most of the other tendon types as well, and it was found necessary to stiffen up the testing system considerably.

Rupture of the rebar bolts was similar to that commonly seen underground, with fairly abrupt necking. Figure 4 shows this appearance at the instant of rupture.

Rupture of smooth bar did not occur, and all energy was consumed in elongation of the steel. On the occasions when bars pulled out of the grout, sliding also absorbed energy. No rupture occurred with the hoist rope strand either. In all cases, slip of the tendons occurred and most of the energy was consumed in sliding.

The 400 kN prestressed cables ruptured in a "brittle" fashion, as shown in Figure 5, when the input energy exceeded the rupture energy.

The tests on the friction rockbolts were not very satisfactory. In a straight hole slip occurred and it was not possible to rupture the bolts. When the bolts had been artificially "fixed" to simulate the clamping that would occur due to dislocation of a hole, rupture was achieved.

The results of the testing programme are summarised in Table 1.

Figure 4. Dynamic failure of rebar tendon

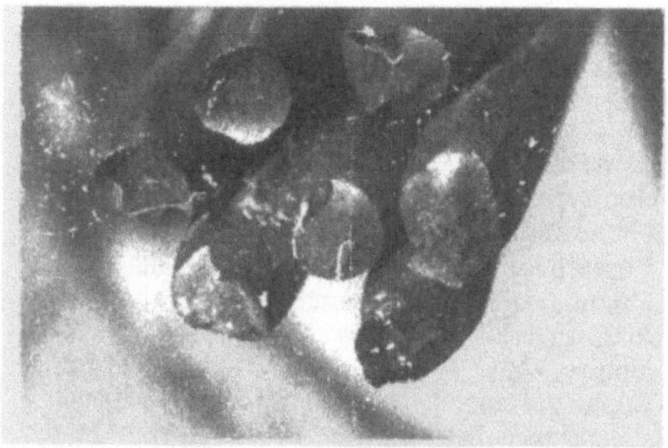

Figure 5. Brittle failure of multistrand cable under dynamic loading

Tendon Type	Rupture Energy kJ	Elongation Energy kJ/100mm	Sliding Energy kJ/100mm
Rebar 16mm	4,8 ± 0,7	-	-
Smooth bar 16mm	-	13,9	$3{,}0 \times l_e - 0{,}15$*
Vaal Reefs hoist rope strand	< 2,4	-	0,8 to 8,2
18mm compact strand cable	17,5 ± 2,0	-	-
39mm diameter splitsets	11,6 to 15	11,6 ± 1,3	3,0 to 6,3
Standard Swellex	4,6 ± 0,5	9,9	3,0 to 5,8
* l_e is initial embedded length in metres			

The summary in the table has been extracted from the results of tests on 58 specimens, often involving multiple weight drops on each specimen. The values in the table provide a set of data that can be used for realistic design of retainment support for large deformation static and dynamic loading conditions.

3.3 *Discussion*

Several interesting and important insights into the likely behaviour of tendons under real dynamic "in-mine" conditions were revealed by the practical difficulties experienced in the actual carrying out of the test programme. The more important insights gained were:

- the very abrupt, even brittle, fracture of fully grouted steel bars, which is often observed in a rockburst damaged tunnel, was not easy to achieve under the simulated conditions. This is because the maximum testing system stiffness was probably well below that of an intact rock mass. Theoretical deficiencies in energy absorption capabilities were therefore not as evident as they would be in explosively deforming excavations in hard rock;
- in practice, reduced system stiffness will result from such factors as:
  - the prior existence of significant fracturing in tunnel walls;
  - commencement of debonding as a result of earlier static yield or subcritical dynamic loading;
  - the presence of "softer" elements in the support system such as non-stiff face plates, poor quality grout, incomplete grouting, etc.

  In contrast, where very strong rock inhibits the development of a softer shell of fractured rock around the runnel, stiff tendons will be vulnerable to acute stress concentrations across joint separation surfaces;
- in uniform straight holes, hollow friction dependent tendons will not break, but will slide even if the fit is very tight in the hole. The full potential tensile capacity of friction type tendons will only be realised if the hole is just sufficiently distorted or dislocated to clamp the tendon without shearing it;
- inadequacies in the terminating arrangements on the tendons (faceplates, the welded ring on split sets, the sealing weld and soft ferrule on Swellex, and the swaged sleeves on the loops of destranded hoist rope strand) often fail prematurely. This means that only part of the potential strength or resistance of the tendon is available as retainment support.

Probably the most striking and certainly the most significant effect that was consistently observed during the successful tests was the paradoxical tendency of tendons, subjected to a first impulse that was insufficient to rupture the tendon, to acquire extra capacity for absorbing more energy later if subjected to several impulses. This extra energy was often considerably more in total than was necessary to cause failure at the first impulse. This type of behaviour could typically occur in practice in situations of poor quality installation of support - poor grout quality, inadequate filling of the hole with grout, slip of mechanical anchors etc. The key issue is that yielding of the support increases its capability of absorbing energy without failing. It is an irony that the enhancement of retainment support performance under dynamic loading conditions may be achieved by poorer quality of support installation. However, it must be well understood that this is by mistake, not by design. The additional amounts of energy consumed in these situations cannot be recognised or utilised in the strict application of a formal design procedure, since the situation that permits the phenomenon is one of conditional stability that cannot be satisfactorily predicted. The aim must be to have a thoroughly designed, well engineered support installation, in which the actual performance corresponds with the desired design performance. It will then be essential for the support to be installed according to a definite specification, and the control of the quality of the installation to be good.

4 CONCLUSIONS

The dynamic test programme that has been carried out has produced a set of data, where none existed before, which will be of use in the design of retainment support for tunnels which are likely to be subjected to dynamic loading. The method of tendon testing proved to be realistic and reliable.

The results of the tests have been presented in terms of energy absorption capability since, under dynamic conditions, this is much more realistic than strength capability. Three energy components have been identified which contribute to the absorption of energy - rupure energy, elongation energy and sliding energy. Failure of the tendons may involve one, two or all three of these components. However, it is often inadequacies in the terminating arrangements such as the face plates, the welded ring on splitsets, the sealing weld and soft ferrule on Swellex and the swaged sleeves on the loops of hoist rope strand, which cause premature failure.

The tests have shown that rebar and prestressed cable can fail in a "brittle" fashion, that smooth bar did not rupture, but absorbed energy in both elongation and sliding, and that lubricated hoist rope strand performed poorly in comparison with the other tendons. The energy absorption values obtained for friction tendons indicated that sliding would be the most likely mode of failure. However, the testing was not considered to be entirely satisfactory in that, in practice, dislocations of boreholes are likely to limit the sliding mechanism and cause rupture, with some elongation, to occur.

ACKNOWLEDGEMENTS

Sponsorship of the drop weight testing of rockbolts by SIMRAC, the South African Safety in Mines Research Advisory Committee, is gratefully acknowledged. The preparation of specimens and the execution of the dynamic tests were carried out by Bruce Soffe and Noel Moya.

REFERENCES

Jager, A.J., Wojno, L.Z. and Henderson, N.B. 1990. New developments in the design and support of tunnels under high stress conditions, *Proc. Int. Deep Mining Conf.: Technical Challenges in Deep Level Mining,* Johannesburg, S. Afr. Inst. Min. Metall., pp 1155-1172.

Ortlepp, W.D. 1994. Grouted rock-studs as rockburst support: A simple design approach and an effective test procedure, *Jl S. Afr. Inst. Min. Metall.,* Vol 94, No 2, pp 47-63.

Ortlepp, W.D. and Stacey, T.R. 1996. The performance of containment rock support such as wire mesh under simulated rockburst loading, *Proc. Geomechanics '96,* A.A. Balkema, pp 81-87.

Ortlepp, W.D. and Stacey, T.R. 1994. Rockburst mechanisms in tunnels and shafts, *Tunnelling and Underground Space Technology,* Vol. 9, No. 1, pp 59-65.

Rock Support and Reinforcement Practice in Mining, Villaescusa, Windsor & Thompson (eds)
© 1999 Balkema, Rotterdam, ISBN 90 5809 045 0

Observations of the dynamic performance of South African tunnel support systems

A.T. Haile
CSIR Division of Mining Technology, Johannesburg, South Africa

ABSTRACT: This paper presents the observations of investigations of rockburst damage to tunnels over a three year period from 1995 to 1998. These investigations have highlighted the performance of the current tunnel support systems under these severe loading conditions. This has enabled analysis of the interaction between the rock bolt reinforcement and the rock mass and the loading of the fabric support systems in relation to the current design considerations. In light of this the importance of a mechanistic understanding of the interaction of the components of a support system within the defined rock mass environment are indicated for improved tunnel support system design.

INTRODUCTION

Off reef excavations are often subjected to ground motions radiated from seismic events, which may result in severe, and sometimes quite erratic damage. This damage will be a function of the site response, which will be controlled by the current state of the support system and rock mass, in relation to the proximity to the seismic event.

In order to evaluate the current performance of support systems within this environment, a series of detailed case studies were conducted over a period of approximately three years (Haile, 1998). These visits concentrated on events that caused significant failure within the rock mass, and of the support system. In this way a detailed picture of the mechanisms of support interaction, and the relative performance of the support systems, could be constructed.

This paper reviews the mechanisms of the deformation of the rock mass in the vicinity of these tunnels. In this way it is considered that insight may be gained into the interaction between the rock mass and the support system. A brief description of a typical South African tunnel environment is given in the following section.

Tunnel excavations are generally square in shape, although the blasting techniques used for excavation often result in a very irregular profile. The tunnel will generally have rockwall dimensions of approximately 3 m to 4 m. At the development face primary support consisting of rock bolts is installed in the exposed sidewall and hangingwall rock. Rock bolts (tendons, anchors) are generally 12 mm to 20 mm in diameter. These may be anchored either mechanically by an expansion shell mechanism at the end of the hole, or by the encapsulation by cementitious or resin grouts, generally over the full length of the rock bolt. In the high stress environments of the deep level gold mines, failure of the rock around the tunnel excavation generally occurs during development. This results in a fractured and highly discontinuous medium in which the tunnel opening must be maintained. The rock bolts act within the rock mass to reinforce this highly discontinuous medium and are also referred to as reinforcement units. Under these conditions high areal coverage support systems are incorporated with the rock bolt reinforcement to maintain the integrity of the rock mass in the exposed skin of the excavation between the rock bolt units. Typical high areal coverage support systems consist of is a combination of steel mesh with an overlying pattern of 10 mm to 20 mm steel ropes, or shotcrete, often with reinforcing fibres.

The condition of these tunnels will be influenced by the stress state in which they are developed and changes in the local field stress due to adjacent mining of the orebody. In addition, the most dramatic influence may be due to the occurrence of a seismic event, which results in the rapid dynamic loading of the rock mass surrounding the excavation. If the event occurs in close proximity to the tunnel, then this loading may result in violent deformations of the rock mass and loading of the support system. It is this environment in which the current support systems are often found to be inadequate, and the investigation of which will give insight into the mechanistic interaction between the components of the support system and the rock mass. This knowledge will allow improved evaluation of the support systems and thus identification of the factors relevant to their design under these conditions.

RESULTS

From the detailed rockburst case studies the following points are highlighted with regard to the performance of the support systems and the mechanisms of failure of the rock mass.

- Bulking of the rock mass between the rock bolt reinforcement was observed at all the rockburst sites. At 70 % of the sites this was observed to have resulted in failure of the fabric support system over a significant area of investigation (Photograph 1). At 30 % of the sites the integrity of the fabric support system was maintained and the unstable rock mass between the rock bolt reinforcement was contained (Photograph 2).

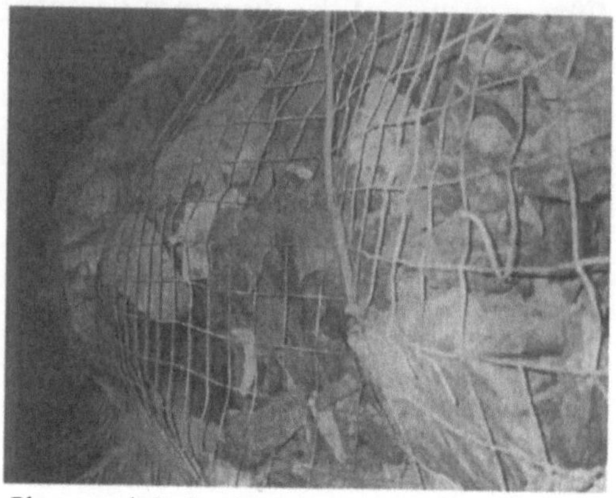

Photograph 2. Rock mass containment by mesh panel.

- Large scale deformation over the whole height of the sidewall was observed at 60 % of the investigations (Photograph 3). This was especially prevalent at investigations conducted in the Klerksdorp gold field. This may be a function of the different geotechnical environment and / or the more prolific use of smooth bar rock bolt reinforcement compared to the use of ripple bar on the West Rand gold field. Only isolated beam type failure was associated with the hangingwall of the tunnel.

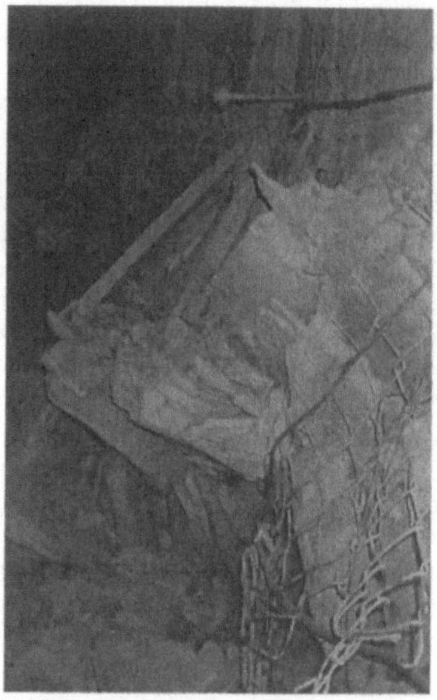

Photograph 1. Failure of mesh panel.

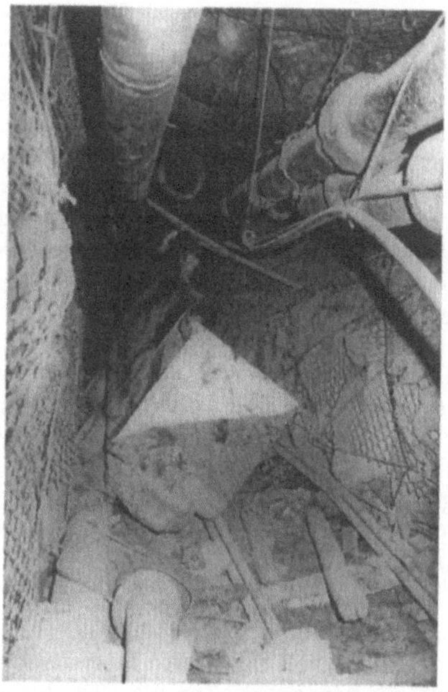

Photograph 3. Large scale sidewall deformation.

- Shear failure of rock bolt reinforcement was observed at 70 % of the rockburst sites (Photograph 4). In all cases this was in the hangingwall of the tunnel excavation.

Photograph 4. Guillotining of hangingwall rock bolt.

- Tensile failure of rock bolt reinforcement was directly observed at 60 % of the rockburst sites (Photograph 5) but of limited occurrence at each site. This was associated with both the hangingwall and sidewalls of the tunnels.

Photograph 5. Tensile failure of hangingwall rock bolt.

Of significance from this investigation is the large proportion of damage associated with the bulking and unravelling of the sidewall rock mass between the rock bolt reinforcement and the prevalence of shear failure of hangingwall rock bolt units. These mechanisms are not considered in the current design procedures. The purpose of this study was to improve the mechanistic understanding of this interaction between the rock mass and the support system to improve the design considerations of support systems in these environments.

The erratic nature of the damage to tunnel excavations with respect to the source of the causative seismic event is indicative of the role that site response has in understanding the failure mechanisms. The main factors that influence the type and severity of damage are:

- the structure and competence of the rock mass prior to the seismic event.
- the previous history of deformation of the support system and thus its remaining capacity immediately prior to the seismic event.
- the orientation and proximity of the excavation relative to the source of the seismic event.

The containment of potential rockburst damage will involve the ability to define high risk excavations with a reasonable degree of certainty and also the implementation of effective excavation and support design strategies. These aspects of design will also require due consideration of the relative cost of these strategies compared to the identified level of risk and the limits of these design considerations.

DISCUSSION

For an anticipated level of rock mass instability around an excavation, the design engineer must be able to select a suitable support system for the given conditions. This will be based on the current and anticipated rock mass condition, anticipated loading conditions and allowable deformation. Also of consideration are the availability of support elements and practicalities of installation, and the cost effectiveness of the support system on the basis of material and installation costs.

Within relatively competent rock mass structures, the role of the rock bolts is to pin potential key block structures by anchorage to stable ground in excess of the defined unstable block geometry. The purpose of this investigation is to develop support design methodologies for highly discontinuous rock mass structures where specific unstable block geometry's can not be identified and the dominant mode of rock mass instability is an unravelling of the rock mass structure.

The principal design methods for support systems are either by containment or structural reinforcement. Containment of the rock mass is achieved by ensuring anchorage outside of the limit

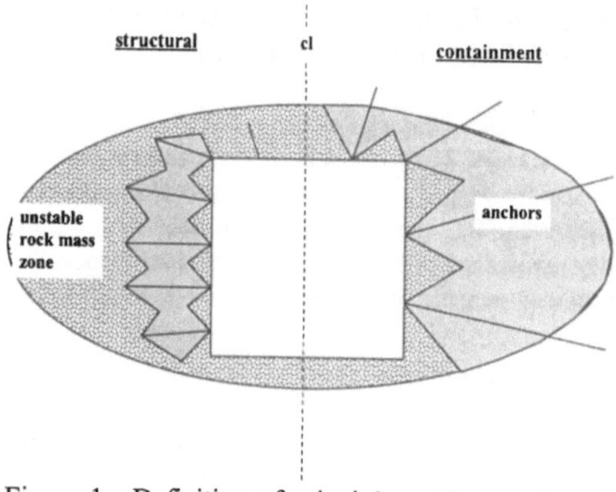

Figure 1. Definition of principle methodologies of excavation stabilisation

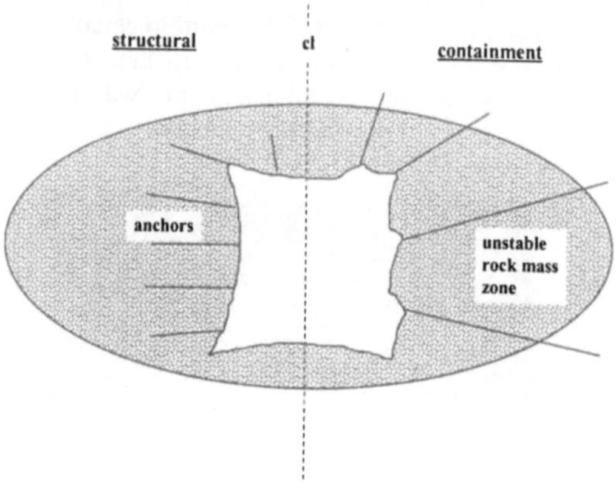

Figure 2. Generalised deformation characteristics for principal support methodologies under high loading conditions.

of rock mass instability, with sufficient capacity within the support system to accommodate the full rock mass loading conditions. Alternatively the support system may act to reinforce the unstable rock mass and thus create a reinforced rock mass structure, again capable of withstanding the envisaged loading conditions. Within most support systems these support mechanisms may be combined to derive an optimum rock mass support system. Such a system may involve relatively short anchors and fabric support combined with long anchors. This would result in reinforcement of the immediate skin of the excavation with suitably anchorage to the deeper rock mass. The principal design considerations as discussed above are shown in Figure 1.

Of importance is an estimation of the depth of rock mass instability around the excavation, to enable the determination of a suitable anchor length. This may be based on data of anticipated depths of instability, empirical relationships, or numerical modelling analysis.

The importance of the influence of the mechanism of the support system interaction with the rock mass, on the deformation characteristics of the excavation, is illustrated in Figure 2.

These deformation mechanisms are clearly illustrated by the rockburst case studies, where deformations often occurred to the extent that failure of components of the support system resulted. A review of the basis of these design methodologies is given below.

The methodology of the creation of a reinforced rock mass structure will be dependent on the interaction of the support system with the rock mass to create a reinforced structure of sufficient capacity to maintain the rock wall stability under the defined rock mass environment. The behaviour of the reinforced structure under loading will result in a more uniform sidewall deformation characteristic. If however the reinforcement is under designed, differential deformation within the structure will occur, causing a loss of capacity of the structure to withstand increased loading. The influence of the fabric support within this methodology will be to maintain the integrity of the near surface discontinuous rock mass between the rock bolt reinforcement. The required load capacity of the fabric will thus be relatively low. However, it is important that relatively stiff fabric systems be employed, such as shotcrete, in order to limit differential deformations within the structure, and thus maintain its overall integrity.

The methodology of excavation stabilisation based on retainment /containment is considered to be a more robust support design methodology, but it is critical that anchorage of the system within stable ground is achieved. Although the engineer should strive to design the support system in order to maximise the inherent strength of the rock mass structure, and thus minimise the support requirements, this methodology will allow consideration of loss of the inherent rock mass strength. This methodology therefore must carefully consider the interaction of the individual components of the support system with the rock mass, and the anticipated demand on these units. To maximise the inherent rock mass strength, for optimum design considerations, the yield capacity of the anchors should be compatible with the envisaged rock mass deformation characteristics. That is, yield of the anchors must be compatible with the dilation of the rock mass between anchor points in order to minimise differential deformations. Incompatibility will result in differential deformation, and thus loosening within the rock mass structure with resultant loss of rock mass strength. The incorporation of high quality, relatively stiff fabric support systems will result in a more even load distribution between the rock mass directly confined by the rock bolts and the potentially unstable rock mass between the rock bolt reinforcement. If the inherent strength of the rock mass is lost due to the degree of rock mass discontinuity, or deformation, then increased demand on the fabric component of the support system must be considered.

Analysis of the case studies clearly indicates the development of characteristic deformation mechanisms and also indicates the shortcomings of the current support systems. Major support system failure was due to loss of anchorage of rock bolts at approximately 60 % of the rockburst sites. This may have resulted from the limited yield and debonding, or snapping of bolts and thus significant reduction in support resistance and rock mass reinforcement; or due to inadequate anchorage depth in relation to depth of rock mass instability, and limited containment of the unstable rock mass volume. Inadequate fabric support (mesh and lacing) capacity and stiffness also resulted in failures of the support system and reduced capacity of a potentially reinforced rock mass structure at all rockburst sites.

The design of the current support system interaction with the rock mass is based on anchorage into stable ground (Anon., 1996). Where tunnels negotiate elevated stress fields due to major stoping abutments, the depth of rock mass fracturing may increase the extent of instability in excess of the length of these anchors. Under these conditions, the basis of support design should be the formation of a competent rock mass structure (beam, arch or shell). The large closures associated with the haulages at these sites may be indicative of low levels of rock mass reinforcement and thus weak structural competency. This may often be a result of the current design procedures not considering this mechanism of support interaction. Under these conditions it is important to ensure sufficient interaction between the reinforcement units as opposed to relying on the fabric support to provide sufficient structural strength to the rock mass. Therefore, rock bolt reinforcement spacing should be reduced to improve rock mass interaction.

The current design recommendations for tunnels in deep level mining indicate the necessity for yielding tendons, particularly under dynamic loading conditions associated with major seismic events, to ensure energy absorption capacity. However, throughout numerous case studies, it was observed that relatively stiff rebar rock bolts, with very limited yield capability and thus energy absorption, would survive the major dynamic deformations and damage (Photograph 6). This is due to the poor interaction between the rock bolts and the rock mass in these environments. This causes the loading of the rock bolts at levels far lower than that anticipated by the design process. Under these conditions dynamic energy associated with the unstable rock mass is dissipated through the deformation of the rock mass contained by the relatively soft mesh and lace fabric support systems, resulting in the often observed bulking profile of the tunnel. This large bulking process between the rock bolt reinforcement, and thus large differential deformations, results in a further reduction of the rock bolt interaction with the rock mass. Thus this mechanism, although not catered for in the design process tends to result in the stability of the tunnel under these loading conditions.

It must thus be appreciated that if the stiffness of the fabric support is increased, to try to limit the deformations of the rock mass to maintain the

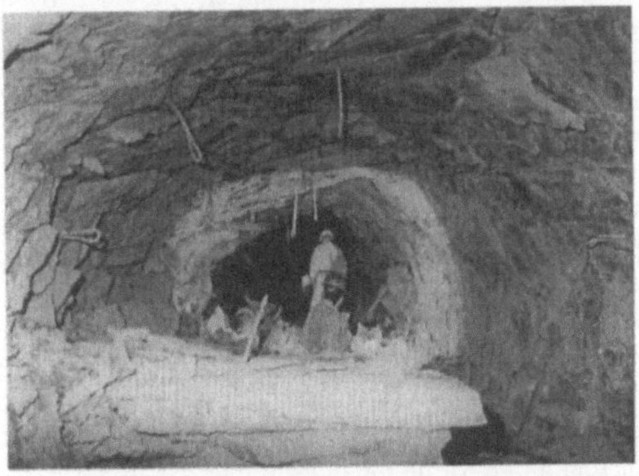

Photograph 6. Rock mass unravelling between rock bolts (background) and rock bolt failure (foreground).

excavation in an operational condition subsequent to such events, then the distribution of loading between the fabric and the rock bolts will change. This will result in increased direct loading of the rock bolt reinforcement to the extent that failure may occur and the total support system unravels. The importance of the yield capacity within the rock bolt system to absorb the dynamic energy is thus heightened.

The design of the support system must therefore carefully evaluate the relative demand on the reinforcing rock bolts and the fabric support. In some areas, particularly where anchorage of the tendons was maintained or had relatively higher resistance than the general peripheral rock mass, excessive bulking of the rock mass between the reinforcing rock bolts was observed. This often resulted in failure of the fabric support. It is thus important to ensure compatibility between load deformation characteristics of the tendons and the fabric support based on an estimation of the relative demand due to rock mass loading. Incompatibility, such as the use of very stiff tendons and soft fabric, will result in large relative deformations within the tunnel peripheral rock mass and associated reduced structural competency. This will lead to subsequent higher loading of the fabric support, due to a further reduction in the interaction of the rock bolt reinforcement. Under dynamic loading conditions it was observed that in 70 % of the rockburst sites the kinetic energy associated with this unstable rock mass volume was in excess of the capacity of the typical mesh and lacing fabric support panels.

Within certain geotechnical areas, the occurrence of highly persistent, and mobile, bedding planes may result in a reduction in confinement provided by a reinforced rock mass structure. This is particularly evident if these bedding planes coincide with the hangingwall or footwall of an excavation and are of low inclination. This was often observed in the case studies conducted within the Klerksdorp gold field, where tunnels are sited in the MB Formation quartzites. These planes of weakness within a reinforced rock mass structure are orientated sub parallel to the typical orientation of the rock bolt reinforcement. This appears to result in low interaction between the rock bolt units and the bulk of the rock mass, and also poor interaction between the rock bolts within the reinforcing system.

The observed increased deformation of the lower sidewall in comparison to the upper sidewall is

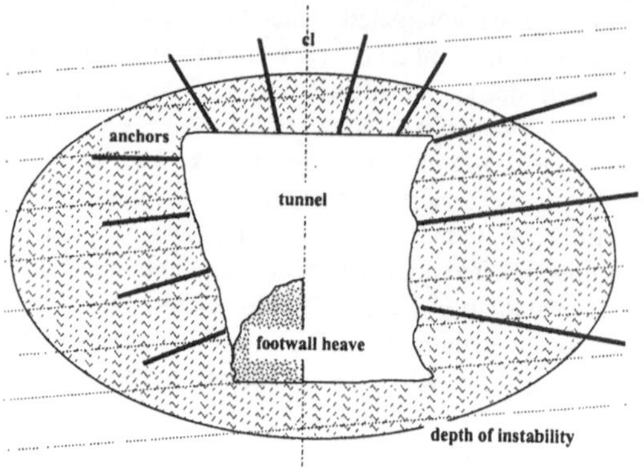

Figure 3. A conceptual model of design methodologies and potential rock mass deformations in the presence of weak bedding planes.

considered to be a function of the presence of these dominant sub-horizontal bedding planes in addition to a lack of continuity of the lower sidewall support into the footwall. The presence of hangingwall support, and the general upward inclination of the upper sidewall rock bolt reinforcement, can create a more competent reinforced rock mass structure and thus greater end constraint to the sidewall reinforced rock mass structure (Figure 3).

Analysis of hangingwall stability often indicated shear failure of tendons in the hangingwall of the tunnel profile. Consideration of the shear capacity of tendons would not be catered for in standard support design practice, but this has been indicated to contribute to hangingwall failures at 70 % of the rockburst case studies. Consideration of the shear capacity of tendons and cables is suggested and that these reinforcing systems be utilised in areas where hangingwall shear, due to existence of prominent bedding and dynamic loading, is anticipated.

The relative intensity of the discontinuities within the hangingwall of an excavation had an influence on the potential of the support system to either accommodate or fail under shear loading. It is considered that in some cases rock bolts were able to accommodate relatively large shear deformations due to the highly discontinuous nature of the rock mass (Photograph 7). This is thought to allow numerous incremental shear dislocations along the rock bolt length, each within the shear capacity of the rock bolt. This results in lower direct shear interaction between the rock mass and the rock bolt and thus reduced dynamic shear loading per shear plane.

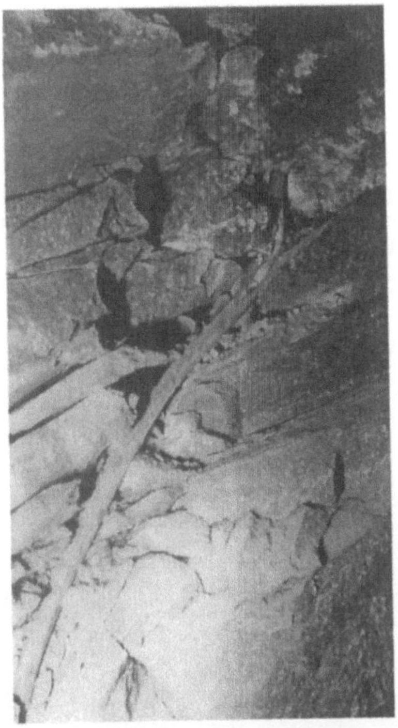

Photograph 7. Shear deformation of smooth bar rock bolt.

The depth of instability within the hangingwall of the tunnel excavations (3.0 m x 3.0 m) has been estimated at 1.5 m to 2.0 m, based on these case studies. With typical support systems as used in the South African mining industry, this implies that hangingwall stabilisation is primarily based on anchorage of the rock bolt reinforcement, and containment of the potentially unstable rock mass rather than structural reinforcement.

CONCLUSIONS

Observations of the performance of the tunnel support systems have allowed a mechanistic evaluation of the current design procedures. Of significance for the improved design of the support systems under high stress rockburst conditions are the following:

- Consideration of the natural depth of instability of the rock mass around the excavation under dynamic loading conditions.
- Consideration of mechanisms of interaction of the rock bolt reinforcement with the unstable rock mass volume.
- Consideration of the loading of the fabric support due to the unstable rock mass volume between the rock bolt reinforcement and the influence of the characteristics of the fabric support.
- Consideration of the shear demand and capacity of the rock bolt reinforcement, particularly in the hangingwall of the tunnel.

ACKNOWLEDGEMENTS

The observations of the case studies as discussed in this paper were conducted under the funding of SIMRAC (Safety in Mines Research Advisory Committee). The support of SIMRAC and the hospitality of the mines of the West Rand and Klerksdorp gold fields visited are gratefully acknowledged.

REFERENCES

Anon., (1996), *Guideline for the compilation of a mandatory code of practice to combat rockfall and rockburst accidents in metalliferous mines and mines other than coal*, Department of Minerals and Energy Ref. GME 7/4/118-AB2, South Africa.

Haile, (1998), *A mechanistic evaluation and design of tunnel support systems for deep level South African mines*, PhD thesis in preparation, Mechanical Engineering Dept., University of Natal, South Africa.

Rock Support and Reinforcement Practice in Mining, Villaescusa, Windsor & Thompson (eds)
© 1999 Balkema, Rotterdam, ISBN 90 5809 045 0

The use of elongate support in the South African Mining Industry

F.J.Glisson, D.H.Kullmann & A.E.Vidal da Silva
CSIR Mining Technology, Johannesburg, South Africa

ABSTRACT: Over the years a large number of timber, timber and steel, and all steel mine props (elongates) have been developed for the support of deep, tabular and narrow gold and platinum mine stopes. These props were developed to meet the requirements of a prop type support, which can yield and thus accommodate the high rates of stope closure. A new support design methodology requires 'standard' performance curves for all support units as input data. The CSIR was commissioned to develop a test procedure and data analysis scheme to provide these standard curves for elongates. This includes the new generation of such props that are pre-stressable and can thus be installed close to the working face without being blasted out. This paper describes the provisional test procedure and provides examples of some initial tests.

1 INTRODUCTION

The South African gold mining industry has, for quite some time, recognized the need for a type of support that was easy and quick to install, light weight, and would also offer the required protection against seismic events as well as falls of ground. In the early days and at shallow mining depths, timber poles were used with a wood wedge as a means of securing and pre-stressing the support. As mining depths increased to over 1500 metres it become apparent that the performance of this type of support was not suitable for mining conditions where stope closure rates were 10 millimetres per day or more. From this evolved many new types of mine support props designed to yield 300 millimetres or more while maintaining an acceptable yield force. The first units developed to this specification were hydraulic props that could yield at one metre per second to accommodate dynamic loading associated with rockbursts. These were followed by timber props, engineered to yield, that have evolved to the present pre-stressable units. Numerous designs of the earlier elongates were patented but only a few were widely used in the mines. These were characterised by their ability to provide the basic functions of: efficient protection for the workers and ease of installation at an acceptable cost. An interim testing procedure has been developed to provide information on the variability of different support units. This information will be used to develop a standard testing procedure for all elongate support units to provide support type performance curves. These curves will be used in the generally accepted support design methodology and will, with a high degree of certainty, establish the minimum support capabilities of particular support types. The test procedure is described and examples of results of two elongate types are presented. The significant difference between laboratory results and results obtained from monitoring props installed underground is highlighted.

2 SUPPORT REQUIREMENTS

Support resistances of 50 kN/m^2 is the rockfall resistance criterion that has been generally used by the industry as a guide for the design of stope support systems in stopes where the major hazards are rockfalls. In stopes subject to seismicity and rockbursts, a general energy absorption criterion has been applied which requires that the support system should be capable of absorbing 60 kJ of energy per square metre of hangingwall. As a consequence the support system must have a yielding capability. The basis of this criterion was a support resistance of

200 kN/m², required to arrest the hangingwall displaced through 0.3 m at an initial velocity of 3.0 m/s during a rockburst and in the process absorb 60 kJ/m² of energy.

This criterion was re-evaluated using the quantified ejection thicknesses for various reefs from the accident database (compiled and maintained by CSIR Mining Technology). Those ejection thicknesses were as follows:

Reef A: 1.5 m
Reef B: 2.0 m

Based on these values, and using the formula,

$$E_a = mgh + 0.5mv^2$$

Where,

m – mass of the rock to be supported
g – gravitational acceleration
h – allowable stope convergence
v – velocity of ejection from roof

The velocity of ejection was taken as 3.0 m/s, and it was also assumed the support system had the required capability of yielding at least 0.2 m.

Based on the ejection thicknesses noted above, the energy absorption criterion would be as follows:

Reef A: 26 kJ/m²
Reef B: 35 kJ/m²

3 TESTING OF ELONGATE SUPPORT

CSIR Mining Technology (1998) has for the last two years, embarked on an extensive SIMRAC commissioned testing programme. The objective of this programme is to evaluate currently used types of elongates.

The testing procedure currently consists of twenty-four tests broken down as follows:

- Five rapid displacement tests at a rate of 3.0 m/s – the units are initially compressed slowly for 50 mm then subjected to 3.0 m/s for at least 200 mm and the test completed at the slow rate until failure
- Ten slow displacement tests at a rate of 15mm/minute
- One test at a deformation rate of 150mm/min
- One test at 15mm/min on a 10° inclined platen
- One test at 15mm/min on a 20mm stepped platen
- One test at 15mm/min on a 50mm stepped platen
- One slow test at a deformation rate of 10mm/day
- One creep test for seven days
- Three underground tests

The results of these tests are then subjected to statistical analysis to provide design curves. A consistency indicator at various degrees of confidence is calculated as well as energy absorption curves.

Examples of these results are shown for two different support units.

Figures 1 and 2 display the comparisons between the average results of several types of elongate for the testing rates as indicated.

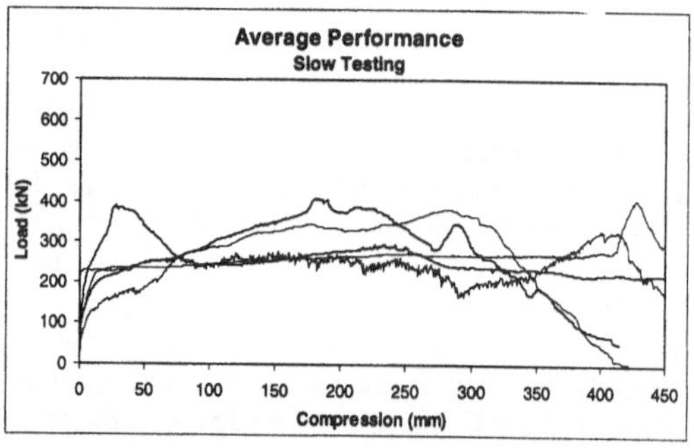

Figure 1. Comparison of the average performance for several elongate types at a rate of 15 mm/min.

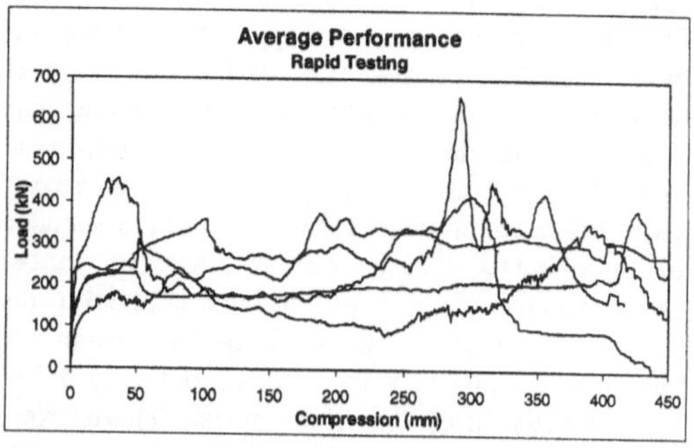

Figure 2. Comparison of the average performance for the same elongate types as presented in Figure 1 when tested at a rate of 3.0 m/s.

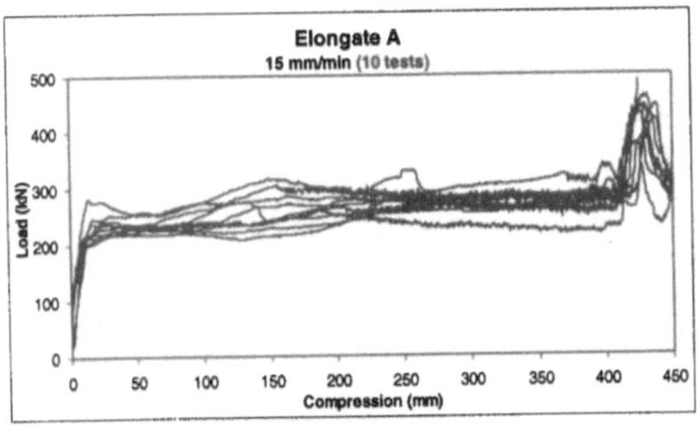

3a)

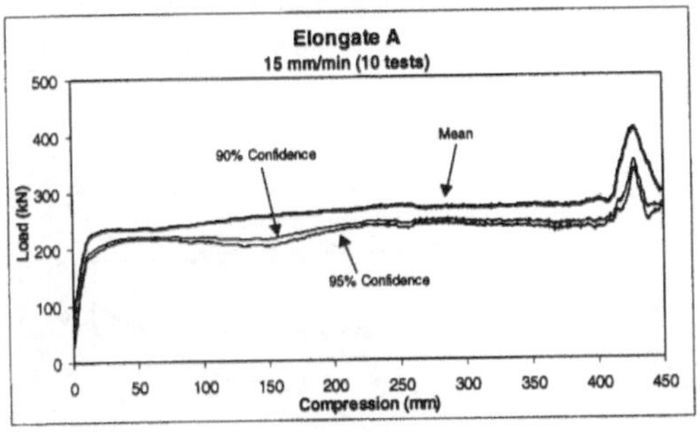

3b)

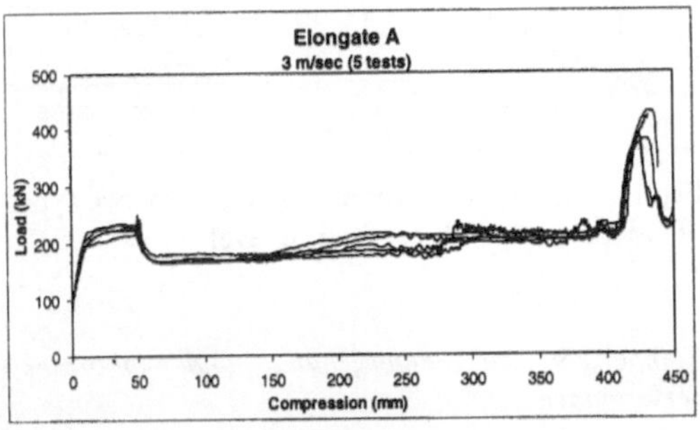

3c)

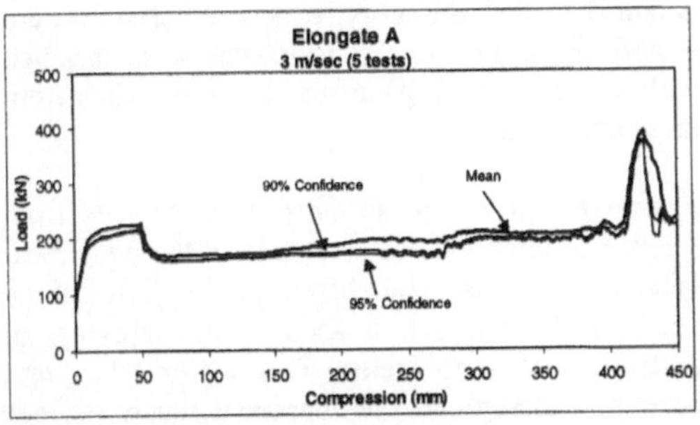

3d)

Figure 3. Variability in performance of Elongate A under slow and rapid compression testing.

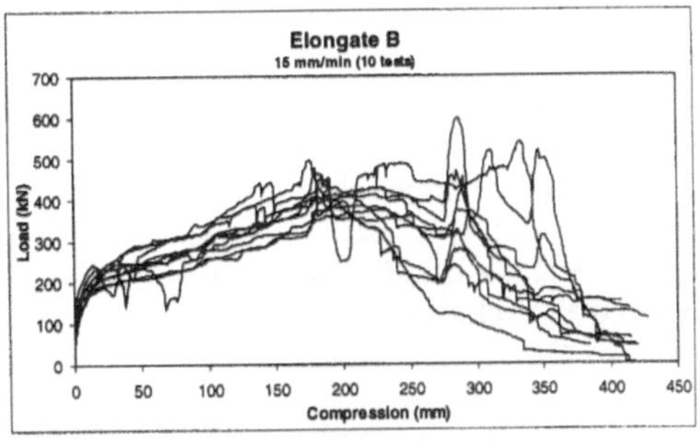

4a)

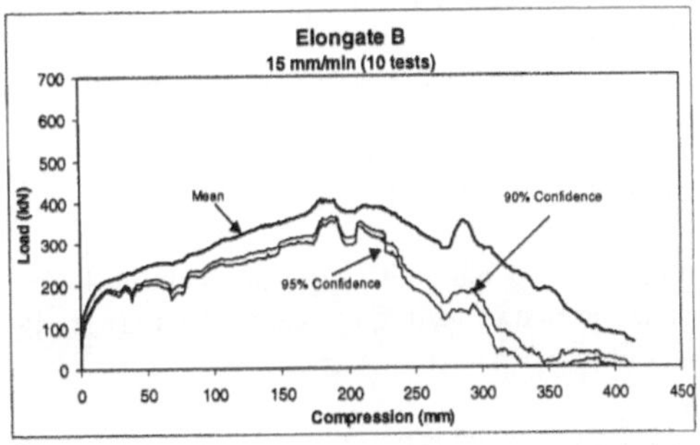

4b)

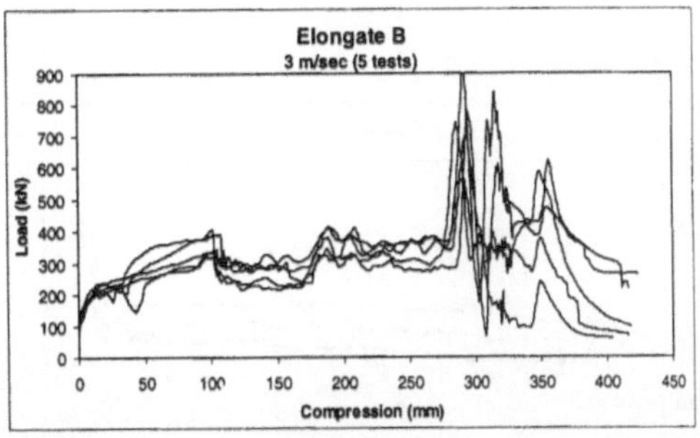

4c)

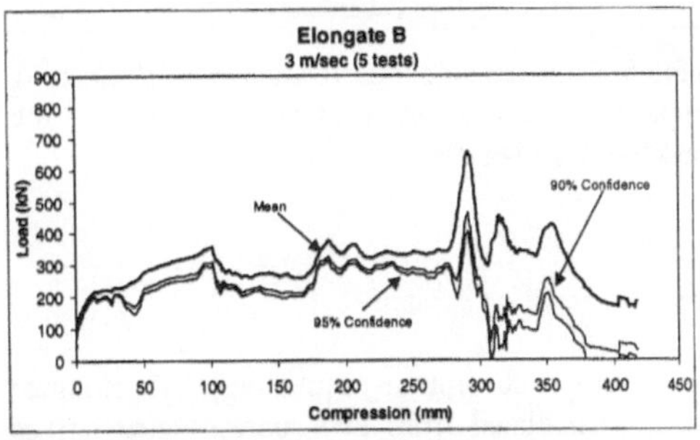

4d)

Figure 4. Variability in performance of Elongate B under slow and rapid compression testing.

Performance variability

Due to the variety of materials used and degree of engineering in the construction of each elongate, the variability of performance between different specimens of each type tested is in itself a point to take into consideration. The one elongate is a well engineered all steel prop, which displays remarkable consistency of performance, see Figure 3.

The other elongate is a timber-based unit with pre-stressing ability and an engineered yielding device (Figure 4). Being a basic prop, with timber as its main constituent it exhibits a greater degree of variability in its performance, as can be seen in Figures 4a and 4c.

Rate dependency

Friction based yield units typically exhibit lower loads as the rate of testing increases. This is clearly seen in Figure 5.

This rate dependency is not followed when timber is combined with an engineered yielding unit as can be seen in Figure 6.

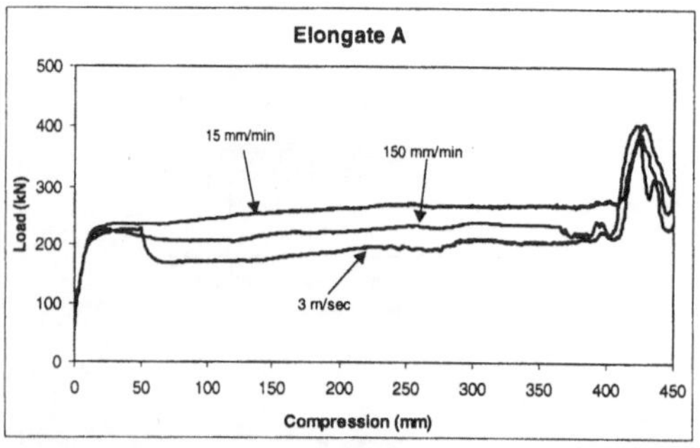

Figure 5. Rate dependent performance of Elongate A comparing average performance curves at the respective testing rates.

Energy absorption capacity

The energy absorption capacities of elongate support, calculated from their performance curves noted in Figures 3b and 4b, at a 90 % confidence level, are shown in Figure 7.

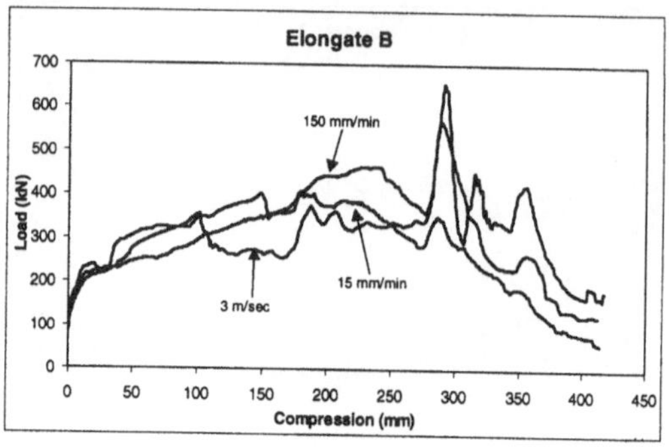

Figure 6. Rate dependent performance of Elongate B comparing average performance curves at the respective testing rates.

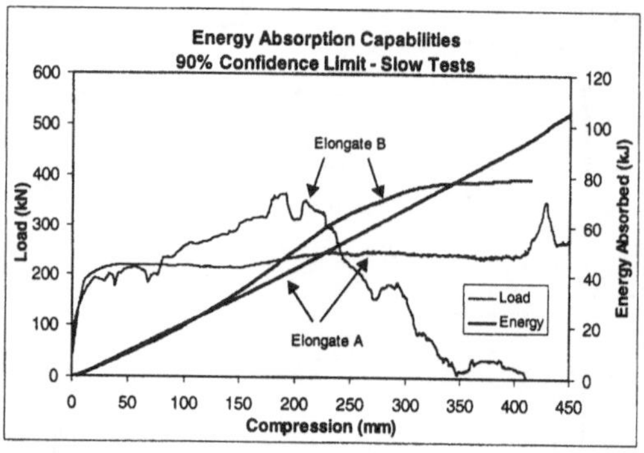

Figure 7. Energy absorption capacities of two types of elongate at a 90% confidence level.

Comparison of underground and laboratory performance

To compare the underground performance with that obtained in the laboratory, experimental sites were established at two mines. Elongates were installed with load cells and regular readings were taken from these and of the convergence.

Figures 8 and 9 indicate the results obtained from the in situ tests and are compared to the equivalent laboratory results. The difference in performance needs to be taken into account by applying an underground usage factor. This factor takes into account underground considerations, such as poor support installation, slow convergence rates and high humidity / temperature as well as timber creep, all of which have been shown to significantly influence support performance (Roberts *et al.*, 1987). Currently, underground field tests are conducted to establish appropriate correction factors for each type of unit, and these will be incorporated in the performance curve calculation as soon as sufficient underground data is available.

This remains a critical issue as all support designs will be required to take this performance discrepancy into account once it can be quantified.

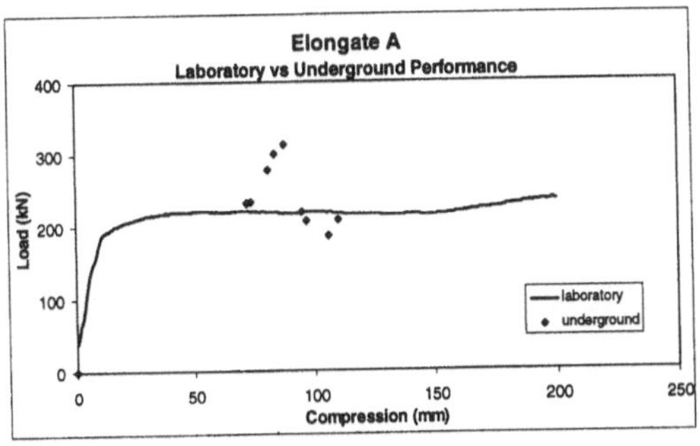

Figure 8. Comparison of laboratory slow test results (90 % confidence) with underground performance for Elongate A.

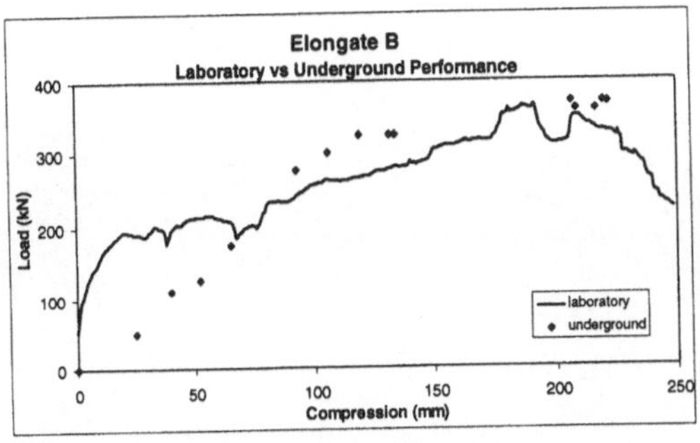

Figure 9. Comparison of laboratory slow test results (90 % confidence) with underground performance for Elongate B.

Stepped and inclined platen tests

These tests were devised to simulate (at least in part) underground installations by evaluating the effects of eccentric loading of the support units. The eccentric loading of support units is a common occurrence underground and has been recognised as a factor adversely affecting the performance of the support units. Three different loading plates were used, of which two were step platens of 20 and 50 mm and the third was an inclined platen of 10 degrees. A limited number of tests have been conducted in this way to determine the validity of the assumption that eccentric loading could compromise the performance of the support units, and promote buckling type failure.

The information obtained to date indicates that this type of eccentric loading may compromise the performance of some units. Due to the limited number of tests conducted, no definite conclusions can be made. It does, however, indicate that more of these tests will be required to establish each unit's buckling potential.

4 CONCLUSIONS

The objective of these tests is to provide a standard testing procedure for all elongate support types. This procedure will provide officially accepted performance curves for each support type, which will in turn be used in the design of support systems.

Following the acceptance of the standard test procedure, support manufacturers will be required to conduct routine quality assurance testing of their support units that will indicate the performance of their production line product. These results can then be compared to the original performance testing results. This will provide for a degree of confidence that the design performance criteria are being met.

ACKNOWLEDGEMENTS

The funding and permission to publish this paper by SIMRAC (Safety In Mines Research Advisory Committee) is gratefully acknowledged.

REFERENCES

Mining Technology (1998). Stope Face Support Systems, SIMRAC Project GAP 330, Final Project Report (to be completed December 1998)

Roberts, Jager and Riemann (1987). The performance characteristics of timber props. COMRO Res. Report NO. 35/87.

Rock Support and Reinforcement Practice in Mining, Villaescusa, Windsor & Thompson (eds)
© 1999 Balkema, Rotterdam, ISBN 90 5809 045 0

Assessment of precast cellular lightweight concrete (C.L.C.) support structures

P.N. Erasmus & J. Smit
Grinaker Duraset – Westorania, South Africa

ABSTRACT: Traditional materials, used in external support structures, like timber and normal concrete have certain fundamental shortcomings. These shortcomings are highlighted in an analysis using defined material and structural properties as the basis. Cellular lightweight concrete (C.L.C.) has been identified as a support material with excellent properties. This paper assesses the performance of a modular precast C.L.C. support system specifically developed for deep level seismically active mining. This CLC technology can be engineered to tailor support to suit a wide range of mining types.

1. INTRODUCTION

External support structures such as timber packs or cribs form a vital part of the support regime in tabular (reef or seam) type of mining operations – particularly roadways and gullies.

These structures have been ineffective in conditions where ground movement is governed by closure and strata integrity. Many attempts have been made to improve the effectiveness of timber support structures with limited success. The fundamental reason for this is the fact that the material and structural properties of timber prevent it from meeting the primary support performance requirements.

Research and development was therefore necessary to find materials and structural designs that will be able to satisfy these primary support performance requirements. Specific material and structural properties required to achieve the ideal load-deformation characteristics are discussed. These properties were used as the basis to analyse different traditional support materials. The same analysis was then used to identify suitable support materials. Cellular lightweight concrete (C.L.C.) reinforced with steel and/or fibre was found to be the most suitable material.

This paper will further show how C.L.C. was developed into a successful precast-modular pack support system. This C.L.C. support system has become the leading support for deep level, seismically active mining during the past five years.

2. PRIMARY SUPPORT PERFORMANCE REQUIREMENTS

In most types of mining, some form of support or reinforcement will be required to control strata and local ground movement. The two factors that govern the amount and extent to which ground movement takes place are closure and integrity of the strata. In most types of tabular (reef or seam ore body) mining, these two factors dictate what the primary support performance requirements will be.

The first factor namely closure depends on the rate and amount of closure. Both have a major impact on what will be required of the support structure. A typical example of closure rate is 4mm to 30mm per day with an increase of up to 3000mm per second during a seismic event in South African gold mines.

To design for closure the primary support performance requirements are:

- The support structure must generate the designed load-deformation characteristics without being adversely affected by the rate or change in rate of closure.
- The support structure must be able to maintain the designed load over the full range of closure that will take place.

The second factor namely strata integrity is determined by the extent of jointing, fractures, bedding planes and rock type (hard or soft).

The primary support performance requirements to maintain the strata integrity are:

- Immediate active support resistance
- Support must yield at a load, which will prevent it from damaging the roof or floor.
- The designed support resistance must be maintained during the mining cycle.

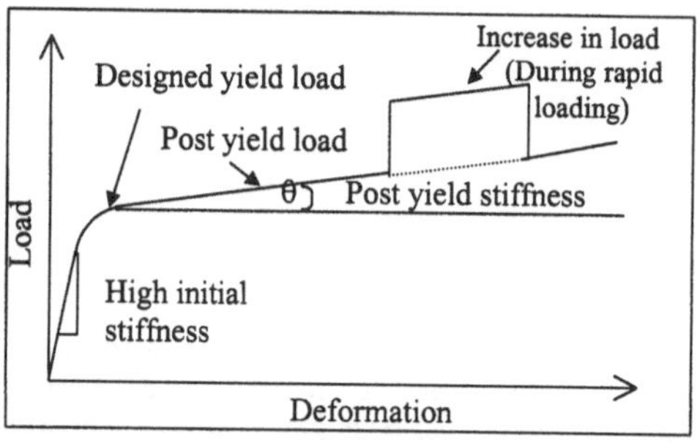

Figure 1. Ideal load deformation characteristics

The primary support performance requirements can be summarised as follows:

1. Not to be adversely affected by the rate or change in rate of closure.
2. Immediate active support resistance. (high initial stiffness – see fig. 1)
3. Ability to yield at a designed load. (see fig.1)
4. Ability to maintain a designed post yield load with a specific post yield stiffness.

These primary support performance requirements are common to most types of mining. Secondary support performance requirements such as the actual designed yield load, post yield stiffness and amount of deformation required will depend on the depth of mining, geology, rock conditions and mining method.

MATERIAL AND STRUCTURAL PROPERTIES	DESCRIPTION AND PERFORMANCE EFFECT
High E-Modulus (stiffness) at typical underground closure rates (1 mm – 50 mm per day).	- This means that the material must generate the required initial stiffness when deformed at slow loading rates. - The stiffness must be high enough to reach the required support resistance after 2-3% closure. - This ensures active support at slow loading rates. - Early maintenance of strata integrity.
Designable yield load.	- This means that the material must be able to be designed to yield at a predetermined load. - This will prevent the support from generating too much load which can damage the roof or floor.
Maintain post yield load during closure.	- During closure the material must maintain the post yield load without load shedding. - This will allow controlled closure of the working place.
Constant low crushing strength during rapid loading over a large deformation range.	- This means the material must have the ability to reduce an impact load (rockburst). The mechanism of dissipating the kinetic energy must be to increase the deformation range rather than an increase in force generated. It must therefore be able to act like a shock absorber during impact instead of acting like a stiff spring. - This will prevent damage to the surrounding rock mass during rapid closure.
The material properties must be homogeneous and non-biodegradable.	- This means that the performance must be repeatable and not be affected by time or hostile environments. - This is to ensure that all support elements and structures behave in the same manner at all stages.
100 % Surface contact between individual units.	- It must be possible to ensure that no point loading occurs in the structure, as this will negatively affect all the above-required properties.
Buckling stability during normal and rapid closure.	- The structure must not be susceptible to buckling during closure.
Fire resistant material	- Must be non-combustible.

Figure 2. Material and structural properties.

3. REQUIRED MATERIAL AND STRUCTURAL PROPERTIES

Specific material and structural properties are required to design a support structure conforming to the ideal load-deformation characteristics in fig. 1. These material and structural properties are discussed in detail in fig. 2.

4. ANALYSIS OF MATERIAL AND STRUCTURAL PROPERTIES.

The following analysis (fig. 3) was used to compare current support materials with C.L.C. The analysis highlights the design and performance shortcomings of timber and normal concrete. It will also be clear from the analysis (fig. 3) why C.L.C was identified as a suitable support material.

REQUIRED MATERIAL OR STRUCTURAL PROPERTY	**TIMBER MATS**	**TIMBER END-GRAIN**	**TIMBER AND CEMENT BRICK**	**CELLULAR LIGHTWEIGHT CONCRETE (C.L.C.)**
High E-Modulus (stiffness) at typical under-ground closure rates (4 mm – 50 mm per day).	Not possible due to natural timber material creep which results in a loss of load carrying capability.	Not possible due to natural timber material creep. The end-grain units where the timber is loaded vertical to the natural grain will induce a slightly higher E-Modulus. Although this is not significant below 30 mm per day.	The cement brick elements increase the overall E-Modulus of the composite structure. However the timber material creep will still induce loss of load at slow loading rates. (Based on combined findings of timber and cement bricks.)	Reliable high E-Modulus which is not dependent on the rate of loading.
Designable yield load.	Due to material creep a flat linear load deformation curve is produced when loaded. The material therefore remains in a state of elastic deformation while building up load during closure.	Slight but unpredictable yield when end-grain units start to shear. The structure remains in a state of elastic deformation while building up load rapidly during closure.	Yielding takes place in an unpredictable manner due to shear failure of the cement bricks once loaded past the bricks compressive strength. This results in the complete structure shedding load later on.	Will yield at a designed yield load. The reliable design is possible due to crushing now being controlled by the strength and size of individual cells. Yielding is therefore induced through a mechanism of controlled cell collapse during closure.

REQUIRED MATERIAL OR STRUCTURAL PROPERTY	TIMBER MATS	TIMBER END-GRAIN	TIMBER AND CEMENT BRICK	CELLULAR LIGHTWEIGHT CONCRETE (C.L.C.)
Maintain post yield load during closure.	Load generation will depend on rate of loading. This combined with wood properties like density, type and condition make the load generation prediction unreliable.	Load generation will depend on rate of loading. This combined with wood properties like density, type and condition make the load generation prediction unreliable.	Loss of load when skeleton structure becomes unstable. Load generation will depend on rate of loading. This combined with wood properties like density, type and condition make the load generation prediction unreliable.	Maintain post yield load over a large deformation range.
Constant low crushing strength during rapid loading over a large deformation range.	Not possible because of natural properties of wood. (Timber reacts like a stiff spring during rapid loading). This results in shear failure of roof and floor around the pack.	Not possible because of natural properties of wood. (Timber reacts like a stiff spring during rapid loading). This results in shear failure of roof and floor around the pack. The end-grain units in these packs contribute additionally to the stiffening up of the structure.	Very high loads will be generated during rapid loading. Catastrophic shear failure of the solid cement bricks induces instability in the structure. This then finally results in total load shedding.	C.L.C. has a low crushing strength during rapid loading over a large deformation range.
Material properties to be homogeneous and non-biodegradable.	Due to timber being an organic material it is biodegradable and therefor not consistent.	Due to timber being an organic material it is biodegradable and therefor not consistent.	Due to timber being an organic material it is biodegradable and therefor not consistent.	Homogeneous and non-biodegradable material properties. The manufacturing process also allows for the production of a very consistent product.

REQUIRED MATERIAL OR STRUCTURAL PROPERTY	TIMBER MATS	TIMBER END-GRAIN	TIMBER AND CEMENT BRICK	CELLULAR LIGHTWEIGHT CONCRETE (C.L.C.)
100 % Surface contact between individual units.	Not possible due to processing methods. Normally adds to low initial stiffness.	Improved trimming process can improve initial stiffness.	Depends on timber quality and accuracy of trimming.	The molding process ensures an accurate and consistent product. This structural property contributes to the achievement of high initial stiffness
Buckling stability during normal and rapid closure.	Becomes unstable when H:W ratio exceeds 2:1. Buckling is mainly due to the elastic load-deformation behaviour.	Becomes unstable when H:W ratio exceeds 2:1. Buckling is mainly due to the elastic load-deformation behaviour.	Becomes unstable when H:W ratio exceeds 2:1. Buckling is mainly due to the elastic load-deformation behaviour. Failure of brick units causes rapid instability of the skeleton structure.	Stable at normal and rapid closure, which is mainly due to the homogeneous deformation process of the cellular matrix. H:W Ratio's exceeding 2:1 have been tested successfully.
Fire resistant material	Combustible	Combustible	Combustible	Non-Combustible

Figure 3. Analysis of material and structural properties of different support materials.

5. C.L.C PRODUCT DEVELOPMENT

Initially C.L.C-support product development commenced in 1990. It was decided at that stage that initial product development would focus on designs for deep level and seismically active mining. A successful product known as Durapak® was developed, which today is one of the leading deep level gully/roadway support systems.

The development programme included many different designs on steel, fibre, geometry and C.L.C.-matrix combinations. The outcome of this development was a modular support system, which makes use of one unit size (600x300x100 mm). This modular unit can be used to build different

Figure 4. Modular C.L.C.-support unit (Durapak®).

configurations, which caters for a variety of heights and support resistance designs.

The unit (see fig. 4) can be described as a precast uniformly dimensioned C.L.C. slab.

6. EVALUATION OF C.L.C. SUPPORT FOR DEEP LEVEL MINING

Since 1992 C.L.C. support technology has been successfully used in deep level seismically active mines.

During this period research was conducted to determine the effectiveness of this technology. The following evaluations are based on the outcome of this research.

6.1. *Testing of load deformation characteristics*

Despite the differences in conditions found underground compared to the conditions at the laboratory, development and comparative testing in the laboratory provides valuable information needed to correlate differences of underground and laboratory results. This information can then be used to interpret underground behaviour.

It is possible to test at the following loading rates in the laboratory:

- Slow loading rate (1 to 3mm/hour)
- Normal loading rate (15 to 30 mm/min)
- Rapid loading rate - Packs (3×10^2 mm/s)
  - Props (3×10^3 mm/s)

Tests at normal loading rates (15 to 30mm/min) were performed at the CSIR laboratories in Johannesburg.

Tests at rapid loading rates (300mm/s) were performed at the Materialprüfungsamt Nordrhein-Westfalen.

Typical underground loading rates can vary from 1mm/day to 3000mm/s - a real challenge for any structure!

6.1.1. *Laboratory test method*

Due to the orders of magnitude difference in loading rates, the tests were done on different hydraulic presses. Normal testing procedures were followed except with the rapid yield test. The loading rate profile followed during the rapid yield tests is shown in Fig. 5.

6.1.2. *In Situ test method*

In-situ tests are performed using hydraulic loadcells (flatjacks) built into the pack structure. Adequate protection to pressure gauges must be ensured to prevent damage due to mining activity.

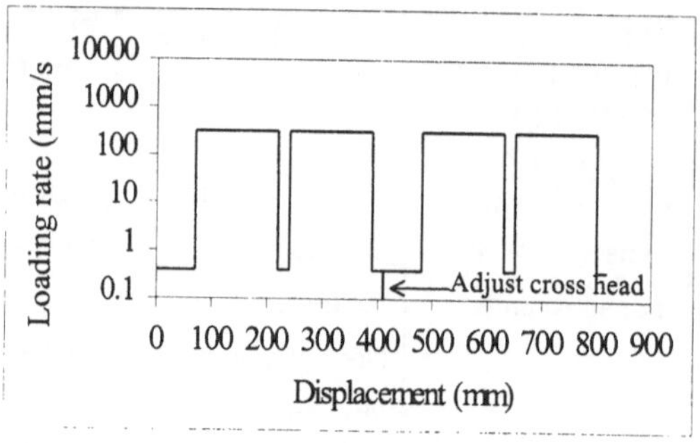

Figure 5. Loading rate profile of rapid yield tests.

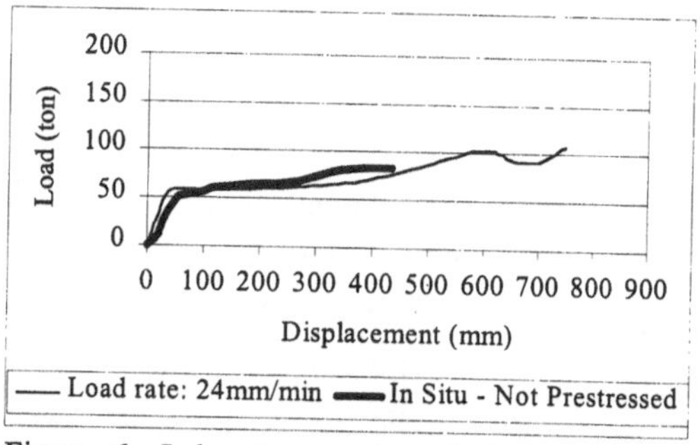

Figure 6. Laboratory and underground tests of C.L.C-support.

A tape measure is used to measure the distance between two fixed points on the hanging wall and the footwall at each corner of the pack.

The measurement interval is a function of expected loading rate and other relevant practical considerations.

6.1.3. *Discussion of results*

a) Effect of different loading rates

Fig. 6 shows a graph of two similar C.L.C. packs tested in laboratory and underground conditions. The laboratory loading rate was 24mm per minute and the underground rate 0.0174mm per minute (25mm per day). The laboratory rate of loading was ~138% faster than the underground test.

The load-deformation characteristics of C.L.C.-support structures are therefore not adversely affected by the rate of loading. A downward adjustment factor of less than 5% can be used to correlate laboratory tests with underground performance

Fig. 7 shows a graph (Roberts, Pienaar & Kruger 1987) of two similar end-grain packs, one tested in the laboratory at 30mm per minute and the other one tested underground at 0.0045mm per minute (6.5mm per day). There is a marked difference in their load-deformation behaviour. The underground test shows a decrease in stiffness and a decrease in load. Various reasons to explain this behaviour have been published of which material creep, bedding-in and point loading are the main contributors

This clearly shows why timber support structures are so ineffective in slow loading rate conditions. Timber support does not meet the first primary support performance requirement of not being affected by a change in the rate of loading. This makes timber support structures unreliable and ineffective.

b) Effect of height to width ratio

Fig. 8 shows a graph of two C.L.C. packs tested in a press at 24mm/min with height to width ratios of 1.67 and 2.5 respectively.

This load deformation plot clearly shows a close correlation in behaviour of the two packs with varying height to width ratios. Due to the controlled plastic deformation behaviour of the packs the potential for elastic induced buckling is limited.

The effect will be a reliable support system even with height to width ratios exceeding 2. Substantial savings are possible with using smaller support structures to replace bigger timber structures.

c) Effect of rapid loading

Fig. 9 shows a graph of two C.L.C. packs tested in a press at 24mm/min and according to the rapid loading rate profile in Fig. 5 respectively.

It can be seen that the increase in yield load from the slow to rapid loading rate is in the order of 57 %. The yield load remained constant during the rapid yield event and went back to the corresponding yield load after the event.

The effect is a support system providing a mechanism for the rock mass to dissipate the kinetic energy during a rock burst or seismic event. The energy dissipation mechanism causes a gradual deceleration of the rock mass and reduces relative movement of the fractured rock. The relative low yield load also ensures that the integrity of the hangingwall and footwall will be preserved.

Fig. 10 shows a graph with a C.L.C. pack and a timber mat pack tested according to the rapid loading rate profile in Fig. 5.

The load generated with rapid loading of the timber pack is 150 % higher than the load with 24mm/min. The load of the timber pack is also increasing during the event. The loading rate during the second event on the timber pack reduced from 300mm/s to 10mm/s due to the inability of the press to keep the required loading rate. At the end of the first event the yield load of the timber pack is 300 %

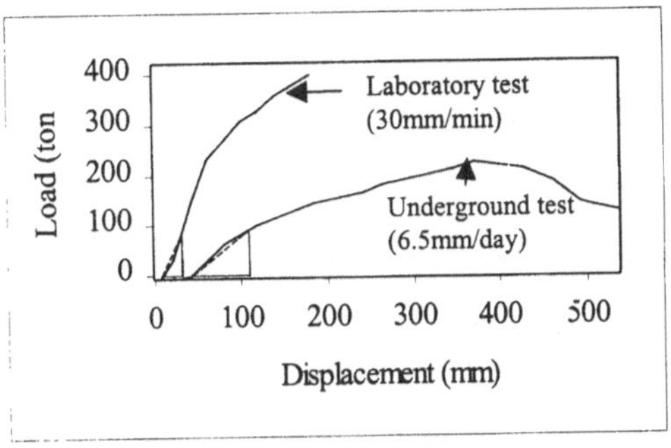

Figure 7. Laboratory and underground tests of timber support.

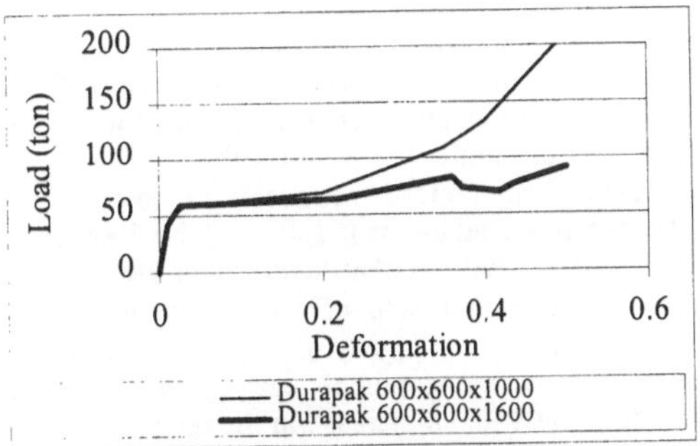

Figure 8. C.L.C height to width ratio tests

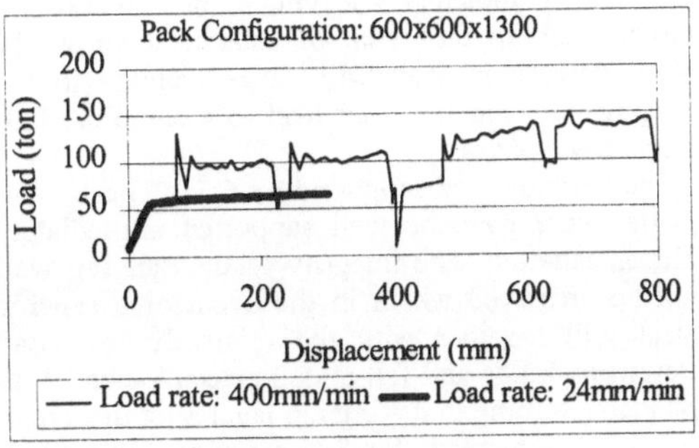

Figure 9. C.L.C rapid loading tests

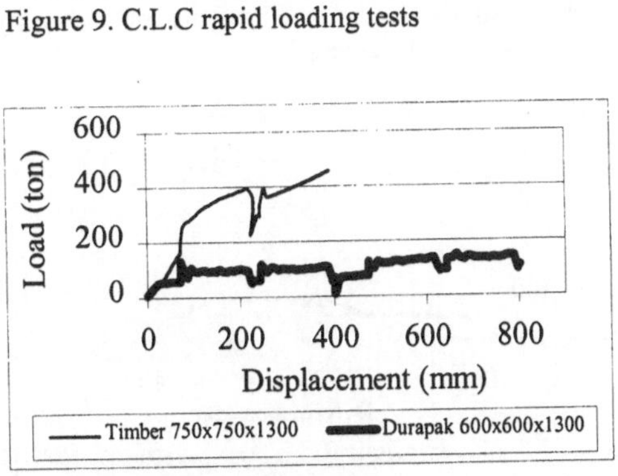

Figure 10. Rapid loading tests

more than the yield load of the C.L.C. pack.

Kinetic energy released during such an event will be absorbed (stored) by the pack structure and not be dissipated. During a seismic event or rockburst, the rockmass rapidly decelerates with subsequent damage to the hanging wall and results in a fall out between packs. The high load generated with a timber pack damages the gully sidewall that leads to non-compliance to support standards, lower grade and production difficulties.

d) Simulation of underground performance

Fig. 11 shows a simulated graph of a timber mat pack and a C.L.C. pack with underground loading rates of 20mm/day for 10 days and a seismic event of 300mm/s for a closure of 150mm.

The load generated by the timber pack is lower than that of the C.L.C. pack for the first 150mm closure. A continuous increase of load then takes place with the timber pack. With the seismic event the load then rapidly increases to exceed the corresponding load on the C.L.C. pack by 300%.

The effect will be that with a timber support network, the initial lack of activeness will cause deterioration of the integrity of the roof. The lack of normal stress development can cause keyblock dislodgement with an increase in the probability of a rockfall occurring. The load can increase to such an extent that the sidewall of the gully can start scaling or fail completely leaving the workplace unsupported. In the event of seismic loading, the rapid increase in load will cause damage to the hanging wall and can even lead to a complete fall-out between packs.

The working area supported by C.L.C. packs will on the other hand be well supported immediately after installation. The integrity of the hanging wall will be preserved which in the event of a seismic event will together with the relatively low load prevent rockfalls and fall-outs between packs. With the load returning to its original level after the event, the pack will be ready for the next event. As a result of the designed loads, the integrity of the gully sidewall will be preserved.

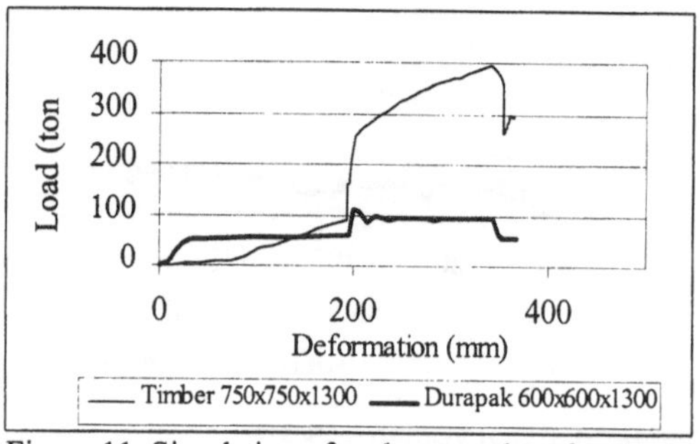

Figure 11. Simulation of underground performance.

6.2. Underground performance.

6.2.1. Choice of pack size

The unique dimensions of this block make it possible to construct different modular configurations, (see appendix). This enables the user to adjust pack sizes for different roof heights and also to achieve specific support resistances. These choices bring about a higher level of cost effectiveness and performance.

6.2.2. Pack construction

The lightweight units transport and handle with ease on the monorope. Manhandling in-stope is quick and easy.

Final construction of the packs is fast and accurate due to the uniformly dimensioned units. One labourer can construct a 60 x 60cm pack 1,5 m high in ± 15 minutes.

Prestressing bags placed close to the top (normally one layer from the top) requires less grout when comparing a C.L.C. pack as with timber packs. This results from 95 % of the pack being immediately solid before prestressing. With timber packs more grout is required to compress the elastic timber structure. The effect of prestressing on a timber pack can also be lost within 24 hours as a result of timber creep.

6.2.3. Pack blast outs

Once a C.L.C. support pack has been properly pre-stressed it will not be blasted out even at a range as close as 50 cm from the face. The surface on the blast side gets roughened up and the scraper might round the corners but structurally the pack remains solid.

The reduction in pack blastouts has an immediate effect on safety and productivity.

6.2.4. Falls of ground

Since introducing C.L.C. support no reports have been received where C.L.C. support was directly responsible for a fall of ground. A visible improvement in the number and direction of hanging wall fracturing indicated the positive effect of active support on hanging wall integrity.

Not only does safety benefit as a result of the reduction in falls of ground but also face advance and stoping width control.

6.2.5. Rockburst Damage

With C.L.C. packs a much higher percentage extraction of shaft pillars has been possible.

Maximum percentage extraction is of utmost importance in medium to high-grade areas.

Most important is the safety of the workers in seismically active areas. It is here where this type of support adds more value than merely being able to achieve high percentage extraction rates. Since the introduction of C.L.C. support, it is possible to continue with mining after a large seismic event. Most of the damage can now be seen on the C.L.C.-pack structures, which through controlled crushing dissipated kinetic energy. This kinetic energy would have damaged the hanging wall, footwall or gully shoulders if timber packs had been used.

6.2.6. Gully Support

In most areas a definite improvement in gully shoulder stability was observed. The two observations were that:

- less undermining underneath the pack occurred and
- Packs were not ejected into the gully itself.

The main reason for this was the fact that the load generated onto the footwall especially on the updip side did not exceed the actual strength of the fractured rockmass beneath the pack.

C.L.C. support offers a far more engineered solution to the gully support problem. The gully is the most important area to protect regarding safety and face accessibility.

6.2.7. Underground Fires

Extensive use of this non-combustible material makes it possible to renegotiate fire insurance premiums. The mere prevention of a major underground fire has obvious safety and production benefits.

7. CONCLUSION

Reliable support systems can be engineered with cellular lightweight concrete (C.L.C) technology. The material and structural properties of C.L.C-support systems ensure that the primary support performance requirements are met. This also allows for accurate designs on support resistance for different mining conditions.

The property analysis and performance evaluation showed that timber support structures are unreliable and ineffective when measured against the primary support performance requirements.

The successful use of C.L.C.-support (Durapak®) technology in South African gold mines since 1992 is clear evidence that the support technology is reliable.

This technology can also be used to develop C.L.C.-support for other types of mining.

8. ACKNOWLEDGEMENTS

We would like to express our sincere appreciation to Grinaker Duraset Mining for giving access to the results of tests performed on their patented C.L.C. support system (Durapak®) and for their approval to present this paper.

9. REFERENCES

Budavari, S. 1983. *Rock mechanics in mining practice.* South Africa: The South African Institute of Mining and Metallurgy.

Chamber of Mines Research Organization. 1988. *An Industry guide to methods of ameliorating the hazards of rockfalls and rockbursts.* Chamber of Mines of South Africa.

Naismith, W.A. 1996. *Risk assessment methodology as a guide for support system design.* Proceedings of the Centennial Symposium. South Africa: Department of Mining Engineering, University of the Witwatersrand.

Peele, R. 1944. *Mining engineers' handbook.* New York: John Willey & Sons, Inc.

Roberts, M.K.C; Pienaar, F.R.P. & Kruger, F.C. 1987. *Properties of pack supports – Comparison of the laboratory and underground performance.* South Africa: C.S.I.R. 122 –145.

Appendix: Modular Durapak Configurations.®

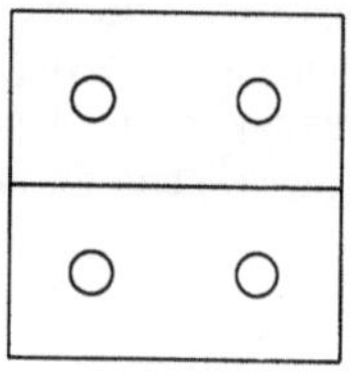

600x600 [mm]

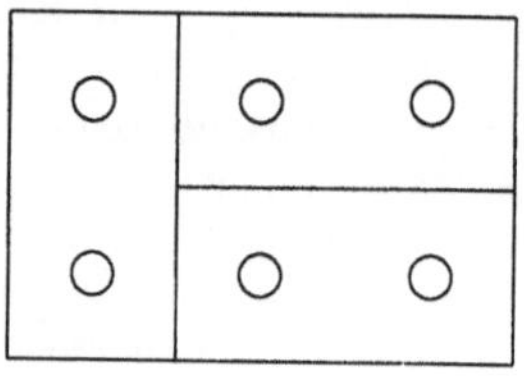

600x900 [mm]

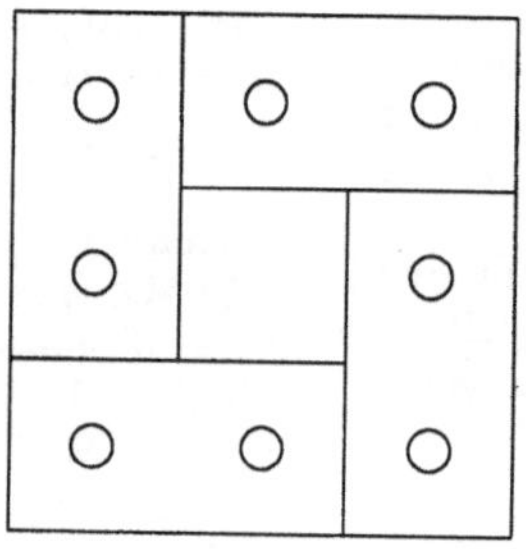

900x900 [mm]

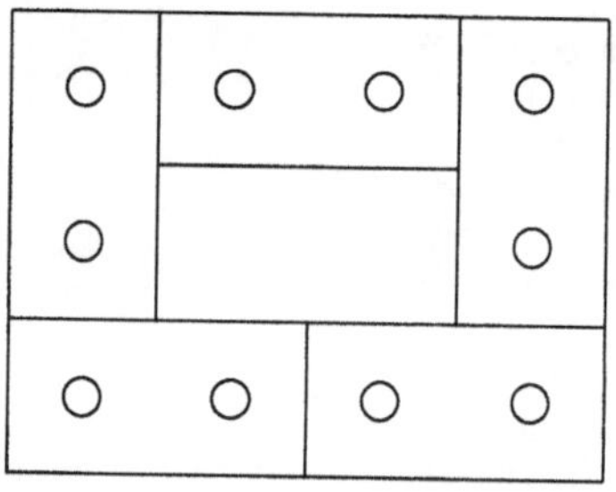

900x1200 [mm]

Rock Support and Reinforcement Practice in Mining, Villaescusa, Windsor & Thompson (eds)
ISBN 90 5809 045 0

Containment support for large static and dynamic deformations in mines

W.D. Ortlepp, T.R. Stacey & H.A.D. Kirsten
SRK Consulting, Johannesburg, South Africa

ABSTRACT: The paper deals with the testing of containment support, under large deformation static conditions and under dynamic conditions typical of the loading experienced during rockbursts. The testing was carried out over the period 1993 to 1998 for various types of wire mesh and wire mesh with wire rope lacing; mesh and fibre reinforced shotcrete; and fibre reinforced shotcrete, with and without wire rope lacing. The wire rope lacing was only used in dynamic tests. The dynamic testing method made use of test samples which attempted to simulate realistic in situ conditions, and the range of loading imposed was equivalent to small to severe rockburst conditions.

The results of the testing provide data which is relevant for the choice of support types appropriate for large static deformations and for situations in which mine-induced seismicity occurs or is expected to occur.

1 INTRODUCTION

Underground mining is progressing to greater depths in many countries in the world. Mining at depths of around 3000m is already common in the South African gold mines, and plans are being put in place to mine to 5000m. Even at shallower depths than these, large deformations of tunnels can result from the induced static and dynamic stresses. The static stresses cause a fractured zone with substantial dilation to occur in the walls of tunnels, and seismicity can cause rockbursts which result in dynamic loading of support elements. Rockbursts manifest as the violent ejection of rock from any of the surfaces of the tunnel, and often in very localised areas. It has been observed that about a metre thickness of rock may be ejected, and that ejection velocities are commonly up to about 10 m/s (Ortlepp and Stacey, 1994).

In such situations, in which failure of rock cannot be prevented, the important requirement is that the failure is contained sufficiently to maintain safety, and to ensure that the tunnels remain serviceable. Therefore, the tunnel support must be capable of maintaining its load carrying capacity even after large movements of the walls have occurred. Figure 1 illustrates two extremes in behaviour after a seismic event. In the immediate foreground the support consists of rockbolts with wire mesh, rope lacing and shotcrete, which is completely intact after the event. Immediately adjacent to this, in the background, rock failure has occurred where only rockbolt support was installed.

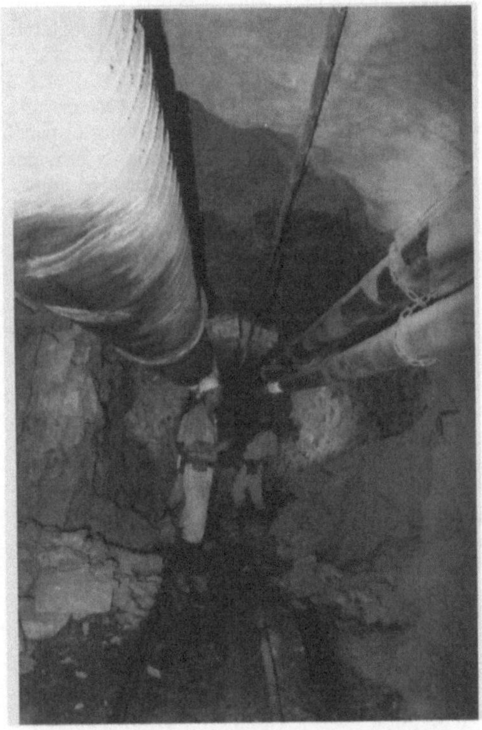

Figure 1. Support behaviour in a rockburst

In this paper the results of tests carried out to determine the static and dynamic performance of a range of tunnel support types are described. The tests include:

- static, distributed load tests on shotcrete reinforced with wire mesh and steel fibres;
- dynamic tests on wire mesh, wire mesh with wire rope lacing, fibre reinforced shotcrete, and fibre reinforced shotcrete with wire rope lacing.

2 STATIC TESTING OF REINFORCED SHOTCRETE

Various shotcrete testing programmes have been carried out previously (Kirsten and Labrum, 1990; Kirsten, 1992; Kirsten, 1993). Recent testing has been aimed at determining whether fibre-reinforced shotcrete could be applicable in the large deformation conditions described above (Kirsten et al, 1997).

2.1 *Test method*

The interaction between shotcrete and rock in a real tunnel is complex, and in the test programme no attempt was made to model this interaction. However, the test method used was aimed at representing, realistically, the loading conditions on the containment support. In this method, which has been described by Kirsten (1992), a 1.6m x 1.6m continuous shotcrete panel is retained by four rockbolts spaced 1m apart.

These bolts define a central 1m^2 area to which a uniformly distributed load is applied by means of a hydraulically pressurised bag. The 300mm of panel outside of the retaining bolt locations provides continuity of the shotcrete across the retention points, which approximates the conditions which occur in practice underground. The test panels were nominally 75mm thick. The use of this method is now well established.

The deformation performance criterion that was set was: after 150mm of central panel deflection, the panel should retain a load bearing capacity equal to or greater than 50% of its peak load capacity. No specific peak load capacity was specified since the most important aspect was considered to be the ability of the shotcrete to continue to provide substantial support, even after significant static deformation, without collapsing. This would also provide the capability of absorbing energy, which is of particular importance under dynamic loading conditions.

2.2 *Materials tested*

A range of fibre types was used in the tests, with various fibre lengths. Consistent fibre content was aimed at, but there was variation due to the shooting process. The results for the fibre types, fibre lengths and fibre contents which were most successful in meeting the specified performance criterion, are presented in this paper. These fibre types were Dramix steel fibres and monofilament polypropylene fibres. Fibre lengths were 40mm in the case of Dramix and 40mm or 50mm in the case of the polypropylene fibres. Fibre contents, by mass, were greater than approximately 2.5% and 0.35% respectively for the Dramix and polypropylene fibres. For comparison purposes, the results for two shotcrete panels reinforced with diamond (chain link) wire mesh of 100mm aperture and 3.2mm strand diameter are included.

2.3 *Test Results*

During the tests the panels first developed a typical orthogonal midspan tension crack at small deflections. This usually occurred between 80% and 100% of the achieved peak load. The first crack was followed by a second similar crack at right angles to the first and as the test proceeded, compression cracks developed diagonally at the bolt heads.

The bands of results for the panel tests are shown in Figure 2. The mesh reinforced panels demonstrate excellent yield characteristics with maintainence of maximum load carrying capacity even at a deflection of 150mm.

The panels reinforced with Dramix steel and monofilament polypropylene fibres also met the deformation criterion as can be seen from Figure 2. Both types of fibres straddled the cracks in the shotcrete as deformation took place, and tended to pull out rather than to break in tension. This probably contributed significantly to the yielding capability of the panels.

It should be noted that all of the panels reinforced with other fibres also retained considerable load bearing capability after 150mm of central deflection even though they did not meet the specified performance criterion.

The peak load capacities of the Dramix reinforced panels are significantly greater than those of the polypropylene reinforced panels, and greater than the wire mesh reinforced panels. As indicated above, this is considered to be of less importance with regard to the performance of the shotcrete in conditions of severe static and dynamic deformation than the ability to maintain residual strength.

3 DYNAMIC TESTING OF WIRE MESH AND SHOTCRETE SUPPORT

3.1 *Test method*

It was considered important to simulate the loading imposed during a rockburst event as realistically as possible. In a rockburst situation the dynamic loading imposed on containment support is in the form of a violent impact by the rock mass, distributed across the surface of the support.

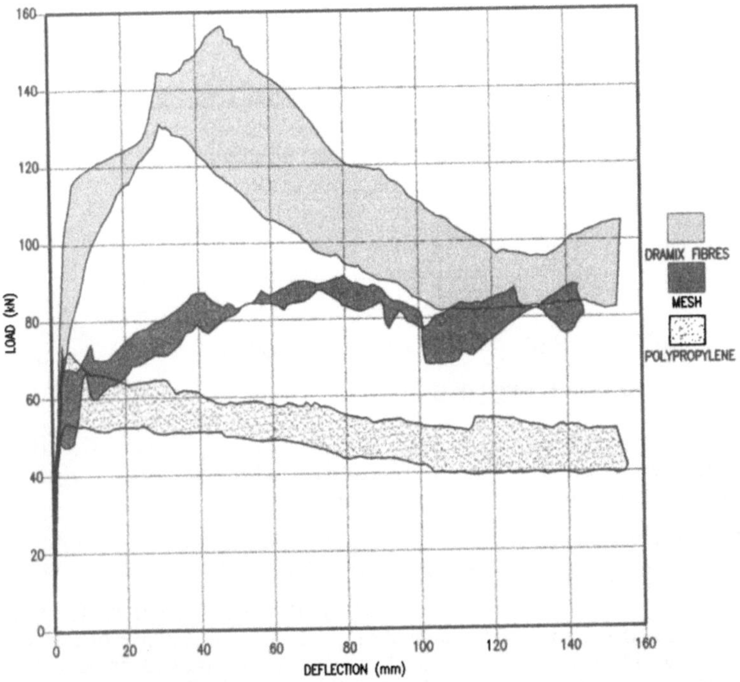

Figure 2. Load deformation behaviour of reinforced shotcrete panels

In this form of loading, the retainment elements consisting of rockbolts and face plates, the containment support itself, and the rock mass all contribute to the support resistance.

Panels of mesh or shotcrete were suspended by means of rockbolts, and an artificial rock mass and pyramid of steel-clad concrete blocks distributed the impact load onto the panel. Tie-back cables provided extended boundary conditions. The aim was to use a method which could apply a series of repeatable loading conditions, and thus allow comparative performance of support to be determined. The dynamic tests were carried out using a drop weight system which could impose an input energy of up to 70 kJ/m^2, and a maximum impact velocity of about 8 m/s.

This testing system has been described by Ortlepp and Stacey (1996). It is significantly different from that used by Tannant et al (1996), and is considered to be more realistic.

3.2 *Materials tested*

The following types of containment support were tested:

- various types of welded wire mesh;
- various types of chain link (diamond) wire mesh;
- various types of wire mesh with wire rope lacing;
- fibre reinforced shotcrete;
- fibre reinforced shotcrete with wire rope lacing.

These types of support were chosen for testing since they are either commonly used in the mining industry or are considered to be attractive options for the future. Prior to the testing no suitable performance data were available for such support.

3.3 *Test results*

The results of the tests are summarised in the plot in Figure 3 in terms of centre deflection of the test panel against the total energy input.

The unreinforced shotcrete has the poorest performance as might be expected. Weld mesh also occupies the lower energy area of the plot, indicating that its capacity to absorb energy is lower. Wire strands broke in almost all of these tests, and in some cases welds broke as well. From the results of the tests on weldmesh, it is estimated that the practical total energy input limit for weld mesh is 10 kJ/m^2.

Diamond mesh performed better, and the practical limit for total energy input is estimated to be 15 kJ/m^2. There was a tendency for the diamond mesh to unravel once a strand has failed, and this allowed the artificial rock mass to spill through.

The performance of fibre reinforced shotcrete is approximately equivalent to that of diamond mesh. Shotcrete with Dramix steel fibres performed slightly better than that with monofilament polypropylene fibres, the former having an estimated energy

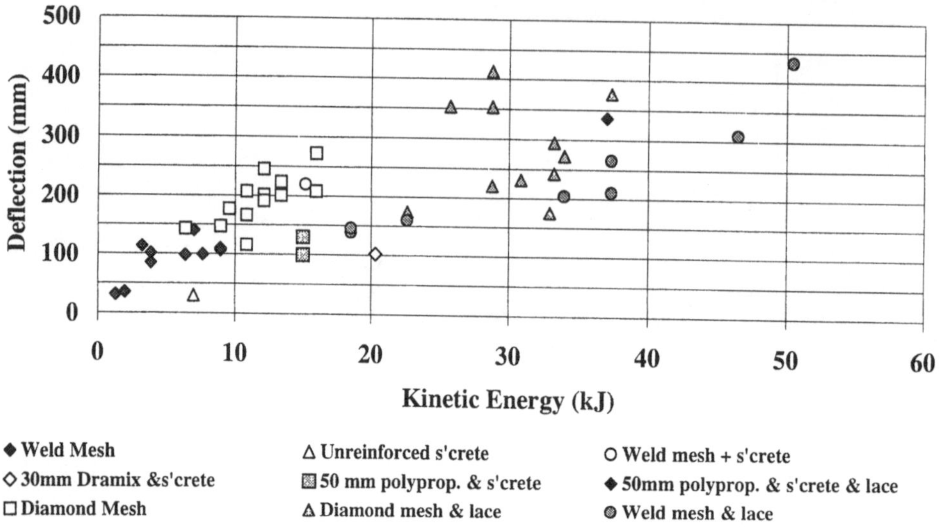

Figure 3. Results of dynamic tests on tunnel support

absorbing capability of 20 kJ/m^2.

With the addition of wire rope lacing to the mesh, the capability of the support system to absorb energy was considerably increased. The effect of wire rope lacing under rockburst conditions is illustrated in Figure 4. In the background the lacing has failed and the rock has been forced into the tunnel, closing it. In the adjacent foreground there appears to be no damage, and no movement of the sidewall.

Compared with the mesh only behaviour of the two types of mesh tested, the performance characteristics were reversed with the addition of lacing - the weld mesh with lacing performed better than the diamond (chain link) mesh with lacing. Failure of a strand of the diamond mesh generally allowed the "rock mass" to spill through. In contrast, the weldmesh contained the rockmass even though some of the strands failed.

The relative energy absorbing capability of the mesh and lacing is at least double that of the mesh only capability, and depends on the size of rope lacing used. The maximum energy input in the testing, which was satisfactorily contained was 51 kJ. The diamond mesh and lacing system failed catastrophically with total energy inputs of 34 and 48 kJ.

Two polypropylene fibre reinforced shotcrete panels were tested with the addition of rope lacing. They absorbed a substantial amount of energy without failing as a support system, and the capacities indicated in Figure 3 are regarded as being conservatively low. This type of support is considered to have significant potential for application in rockbursting conditions.

Several tests were carried out using weld mesh modified to allow yield, with wire rope lacing in which a yield capability was introduced. This support withstood, without failing, the maximum amount of energy of 71 kJ that the testing system could apply. This amount of energy is representative of a severe rockburst, and the result demonstrated the possibility of containing this damage satisfactorily.

4 LONG TERM PERFORMANCE OF SUPPORT

Many mine atmospheres are extremely corrosive and the mine water can be significantly acidic. The rusting of steel support elements can be seen in many operations, and can lead to complete failure of the support (Durrheim et al, 1998). Very rapid corrosion of steel fibres was observed at Oryx Mine (Venter and Gardner, 1998), to the extent that they no longer contributed to the integrity of the support system. Subsequent use of shotcrete reinforced with monofilament polypropylene fibres proved to be very successful.

There does not appear to be any satisfactory information on the long term performance of steel fibre in a cracked shotcrete or concrete environment. This is a problem area that must be solved before such fibres can be used with confidence in mines as reinforcement in heavily deforming shotcrete.

5 CONCLUSIONS

The test results that have been presented allow the following conclusions to be drawn regarding the performance of wire mesh and shotcrete containment support (with and without wire rope lacing) when subjected to large static and dynamic deformations:

- shotcrete reinforced with 40mm long Dramix fibres, or with monofilament polypropylene fibres

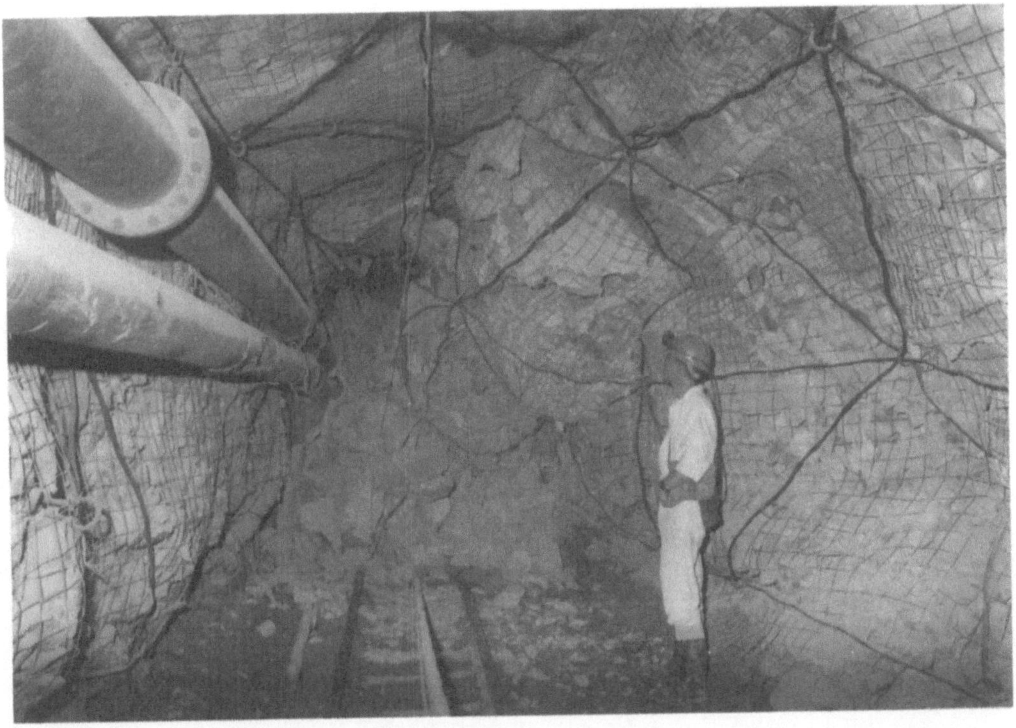

Figure 4. Wire mesh and wire rope lacing behaviour

40mm or longer, retains at least 50% of its peak strength after 150mm of deflection;

- Dramix reinforced shotcrete has a peak strength significantly greater than that of polypropylene reinforced shotcrete;
- diamond (chain link) mesh performs better than weldmesh in a mesh and rockbolt system. With the addition of wire rope lacing, the weldmesh and lacing combination has the superior performance;
- fibre reinforced shotcrete performs as well as wire mesh in the absorption of energy;
- the capability of absorbing the large amount of energy released in severe rockbursts has been demonstrated;
- the corrosion of steel support elements remains a problem that requires solution.

ACKNOWLEDGEMENTS

The drop weight testing was sponsored by SIMRAC, the South African Safety in Mines Research Advisory Committee. The static tests on fibre reinforced shotcrete panels were sponsored by the Shotcrete Working Group.

REFERENCES

Durrheim, R., Haile, A.T. and Ortlepp, W.D. 1998. Observation of shotcrete behaviour, School: Shotcrete and its application, *S. Afr. Inst. Min. Metall.*, Johannesburg.

Kirsten, H.A.D. 1993. Equivalence of mesh- and fibre-reinforced shotcrete at large deflections, Can. *Geotech. J.*, Vol. 30. pp. 418-440.

Kirsten, H.A.D. 1992. Comparative efficiency and ultimate strength of mesh- and fibre-reinforced shotcrete as determined from full-scale bending tests. *Jl S. Afr. Inst. Min. Metall.*, vol. 92, no. 11/12. pp. 303-323.

Kirsten, H.A.D. and Labrum, P.R. 1990. The equivalence of fibre and mesh reinforcement in the shotcrete used in tunnel-support systems. *Jl S. Afr. Inst. Min. Metall.*, Vol. 90, No. 7. pp. 153-171.

Kirsten, H.A.D., Ortlepp, W.D. and Stacey, T.R. 1997. Performance of fibre reinforced shotcrete subjected to large deformations, *Proc. 1st Southern African Rock Engineering Symposium, S. Afr. National Group of ISRM*, Johannesburg, pp 301-307.

Ortlepp, W.D. and Stacey, T.R. 1996. The performance of containment rock support such as wire mesh under simulated rockburst loading, *Proc. Geomechanics '96*, A.A. Balkema, pp 81-87.

Ortlepp, W.D. and Stacey, T.R. 1994. Rockburst mechanisms in tunnels and shafts, *Tunnelling and Underground Space Technology,* Vol. 9, No. 1, pp 59-65.

Tannant, D.D., Kaiser, P.K. and McCreath, D.R. 1996. Rockburst support research in Canada, *ISRM News Journal,* Vol 3, No 3, pp 15-18.

Venter, D.P. and Gardner, L.J. 1998. Application of shotcrete at Oryx Mine, School: *Shotcrete and its application, S. Afr. Inst. Min. Metall.,* Johannesburg.

6 Safety and training

Rock Support and Reinforcement Practice in Mining, Villaescusa, Windsor & Thompson (eds)
© 1999 Balkema, Rotterdam, ISBN 90 5809 045 0

Development and implementation of new ground support standards at Mount Isa Mines Limited

Y.Potvin
Australian Centre for Geomechanics, Perth, W.A., Australia

D.B.Tyler, G.MacSporran, J.Robinson, I.Thin, D.Beck & M.Hudyma
Mount Isa Mines Limited, Qld, Australia

ABSTRACT: Ground support standards for MIM underground hardrock mines were reviewed and updated in 1996. The review was conducted by a cross section of the workforce and focussed on ground support selection, ground support design, installation procedures and quality control management. A wide range of bolt types were investigated and trialed, with the cement grouted rebar chosen as the preferred primary rockbolt. A key component in the ground support standards review was the development and implementation of standard work instructions for installation of all ground support. Quality control is ensured through routine monitoring and periodic audits by rock mechanics engineers. The ground support standards are constantly monitored to look for means of improvement, and are regularly updated to reflect changes.

1 INTRODUCTION

Ground support is used in underground mines to protect personnel and equipment from ground falls. In many cases, it is a significant portion of the total mining costs. The reliability and efficiency of the support system is therefore critical to ensure that mine workers can operate in a safe environment and also to ensure that this large expenditure is not wasted.

There is a wealth of knowledge and experience in ground support practices acquired through 75 years of underground mining experience at Mount Isa Mines (MIM). However, ground condition changes, new products and new technologies are fundamentally changing the nature of ground support. In order to maintain and improve a high reliability from the ground support systems used in the four operating mines and the two developing projects, new ground support standards were developed and implemented.

The process of developing and implementing the new standards started in 1996 and is still on-going. A cross section of the workforce is involved, including mine workers, supervisors, management and rock mechanics engineers. The work focused on four main areas critical to the success of ground support systems:

- The selection of ground support devices;
- The design of the support patterns;
- The installation procedure;
- The control of quality.

This paper reports on some of the work performed and the progress made towards achieving best ground support practices.

2 GROUND SUPPORT SELECTION

The ideal long term ground support system for the requirements of Mount Isa Mines would have the following characteristics:

- Full encapsulation and full steel bar to ensure adequate corrosion resistance
- Simple and robust installation; also compatible with mechanised bolting rigs.
- Low cost
- Minimal (long term) strength of 10 tons per bolt
- Provide support immediately after installation
- One pass bolting installation

Cement grouted rebar is the preferred ground support system at MIM. It has been used extensively and its reliability and efficiency has been proven through decades of continuous used.

Traditionally, rebar have been installed at MIM from a mobile platform. More recently, six fully mechanised bolting rigs have been purchased in order to improve the safety and productivity of ground support installation.

Grouted rebar met all but the last two criteria listed above. However, with the deeper mining activity, the higher stresses and the larger headings, the importance for the ground support to provide immediate support action has become increasingly critical. As a result, alternative ground support systems are being investigated.

All support systems involving hollow bars such as Split Sets, Swellex and Hollow Groutable Bolts (HGB) have been tried during the last decade and rejected on the basis that their resistance to corrosion is inadequate. Split Sets are the only hollow bolt still used in specific short term applications, and they have been particularly efficient in the laminated hanging wall of the Lead Mine. Mechanical anchor bolts are also used occasionally in short term applications.

Although relatively expensive, it was thought that the CT Bolt would meet all the technical requirements. An extensive trial has shown that the complexity of the installation procedure and the difficulty in using thick grout resulted in an unacceptable proportion of poorly installed bolts. Specific installation problems in poor ground conditions were encountered. Some CT bolts were spoiled during installation while a significant proportion were not fully grouted.

Resin bolts have also been tried in the mid 80's at MIM. Difficulties in controlling the quality of the resin resulted in ground falls. Grice (1984) has identified the following problems with the resin rebar:

- Over spinning
- Under spinning
- Incomplete encapsulation due to the lost of resin in cracks
- Expired resin (poor stock management)
- Heat

However, resin rebar can potentially meet all the technical requirements for ground support at Mount Isa providing that the quality control problems can be resolved. The quality of resin has improved over the last 10 years and it is possible that mechanised bolting rigs can overcome some of the other quality control issues mentioned above. New trials with resin rebar, under strict quality control measures, are currently being conducted at the Lead Mine and at the George Fisher project. Results from these trials were not available at the time of writing this article.

A technique combining the traditional wedge bolt and the grouted rebar has been developed at the Kiruna Mine in Sweden. The rebar, which is inserted after filling the hole with a thick grout, is "hammered-in" using the drifter of the jumbo (or the bolting rig). The wedge opens the end of the bolt, providing immediate anchor support until the grout cures. The plate can be installed immediately after hammering the wedge and a productive one-pass bolting (grouting, inserting and plating the bolt in a single operation) is achieved.

The "Kiruna Bolt" was tested both on surface and underground. The wedge anchor design (shown in Figure 1) was found to be not reliable in typical Mount Isa Mines ground conditions. Pull testing has shown that this technique is highly sensitive to installation procedures and ground conditions.

Several modifications to the wedge shape were attempted in order to improve the reliability of the anchoring mechanism (see Figure 2). Although some improvement was achieved, the wedge mechanism still did not meet the high standard of reliability sought.

A modified expansion shell anchor has been developed to improve on the unreliability of the wedge. Past attempts to push expansion shells into cement grout filled holes have not be successful. Trimmed point anchor expansion shells have been machined in order facilitate pushing of the shell into a cement grout filled hole. Installation trials with the tappered shells have been successful. There has been no significant effect on the cement encapsulation of the rebar. A full scale field trial of this system is currently underway.

3 GROUND SUPPORT DESIGNS

There is no significant issue with ground support design at MIM. Standard designs for the different drive size and shapes have been developed. The philosophy is to implement a "blanket support" capable of supporting the largest possible wedge, as estimated by wedge analyses (safex or unwedge). Figure 3 shows a typical ground support pattern for the Deep Copper Mine.

Regular inspections by the superintendent, supervisors, geologists and rock mechanics engineers ensure that unusual ground conditions are identified as soon as possible. Modification to the blanket support can then be made to account for local geology variation, or anticipated stress changes.

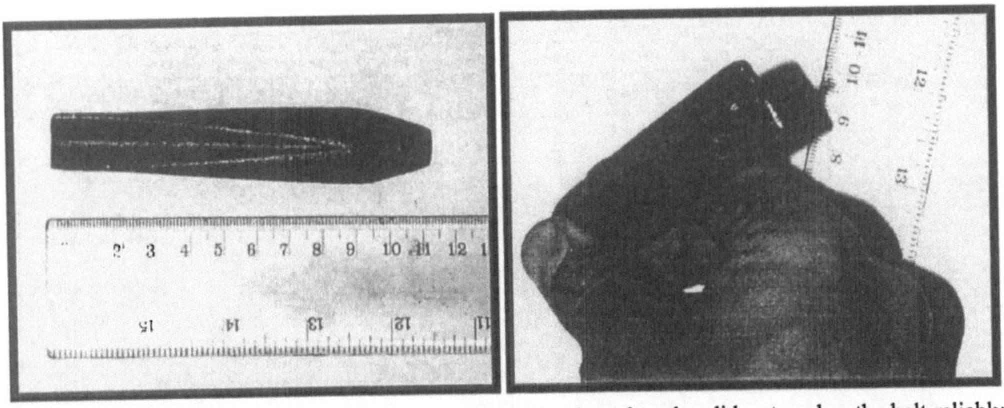

Figure 1. "Kiruna bolt" wedge trialed at MIM. The cross shaped wedge did not anchor the bolt reliably in varying ground conditions.

Intersection of drives results in larger back exposures and therefore, larger potential wedges. Standard drive support does not account for this. Cable bolt support is systematically installed according to the pattern shown in Figure 4 in most intersections. There is a provision for not installing the cable bolt support if the ground condition is good and approval is obtained from both the rock mechanics engineer and the underground manager.

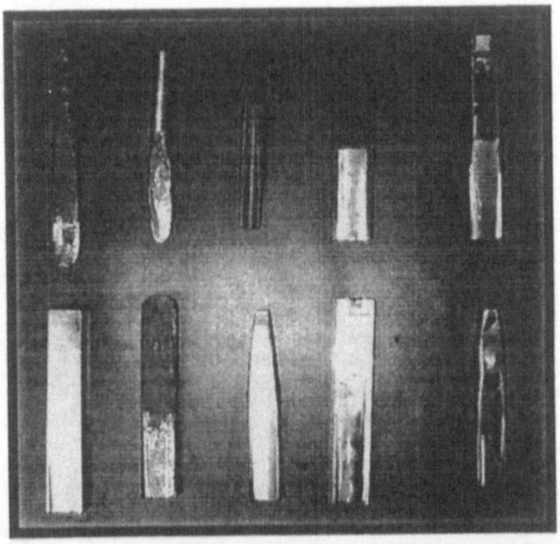

Figure 2. Various designs of wedges trialed to maximise the point anchor load of slot and wedge bolts. The angle, thickness, and length of the wedge are all key design parameters.

4 INSTALLATION PROCEDURE

The process of installing ground support may induce falls of ground. Often, the ground has recently been disturbed by blasting and is submitted to drilling vibration and water lubrication. As a result, the risk of ground falls must be carefully managed during manual installation of rock bolts. The key control measures are:

- training and standard work instruction (SWI)
- Timing of ground support and
- Scaling

All mining crews are trained on ground control and ground support issues. The training is repeated annually. To ensure that safe procedures are implemented at the working face, standard work instructions (SWI) have been developed on a single laminated sheet for each task and must be available at the job sites at all time. The SWI contains simple instructions and sketches describing how to carry out ground support tasks safely.

In theory, it is preferable to install the ground support as soon as possible after the excavation has been made. However, in many cases at Mount Isa, the ground conditions allow for several cuts to be exposed safely before ground support is installed. Nevertheless, the new ground support standards requires that no more than 2 cuts be exposed before the ground support is installed. This has been implemented to reduce the potential for judgement error and reduce the risk of rock falls.

When less favourable ground conditions are encountered, the bolting pattern is intensified and the excavation is supported after each cut.

Once the ground support process has started, it must be completed in its entirety before any

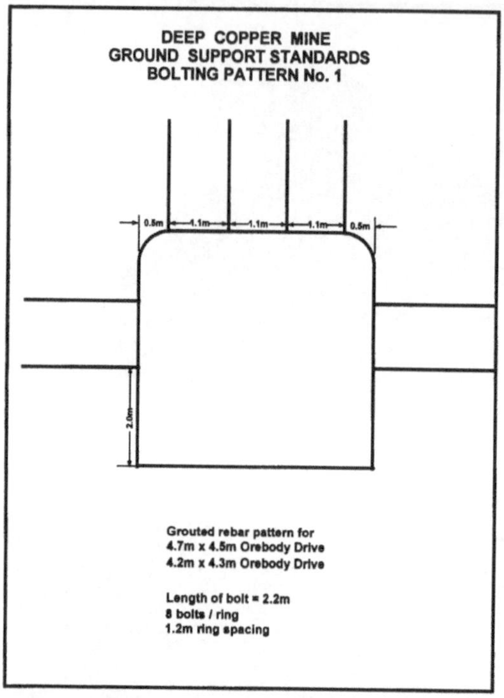

Figure 3. Typical primary ground support pattern used in the Deep Copper Mine.

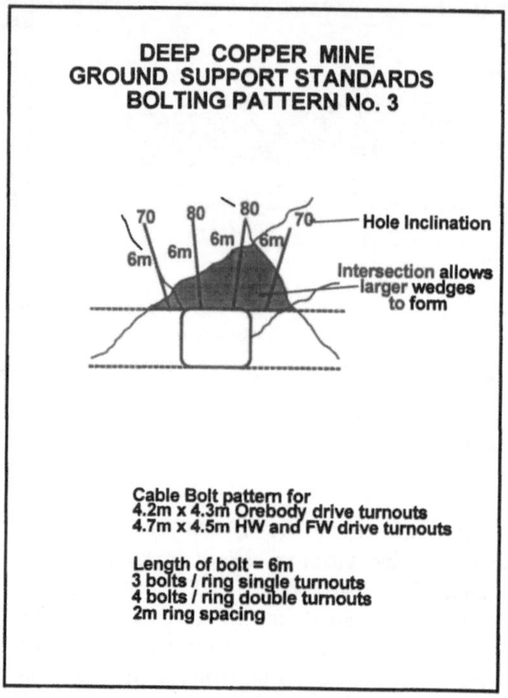

Figure 4. Typical cablebolt pattern for development intersections.

blasting activity takes place in the area. This is to minimise local ground disturbance induced by blast vibration during the installation process.

The back and walls are to be scaled manually and mechanically prior to boring holes for rockbolt installation.

5 QUALITY CONTROL

Auditing is an essential component of any quality control program. The quality of ground support is audited informally on a continuous basis by supervisors and rock mechanics engineers during their daily underground tour.

From these regular observations, a table of ground support and ground rehabilitation jobs with a priority assigned to each job is kept up-to-date by the rock mechanics engineers and distributed to the relevant production personnel on a regular basis.

Formal audits are also conducted every six months, as specified in the new ground support standards.

The large majority of ground support at MIM uses cement grout and since the load bearing capacity of grouted support is sensitive to the grout quality,

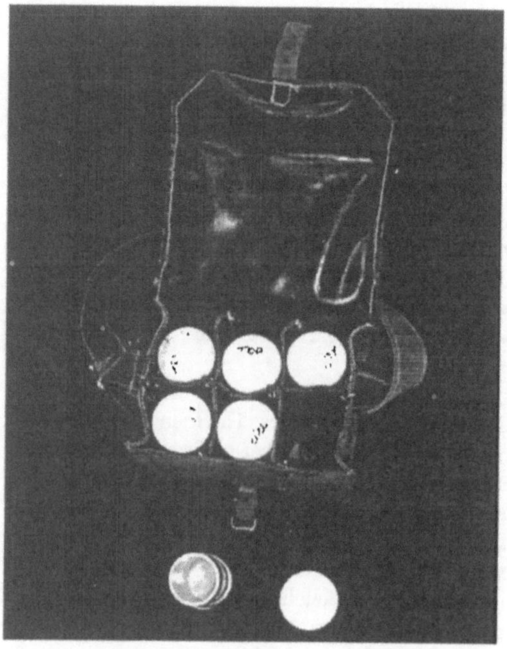

Figure 5. Grout sampling kit used to collect cement grout samples from ground support crews.

monitoring of the grout quality is essential.

Random cement grout samples are collected underground (without giving any warning to the crews) using a sampling kit developed in-house (see Figure 5). The cylindrical grout samples are brought to surface and are cured for seven days before being tested using a standard ASTM uniaxial compressive strength test. This testing procedure is used to identify weak grout and the ground support crews responsible for it. Their grouting practices can then be reviewed and corrected if necessary.

The introduction of a new ground support product may have disastrous implications if the control is inadequate. For example, it is often after several months of using a product, when a ground fall occurs, that the problems related to a new product are identified. The mine must then face the situation of having unknown ground support quality in extensive areas of the mine until a costly rehabilitation program is completed.

At MIM, prior to proceeding with a new ground support product, standard short embedment length (0.5m and 1m) pull tests are performed in the mining research laboratories. The mechanism of failure and the quality of grout (when relevant) is assessed visually by cutting the pipe containing the support element tested.

When a cement additive is tested, the sensitivity of the dosage is investigated by preparing samples using half and double the recommended quantity. If the strength of the cement is sensitive to gross mistake in the dosage (which is likely to happen in the mine), it is then rejected.

When the laboratory tests confirm the suitability of new products (new bolts, additive, resin, etc), underground trials are then undertaken. Attention is paid to the installation procedure and on identifying potential flaws. If the trials are successful, the new product is then introduced to a limited number of crews with frequent quality control checks. It is only after this last step is completed and the new product is proven reliable in a production environment that the new product is approved for mine wide use under the regular quality control program.

6 SUMMARY AND CONCLUSION

A safe and economic underground mine must have a reliable and efficient ground support system. This is achieved by selecting appropriate support elements for the conditions prevailing at the mine, by:

- designing adequate bolting patterns,
- following rigorous installation procedure
- and implementing a good quality control program.

The ideal ground support system at MIM is a one-pass mechanised installation of a fully encapsulated solid steel bar, which provides immediate support at a low cost.

The design bolting patterns are based both on experience and wedge analyses. They are well documented and systematically implemented.

The risk of ground fall must be controlled and managed during ground support installation. An emphasis on training and ensuring that standard work instructions are available at the job site help to reduce this risk.

The implementation of a quality control system is also imperative to ensure that the ground support system will work when required.

The approach of the new ground support standard at MIM is holistic. It does not focus on only one particular aspect of ground support, but to the whole process. It allows for the detection of human error and for continuous improvement.

This effort is on-going and the results are positive. The number of ground falls have decreased significantly since the work on developing and implementing new ground support standards has started.

REFERENCES

Grice, A. G. (1984) Ground support review at Mount Isa Mines Limited. Internal MIM report.

Rock Support and Reinforcement Practice in Mining, Villaescusa, Windsor & Thompson (eds)
© 1999 Balkema, Rotterdam, ISBN 90 5809 045 0

Practical considerations in the training of miners on rock bolting

J. Hadjigeorgiou, K. Hammond & F. Lemy
Department of Mining and Metallurgy, Université Laval, Ste-Foy, Que., Canada

ABSTRACT: This paper summarizes recent experiences in developing a multimedia training tool for miners. The developed interactive multimedia instructional system covers rockbolting for underground excavations.

1. INTRODUCTION

Training is an essential component to ensure the safe and economic functioning of mining operations. Consequently there has been a renewed focus in Quebec and elsewhere on improving the training of underground mining personnel. This has been the driving force behind the developed multimedia tool. This aims to contribute to improved knowledge and skill levels, thus enhancing on-the-job performance of all those involved in rock bolting.

Rock bolting is by far the more popular support system in Quebec underground mines. The interactive multimedia instructional system developed provides technical data on the choice and characteristics of rock bolt systems, as well as the methodology for correctly installing these different support systems. Throughout the multimedia, the user is provided with recommendations and practical advice on the correct application of the support systems. It can be followed as a complete course in a linear sequence or the user can reference a specific topic quickly through accelerated navigation options.

A multimedia format was chosen as it was deemed appropriate for this subject matter. The subject can best be conveyed through the use of many different media to facilitate learning and reinforce concepts in a dynamic and motivating context. Multimedia offers many advantages, as it:

- provides for visual and auditory reinforcement of information, in addition to textual content
- can accommodate different learning styles through a choice of options available to the user
- offers guidance, while allowing learners to progress at their individual pace
- can be integrated with traditional teaching methods, to enrich the learning experience
- provides uniform training
- offers on-site training which is timely and easily accessible when needed
- can be customized to accommodate specific needs
- provides an interactive resource that can be referenced as frequently as required
- delivers training using the latest technology and advances in pedagogical methods
- can be easily updated.

The developed interactive multimedia instructional system can be employed at different levels. The use of graphics and videos, along with a simple and straightforward navigation system, make it accessible to people without any prior computer experience. The user is free to choose his own path through the content, depending on his learning style and the knowledge he wishes to acquire. For those who prefer a more guided approach, a sequential order is suggested, beginning with an overview and then followed by specific technical and practical information. For those with specific information needs, a rapid navigation screen provides quick access to any given topic and tracks completed sections. Both linear and rapid navigation options are available at all times and are intended to help the student use the system effectively, according to his level of expertise and particular needs.

The developed multimedia is intended as a complement to classroom training. The goal is not necessarily to completely replace other training forums, but rather to render the whole training process more efficient by using computer technology to present those concepts which can best be explained through a multimedia approach. A versatile instructional system, the multimedia can be followed as a course or used as a pedagogical resource during problem-solving exercises.

One suggested way to use the multimedia for the training of miners, is for the training coordinator to animate the training session and review the different modules with the trainees. Afterwards, each miner can follow the course or elements of the course at will and at his or her own pace.

The developed multimedia involved the cooperation and skills of several people, including technical experts, a pedagogical designer and a programmer. A needs analysis was conducted during the early stages of the project. The technical data and information were collected, analyzed and verified in the field. The structure of the final product was designed based on pedagogical considerations, with each screen individually grafted. Programming involved the smooth integration of the different modules. Finally, to arrive at the finished product, it is the recommendations of the intended audience that provided many of the guidelines.

2. THE SYSTEM

From a technical point of view, the system covers the following modules:

- a general overview of rock bolting
- a section containing pertinent information on the different types of rockbolts
- a module including useful data on rock supports used in combination with rockbolts

The general overview describes the mechanics of rockbolting, its usefulness, applications and functioning principles. Furthermore, a subheading on scaling has been integrated into this module since it is an essential operation to ensure safe operating conditions and the efficiency of rock reinforcement and support.

The bulk of the technical content of the multimedia training tool is found in the section concerning the different kinds of rockbolts used. The presented rockbolts, as well as their applicability, were selected based on our field experience as well as by a comprehensive survey undertaken in fifteen underground hard rock mines in Quebec and New-Brunswick. The rockbolts described are: mechanically anchored, resin grouted rebars, Split Set, Swellex and tubular bolts. For each of them, a complete description, technical data and a detailed installation procedure are given. The multimedia provides an extensive section on installation procedures. Each step of the installation is described with narration, text, photos, images, animations or videos. These steps are accompanied by practical recommendations and safety instructions. Furthermore, different procedures are proposed according to the installation equipment used, such as jacklegs, stopers, drifters and fully mechanized bolters. Finally, special attention is paid to quality control of the rockbolts, using simple visual signs or easy tests, in order to verify whether the reinforcement is properly installed or remains efficient afterwards.

The last section of the software describes complementary rock supports used with rockbolting in the Quebec mining industry. These include welded and chainlink wire meshes, shotcrete and straps.

3. SUPPORTING LEARNING THROUGH THE USE OF PEDOGOGICAL STRATEGIES

It goes without saying that one must look beyond the technology itself in order to create effective learning systems with multimedia technology. Pedagogical strategies, based on proven results from educational research, are the key to successful multimedia instructional systems.

A needs assessment is always the first step in any training project, Rossett (1987). The identification of concrete learning objectives, based on desired on-the-job performance outcomes, helps to determine the content and the most effective instructional strategies. This planning stage is crucial in developing a product tailored to the specific characteristics of the target audience and ensures that pertinent knowledge and skills are transmitted to the learner. Usually this phase is accomplished through a combination of interviews with site personnel, on-the-job observations of specific tasks being performed, and referencing existing documentation.

Various factors affect the learner and his or her capacity to learn. The motivation of the learner and their interest towards the subject matter remains a key element in the learning process, Brien (1994). In the multimedia developed, several strategies were used to elicit learner interest, such as:

- clear objectives stated at the beginning of the course, with links established between these objectives and their pertinence to the daily tasks at a mine.
- a logical and easy-to-use structure, which minimizes the efforts required by the learner and typical obstacles that could result from the technological aspect of the learning system (such as the level of comfort of the learner, navigation problems which could hinder learner progress, etc.).
- the use of a variety of stimuli (symbols, images, animation, sound) representative of the professional interests of the target audience.
- different navigation options, adaptable to individual learner needs.

How the information is presented to the learner is another important factor in effectively communicating the message, Marton (1994). In the multimedia developed, certain strategies were employed in an effort to render the communication between man and machine interesting and effective:

- a uniform structure, where each rockbolt type follows a similar format, allowing the learner to become quickly familiarized and comfortable with the product. At the same time, a variety of stimulating and dynamic elements, within the standard format, stimulate interest and promote learner perception.
- the metaphor of a mine was used to create a user-friendly environment. This was done through the inclusion of an abundance of photos of in-situ rock support applications, many screen backgrounds were created from photos of actual rock surfaces in Quebec mines, and the rapid navigation screen was designed in the format of a network of underground tunnels typical of a mine.
- different media were used to stimulate all the senses and help the learner assimilate concepts. The use of icons and symbols appropriate to the subject matter help to situate the learner.
- in order to present information effectively and facilitate assimilation of concepts, a modular approach was adopted, Horn (1989). The content was analyzed and organized into a modular format. More complex concepts were regrouped under sub-headings, allowing for smaller, more digestible quantities of information to be presented at one time.

4. EVALUATING THE MULTIMEDIA

Two formative evaluations were conducted during the development phase. The method used to carry out these evaluations consisted of observations of users during their interaction with the system, followed by semi-structured interviews. The main goals of the evaluations were to test the appropriateness of the system for the target audience and to verify the proposed navigation system, help facility, menus, as well as obtain general feedback and suggestions. After each evaluation, corrections were made to the multimedia. Where appropriate, suggestions of participants were integrated.

Several items were highlighted as a result of the formative evaluations. These included the visual aspect of the multimedia, the navigation system and the help facility. The quality of photos and animations were greatly appreciated and considered a strong point of the system by those who participated in the evaluation. The rapid navigation option was by far the most popular method of going through the system, particularly for those with greater expertise. It was noted that although all users read the initial sign-on instructions, none of them could remember them when following the course. This therefore prompted us to remove most instructions from the beginning and display them contextually, at the needed moment.

A more comprehensive evaluation was carried out at the end of the project, to evaluate the multimedia on both cognitive and affective levels. That is, to evaluate the actual learning achieved, along with satisfaction levels of learners. To determine learning, a pre-test was administered. Students were then left alone to go through the multimedia at their own pace. An identical post-test was administered afterwards, in order to compare the results of the two and determine how much learning actually took place. This was followed up by an interview with each participant to determine satisfaction levels and solicit further suggestions.

The results of this comprehensive evaluation demonstrated that most of the training objectives were achieved. On a satisfaction level, the system was well received. The multimedia was cited as being very user-friendly, interactive, with ample visually appeal.

5. ADAPTING CONTENT TO MULTIMEDIA

One must respect the limitations of the technology, while reaping the benefits of what it has to offer. A limited amount of text can be presented at one time on a computer screen. This becomes a constraint, unless the information is carefully broken down into smaller segments. One must also decide on how related topics will be accessed (i.e.: through buttons, narration, visuals etc.). Multimedia implies the use of several media to convey the message. Information that was originally in text format can often be translated into other media such as narration or visuals, thus creating a more dynamic and multi-sensorial learning experience.

One of the greatest advantages of multimedia is its dynamic and interactive nature, with its potential to animate the subject and provide feedback to the learner. As much as possible, one should capitalize on these advantages, stimulating the senses with a combination of sound, motion, and visuals. The learner should be an active participant in the learning experience, able to control his own progress, and receiving guidance only when needed.

Finally, it is important to test the multimedia using sample populations of typical users, during and at the end of development. This allows for potential problems to be identified and corrected early on, and before the final product is distributed. Often, some of the best ideas come from potential users through comments received during these evaluations.

6. CONCLUSIONS

The evaluations conducted received very favorable comments and demonstrate the potential of multimedia for training on technical topics in mining applications, such as rockbolting.

The integration of several different media into one pedagogical resource permit easier and more efficient access to information. Flexible and rapid access to information seems to best suit the needs of most users.

This multimedia tool does not replace the trainer or instructor, but it provides him with a useful teaching tool. Consequently more time will be available to guide and advise the workers and to verify the efficiency of training in the field.

7. ACKNOWLEDGEMENTS

The authors would like to acknowledge the financial support of the Institut de recherche en santé et en sécurité du travail du Québec. Several people have contributed in developing the technical elements of the training tool as well as providing constructive comments during the evaluation trials. Frederic Flament did the programming.

REFERENCES

Brien R. 1994. Science cognitive et formation. Presses de l'Université du Québec, Sainte-Foy.

Horn R.E. 1989. Mapping Hypertext. Lexington, Massachussetts. The Lexington Institute.

Marton P. 1994. La conception pédagogique de systemes d'apprentissage multimedia. III Rencontre International de communication éducative, Université autonome de Mexico, pp. 5-9.

Rosset A. 1987. Training Needs Assessment. Educational Technology Publications.

Rock Support and Reinforcement Practice in Mining, Villaescusa, Windsor & Thompson (eds)
© 1999 Balkema, Rotterdam, ISBN 90 5809 045 0

Geotechnical mining regulations in Western Australia

A.M.Lang
Department of Minerals and Energy, Perth, W.A., Australia

ABSTRACT: The Mines Safety and Inspection Act 1994 of Western Australia includes in its associated Mines Safety and Inspection Regulations 1995 geotechnical regulations for underground and open pit mining. The intent of the geotechnical regulations is briefly discussed. Recent safety performance of the Western Australian underground mining industry is reviewed and compared with that in Ontario and South Africa. The general approach taken with underground metalliferous mining geotechnical regulations in Ontario, South Africa and Western Australia is reviewed. The audit methodology developed to determine the level of compliance with the underground geotechnical regulation in Western Australia is described. The level of compliance achieved by industry is discussed.

1 INTRODUCTION

The Department of Minerals and Energy, Western Australia (DME) is the regulatory authority for the minerals and energy industry of Western Australia (WA). The minerals industry of WA is regulated by several pieces of legislation including the Mines Safety and Inspection Act 1994 (Act) and Mines Safety and Inspection Regulations 1995 (Regulations). The current Act and Regulations came into operation on 9 December 1995. Drafting of the Act commenced in September 1993 and involved extensive consultation with representatives from the mining industry, unions and government (Torlach 1996). The Act received bipartisan support in parliament and was enacted in November 1994. Similarly, drafting of the new Regulations commenced in December 1994 and benefited from extensive tripartite input during the following twelve months. The Act and Regulations introduced a change in style of mining health and safety legislation in WA.

The previous regulatory framework was highly prescriptive, relatively "black and white" and told the employer precisely what was required to comply. The rapid introduction of new mining technology required constant extensions or elaboration of the regulations that had been framed when more labour intensive underground mining methods were the norm. The regulations left much to the decision, discretion or direction of the inspector. Expressions such as "... as approved by the inspector ..." were common. Increasing involvement of the inspector in mining operations gradually led to the view that the responsibility for safety in mining lay with the inspector rather than the mine management.

The Act and Regulations proclaimed in December 1995 embodied what is generally known as the "Robens-style" of legislation, after Lord Robens, the Chairman of a Committee of Inquiry into safety in the workplace appointed by the British Government. The historical background to mining legislation in Western Australia and the development of a new approach to health and safety legislation has its origin in developments in the United Kingdom (Knee 1995). In essence the "Robens-style" legislation "clearly spelled out the general principles of responsibility for safety and health and set up a framework for both statutory regulation prescribing general requirements, and non-statutory standards and codes of practice to provide guidance in some detail on particular aspects of safe and healthy operating practice".

2 MINES SAFETY AND INSPECTION ACT

It is important to recognise that the Act and the Regulations must both be consulted to fully appreciate what is required to comply. The Regulations should not be read in isolation from the Act. The Act contains very important information about the duties of employers, employees, self-employed persons and manufacturers in Sections 9,

10, 12 and 14 respectively. These sections of the Act embody in enacted law the common law duty of care that is expected of employers, employees, self employed persons and manufacturers, suppliers and erectors of plant. The importance of fully understanding the duty of care that is expected to be exercised by all parties involved cannot be over-emphasised.

2.1 *Duty of care*

The common law duty of care has been brought into the legislation. In brief, the duty of care required to be exercised by employers for their employees embodies the following fundamental issues:

1. The identification of hazards.
2. The management of risks.
3. The consideration of what is "practicable".

A hazard in relation a person, as explained in the Act, means anything that may result in injury to a person or harm to the health of a person. Risk in relation to any injury or harm, as explained in the Act, means the probability of that injury or harm occurring. The concept of what is "practicable" means reasonably practicable which may vary from mine to mine and also within each mining operation. The variety of access methods, equipment, ground conditions, mining methods and mining history, to mention a few, should provide considerable scope for innovative solutions to the challenge of practicability.

It is recommended that a formal hazard identification process (AS/NZS 4360:1995, DMR 1997) should be implemented to identify and rank the hazards to which employees may be exposed. The prevention of employee exposure to significant hazards in the workplace requires that the appropriate controls be selected, put in place and maintained to ensure their effectiveness with time.

The employer's duty of care may be summarised (Torlach 1996) as provision of:

1. a safe place of work;
2. a safe system of work;
3. competent and properly trained staff and supervisors; and
4. fit-for-purpose plant and equipment.

The duty of care obligations for employees require that they take reasonable care to ensure their own health and safety at work and that of others, co-operate with their employer and comply with instructions to maintain a healthy and safe work environment. Active employee involvement in health and safety matters is encouraged by the election of "safety and health representatives".

As noted above, these obligations are all qualified by the term "reasonably practicable" which takes into account:

1. the severity of any potential injury or harm and the degree of risk;
2. the state of knowledge about the hazard, the risks and the means of mitigation;
3. the availability, suitability and cost of the means of mitigation.

2.2 *Due diligence*

The expression "due diligence" is a generic term that is related to duty of care. Due diligence is generally used in the corporate world in relation to commercial and business matters. Its use in the mining industry is generally restricted to mineral resource and ore reserve issues. It is a useful term that may assist a person to judge how well the hazards present in a particular situation have been identified and treated. One measure of due diligence being whether one would be comfortable explaining the approach taken to an employee's family, the Inspector of Mines or possibly a Court following a serious accident. The formalisation, at corporate level, of company "due diligence standards" for risk management would provide mine site management with consistent guidance as to the standard of care expected of them by corporate management.

3 WA GEOTECHNICAL REGULATIONS

The Regulations contain two regulations that deal with geotechnical considerations in mines. Regulation 10.28 deals with underground mines, while Regulation 13.8 deals with open pit mines. This paper will only deal with Regulation 10.28 and its application to underground mines. The general penalty upon conviction for an offence, which is imposed under Regulation 17.1, is also stated for reference.

3.1 *Underground geotechnical regulation*

Geotechnical considerations

10.28. (1) The principal employer at, and the manager of, an underground mine must ensure that geotechnical aspects are adequately considered in relation to the design, operation and abandonment of the mine.

Penalty: See regulation 17.1.

(2) The principal employer at, and the manager of, an underground mine must ensure that the following things are done in relation to workplaces, travelways and installations underground in the mine-

(a) due consideration is given to local geological structure and its influence on rock stability;

(b) rock damage at the excavation perimeter due to blasting is minimized by careful drilling and charging;

(c) due consideration is given to the size and geometry of openings;
(d) appropriate equipment and procedures are used for scaling;
(e) appropriate measures are taken to ensure the proper design, installation and quality control of rock support and reinforcement; and
(f) the installation of ground support is timed to take into account rock conditions.
Penalty: See regulation 17.1.
(3) The principal employer at, and the manager of, an underground mine must ensure that the following things are done in relation to all development openings and stoping systems underground in the mine-
(a) geotechnical data (including monitoring of openings when appropriate) is systematically collected, analysed and interpreted;
(b) appropriate stope and pillar dimensions are determined;
(c) rationale for sequencing stope extraction and filling (if appropriate) is determined;
(d) there is adequate design, control and monitoring of production blasts; and
(e) rock support and reinforcement are adequately designed and installed.
Penalty: See regulation 17.1.

3.2 *Penalty regulation*

General penalty
17.1. The penalty for contravention of a provision of these regulations that refers to this regulation is-
(a) in the case of an individual, $5 000; and
(b) in the case of a corporation, $25 000.

4 DISCUSSION OF REGULATION 10.28

Regulation 10.28 is essentially a performance based standard that states the result to be achieved rather than a detailed prescriptive methodology. This is consistent with the self-regulatory style of the Regulations.

The Regulations support the duty of care in the Act by outlining in more detail how the general duty can be complied with in this aspect of the operations. The duties are imposed on the principal employer (not on contractors) and the mine manager as these are the people most able to influence the matter. The three sub-regulations are briefly discussed below.

4.1 *Regulation 10.28(1)*

This sub-regulation is a general statement that, in effect, requires the "geotechnical aspects" at a mine to be identified and addressed during the design, operation and abandonment of the mine. It was intended to be general and encompass the full range of geotechnical aspects that may be encountered in all underground mines. Geotechnical aspects includes the various geotechnical issues or hazards that are unique to an individual mine, eg geological structure, high or low stress, high water pressures, soft rock conditions, hard rock conditions and seismic rock conditions (DME 1997a). When read in conjunction with the employers' duty of care requirements under Section 9 in the Act, it will be readily apparent that the geotechnical hazards should be identified, ranked and treated.

It will be recognised that the geotechnical aspects may not initially pose a hazard to the underground workforce, however, continued mining, possibly without a full appreciation of a particular aspect (eg seismicity), may eventually pose a more serious risk to the workforce as mining becomes more extensive and workforce exposure increases. Modern risk management practice (AS/NZS 4360:1995) requires that such geotechnical aspects are adequately considered (ie identified, ranked and treated) in a manner appropriate to the hazard posed. The principal employer and manager have, in effect, a duty to ensure that the geotechnical aspects are addressed in a manner appropriate to the risk profile of the mine. The management and employees of a particular mine are the people that should be in the best position to identify the particular hazards that are unique to that mine.

Design is considered to mean the appropriate engineering design of all the sub-systems in the mine including drilling, blasting, loading, haulage, services reticulation, ventilation, pumping, fill systems, rock support and reinforcement and stope and pillar design, etc. Planning is considered to be an integral part of the "operation" of a mine. Mine planning deals with the correct selection and coordinated operation of all the sub-systems in the mine to produce the required outputs, both safety and financial. Planning necessarily includes consideration of mine production capacity, workforce numbers, equipment selection, scheduling and budgeting. Thus, design and planning are not the same, but complementary parts of the engineering method.

4.2 *Regulation 10.28(2)*

This sub-regulation refers to the geotechnical issues that may be present in relation to workplaces, travelways and installations. These areas are where the workforce is exposed to the potential hazards that may exist. An appropriately trained workforce can, to a certain extent, exercise some influence or control over the prevailing ground conditions. However, the ultimate responsibility to ensure that these issues are appropriately addressed lies with the principal employer and mine management.

Geological structure (planes of weakness in the rock mass) and their influence on rock stability is a basic issue that should be recognised and acted upon. Data from the DME AXTAT accident and incident database confirm that the largest single cause of lost time injury to the underground workforce is rock falls. Experience indicates that geological structure and "less than adequate" mining practice can contribute significantly to the prevalence of rock fall related injuries. The use of inappropriate drilling and blasting practice has the potential to significantly degrade the ground conditions in what, otherwise, would have been a strong competent rock mass. The selection of the size and geometry of the mine openings, without properly addressing the geological structure, rock stress field and other voids in the vicinity has the potential to produce a hazardous work environment. Incremental increases in the excavation span of entry mining methods, without a review of rock support and reinforcement requirements, is unacceptable.

It is noted that the rock stress field, amongst other factors, has not been explicitly included in the regulation. Rock stress and other factors (eg ground conditions, rock strength) are generally considered to be essential inputs in many methods of geotechnical analysis. It is simply not possible to ignore the rock stress field when using some of the design techniques that are available (eg numerical stress analysis).

The last three parts of the sub-regulation deal with the some of the operating procedures that are necessary to minimise the hazards from rock failure. Hand scaling, with an appropriate bar, is a basic mining skill, the simplicity of which, may make its routine application seem out of place in a mechanised mine. Routine inspection of the backs, sidewalls and face of an active workplace, with the aid of a scaling bar, is a simple and very effective way of removing high potential hazards (ie loose rocks).

Rock failure may be controlled by static loading considerations, as in a joint-bound block and its fall under the action of gravity. Rock failure may also be controlled by a more dynamic loading process due to seismic events in close proximity to the mine. The design of the rock support and reinforcement presumes a knowledge of the loading conditions to which it will be subjected. The design methods, installation procedures and quality control measures are not detailed in the regulation as these will obviously vary depending on the ground conditions, loading conditions and mining geometry. The important point is that the principal employer and mine manager must be able to demonstrate that the selection and use of a particular rock support and/or reinforcement system is based on sound and defensible geotechnical engineering practice.

4.3 *Regulation 10.28(3)*

This sub-regulation refers to the geotechnical issues that may be present in all development and stoping systems. These geotechnical issues, and other issues, are essential inputs to the mine planning and design process. In this situation the workforce can have very little direct involvement. The principal employer and mine management, obviously, have the responsibility to ensure that the appropriate geotechnical factors are adequately incorporated into the mine planning and design process.

Geotechnical data should be collected during the mining process to monitor the performance of excavations and to confirm any assumptions that were made during the design phase. The actual type of geotechnical data to be collected is not specified. It should obviously be consistent with the identified geotechnical hazards and needs of the mine. It may range from the systematic recording of visual observations of ground conditions and support behaviour, made while regularly walking around active mining areas (and inactive areas on a less frequent basis), through to the use of a fully integrated mine-wide monitoring system of seismic transducers and other geotechnical instrumentation.

The stope and pillar dimensions in use at the mine should have been determined by an appropriate rational geotechnical engineering design process. The incorporation of local mining experience into this process is obviously important. The ability to successively refine the design process, as a result of the "learning opportunities" that can occur, is an important aspect of optimising mining efficiency. A range of empirical and numerical stress analysis design approaches are available (DME 1997a).

The sequence of stope extraction should be such that the remaining ore reserves are in the "best condition" for future mining. The best condition referred to may include: acceptable rock stress field, acceptable and consistent levels of seismic activity, access requirements, fragmentation, dilution, ventilation, etc. If the mining method requires the use of fill, this will necessitate the consideration of fill material: strength properties, supply, distribution, placement, quality control and stability. Stope sequencing is strongly influenced by the in situ and induced rock stress regime, ground conditions, stope and pillar dimensions, seismicity, fill exposure capabilities, access and ventilation requirements. Appropriate stope sequencing can be aided by the use of appropriate numerical stress analysis methods.

The rock support and reinforcement requirements for large permanent excavations and stopes can be addressed using similar techniques to those discussed in the previous sub-regulation. The support and reinforcement requirements for non-entry stoping methods will focus primarily on the prevention of large-scale stope collapse, rather than on the higher level of security required for entry stoping systems.

4.4 *Summary*

It will be recognised that the regulation lists a number of the individual outputs or end points of an overall geotechnical engineering process. The inputs to that process may change with time as more knowledge becomes available. However, the outcomes of the process are essentially the same. The primary outcome of the process being a working environment in which the hazards are identified and treated in such a manner that the health and safety workforce is maintained at a socially acceptable level.

Regulation 10.28 prescribes a series of actions or end points that, if taken together, should form a coherent geotechnical engineering process by which mine management may identify and proactively manage the unique set of geotechnical hazards present in a mine. A paper (Lang 1995) in the Sixth AusIMM Underground Operators' Conference volume, discussed many of the above issues.

5 GUIDELINES

To encourage discussion and debate of these important geotechnical matters, the DME has produced the following:

1. Guidelines: geotechnical considerations in underground mines (DME 1997a).
2. Guidelines: underground barring down and scaling (DME 1997b).

These guidelines together with a number of others are available via the Internet at the DME web page in the EXIS (external information system) database. The DME's EXIS-on-the-web can be accessed via http://exisweb.dme.wa.gov.au/exis/Exisonweb.nsf.

These guidelines have been approved by the Mines Occupational Safety and Health Advisory Board (MOSHAB), a tripartite board established under the Act to advise the Minister for Mines on mining industry occupational safety and health matters.

The above guidelines provide a summary of the state of the art of geotechnical engineering, in June 1997, as it applied to underground mining. A number of references are provided to publications available in the public domain that may be of interest to those wishing to pursue various aspects in more detail. The guidelines are not intended to be a detailed exposition of all the geotechnical issues that could conceivably exist. They do not present a highly prescriptive methodology, which, if followed, would ensure compliance with the regulation. This would be contrary to the underlying concept of self-regulation present in much of the legislation (Torlach 1996).

The general aim of the guidelines is to encourage the application of current geotechnical engineering knowledge in the pursuit of safe, cost effective mining. Where it can be demonstrated that the existing level of geotechnical knowledge is inadequate, impetus is generated for further research and development.

6 SAFETY PERFORMANCE

6.1 *Fatal accidents*

A comparison of the fatal injury rate per 1000 underground metalliferous mine employees for Western Australia, South African gold mines and Ontario is given in Figure 1. The data for Figure 1 were obtained from the Ontario Natural Resources Safety Association (ONRSA), the Chamber of Mines of South Africa (CM) Internet web site and the DME AXTAT accident and incident database. The data in Figure 1 are based on ONRSA's best estimate of the size of Ontario underground workforce (Lang 1997).

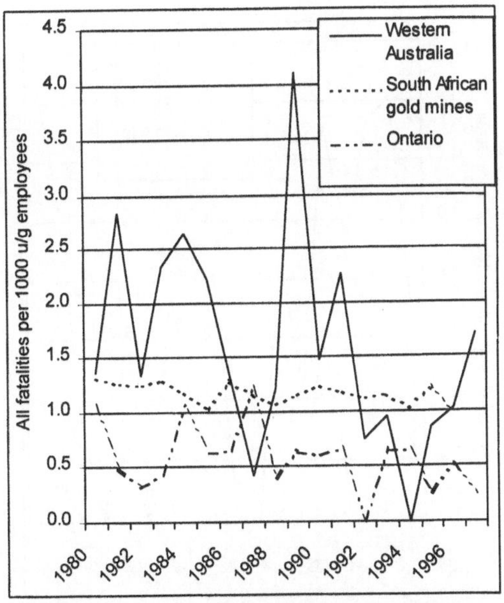

Figure 1. Graph of all fatal injuries per 1000 underground employees in metalliferous mines for Western Australia, South Africa and Ontario (data provided by DME, CM and ONRSA).

Figure 1 demonstrates that the WA fatal injury rate has shown considerable fluctuation during the period from 1980 to 1997. The data for WA in 1989 are strongly influenced by a single event that claimed the lives of six people. The largest single cause of underground fatalities in WA during the period from 1980 to 1997 was rock falls which accounted for 45% of all fatalities. The encouraging downward trend in WA for the first half of the 1990's has not been sustained during recent years. The Ontario underground fatal injury rate has generally been the lowest of the three shown. It has displayed considerably less variation than that for WA during the same period. The underground fatal injury rate for the South African gold mining industry has been relatively steady in the range of 1.3 to 1.0 from 1980 to 1996.

It is noted that the Ontario underground fatal injury rate has generally been lower than Western Australia, with the exception of two years. Figure 1 demonstrates that, while there has been an improvement in the Western Australian underground fatal injury rate, it has not achieved the consistency displayed by Ontario or South Africa. This is partly explained by the smaller size of the Western Australian underground workforce.

A comparison of the rock fall fatality rate per 1000 underground metalliferous mine employees for Western Australia and Ontario is given in Figure 2. It is noted that a major inquiry into ground control and emergency preparedness in Ontario mines occurred following a number of fatalities and rockburst incidents at a number of mines in 1984 (Stevenson 1986). The data for Ontario has been fairly consistent at or below 0.6 fatalities per 1000 underground employees and has shown a commendable downward trend since 1984. The data for Western Australia has shown considerably more variation, particularly during the past few years.

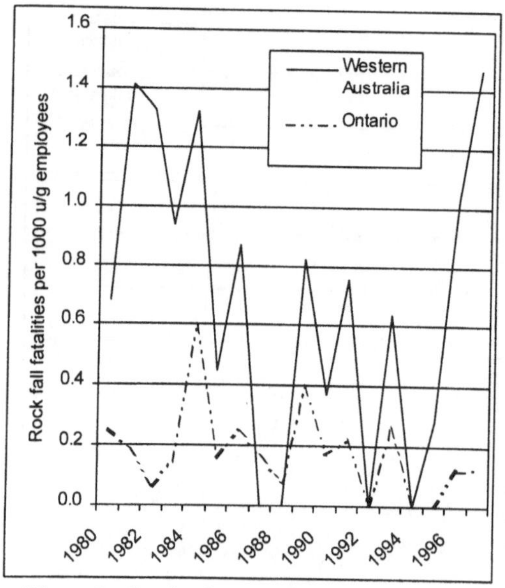

Figure 2. Graph of rock fall fatalities per 1000 underground employees in metalliferous mines in Western Australia and Ontario (data provided by DME and ONRSA).

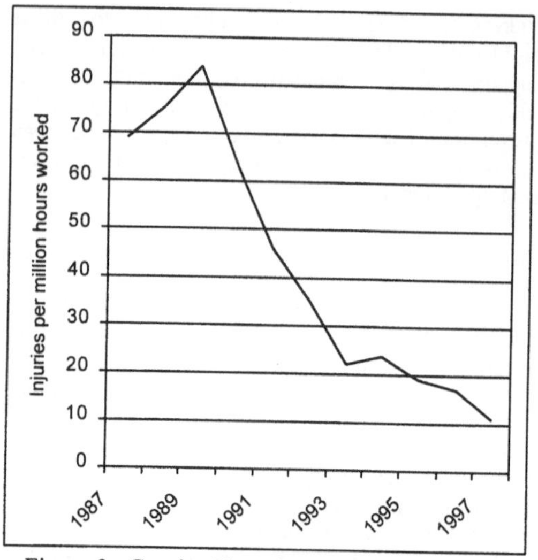

Figure 3. Graph of lost time injuries per million hours worked in Western Australian underground mines (AXTAT data).

6.2 *Lost time injuries*

A lost time injury is a work injury which results in inability to work for at least one full day or shift at any time after the day or shift on which the injury occurred. Figure 3 is a graph of the lost time injuries per million hours worked in WA underground mines. This graph shows a significant reduction in number of lost time injuries during the period from 1989 to 1997.

Figure 4 shows the distribution of the most common accident types resulting in lost time injuries from 1987 to 1997 in Western Australian underground mines. The data in Figure 4 demonstrate that rock falls have been the most common single accident type resulting in lost time injury during the period from 1987 to 1997. Analysis of AXTAT data for individual years during this period also demonstrates that rock falls have been the most common type of accident resulting in lost time injuries for each individual year.

From the data in Figure 4, it may be concluded that rock falls are the one of the most significant hazards in underground mining.

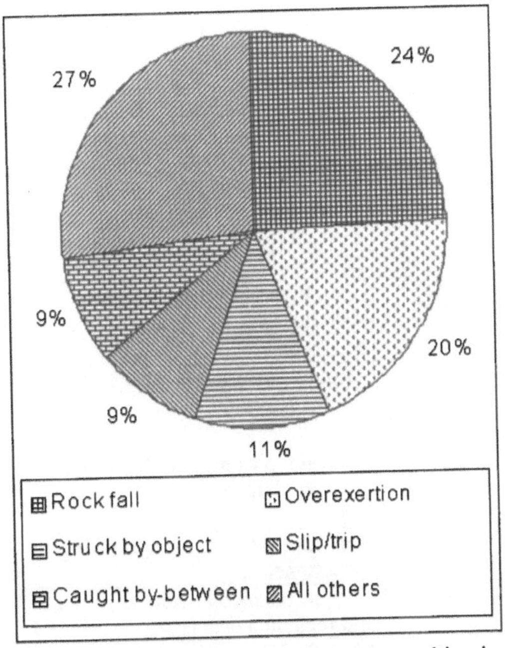

Figure 4. Distribution of accident type resulting in lost time injuries from 1987 to 1997 in Western Australian underground mines (AXTAT data).

7 PREVENTION OF MINING FATALITIES TASKFORCE

In September 1997, the WA Minister for Mines instructed MOSHAB to establish an inquiry into mining fatalities following a series of fatalities culminating in a double fatality caused by a rock fall in an underground mine. The report of the Prevention of Mining Fatalities Taskforce was completed in December 1997 (MOSHAB 1997), and contained a number of recommendations dealing with geotechnical matters. Four of the nine priority one recommendations related to geotechnical considerations, as follows:

1. The management of all underground mining operations to ensure a high standard of compliance with the obligations of Regulation 10.28 Geotechnical Considerations.
2. The DME to target its resources to ensure compliance with the full obligation of Regulation 10.28 Geotechnical Considerations.
3. The Chamber of Minerals and Energy and the DME actively promote the adoption of the requirements of the Department's guideline "Geotechnical Considerations".
4. MOSHAB develops a code of practice for securing backs in headings of extended height and width with continuous meshing, shotcreting or other surface treatment.

To implement the geotechnical recommendations of the report the DME has completed the following:

1. Developed and implemented a geotechnical audit for underground mines.
2. Distributed copies of the two guidelines listed above to WA mining operations and made these guidelines (and others) available on the Internet.
3. Developed a draft Code of Practice for Surface Rock Support for Underground Excavations (MOSHAB 1998).

7.1 *Geotechnical audit*

The DME has adopted an audit approach to determine the extent to which the management systems that companies have in place, at their mine sites, conform to what might be reasonably expected of a well managed mine. Two general types of audit are used by DME:

1. Management systems audit.
2. High impact function audits.

The management systems audit looks at the overall health and safety management systems that have been implemented on the site. The high impact function audits look at those aspects of the operation that may pose a serious hazard to the workforce and include amongst others: cyanide, emergency plan, explosives, high headings, geotechnical considerations and ventilation. The audit documents are, at present, essentially internal working documents of DME and will evolve with time. It is planned to make them widely available as part of a self-audit system in which results would be reported to DME. While the audits are comprehensive, they may not necessarily be totally inclusive of all hazards relevant to all mining operations. Each mining operation is encouraged to develop its own audit systems following the systematic identification and documentation of all relevant hazards.

It is expected that an increasing level of effort will, in future, be devoted to the proactive auditing approach as distinct from a more reactive style of mine management. Considerable work has already been undertaken at some mines in this area.

From the mines that have been audited, the following points are noted:

1. Considerable resources are devoted to short term production planning and contractor management (where a contractor is present).
2. Formal, multi-disciplinary mine planning and design process, for the medium to long term, is generally ad hoc and/or not well documented.
3. Hazard identification tends to be reactive (ie after the event) and not proactive (ie identify and rank the hazards and then put the appropriate controls in place prior to the occurrence).
4. Adequate characterisation and documentation of the ground conditions is generally lacking.

5. There is a tendency to rely on a "standard" support pattern that may not be matched to the ground conditions in each individual area of the mine.

6. Reasons why particular rock support and reinforcement systems are being used are not well documented.

7. Rock support and reinforcement quality control testing has improved, albeit from a very low base, however continuing work and vigilance are required.

8. Equipment being used to install rock support and reinforcement is generally not purpose-designed and built for the task.

8 OTHER APPROACHES TO GEOTECHNICAL REGULATION

It is instructive to review the approaches that have been adopted in some other areas of the world where there is a well developed underground metalliferous mining industry. Ontario and South Africa and have been selected on the basis of an extensive mining history in challenging seismic rock conditions generally at considerable depth. Ontario is recognised as one of the world leaders in the area of underground mining occupational health and safety. South Africa has made a steady improvement in the occupational health and safety aspects of its underground mines, despite generally increasing rock stress levels and virgin rock temperatures associated with mining at increased depth.

8.1 *Ontario geotechnical regulations*

The Ontario Ministry of Labour provided a copy of the Occupational Health and Safety Act and Regulations for Mines and Mining Plants 1990 (OHSA&R 1996). Regulation 854 of the above Act has 11 separate regulations or sub-regulations that deal with or refer to geotechnical matters. These regulations are listed in Table 1.

The regulations tend to set out in some detail what is required of the mine management to comply with the regulation. The regulations appear to require a high level of geotechnical engineering effort at each mine.

8.2 *South African geotechnical regulations*

The Department of Minerals and Energy in South Africa provided a copy of the Mine Health and Safety Act 1996 (MHSA 1996) and the draft Regulation relating to geotechnical engineering matters. Regulation 14, in draft form, currently has the title "Proposed Regulation to combat rockfall and rockburst accidents in mines". This regulation has 12 parts that deal with the various aspects of geotechnical or rock engineering, as listed in Table 2.

The comprehensive nature of the draft geotechnical regulations proposed for South African underground metalliferous mines is evident from Table 2. Each mine is required by the Regulations to develop its own code of practice to combat rock falls and rockburst accidents. It is understood that the code of practice for a particular mine has the same legal standing as the Regulations. A mine will be expected to have its own rock engineering section, staffed by appropriately qualified people.

Table 1. Geotechnical topics in Ontario Regulation 854

Regulation	Topic
6	Mine design
19	Boundary pillars
21 (5)	Injury notice (rock failure size)
65 (1)	Communication program
66 (1)	Workplace examinations
67	Procedures for ground control
67.1	Scaling
69(1)&(5)	Illumination for ground assessment
70	Monitor rock movement
72	Rockburst record
73	Ground support quality control

Table 2. Geotechnical topics in draft South African regulations

Regulation	Topic
14.1	Appointment of rock engineering practitioner
14.2	Code of practice to combat rockfall and rockburst accidents
14.3	Mine design
14.4	Work rules, standards and procedures
14.4.1	Examination and control
14.4.2	Monitoring
14.4.3	Installation of support
14.4.4	Hanging wall support removal
14.5	Support testing procedure and accreditation of support testing facility
14.6	Ground support quality control
14.7	Seismic monitoring
14.8	Rock related accident/incident investigations
14.9	Rock engineering management programme

8.3 *Comparison with Western Australia*

It would not be appropriate for the author to comment in detail on the Ontario or the draft South African approach to geotechnical regulation. However, it is noted that many of the topics covered in the Ontario and draft South African regulations are also mentioned in the WA Regulations. The WA Regulation (10.28) is probably less prescriptive and more self-regulatory in nature. However, this is in keeping with the overall tone of most of the WA legislation. The self-regulatory approach adopted in WA places great demands on mine management to establish and enforce their own mine standards.

9 LEADERSHIP

Senior mine management, at the mine site and corporate level, will recognise the substantial economic benefits that can accrue from sound mining engineering practice. Strong leadership is required from senior industry management to ensure that the necessary human and engineering resources are marshalled to address the challenges of underground mining, particularly in rockburst prone ground conditions. These challenges can be successfully overcome through knowledge of the ground conditions and the use of appropriate geotechnical engineering design and monitoring tools (DME 1997a) that have become available during the past twenty or so years. Senior mining industry management are now in a good position to facilitate and require mine operators to implement consistent, coherent, achievable and cost effective plans to reach mine safety and production targets. Anything less could not be described as sound risk management.

10 CONCLUSIONS

1. The Act and the Regulations both need to be consulted to understand what is required to fully comply.
2. The duty of care embodied in the Act requires that each mine identify, rank and treat all hazards, including geotechnical hazards, that occur at the mine.
3. The underground geotechnical regulations in the Western Australian Mine Safety and Inspection Regulations have generally raised the profile of geotechnical engineering within the mining industry and helped to direct industry attention to these important matters.
4. The regulatory approach adopted in Western Australia has tended toward a self-regulatory style.
5. A self-regulatory regime requires a proactive management approach, such that the mine management establishes and enforces the appropriate mine standards.
6. Rock falls have been the most common accident type resulting in lost time injuries during the period 1987 to 1997 in Western Australian underground mines.
7. A considerable corpus of geotechnical engineering knowledge, relevant to the mining industry, has been accumulated during the past 20 years.
8. The geotechnical regulation seeks to encourage the application of current geotechnical engineering knowledge.

REFERENCES

AS/NZS 4360:1995. *Risk management*. Sydney: Standards Australia.

DME 1997a. *Guidelines: geotechnical considerations in underground mines*. Perth: Department of Minerals and Energy (Western Australia).

DME 1997b. *Guidelines: underground barring down and scaling*. Perth: Department of Minerals and Energy (Western Australia).

DME EXIS-on-the-web at the following address http://exisweb.dme.wa.gov.au/exis/Exisonweb.nsf

DMR 1997. *Risk management handbook for the mining industry*. MDG 1010. St Leonards: Department of Mineral Resources (New South Wales).

Knee, M. 1995. Changes in legislative approach – the implications for the mining industry. *Proc AusIMM Annual Conference*:285-291. Melbourne: The Australasian Institute of Mining and Metallurgy.

Lang, A.M. 1995. Ground control management in underground mining. *Underground Operators' Conference*:45-53. Melbourne: The Australasian Institute of Mining and Metallurgy.

Lang, A.M. 1997. *Aspects of mining in rockburst prone ground conditions*. Internal report GE005. Perth: Department of Minerals and Energy (Western Australia).

MHSA 1996. *Mine Health and Safety Act 1996* and draft Regulations. Draft Regulation 14: Proposed Regulation to combat rockfall and rockburst accidents in mines, dated 20 January 1998 (5th version). Pretoria: Department of Minerals and Energy.

MOSHAB 1997. *Report on the Inquiry into Fatalities in the Western Australian Mining Industry*. Perth: Mines Occupational Safety and Health Advisory Board.

MOSHAB 1998. *Surface rock support for underground excavations*. Draft Code of Practice. Perth: Mines Occupational Safety and Health Advisory Board.

OHSA&R 1996. *Occupational Health and Safety Act and Regulations for Mines and Mining Plants*, Revised Statutes of Ontario, 1990. Binder Edition. Sudbury: Queen's Printer for Ontario.

Stevenson, T. 1986. *Improving Ground Stability and Mine Rescue, The Report of the Provincial Inquiry into Ground Control and Emergency Preparedness in Ontario Mines*. Ontario: Ministry of Labour.

Torlach, J. 1996. New legislative directions. *Minesafe 1996*:117-122. Perth: The Chamber of Minerals and Energy of Western Australia Inc.

Rock Support and Reinforcement Practice in Mining, Villaescusa, Windsor & Thompson (eds)
© 1999 Balkema, Rotterdam, ISBN 90 5809 045 0

Elimination of fatalities taskforce – Underground rockfalls project

S.M.Harvey
WMC Resources Limited, Perth, W.A., Australia

ABSTRACT: This paper describes the development of the Underground Ground Control Standard by the Elimination of Fatalities Taskforce for WMC Resources. It describes each stage of development of the standard , the implementation and the indicative results to date.

1 BACKGROUND - EOFT

The Elimination Of Fatalities Taskforce (EOFT) was established in June 1996 by the senior management of WMC Resources Ltd (WMC) to address the issue of fatal incidents within the company.

Over the past six years the accident injury statistic rates have been falling. WMC has reduced the Lost Time Injury (LTI) and Medical Treated Injury (MTI) frequency rates. However the rate of fatal incidents has remained relatively constant.

The safety objective set by the Managing Director was zero fatalities. This objective had not been achieved. Major improvements have been made in many areas of safety, throughout WMC. These improvements have been achieved by the good work carried out at management and operation level.

The Managing Director, Hugh Morgan, along with the Executive General Manager Nickel and Gold, Peter Johnston, decided to form a group to focus exclusively on the fatality issues to eliminate them from the workplace within WMC. The group was sponsored by Hugh Morgan and a steering committee consisting of all the Executive General Managers (EGMs) was formed. A taskforce was formed from a vertical cross section from the company and consisted of ;

- Executive General Manager Nickel and Gold; EOFT Leader
- U/G Supervisor - St Ives
- Airleg Miner - Kambalda
- Maintenance Technician - Nickel Refinery
- Process Technician - Nickel Smelter
- Mining Safety Systems Coordinator - Gold
- Scraper Operator - Central Norseman
- Mining Manager - Leinster
- Blister production Coordinator - Copper Smelter
- Operations Manager - Gold
- Underground Operator - Olympic Dam
- SEQ Manager - Gold.

There were several advisers to this taskforce. These were;
Group Manager OH&S
Principle Safety Adviser - Exploration Division
Environmental and Safety Manager - Petroleum
External advisers - ALARA Risk Services

The taskforce initially convened on 17 July 1996, where they were briefed on the task scope and the guidelines for the group were developed. A strategy to review the fatality issue was determined and a perception survey was sent to all sites for them to indicate what were the five highest hazard or risk areas at their operation.

The taskforce reviewed the survey results and developed a preliminary list of major hazards within WMC. The taskforce was introduced to

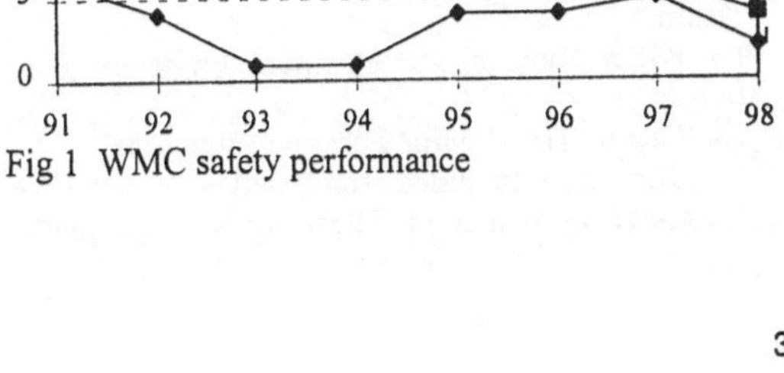

Fig 1 WMC safety performance

several methodologies that could be utilised for the purpose of analysing the fatalities issues. These were;
- ISRS - International Safety Rating System.
- WRAC - Workplace Risk Assessment and Control
- DuPont
- Petroleum's Safety Case
- Alcoa's approach to this issue of eliminating fatal accidents from the workplace.

From this meeting a strategy and action plan to tackle the problem was developed. The strategy was grouped into two parts.

Part A targeted the high priority areas to tackle in the short term. The high priorities were;
- rockfalls,
- falls from heights,
- underground fires
- heavy equipment collisions and
- molten metal.

Part B of the strategy was targeted at the long term management and cultural issues such as planning, training, remuneration and communication. These issues were seen as essential to achieve a permanent cultural change in regard to safety. This change would be achieve by the initiatives flowing from the Part A strategy.

The Alcoa approach was chosen as the foundation to address the Part A issues. The immediate focus would be on the highest priorities and separate project teams were set up to review each major hazard. The teams reviewed the hazard, its root causes and all available measures to control the hazards. Once the controls had been developed by the team, a corporate Major Hazard Standard was formulated prior to site implementation.

These standards are to be audited by an external party to ensure the required compliance was being achieved. The ISRS audit framework was to be used for these initial audits.

A task of this nature is going to be very taxing on the resources of the company and the Managing Director has put his weight behind it by giving clear notice to all that the EOFT project is of the highest priority by stating - "there is no restraint on the imagination of the taskforce or any employee in our quest to fulfil our commitment to the elimination of fatalities at WMC workplaces".

The hazards to be addressed were chosen following an analysis and risk ranking by each operation including an exposure factor. The ranked hazard list were then sorted into the following groups:

1. Major hazards for which corporate standards could be developed.

2. Broad hazards which would be priority issues for standard work procedures. Eg. Manual handling and electricity.

3. Process management issues which should be addressed at local site level.

4. Hazards that could be referred to specialist bodies within WMC. Eg. drugs, alcohol, hygiene

Hazards that were common to two or more sites were categorised according to the above classifications. This resulted in a list of 19 major hazard areas that would require corporate standards being developed. The areas are:
- underground rockfalls
- underground mine fires
- aircraft use
- molten metal
- vehicle transport (off-site)
- isolation (permits, tagging etc)
- falls from height
- heavy equipment - underground
- heavy equipment - surface
- surface fires (buildings etc)
- guarding (contacting machinery)
- chemical handling
- explosives
- natural disaster
- pit wall stability
- remote areas
- underground ventilation
- facility design, construction and maintenance
- tyre explosion

2 ROCKFALLS PROJECT

The highest priority area identified by the EOFT taskforce was rockfalls. From the WMC statistics 37% of all underground fatalities were due to rockfalls. Hence rockfalls was the first issue to be tackled, followed by aircraft, underground fires and molten metal.

The Rockfalls project team was made up of a cross section of personnel who had exposure to rockfalls including personnel from the Nickel, Copper /Uranium and Gold business units. As a recognition of the high priority of the rockfalls issue a large group was assembled.

This team included representatives from management, workforce and contractors with experience in ground control and geomechanics. The members were:
- Brendan Cope, Senior Mining Engineer; Olympic Dam
- Peter Crooks, Airleg Miner; Otter-Juan
- Steven Harvey, Mining Safety Systems Coordinator - QV1, Team Leader
- Jock Lewis, Jumbo Operator; Bronzewing - Eltin
- Nigel Thomas, Development Engineer; Perseverance
- Dennis Treen, Mine Foreman; Junction

The team included some advisers who were 'experts' in their field. These advisers also participated in developing the scope and the risk assessment exercise. The advisers were;

- Peter Teasdale, Principle Mining Engineer - WMC Resources Gold Business Unit
- Adrian Lang, Geotechnical Engineer - Department of Minerals and Energy; WA
- Brian McCowan, Manager - Mining NSW; ACIRL
- Doug Morrison, Senior Mining Consultant - Golder Associates; Canada
- Andrew Murie, Senior Rock Mechanics Engineer - Golder Associates; Perth
- Chris Windsor, Senior Research Engineer - Rock Technology; Perth

Rock Technology and Golder Associates have been involved in conducting reviews of ground support and compliance to regulation 10.28 of the Mine Safety and Inspection Regulations 1995 for the WMC Gold Business Unit. Andrew Murie, Chris Windsor and Brian McCowan were involved in developing the scope and the initial set up meetings. Peter Teasadale, Doug Morrison, Chris Windsor and Adrian Lang were involved in the risk assessment exercise.

The development of the scope and the risk assessment exercises were facilitated by Jim Joy and Bob Logan of ALARA Risk Services.

The project team initially convened on the 4th October 1996 to scope the task to be undertaken, set the ground rules and a preliminary agenda for the information gathering exercise. The team defined a rockfall hazard as an uncontrolled displacement (due to gravity or stress) of material (rock or support elements) from the surface of an excavation.

This was to ensure the team did not get sidetracked by unrelated issues that were important, but unrelated to the rockfall issue. The team developed the scope of the project using this definition. This included all aspects of ground control;

- planning and design
- geomechanical/geotechnical issues
- support systems and practises
- work procedures and training
- monitoring and control

A two day seminar was conducted for the team to bring everybody up to the same level of understanding of the issues identified in the scope. Andrew Murie, Chris Windsor and Brian McCowan briefed the group on ground support systems, the requirements to achieve compliance with Regulation 10.28, ground support systems being used and some of the conditions to look for when assessing rockfall hazards.

A list mines that the group would visit was drawn up with a timetable for the visits. A timetable for the risk assessment process was also finalised.

The program was scheduled to take eight weeks, in two stages. The initial stage was to look at other sites to see what they were doing to get a picture of what current "Best Practice" was available. This took five weeks and included site visits to twenty two mines and talks with three specialist groups. The second stage was the risk assessment and drafting of the standard.

3 INFORMATION GATHERING - SITE VISITS

In the time available the team could not visit all underground mines within Western Australia and Australia. The team picked the best sites to visit. The choice of sites was based on a combination of factors. These were;

- safety record and training
- ground support practices
- ground conditions and geotechnical issues
- systems and processes
- planning and mining systems
- reputation

The main criteria were that the mines had a good safety record with regard to rockfall related incidents or they had unusual ground conditions or both.

The team visited all WMC sites with underground mines and other relevant sites. The mines visited were;

- Perseverance
- Rockies Reward
- Redeemer
- Mt Keith
- Otter Juan
- Olympic Dam
- St Ives complex - Junction
- KNO complex - Several mines
- CNGC - OK/Bullen
- H50

The mines visited within Western Australia in addition to WMC mines were;

- Cassidy Shaft - KCGM
- Big Bell and Scuddles - Normandy Group
- Kanowna Belle - North Ltd
- Telfer - Newcrest Mining Ltd

For the visits to mines in the eastern states the team broke into two. One team went to NSW and the other team went to Queensland.

The NSW team went to:

- Broken Hill - Pasminco
- Northparkes - North Ltd
- Woodlawn - Woodlawn Mines
- West Cliff Colliery - CRA

The Queensland team went to:

- Mt Isa - MIM
- Osborne - Placer Pacific
- Cannington - BHP

The team also spoke with three separate special interest teams. These were;

- ICI Explosives
- JKMRC at Qld University

– Queensland Department of Mines

From the site visits many good things were observed that the team saw could be incorporated into the standard or brought back as good ideas for use within the team members' own workplaces. To ensure the team collected the same information from each site and for easier comparison a standard questionnaire was used by all members of the team.

4 RISK ANALYSIS AND STANDARD DEVELOPMENT

The risk analysis was conducted over one week and a second week was devoted to developing controls and the draft standard. The risk assessment method used was determined by the team after advice from the facilitator. It was one of several methods available that are widely used throughout industry.

The method used was the Workplace Risk Assessment and Control (WRAC) method. There is a lot of research into this form of risk assessment and it shows that the output from a diverse group of varying experience is very powerful in controlling risks.

The risk assessment process is not "rocket science", but was lengthy and involved. The process identifies what should occur and all the hazards in that process. All these hazards are then assessed for their risk and controls put in place for each hazard. Each hazard can be treated by either eliminating it, treating it or tolerating it. Action needs to be taken according to how the hazard is to be treated and it will need to be monitored for changes and performance. See figure 2.

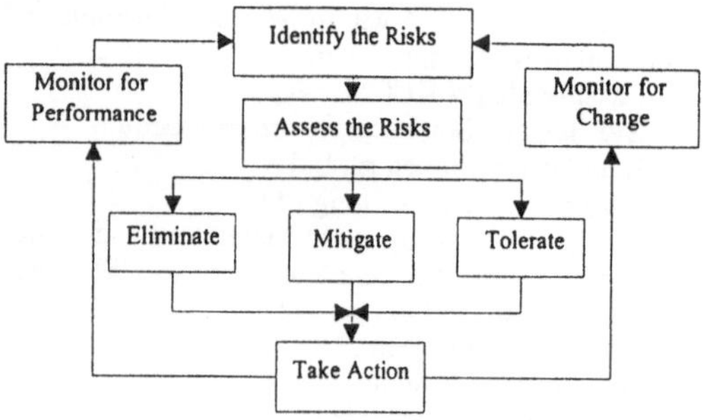

Figure 2 Basic risk management system

The ground control process was broken down into five areas;
- planning
- preparation/resourcing
- operations management
- ground control work/task execution
- monitoring/auditing

Figure 3 shows the process flow for ground support. It starts with the planning and design of the mine and the ground support. Then the necessary resources, (personnel, financial and physical), must be gathered to install the ground support. The operation management is the day to day managing and organisation (planning and resourcing) that is required to ensure the ground support is installed. All aspects of installing ground support should have Standard Work Procedures, (SWP's) so that uniformity is achieved when installing ground support safely. The whole process must be monitored including testing and checking of the installed system to ensure that it is the correct system for the purpose.

As the greatest exposure to personnel is in the actual "doing" stage of the work cycle the team focused on this area for the development of the hazard exposures. All the tasks that exposed personnel to a rockfall hazard were listed. Then against all these tasks, scenarios or risk exposures were identified. A list of nine tasks was identified as exposing personnel to rockfalls:

1. visual inspection of areas
2. watering down
3. scaling
4. bogging out
5. install ground support
6. survey and mark up
7. sampling and mapping
8. digging out lifters
9. drilling

Visual inspection is something that all underground personnel are exposed to. Some of the tasks are targeted for a very select audience, eg. Survey & mark up, sampling & mapping and digging out lifters. All the other tasks, while not applying to all personnel, apply to a large percentage of the u/g workforce directly.

The nine areas were expanded and some 100 plus exposures to rockfalls were identified. These were ranked, based on risk according to the WMC Standard - Development, Implementation and Maintenance of Major Hazard Standards developed for the EOFT process.

Risk = Likelihood x Consequences

Where likelihood is the probability that the event will occur and consequence is the severity of the outcome. The risk is the result of the product as expressed in the matrix shown in Table 1.

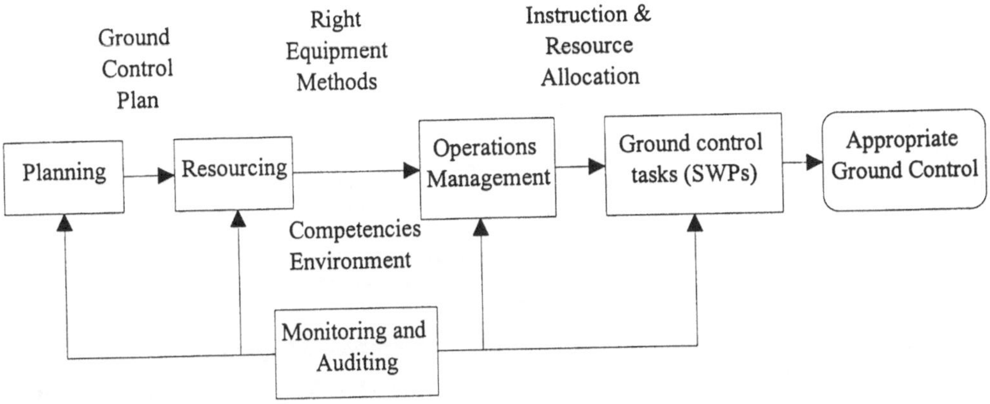

Figure 3 Ground control process flow chart

Table 1 Risk assessment matrix

Consequence / Probability	MTI	MTI	LTI	Fatality	Multiple Fatality
Highly Likely	15	10	6	3	1
Likely	19	14	9	5	2
Moderate	22	18	13	8	4
Unlikely	24	21	17	12	7
Rare	25	23	20	16	11

A ranking of 1-8 is a high risk and must be addressed immediately either by eliminating the risk or treating it. A ranking of 9-16 is a moderate risk that must be addressed but these are of lower priority. The last group 17-25 is of low risk and can either be tolerated or ignored as the risks are low and are already being controlled or no longer need to be controlled eg. work completed or eliminated.

All the risks were ranked based on the team's knowledge, opinions, views and the information gathered. The next job was to determine controls for these. Again the team's combined knowledge was used to determine the controls for each of the risks.

Only risks ranked below 16 were given controls as it was deemed that risks ranked above this would be covered by other controls, systems or the likelihood of occurrence was so low as be tolerable.

Once all the risks were ranked controls were put in. For each risk there may have been several controls. Most of these controls however were repetitive so the total list was not unwieldy.

The controls were then grouped into the various stages of the process as the controls for all the tasks are not necessarily directly related to the SWP or installing the ground support. Many were in the planning and resourcing stages.

An old proverb is "Prior Planning Prevents Poor Performance".

From this a draft standard was developed and reviewed by the team several times. The team debated many sections within the standard and from these discussions the standard was refined.

This reviewed draft, (draft 4), was then sent to all the Resident Managers and Mining Managers for comment. A meeting was held on December 13 to discuss the standard. The standard was received well in principle, but there were some changes made.

From here the final version of the Underground Ground Control Standard (UGCS) was developed following input from sites, management and the workforce.

A professional technical writer was contracted to put the standard into a standard format that will be used for all EOFT standards. This process also involved editing the standard to ensure it was logical and clear.

5 RECOMMENDATIONS FOR IMPLEMENTATION

Distinct from the standard the team made a number of recommendations. These can be broken into separate groups. Recommendations for immediate implementation, recommendations with a long term focus and recommendations for the implementation process for the standard. Guidelines were also established for the development of the Ground Control Plan.

5.1 *Recommendations for Immediate Implementation*

The recommendations for immediate implementation were seen by the team as being crucial to stopping the vast majority of fatalities and could be implemented quickly and relatively easily. All of the recommendations were extracted directly from the standard and as such they formed an integral part of the implementation program. These recommendations allowed the mines to make immediate moves on the implementation of the standard.

These recommendations were;

- Each mine must develop standards for ground support in all relevant mine situations.
- A Current Support Evaluation Survey must be carried out at all mines.
- No person is to go beyond supported ground (ground supported to at least the minimum.)
- Scaling Standard Work Practices are to be developed and put in place at each mine. The SWPs must consider the need to work from under supported ground, as well as the length, rigidity and maintenance of scaling bars.
- The face must be scaled and made safe prior to undertaking any other tasks.
- Boggers are not to be used for scaling
- Hand held drills are not to be used for scaling
- Jumbo scaling cannot be the sole or final method of scaling. Manual scaling and sounding is the final method, unless the Ground Control plan does not allow hand scaling due to the risk.
- Persons are not to approach past the rear of a Jumbo and other ground support equipment while it is operating without the prior knowledge and consent of the operator.
- All underground personnel must be aware of the general layout of the mine and the specifics of their workplace.
- All mines must have Shift Plans.
- There must be a documented system on the surface to record the location of all underground personnel.
- There must be a mine map located on the surface.
- If a person is not "fit for duty" (to be defined in the Plan) they are not to go underground.
- No one to enter a working area without reporting to the supervisor or the location control system.
- No repairs or maintenance on equipment is to be conducted which may expose persons to unsupported ground.
- Surveyors, geologists, samplers and other professionals must be familiar with the Shift Plan including relevant information on local conditions in the area where they will travel and work, before approaching any work location.
- When charging up, charge from the top down and only on one level at a time.
- If backs are at a height of more than 3.5m, an inspection platform must be used to identify and control rockfall hazards.
- Review new and available information on blasting technology and profiles to reduce rockfall risks.
- Hand held lights or spotlights made available and used for hazard identification and control where cap lamps do not provide adequate visibility.
- Install signs identifying mine locations and hazards, all signs are to be to Australian Standards.

5.2 *Recommendations for further investigation*

These recommendations were seen as longer term issues and would require further investigation by the company as a whole or by specialist groups within the company. The team recommended that there should be a group or several groups given the task of conducting further investigation into

- The best practice methods used in both the coal industry and the Canadian mining industry.
- The availability and practical application of ground control equipment such as dedicated, purpose built rock bolting apparatus, scaling machines, cable bolting equipment and shotcrete technology with regard to the requirements of WMC.
- Platform design to ensure a stable, safe work area.
- The hydraulic design of platform lifting equipment should minimise any uncontrolled fall of a platform if the hydraulics are lost.
- The regular medical monitoring of shotcrete crews and monitoring of the chemicals used.
- The Quality Assurance requirements for ground support materials suppliers.

5.3 *Recommendations for Implementing the Standard*

For the implementation of the standard ALARA Risk Services developed a framework that all the mines could follow.

6 IMPLEMENTATION

It has now been 18 months since the standard was approved and released. Implementation of the standard has been a long process.

The process of implementation has been as follows:

- Conduct a Gap Analysis, reality verses the requirements of the standard.
- Determine what the gaps are and what needs to be done.

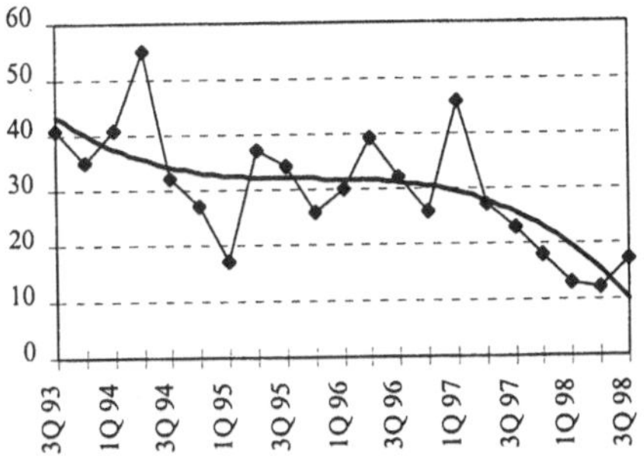

Figure 4 WMC rockfall injury trend (WA only)

- Rank these requirements in order of priority based on risk assessments of these requirements. This has been based on probability of an occurrence, levels of exposure, most likely outcome of an incident. Other factors that can be considered are resources required, time to implement, expected effectiveness and cost.
- Develop an action plan to assign a time frame and a responsible person for each task.
- After the first 12 months a formal audit was conducted by an external person. The audit protocol format was developed based around the DNV format.

All current sites having been audited once by an external person to the audit protocol. This has been beneficial in several areas;

- Highlighting areas for improvement in the standard.
- Highlighting areas for improvement in the audit protocol.
- Highlighting areas for improvement at individual sites.

Another initiative to aid implementation was to develop common guidelines to specific parts of the standard to ensure interpretation would be the same at all sites throughout WMC.

The UGCS is currently undergoing the first revision.

7 CONCLUSIONS

Though implementation has been slower than expected the impact to date has been significant in many ways.

It has lead to increasing awareness at all levels of the need for good ground control systems, improved planning and training. Already at the mine sites there have been changes in the ground control systems installed and changes in attitudes, which is a key to improving safety in the WA mining industry.

From looking at the numbers of reported rockfall injuries (see Figure 4) there is a definite downward trend in the reported rockfall related injury numbers for the WA operations corresponding to the time period that the UGCS was introduced and implemented. The draft UGCS was released in the last quarter of 1996 and implementation has progressed since then.

It cannot be conclusively stated that the introduction of the UGCS has lead to this decline in rockfall injuries, as it is one factor amongst several. Certainly the focus given to the UGCS and ground control issues over the last 12-18 months has lead to an increasing awareness and potentially lead to the decreases in injury numbers.

8 ACKNOWLEDGMENTS

In researching and developing this standard and paper many people and organisations contributed their valuable time and effort. The author wishes to thank those companies and personnel who participated and gave their assistance.

Special thanks goes out to

- The members of the development team.
- The advisers to the team.
- All those who assisted with reviewing the standard.
- Those companies and mines who made themselves available to be apart of this process.
- Those who made all the arrangements for site visits and travel.
- WMC for allowing the preparation and presentation of this paper.

Rock Support and Reinforcement Practice in Mining, Villaescusa, Windsor & Thompson (eds)
© 1999 Balkema, Rotterdam, ISBN 90 5809 045 0

Continuous improvement in geotechnical design and practice

T.Li
WMC Resources Limited, St.Ives Gold, Kambalda, W.A., Australia

ABSTRACT: Geotechnical design and practice has been recognised as a key issue to be addressed by the mining industry in Western Australia in order to eliminate the unacceptably high number of fatalities and to reduce injuries and incidents. A sharp increase in the geotechnical efforts committed by the underground mines has been experienced. Consequently, the incident rates at the mine sites have been significantly reduced. In this paper, several key features of the effort to improve geotechnical design and practice are discussed and cases presented. A number of successful initiatives are presented to highlight the progress that has been made. The geotechnical challenges facing the underground mines are also discussed.

1 INTRODUCTION

Rockfall related fatalities accounted for almost 50% of all fatalities in Western Australian underground mines in the last several years. The lack of sound geotechnical practice, has been identified by the companies and the mine inspectorate as the key issue that must be addressed by the industry to minimize the incidents. The aim of this paper is to overview the geotechnical design and practice, including rock support and reinforcement, as carried out at some mine sites to address this issue.

The cases in this presentation are mainly from the underground mines at St Ives Gold, Western Mining Corporation (WMC). St Ives Gold consists of two underground mines and three open pits, producing about 3.1 million tonnes of ore grading 4.5g/t. The total reserves are 11.3 mt at 6.1 g/t underground and 13.6 mt at 3.0 g/t open pits; and resources of 12.8 mt at 3.9 g/t underground and 27.4 mt at 2.3 g/t open pits (WMC 1997). St Ives Gold is a very dynamic operation in that there is generally a large turn-over of small to medium sized underground mines and open pits. To date, the requirements for geotechnical input have been for on-going technical support in the existing mines, and geotechnical studies for new mines or mine expansion projects.

The aim of this paper is to demonstrate the commitment and the approaches taken by the management and the geotechnical professionals in implementing a wide range of geotechnical programs. These programs are aimed at improving the geotechnical design and practice on a continuous basis.

2 GEOTECHNICAL EFFORTS AND EFFECTS

Adequate geotechnical input into mine design and operational practices has been recognised as an integral part of the overall drive to improve safety and efficiency in underground mines. There has been a sharp increase in the geotechnical efforts undertaken by the mining industry, especially in the last a few years since the release of Regulation 10.28 in 1995 (Lang 1999).

WMC set up a corporate wide Elimination of Fatality Taskforce (EOFT) in mid 1996, with the objective to develop and implement standards and procedures that will eventually lead to fatality free operations (Harvey 1999). By mid 1997, the first of the 20 standards: Underground Ground Control Standards (UGCS) was finalised and implemented.
One measurement of the increased geotechnical effort is the increasing number of geotechnical professionals, in particular the site based geotechnical engineers. For instance, St Ives Gold has increased the geotechnical staff from 2 part-time in 1996 to 3 full time and 1 part-time at

present. WMC has increased its site based geotechnical staff four times, from 4 in 1995 to 16 in 1998. Another measurement of the increased geotechnical effort is the increased geotechnical training courses conducted by educational institutions and the mines.

Most importantly, a number of geotechnical programs are being initiated and implemented at the mine sites. The geotechnical programs can be summarised as the following: the systematic collection of geotechnical data and the processing of this data, the use of geotechnical modelling and design parameters, the optimisation of ground support/reinforcement methods and procedures, the implementation of monitoring and testing, the back-analysis of stope and pillar performance, and geotechnical training.

Rockfall related incident rates and severity, on the other hand, have been reduced significantly, largely due to these measures. Figure 1 illustrates the trend of the rockfall related incident rates at St Ives Gold. The actual consequences of the incidents were recorded as fatalities, lost time injuries (LTI), medical treated injuries (MTI), and minor injuries (MI). The potential consequences of some incidents were re-assessed as the incidents that could have resulted in more severe or even fatal injuries. It is shown that the severity of the potential consequences of the incidents has also decreased significantly.

The decrease observed in the incident rates since 1997 coincided with the appointment of two full-time geotechnical engineers; and the implementation of the EOFT Underground Ground Control Standard (UGCS), in particular the 20 key elements of the UGCS (Harvey, 1999).

Decreasing incident rates and high fatalities seem to be a trend across WA underground mines. One possible explanation of this seemingly unmatching trend is that in the past, a number of mines might not have had adequate geotechnical input to the design and operation at an early stage. While this may not have caused problems in the past, the inadequacy or the lack of geotechnical input may become apparent later in the mine life. If this is the case, no simple geotechnical measure can rectify the problems.

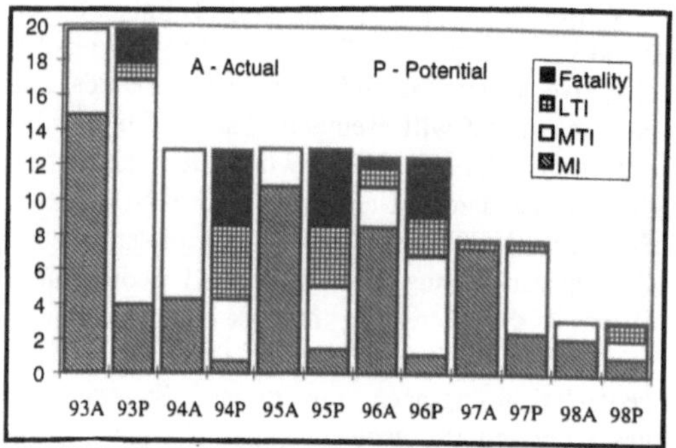

Figure 1. The trend of the rockfall related incident rates at St Ives Gold.

Some geotechnical measures can have immediate impact on an operation, such as changes in development heading design and procedures, and alterations to the ground support methods and procedures. Some others may have an impact long after they have been implemented. For example, the effectiveness of a mining sequence or a pillar design strategy would only be evaluated years after the strategy was adopted.

More critical issues are the understanding of the geotechnical conditions at the current and future stages of mining, the awareness of the potential consequences of not following sound geotechnical approaches, and consciously striving for quality implementation of sound geotechnical measures. This is applicable from the operators at the development headings to the corporate level, and from daily operations to the life of mine planning.

It is this cultural change combined with technical excellence that can minimise incidents, eliminate fatalities, and sustain excellent performance. This fundamental change is taking place industry wide in WA. The experience in North America suggests that there could be several years after the systematic efforts have been directed to addressing the fatalities and incidents that significant results would be forthcoming.

3 MINE SITE BASED GEOTECHNICAL MANAGEMENT SYSTEM

A relatively strong geotechnical presence at mine sites has been in place since 1996 at St Ives Gold. The positive impact of sound geotechnical work has always been recognised. Initially, the role of the geotechnical engineers was mainly confined to the short term issues, such as ground support/reinforcement and other daily operation input. Longer term issues, such as geotechnical design were undertaken mainly through external consultants. While this might be appropriate given the level of geotechnical resources available and the prevailing conditions at the time, it is questionable that such approach can achieve an effective and long lasting geotechnical practice.

In late 1997, a geotechnical group was established at St Ives Gold. Geotechnical programs consisting of long term and short term projects and goals were developed and approved. The core of the program is a geotechnical management system that features a number of points including:

- Adequate geotechnical resources, especially at mine sites.
- Systematic geotechnical data collection, processing and application.
- Geotechnical design and evaluation methods and procedures.
- Geotechnical training for the geotechnical engineers and all the other personnel.

The thrust of this approach is to deliver quality and timely geotechnical services for the short and long term needs of the mines. The short term geotechnical routine practices and the long term projects are integrated. The following are a few examples of the geotechnical programs introduced and implemented at St Ives Gold.

3.1 Geotechnical data collection and processing

Up to few years ago, geotechnical data collection had not been a routine practice in our mines. Efforts by the geotechnical engineers and geologists in the last two years has seen significant progress towards systematic data collection and processing. At present, a geotechnical database has been set up in the Kambalda Nickel Operations – St Ives Gold Geobase. Geotechnical data, mainly core logging, is stored in the Geobase through the Data Entry System installed on standard PCs. Attempts have been made to establish procedures for 3D modelling of the geotechnical data from borehole logging and mapping with encouraging results.

The development and implementation of geotechnical data processing make the whole process from data collection to geotechnical information streamlined. The 3D distribution of geotechnical domains, and the interpreted and interpolated geotechnical conditions within each domain are delineated following standard methods and procedures.

3.2 *Optimisation of development and ground support methods and practices*

The Junction Mine at St Ives Gold offers a good example on the continuous improvement in the design of development headings and ground support standards. Before 1997, the standard decline design

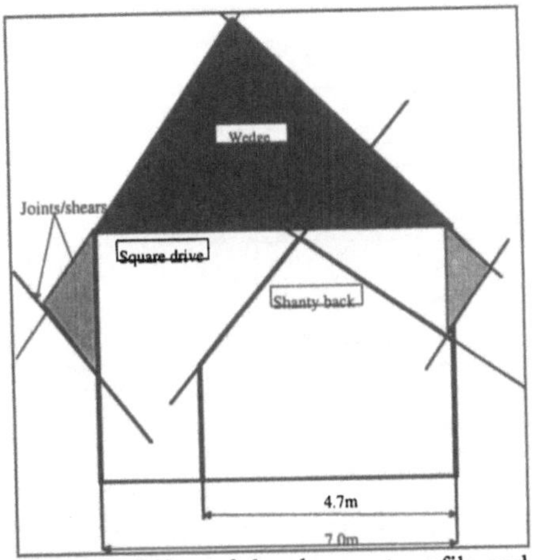

Figure 2. Effect of development profile and size on rockfall potential.

at Junction Mine was 6m by 6m square profile, while the ore drives were mined to shanty profile at full width of the orebody, which can range from 5m to 15m wide. Figure 2 illustrates the effects of development sizes and shapes on the rockfall potential. The geological discontinuities which can be present in and around the orebody have the potential to define large wedges at the back of the ore drives. The wider the drive, the larger the potential wedge. The square shaped drive backs tend to be more prone to wedge formation. The shanty back shapes are likely to minimize the potential for large wedges.

Two major changes in development heading design were introduced in mid 1997, reducing the sizes of the drives and declines, and adopting the arched back profile for declines and modifying a shanty back profile for ore drives. The size of the decline was reduced from 6m by 6m flat back to 5.5m by 5.8m arched back. The size of the ore drives was reduced from full orebody width to 4.7m wide by 5m high with shanty backs (Figure 2).

These simple changes have resulted in several benefits. Development rates are higher, as the development cycle time is reduced due to the reduction in drilling, charging, mucking, and ground support. Ground support/reinforcement is more effective as the load distribution over the support elements is more even. The risk of large wedge rockfalls has been removed significantly, and there has not been a single large wedge rockfall since the implementation of these geometrical changes.

The development was further improved by the

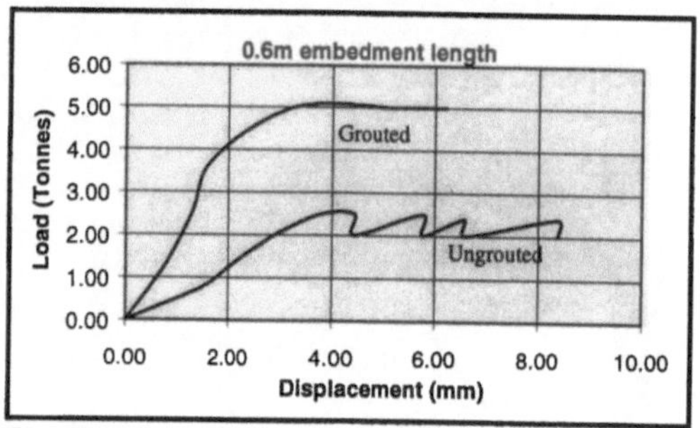

Figure 3. A comparison of the loading capacity of grouted and ungrouted split sets.

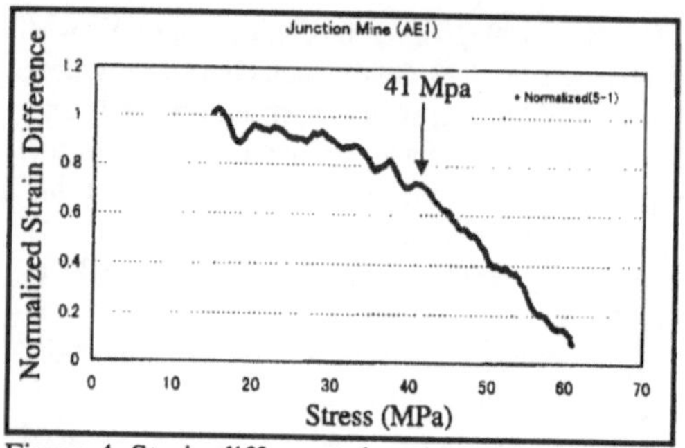

Figure 4. Strain difference for a cored sample from the Junction Mine

implementation of perimeter blasting techniques which reduced the blast induced rock mass damage to the immediate excavation backs and walls.

The improvement in development size and profiles were accompanied by the optimisation in ground support/reinforcement methods and procedures. Extensive trials and tests were performed by the site geotechnical engineer over a wide range of support/reinforcement elements and schemes.

Figure 3 shows the comparison of the loading capacity of grouted and ungrouted split sets as determined by in-situ pull testing. The cement grouted split sets generally doubled the loading capacity over the ungrouted split sets. The loading capacity of the grouted split sets was found to be approximately 10 tonnes per metre of bolt embedment. The ungrouted bolt strength was found to be approximately 5 tonnes. This increase in loading capacity has been confirmed by other test results (Villaescusa & Wright 1997).

The results also show that grouted split sets are significantly stiffer and therefore their application may not be suitable for some ground conditions.

3.3 *Stress measurement and monitoring*

The two key parameters that measure rock mass behaviours are stresses and deformation. Up to now, Hollow Inclusion (HI) cell stress measurement techniques have been the major method of in situ stress determination. A renewed interest on other alternative stress measurement methods has emerged due to various reasons. The determination of in situ stresses by the Acoustic Emission technique using a piece of diamond drill core, for example, has been compared well with that using the HI cell method (Seto & Villaescusa 1998). The benefit of the Acoustic Emission technique is that it only requires oriented diamond drilled cores and therefore the in situ stresses could be determined well before any access to the underground excavations is available. Figure 4 presents the relation between differential strain and stress for a core sample from the Junction Mine. A bending point identifying the maximum principal stress can be clearly recognized at approximately 41 MPa.

Monitoring of stress changes due to mining has also been implemented at an increasing number of mines. Apart from the HI cell method, other methods, such as vibrating wires, have been used for stress change monitoring (Brown 1993). Stress measurements from cored rock offer a very promising alternative regarding stress change determination in isolated pillars.

The increased effort in geotechnical monitoring is probably better demonstrated in the case of mining induced seismicity monitoring. The information derived from the monitoring can be used to assess the rock mass behaviour and allow the geotechnical engineers and mine managers to make informed decisions on mine design, planning, and operational issues.

Conventional observational methods and simple devices, such as extensometers and crack monitors, can also provide reliable information about the likely rock mass behaviour.

3.4 Stability performance measurement and monitoring

St Ives Gold has used the Cavity Monitoring System (CMS) survey method to determine open stope voids, along with a number of mines in Australia. The CMS survey, for the first time in the history of underground mining, is able to quantify the performance of stopes and pillars. The amount of overbreak and underbreak, dilution, and ore losses can be determined to a reasonable accuracy. Analysis of the CMS survey results can be used to highlight the potential problems in stope design, to assess the reinforcement effectiveness, and also to optimise the drilling and blasting practices.

4 BACK-ANALYSIS TO IMPROVE THE GEOTECHNICAL DESIGN

Precedence of stope stability performance has been the practical design method for stope walls and pillars. In recent years, methods based on back-analysis of the stope walls and pillars performance have been developed and used extensively (Potvin et al 1989, Nickson 1992, Villaescusa et al, 1997). These back-analysis based methods typically incorporate geological conditions, stope geometries, and stress analysis to quantify the impact of geotechnical parameters and provide guidelines for the acceptable range of design parameters, such as exposed hangingwall span dimensions.

4.1 Modified Stability Graph

The Modified Stability Graph method was developed based on a large set of data from Canadian open stope mines (Potvin et al 1989). This method has been extensively used in WA by mine geotechnical engineers and consultants for the determination of stope dimensions. St Ives Gold has been using this method in the last two years. Since early last year, an attempt was made by the geotechnical engineer at Junction to develop a local stability chart using similar principles and methodology such as those used in the Modified Stability Graph method. Extensive back-analyses of stope performance have been carried out using a 20 stope hangingwall database. It is envisaged that such a local, stope performance based stability chart would improve the geotechnical design.

4.2 Hangingwall Stability Rating

Disruptive bench hangingwall failures during the extraction of initial benches in weak rock led to the development of an empirical model of hangingwall stability at the Lead Mine in Mount Isa (Villaescusa et al 1997). The method is based on the quantified stope hangingwall performance, using observations and CMS surveys to establish a stability chart which links the factors controlling ground behaviour and excavation geometry for each of the orebodies at Mount Isa. The Hangingwall Stability Rating (HSR method), assumes that the geological discontinuities, induced stresses, blast damage, and excavation geometry are the main factors controlling hangingwall stability (Villaescusa et al 1997).

The method has been calibrated and used as a predictive tool in mine planning to optimise new mine block designs. In the Lead Mine, for example, greater bench heights could be planned for better quality orebodies, with significant reduction on orebody drive development costs. Alternatively, when a particular bench height is fixed, the stability chart may be used to calculate the maximum (unsupported) stable length that can be safely exposed. Successful application of this method to open stopes in similar ground conditions has also been achieved (Villaescusa et al 1995, Harris & Li 1995).

The continued improvement in back-analysis methods and the application of these methods will gradually lead to better understanding of the stope performance and high quality stope design and sequencing.

5 GEOTECHNICAL TRAINING

The geotechnical training outlined mainly refers to the training of managers, engineers, and operators at a mine site. Although the level of understanding of geotechnical design and practice is different for different groups of personnel, there is a recognition that a geotechnical culture needs to be established to maximise the impact of geotechnical input.

A geotechnical culture can be simply defined as an environment where all personnel are consciously pursuing the understanding of geotechnical conditions, value the geotechnical input and play their respective roles for the implementation of geotechnical programs.

This cultural change can only be brought about by a systematic and structured approach to conducting geotechnical programs and in particular, geotechnical training. The approach pursued by St Ives Gold has both formal and informal components. The formal geotechnical training consists of selected training courses for different personnel. For example, two geotechnical design workshops were delivered by the Western Australian School of Mines to about 50 WMC managers, engineers, risk control advisors, and geologists based in Kambalda and Norseman. A one day workshop on ground support/reinforcement was delivered by Rock Technology to GBF, a mining contractor working at St Ives Gold. Geotechnical awareness training has been delivered by the geotechnical engineers to all operators working at the underground mines at St Ives Gold. Geotechnical awareness training has become part of the standard site induction program.

The informal training has been conducted by the site geotechnical engineers by engaging the relevant personnel in geotechnical data collection, ground support/reinforcement trials and monitoring, and geotechnical design. Most of the site geologists now have acquired the basic skills in geotechnical logging and mapping. These skills help them identify potential hazards when they are undertaking their routine geological work. This increases safety and helps to rectify any potential hazard.

6 GEOTECHNICAL CHALLENGES

Two major geotechnical challenges facing the industry have been identified (Brown 1994). These relate to technical and education/training challenges. These are still the issues that the WA mining industry has to address today in order to ensure safe and economical mining operations.

6.1 *Increasingly difficult geotechnical conditions and high expectations*

A number of technical challenges are currently being experienced, like deeper mines, and remnant mining. Whilest more deposits are hosted in more complex geological and geotechnical settings as exploration targets new areas and geological structures.

Economic conditions dictate that the current mining methods and sequences are designed to extract orebodies cheaper and quicker. These methods and sequences must be also justified to be viable through geotechnical evaluation.

Industry leaders have realised that mining is no longer being accepted and tolerated as a high risk activity (Morgan 1997). Licences to operate could be at stake as community sentiments change. The values of mining houses have also shifted from the sole profit generating for the shareholders to the well-beings of all stakeholders and the society at large.

All of the above changes suggest an increased demand for quality geotechnical input to bring about safer and more efficient operations. The technical efforts should be accompanied by efforts in training and cultural changes.

6.2 *Overcoming the shortage of qualified geotechnical professionals*

A typical geotechnical/rock mechanics engineer has been someone with a mining, geological engineering or geology degree and several years of mine site based geotechnical experience. A large number of these professionals have undertaken postgraduate training through full-time or part-time studies.

The rapid expansion of the mining industry and the record number of mining and geological related graduates in the last several years has not resulted in an increase in the number of postgraduates in geotechnical engineering.

Currently, there is a severe shortage in the qualified geotechnical/rock mechanics engineers in Australia. This has been overcome to a certain degree by employing overseas-trained geotechnical professionals. However, this process has limitations and delays that often make it impractical for small mining companies.

There are initiatives that could provide long term solutions to this shortage. A graduate geotechnical engineer training program has been established by WMC operations. Postgraduate training through industry and government funded research programs has been implemented, such as that at JKMRC (Julius Kruttschnitt Mineral Research Centre) of the University of Queensland. There is also a successful Master's degree in Mining Geomechanics program offered by WASM on part-time basis since 1993. ACG (Australian Centre for Geomechanics) has been offering various geotechnical workshops or courses.

"On the job" training has been an effective way of training geotechnical professionals for large operations with a strong technical group where more experienced senior rock mechanics engineers are available and geotechnical design and practice standards and procedures are well established. Geotechnical consultants have also been playing a training role for the site based geotechnical engineers or engineers/geologists with geotechnical responsibility.

Geotechnical engineers, unlike geologists and mining engineers, do not normally have a defined career path within a mining organization. It is a challenge for the mines to attract and keep experienced geotechnical engineers for long periods of time. Experienced geotechnical staff on site is required for sound geotechnical design and practice, which leads to sustained improvements on mine safety and efficiency.

7 CONCLUSIONS

An increased effort on geotechnical design and practice has been experienced in Western Australia. A systematic approach to sound geotechnical design and practice has demonstrated that positive impact can be achieved to reduce the number of injuries, incidents and fatalities. There are a number of geotechnicl challenges facing the mining industry and the geotechnical professionals. The magnitude of these challenges requires a continuous improvement on all aspects of geotechnical endeavours.

ACKNOWLEDGEMENTS:

The permission by WMC management to use the materials contained in this presentation is appreciated. The author is grateful to Mr Luke Tonkin, Mr Dave Finn and Mr Peter Crooks of St Ives Gold and Dr Ernesto Villaescusa of WASM for reviewing this document.

REFERENCES:

Brown, E. T. 1993. Geotechnical monitoring in surface and underground mining - An overview. *Proc. Geotechnical Instrumentation and Monitoring in Open Pit and Underground Mining*. Szwedzicki (Ed.) Balkema, Rotterdam. pp3-11.

Brown, E. T. 1994. Australian mining geomechanics - development, achievements and challenges. Short Course Note: Acceptable Risks and Practical Decisions in Rock Engineering Hoek, E. (Ed.). August 1994.

Harris, A. and T., Li. 1995. Mining of the southern 1900 Orebody at Mount Isa. *Proc. 6th Underground Operators' Conf.* Kalgoorlie. 1995.

Harvey, S. 1999. Elimination of fatalities taskforce – underground rockfalls project. International Symposium on Rock Support and Reinforcement Practice in Mining, Kalgoorlie.

Lang, A. 1999. Geotechnical mining regulations in Western Australia. International Symposium on Rock Support and Reinforcement Practice in Mining, Kalgoorlie.

Morgan, H. M. 1997. WMC's number 1 safety objective: Elimination of Fatalities. *Presentation to the Queensland Mining Industry Health and Safety Conference*, Yeppoon, 10 September 1997.

Nickson, S. D. 1992. *Cable Support Guidelines for Underground Hard Rock Mine Operations.* M.Appl.Sci. Thesis. The University of British Columbia. 12/1992.

Potvin, Y., M. Hudyma & H. Miller, 1989. Design Guidelines for Open Stope Support, *CIM Bulletin.* **82**. pp53-62.

Seto, M. and Villaescusa, E. 1998. In situ stress determination by acoustic emission techniques from McArthur River Mine Cores. To be presented *on 8th ANZ Geomechanics Conf.* Hobart. 1998.

Villaescusa, E., Karunatillake, G. and Li, T. 1995. An integrated approach to the extraction of the Rio Grande silver/lead/zinc orebodies at Mount Isa. *Proc. 4th Int. Symp. On Mine Planning and Equipment Selection.* Calgary.

Villaescusa, E. D. Tyler and C. Scott, 1997. Predicting underground stability using a Hangingwall Stability Rating. Proc. 1st Asian Rock Mechanics Symposium, (H.K. Lee, H.S. Yang and S.K. Chung, Editors), Seoul, 171-176.

Villaescusa, E. and J. Wright, 1997. Permanent excavation support using cement grouted split set bolts. *The AusIMM proceedings*. 1997.

WMC. 1997. Annual Report to Shareholders. 1997.

7 Strata control in coal mines

Rock Support and Reinforcement Practice in Mining, Villaescusa, Windsor & Thompson (eds)
© 1999 Balkema, Rotterdam, ISBN 90 5809 045 0

Keynote lecture: Roof strata reinforcement – Achievements and challenges

P.G. Fuller
BFP Consultants Pty Limited, Richmond, Vic., Australia

ABSTRACT: Developments in coal mine strata reinforcement technology, hardware and installation practices are outlined in the paper. It follows the progression from early concepts of roof bolts suspending layers through to reinforcement of a multi-layered roof beam to withstand both vertical and horizontal applied loads.

Advances in strata reinforcement have occurred in most countries where strata-bound deposits are mined but the paper focuses on those made in Australia and only major developments from other countries. In spite of the many achievements made, there are still some challenges in strata reinforcement which have been identified in the paper. Many of these relate to the need for increased reinforcement efficiency as part of the overall roadway development process which is continually being improved without compromising safety.

1 INTRODUCTION

Stabilizing coal mine roof strata by roof reinforcement techniques has its origin in the concept of arching action in earthen materials (soils) originally contemplated by Terzaghi & Peck 1948. In that early work, the existence of cohesion was recognized as being essential for a stable arch to form. Applying this concept to rock mechanics meant that increased undercut spans were possible if the shear strength of the rock mass was sufficient. Evidence of natural rock arches goes back centuries but the idea that stable spans could be increased by increasing the shear strength of the rock mass are comparatively recent.

This paper is intended to review the developments that have taken place with reinforcement techniques to increase the shear strength of roof strata encountered in underground coal mines. Some have regarded the stabilization of laminated strata as being an easier problem to tackle than stabilizing rock masses in general. However, both history and experience has shown that neither issue is straightforward and that perhaps the only simplification provided by strataform deposits is that they are easier to represent in two dimensional diagrams and the mechanics of their behaviour may be a little clearer to visualize.

Strata reinforcement has been an issue in all countries where high productivity underground mining of strata-bound resources has taken place. Advancements in the technology have been made in each location depending on the strata conditions encountered and the mining methods employed. While this paper is intended to cover the development of strata reinforcement in the broadest possible context, it has been necessary to limit its scope to advances made in Australia and only major developments from other countries.

2 EARLY DEVELOPMENT

As with many technologies associated with mining, strata reinforcement evolved from a mix of in-situ trials and physical models that were intended to provide a basis for understanding the action of the reinforcement. From the literature, there appear to be only limited applications of the technique before 1948 but by the mid 1950's more than 700 coal mines in the US had adopted roof bolting as a systematic method of roof support (Panek 1956). However, this change from timber prop and lid supports to roof bolting had occurred with only limited knowledge of the reinforcement principles involved.

Roof conditions in Australian and UK coal mines are more variable than in the US and consequently the change to roof bolting was much more gradual. Although the method was trialled at many pits in the UK, it was not regarded as a viable alternative to steel arches for longwall gateroad support. Most Australian operations at the time were bord and pillar with secondary pillar extraction to maximize recovery. Early trials with roof bolting showed the method to be generally successful in more massive roof conditions particularly where relatively thick sandstone and siltstone bands were present in the immediate roof. In conglomerate roof and where layers were much thinner such as with laminate roof, bolting was less successful but nevertheless it became the industry standard in Australia for roof support of main headings and solids development. Pillar extraction using the Wongawilli method relied heavily on the controlled failure of the roof and from experience it was found that a low density of roof bolting combined with timber props provided the necessary temporary support. Props also provided a visual indication of roof movement and their failure acted as a warning of imminent roof failure.

2.1 *Theoretical Analyses*

The laminated nature of coal mine roof strata allowed theoretical analysis of its stability using beam theory (Panek 1956). Beams were assumed to have fixed ends and were subjected to driving forces equivalent to their self weight. A single member beam analysed in this way closely resembled the behaviour of the lowermost bed in a coal mine roof provided it was not loaded by an overlying bed or supported from below. Stiff end constraints were needed from pillars to limit the potential for end rotation to develop.

Multiple layers were then analysed assuming no cohesion existed between layers and this was compared with a single beam of equivalent thickness to demonstrate that increasing the interlayer shear strength by bolting could substantially reduce both the bending stresses and deflection (Panek 1956). Early beam analyses showed the potential that bolting had to improve the stability of layered strata but because of the inherent assumptions in the analyses concerning:

- end constraints,
- driving forces involved and how these changed with downward movement of the beam, and
- lateral homogeneity within layers,

many regarded them to be of limited real value as a basis for roof support design.

It was unclear from early beam analyses whether the roof bolts were acting to reinforce a laminated beam or whether lower layers were simply suspended by roof bolts from more competent bands above. The reinforcement effect involved holding layers in contact as shown in Figure 1 to build a thicker beam.

Physical model tests produced results that indicated the overall stabilizing effect of bolts on layered strata to be a combination of both reinforcement and suspension (or support). As a result, the direction of analysis work moved to consider the combined effects of support and reinforcement of roof bolts (Panek 1963). The support effect was found to be substantial when bolts were anchored into a thicker and more rigid layer and when the bolts were pre-tensioned on installation, as was usually the case because mechanically anchored bolts were being used at that time. Reinforcement of the strata by increasing interlayer friction led to reduced bending and hence reduced downward deflection of all layers. The relative effects of support versus reinforcement on the deflection of a particular layer were found to depend on the sequence of layer thicknesses present and roof bolt pre-tension and were difficult to generalize. However, the net influence on layer stability was represented by the support effect multiplied by the reinforcement effect rather than by the sum of these two factors.

Beam analysis studies culminated in a bolting design for stratified roof (Panek 1964). Despite its title, this work appeared very theoretical and represented more an attempt to transfer the results of previous studies on support and reinforcement effects into practice than to provide a useful roof bolt design manual. The simplifying assumptions on which the work was based were adequately stated but not included as limitations to the use of the design. As a result, the design approach was rarely used. It was regarded by those responsible for specifying roof bolting as a theoretical statement of roof stabilization mechanics and in that sense it was an achievement, but not one that could be easily put into practice.

By the mid 1960's it became apparent that the early beam analyses of layered structures could only partly explain the experience of bolted coal mine roof. These had assumed fixed end conditions and had ignored any effects of lateral stress, either active or passive due to arching. With stiff coal pillars beam behaviour could be approximated by fixed end

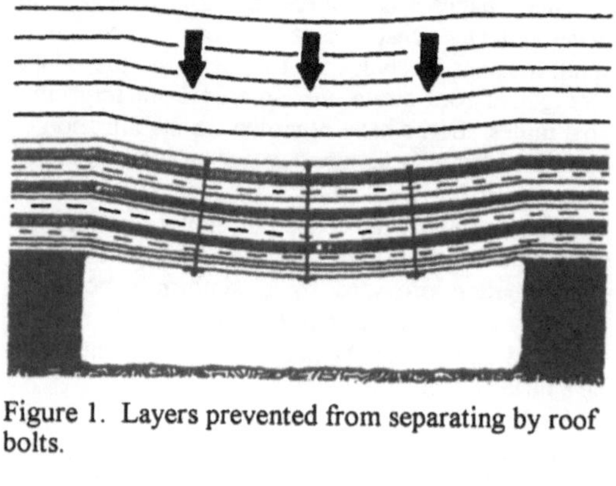

Figure 1. Layers prevented from separating by roof bolts.

formulae. With softer coals and where cleat was sub-vertical, behaviour ranging between fixed end and simply supported was more representative.

Lateral stress effects only started to be taken seriously in the early 1970's (Wright 1974) and the additional influence of both lateral confinement and applied lateral stress on beam behaviour was reflected in the equations in the SME Mining Engineering Handbook of that era. No direct implications on roof reinforcement requirements were foreshadowed from this broadening of the initial beam studies to include lateral stress but later experience suggested that there should have been.

2.2 *Roof Bolt Hardware*

Bolts used in initial trials to reinforce stratified coal mine roof had a slit end into which a steel wedge was driven during the installation. It appears that the requirement for the bolt to be active immediately on installation precluded the use of any continuous bonding medium such as cement to achieve an anchorage with the strata. Bolt diameters were relatively large (usually nominal 25mm) and of low carbon mild steel. A reaction at the exposed roof was achieved with a flat, mild steel plate and a nut, which allowed bolts to be pre-tensioned to a nominal 3 to 5 tonnes.

Early experience with this type of slot-wedge anchor was that it provided inadequate anchorage in soft strata conditions which were frequently encountered. This deficiency was overcome by an expansion shell anchor but the operation of the bolt as a point anchored bar remained the same. These bolt types matched the requirement of early theoretical analyses in which bolts were to be pre-tensioned to the highest possible loads not exceeding 60% of the bolt yield load to maximize clamping forces and interlayer friction. At that stage the need for dowel action to laterally pin layers together was not recognized as a need for reinforcement.

Bolt anchorage remained the main practical issue during the 1960's. In coal roof support applications flat steel plates were found to be adequate for most applications. Occasionally dished plates and spherical washers were used to handle any plate-bolt misalignment but the stiffness of these was never regarded as being an issue. Wooden blocks were used under flat plates to increase the bearing area. These reduced the compressive stiffness of the overall bolt to strata interaction but this was not considered in early applications of the method. Users and suppliers of roof bolt products at that time were focussed on load capacity of each installed bolt and hence the total support load capacity of an installed array of bolts.

By the late 1960's, roof bolting had become a proven alternative to timber props in all but the worst roof strata conditions and in pillar recovery headings. Designs were based on providing support load capacities at least equal to the weight of the bolted beam created by the bolts and this was found to work adequately in the majority of conditions. Although reinforcement of the strata was recognized as part of the function of the bolting, this could not be incorporated into any practical and useable design.

The above attitude and approach had become entrenched in the industry and many operators considered any attempts to develop the technology further to be futile. However, roof falls continued to occur and it was clear that despite the general success of the method, there was a lot that was still not understood.

3 MAJOR ADVANCES IN STRATA REINFORCEMENT

Development of strata reinforcement from the initial experience of the 1950's and 60's has been driven by a combination of:

- the need to reduce the incidence of roof falls, and
- a desire to maximize roadway development rate productivity which is strongly dependent on the amount of roof bolting that needs to take place.

Poor connection between the roof bolts and the strata was recognized as a major problem that was responsible for many roof falls. Economic pressure on the industry meant that each installed roof bolt had to be more effective in stabilizing the roof. Thus an environment was created in which a better understanding of strata stabilization was required and the products available needed to be more efficient in their function.

The early success of roof bolting in Australia had been due in part to the level of conservatism that was inherent in the initial trials. An example of the transition from timber props to roof bolting is shown in Figure 2 (Samuel 1980).

Conservative designs became part of the roof support rules set for each colliery by the registered manager and were to some extent "cast in stone" unless there were well founded reasons for change.

Figure 3 shows an example of such support rules. An analysis of operating continuous miner panels for one year (1979-80) involving the twenty continuous miners showed that on average 30.8% of available machine operating time was lost due to roof bolting (Grant 1980). Thus, there was significant scope to improve overall roadway development rate productivity by reducing roof support densities and reducing the time spent on roof bolt installation. An example of the need for improvement was expressed by Pearce 1976 as follows:

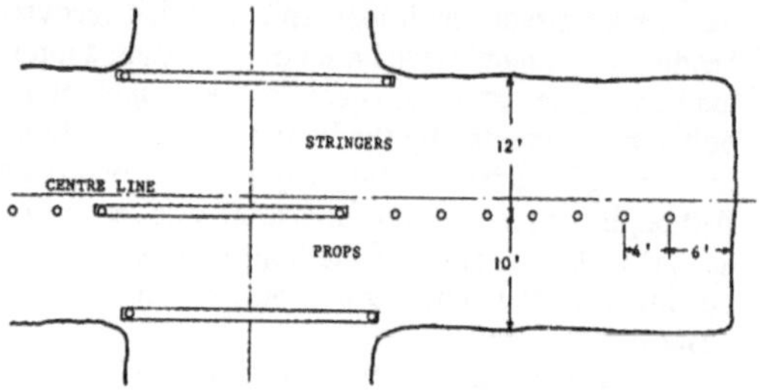

Figure 2 (a) before 1957.

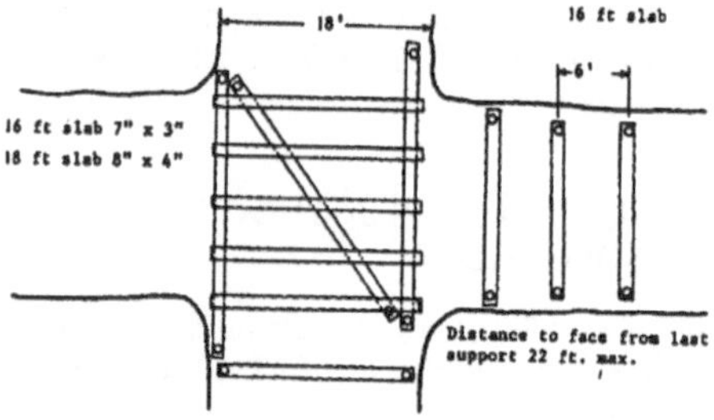

Figure 2(b) early 1969

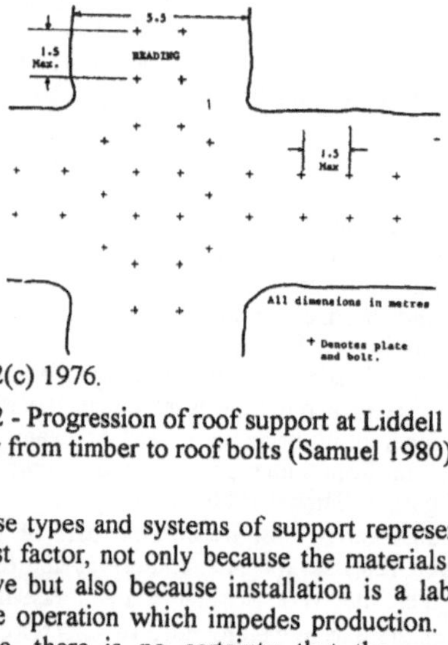

Figure 2(c) 1976.

Figure 2 - Progression of roof support at Liddell Colliery from timber to roof bolts (Samuel 1980).

"These types and systems of support represent a high cost factor, not only because the materials are expensive but also because installation is a labour intensive operation which impedes production. At this time, there is no certainty that the support system chosen is more or less than required for the conditions."

In addressing these needs significant advancements in strata reinforcement have been made in three areas:

- Products
- Reinforcement technology
- Installation methods

Developments within all three have been interactive and there are many examples where improved understanding of the technology has prompted the design of a new product that is amenable to high speed installation.

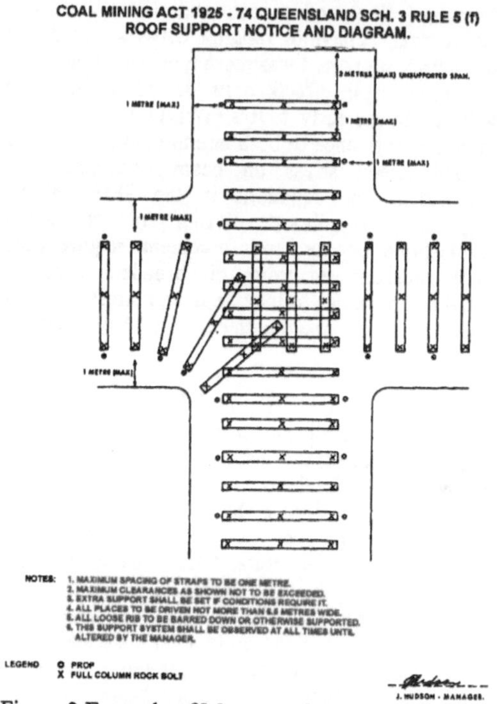

Figure 3 Example of Managers Support rules (Sleeman 1980).

3.1 Reinforcement products

Since the early days of roof reinforcement where only a very basic range of reinforcement products was available, a steady flow of new components and systems has become available. One of the first was the polyester resin anchor to improve the point anchorage capacity of roof bolts in weak strata. The main advancement was not the resin but the two part packaging of polyester resin and filler with catalyst in pre-defined proportions to allow simple and rapid achievement of a short bonded length of the roof bolt to the strata. Anchor tests showed that bond lengths as short as 200mm in some conditions was sufficient to at least develop the yield load in standard M24 steel bars. Hole size was found to influence anchor strength and 5-6mm difference between hole and bolt diameters provided highest anchorage efficiency with most bar type bolts.

Roof bolt design has progressed steadily and the number of alternatives has increased particularly since 1990. Bolts are now available with many external shapes to achieve improved resin mixing and bond strength.

In addition most roof bolt manufacturers now offer a range of steel grades for higher volume product sizes. Demand for higher bolt tensile capacity (both yield and ultimate) has encouraged manufacturers to develop larger diameter bolts.

Bolts are now available in the following nominal sizes: 16mm, 20mm, 24mm and 28mm.

Roof bolt accessories have increased in line with the greater range of bolts. Plates now come in flat and preformed profiles, the latter providing increased stiffness and a greater range of angular misalignment (Gray 1998, Rataj et al 1998).

Cable bolts were introduced into Australian coal mining in the early 1980's (Windsor 1992) primarily as secondary or supplementary roof reinforcement for roadway intersections and in regions of longwall gateroad where poorer roof conditions were encountered. Initially these were either single or twin 15.2mm diameter strands grouted in 50mm diameter boreholes with cement grout as in metal mining. Cable bolts have the advantage over conventional bar bolts that their flexibility allows longer bolt lengths to be installed where the mined height is usually much less than in metal mine development. Standard cable bolt installation was time consuming and frequently interfered with transportation requirements of a roadway so it tended to be used only in potentially unstable areas. Where poor roof conditions were encountered due to either high horizontal stress or weak, highly laminated roof strata, two new forms of cable bolts; Flexibolt and Megabolt can be used. These were developed in Australia to be installed as part of the primary or secondary bolting cycle with resin anchors and to become active at the same time as conventional bar type bolts. Flexibolt is a 23.5mm diameter 21 wire strand cable of 55 tonnes nominal tensile strength with a threaded end to take a drive nut of the same external dimensions as a 24mm roof bolt (Fuller & O'Grady 1993). Even higher tensile capacities are possible with the Megabolt. It is constructed from multiple 7mm diameter high tensile wires and the four types available have nominal tensile strengths of 50, 63, 84 and 90 tonnes (Hutchins et al 1997).

Although long lengths can be installed with either of the above cable types, there are limits to the length which can be fully encapsulated with resin. Cables longer than 4m usually require secondary grouting with cement and a simple and quick method of achieving this remains a challenge. However, there is no doubt that their development has enabled mines with very heavy roof conditions to continue to operate with roof reinforcement and not have to revert to timber cribs within the opening (McCowan 1994, Fuller et al 1994).

Stress corrosion cracking of roof bolts has been recently identified as a potential problem with partly encapsulated roof bolts subjected to stress in a corrosive environment (Gray 1998). The ribbed surface of most bar type bolts can create stress raisers on the surface from which stress corrosion cracks can initiate in the right environment. Bolts more resistant to stress corrosion cracking can be produced but a challenge for manufacturers is to provide a product range where at least some bolts can be used in more highly corrosive environments.

Development of polyester resins has been a significant factor in the progressive improvement of the reinforcement effects of roof bolts and cables. The ability for bolts to be partially or fully encapsulated in a single pass operation has only been possible because of defined set time resins being available and appropriately packaged. A challenge for resin anchor suppliers is to be able to quantify the effects of shelf life at given shelf temperature and resin temperatures on resin mix and hold times. Operators in some high ambient temperature areas are often plagued by resins not performing as expected.

3.2 Reinforcement Technology

Two major advances in the understanding of roof strata mechanics were made in the 1970's as a result of both in-mine experience and measurement of downward movement in reinforced strata. The first was that minimization of strata separations and roof movement was essential to maintain the structural integrity of mine roof (Karabin & Debevec 1976). This was a significant move away from the notion that bolts needed to have the potential to support at least the weight of the bolted roof, regardless of its deformation.

The second was that field experience indicated that untensioned fully encapsulated bolts performed much better than mechanically anchored bolts in supporting layered roof (Fairhurst & Singh 1973). The reason for this was initially thought to be due to the increase in interlayer shear strength due to a dowelling effect from the bolts. It was soon realized that bedding plane shear resistance could develop with untensioned bolts because the axial bond strength of the resin bolt combination was so high that driving forces due to gravity were insufficient to cause parting between layers.

The mechanics of this new way of achieving strata reinforcement without the need to pre-tension bolts took some years to be clearly understood. Fully bonded bolts were able to provide both axial restraint and shear resistance to layered strata as shown in Figure 4. Theoretical work (Fairhurst & Singh 1974) with a simplified shear model for the dowelling effect of each fully bonded bolt confirmed that significant shear resistance between layers could be developed by this system and that the improved roof conditions observed in practice were due to the more effective reinforcing properties of these bolts. However, this work could not be used for structural design of a reinforced layered system but it provided a basis for fine tuning a bolting pattern that had been judged as successful in a mine.

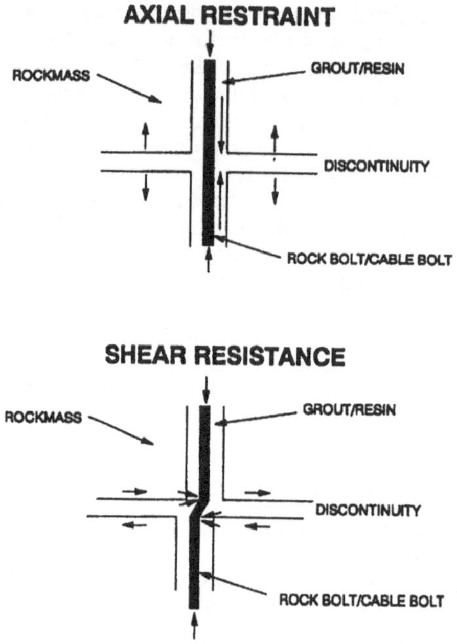

Figure 4. Axial and shear reinforcement from fully bonded bolts.

The above advancement in strata reinforcement indicated both the axial and shear stiffness of fully bonded bolts to be important. Bedding plane shear movements were predicted (Fairhurst & Singh 1974) to only develop a few tonnes force in each bolt so there was considerable potential to increase the reinforcement effect by pre-tensioning the bolts prior to setting of the grout column. This conclusion was virtually ignored for 15 years during which fully encapsulated bolts were considered to perform adequately without being pre-tensioned. Fully bonded bolts required a high axial stiffness as highlighted by Gale et al 1988. Figure 5 (Fabjanczyk & Tarrant 1992) shows that high bond stress is required to utilize the capacity of higher strength bolts and to guard against high plate and hole collar loads.

Pull out testing of fully bonded bolts was used to define the comparative performance of different bolt-bonding medium-borehole combinations. There was an initial pre-occupation with load capacity from these tests but as the issue of displacement became more prominent, so the complete pullout load-displacement curve became the standard means of presenting bond strength results. The question of what bonded length should be tested was a topic of debate during the 1980's and because pullout load-displacement performance has a strong dependence on bond length, there was need to standardize test lengths. A 50mm long push test (Fabjanczyk & Tarrant 1992) was intended to become such a standard but many practitioners had reservations about how representative a push test with independently mixed resin would be.

Shear behaviour of fully bonded bolts has received much less attention than axial performance and is more difficult to describe because of the number of variables involved. Bolt angle to the shear surface is known from shear testing to significantly affect shear behaviour (Bjurstrom 1974, Haas 1976, Windsor & Thompson 1993). This has highlighted the potential increases in shear stiffness that could result from angled bolting across bedding planes. Theoretical analyses of shear behaviour of fully bonded bolts and cables (Fuller & Cox 1978, Dight 1982) showed that bolt bending and its alignment in the shear direction affects shear stiffness. Bonding efficiency also influenced shear stiffness (Dight 1982). More recent work (Goris et al 1996) has shown significant increases in shear resistance from a resin bonded single strand cable bolt installed normal to both rough and smooth joints. At 5mm shear displacement the shear resistance was increased by 60% due to the cable.

Instrumentation of roof bolts has been significant in improving the understanding of load profiles that develop in a pull out test and in arrays of bolts in-situ. Direct strain gauging of bolts to detect tension and bending at points along the bar (Karabin & Debevec 1976, Wade et al 1976, Gale et al 1988) has identified the early onset of yield in fully resin encapsulated roof bolts and the need for high yield strength bolts in applications where driving forces are high. The effect of a range of roof strata conditions on bolt load development has only been possible to determine with strain gauged bolts. Problems with bonding of resin to strata in weak ground has shown up as low load development in bolts and hence poor utilization of installed reinforcement capacity. The behaviour of arrays of bolts has been examined with instrumented bolts and asymmetry in load development across the roof has been commonly observed (Gale et al 1988). Load cells have occasionally been used on bolts but these only provide realistic results if their load-deformation response is very stiff and bolts are partially grouted.

The variability of bolt load and plate load measured with the above techniques highlights the difficulty in providing an engineering based design for a roof reinforcement pattern. Clearly this represents an on-going challenge for strata reinforcement.

The effect of multiple layer separation on axial load development in bolts and cables which are fully or partially bonded has been simulated in a "PC Windows" based package (Fuller et al 1996). This required the roof displacement profile and bolt

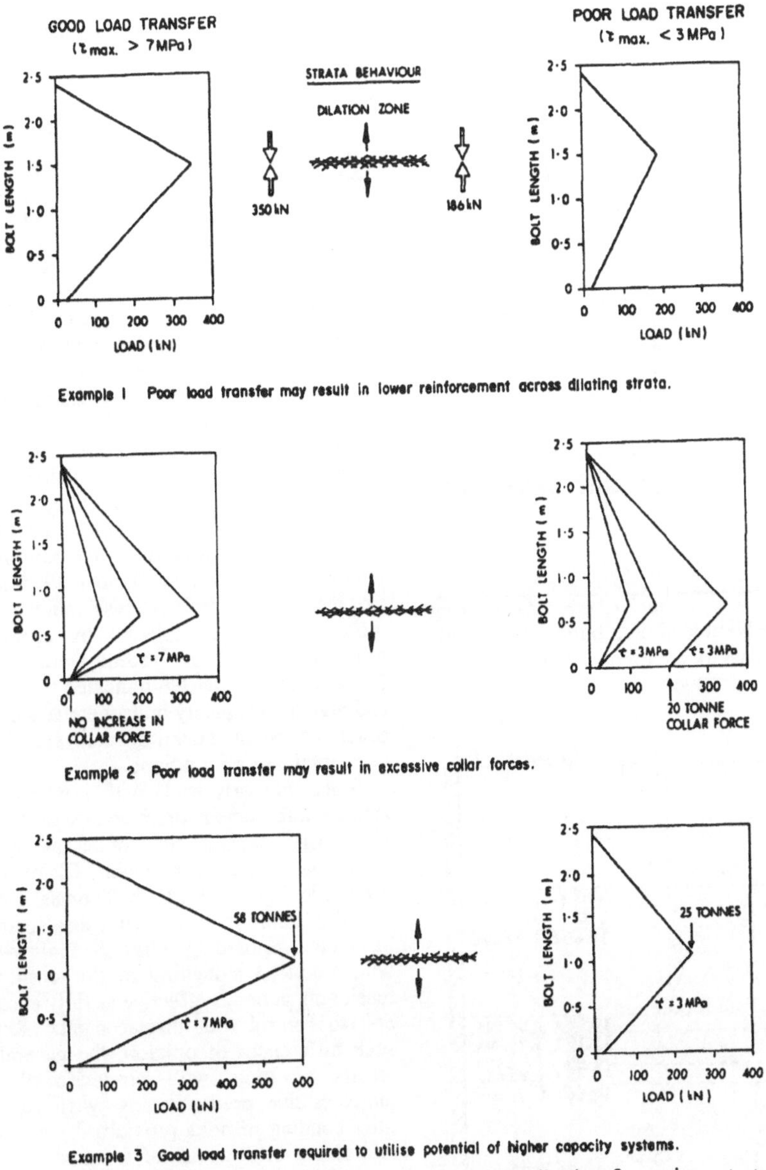

Figure 5. Effect of bond strength on bolt load distribution for various strata separation (dilation) zones. (Fabjanczyk & Tarrant 1992).

bonding properties as input and shows the relative effects of different bolt types and location of separation planes on the bolt load profile. An example of bolt load in a 2m thick bolted horizon is shown in Figure 6 where 14mm total separation has taken place at three locations through the bolt. Note that a plate load of 1.4 tonnes is also predicted. If the same displacement occurred on multiple separation planes the bolt is more evenly loaded and peak loads are smaller (Figure 7).

In-situ field testing and performance measurement, laboratory testing of shear and axial bolt properties and analytical simulation have all contributed significantly to a much improved understanding of bolt mechanics and their reinforcement function.

Until the early 1990's, pre-tensioning of roof bolts was only considered essential for point anchored bolts which had all but disappeared from use as strata reinforcement. Some minor post-tensioning of

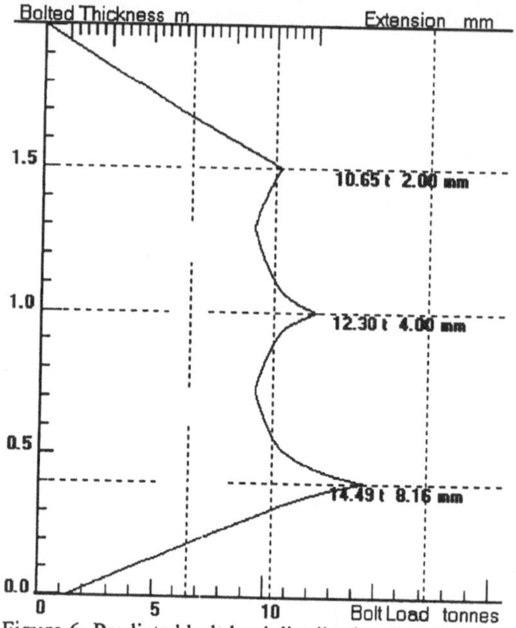

Figure 6. Predicted bolt load distribution (bold line) for 14mm strata separation confined to three horizons. (Fuller et al 1996).

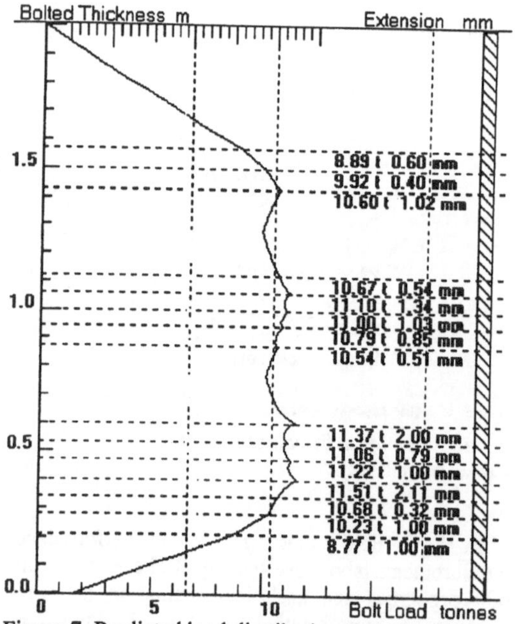

Figure 7. Predicted load distribution (bold line) for 14mm strata separation on multiple planes. (Fuller et al 1996).

fully bonded bolts occurred when plates were attached but this was more a consequence of tightening the nut to seat the plate rather than any intended design. In the last few years, tensioning of roof bolts can cables has re-emerged at least in the eyes of some practitioners as a practically feasible method of increasing strata reinforcement for minimal extra cost. It also had the potential to allow roof bolting densities to be reduced and for necessary improvements in roadway development rate productivities to be achieved (Frith & Thomas 1994). Higher installed roof bolt tensions have been possible by installing a thrust bearing between the nut and domed washer in Figure 8 (Frith & Thomas 1994). Trials of pre-tensioned roof bolting at a number of Australian collieries produced results interpreted in terms of a more competent beam in the immediate roof which was better able to withstand horizontal stresses without low angle shearing or guttering occurring (Frith & Thomas 1995). One consequence of this "beam building" was an increase in overall roof movement before any adverse signs of imminent roof instability were observed. This was interpreted by Frith & Thomas 1995 as indicating that the "effect of bolt pre-tensioning is to make roof more displacement tolerant by promoting better beam action". More recent work in the US by Iannacchione et al 1998 indicated that high stiffness and high load capacity bolts were needed to limit the onset of beam buckling and shearing in high horizontal stress conditions.

Since the early trials with pre-tensioning, some strata reinforcement practitioners in Australia have vigorously debated its value as a means of improving roof reinforcement. Claimed benefits of pre-tensioning by Frith & Thomas 1998 and the resulting improvement in roof displacement control have been refuted by Gray & Finlow-Bates 1998 who produced modelling results to show that pre-tensioning is not as effective as fully encapsulating a bolt to control roof displacement. Regardless of such differences of opinion, the generally positive results of trials with pre-tensioned roof bolts indicates that pre-tensioning when combined with fully bonding of bolts provides the maximum strata reinforcement potential of a particular roof bolting array. This is consistent with the previous predictions of Fairhurst & Singh 1974 and is probably due to fully bonded pre-tensioned bolts offering the maximum possible resistance to strata separation and shear. If the results to date continue to be substantiated in the long term, the techniques to develop high installed pre-tension in fully bonded roof bolts will be a significant strata reinforcement achievement.

Cable bolts and in particular Flexibolts, the Hi-Ten variant and Megabolts can potentially be pre-tensioned to high load levels without the risk of tensile failure due to overload. Fuller et al 1998 and Rataj et al 1998 have shown that 6-8m long resin

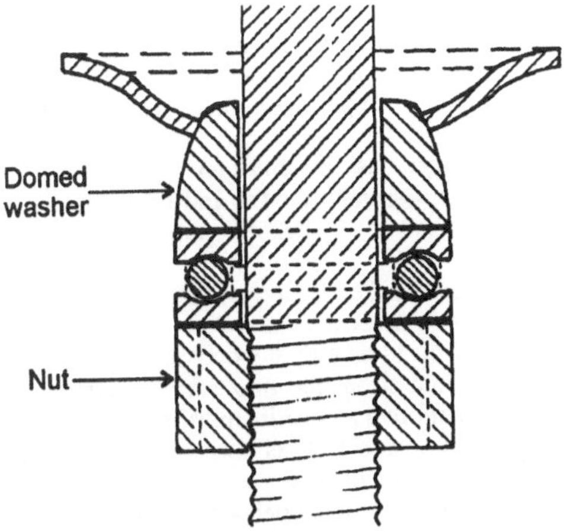

Figure 8. Schematic bolt head assembly with thrust bearing (Frith & Thomas 1994).

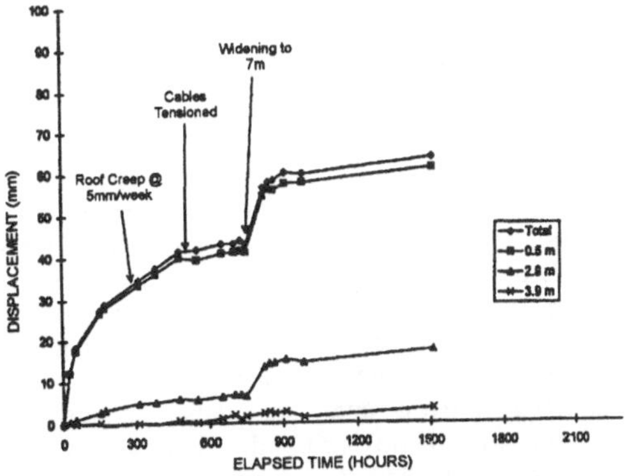

Figure 9. Roof displacement at three horizons; LW307 installation roadway, German Creek Coal Operations (Rataj et al 1998).

anchored high load capacity cables in a primary bolted roof can significantly decrease the rate of roof displacement (Figure 9). This is an example where a combination of point anchored and fully bonded reinforcement technologies has improved the mine operator's ability to limit the rate of roof displacement and thereby prevent its subsequent deterioration. A higher stiffness reinforcement system would be possible if the cables could be pre-tensioned and fully bonded. Limited trials of this combination have been undertaken at Wyee Colliery (Fuller et al 1998) with inconclusive results and at this stage, its practical application remains an operational challenge.

3.3 Installation Methods

Early reinforcement installation is critical in terms of its ability to control movement in laminated roof subjected to high horizontal stress. This has been made possible by significant advancements in roof bolting rigs mounted on the continuous miner. Initially two side mounted rigs and one central bolting rig (Sleeman 1980) allowed roof bolting to within approximately 2m of an advancing face. Bolt installation took place between face cutting cycles and consequently had a negative impact on machine productivity and roadway advance rate. A major advance was made in the early 1990's with the Voest-Alpine ABM-20 continuous miner which incorporated up to four machine mounted roof bolters and two machine mounted rib bolters (Melrose 1994). This machine enabled simultaneous coal cutting and roof and rib bolt installation. Multiple bolts could be installed simultaneously which decreased the roof bolt cycle time. Variants of this concept are now available from other manufacturers and this more mechanized approach to bolting is now common in longwall mining operations.

Secondary bolting with cables usually takes place with a dedicated crew operating stand alone equipment. Recent improvements have been developed in both longhole drilling equipment and cable bolt placement machines to reduce the amount of manual labour involved and to increase the overall speed of installation. Grouting remains a challenge not because of any difficulties but rather a lack of clear direction of industry requirements in terms of what would be acceptable in practice.

3.4 Reinforcement Design

In spite of the many technical advances that have been made in reinforcement technology and methods, its design in a formal sense remains a challenge. Gale et al 1994 have devised a design procedure based on computer modelling of strata behaviour and Seedsman 1994 has incorporated a roof bolt model into a layered beam model to create an analytical design method. Seedsman 1997 and Smith & Seedsman 1998 represent the most recent attempts at formulating a rational design method but like all previous methods, these have not gained any acceptance by the coal mining industry. The reasons are that end users are sceptical of anyone's ability to adequately define the driving forces involved and roof deformation and failure processes in any model for design.

Formal design that is used in practice will probably remain a challenge and it is debatable whether it is needed anyway. It can be argued that the most responsible approach to defining strata reinforcement would be to use experience from similar sites with similar conditions, to deliberately over-engineer a bolting array and to monitor its performance after installation. Refinements would then be made based on the results following face advancement and any secondary stress effects such as a retreating longwall. It is a more empirical approach and one which takes account of driving force changes, variability in bolt installation quality and local variability in strata geometry and mechanical properties. This approach needs some rational basis for making the refinements. An example might be that the height of softening in the roof should not exceed some limit before primary bolting is supplemented by secondary reinforcement. It may also be possible to relate the height of softening to downward displacement of the reinforced horizon and this could be the key indicator of stability and the trigger for any secondary reinforcement.

From an operations perspective, such an empirical approach to strata reinforcement design would be more cost effective and reliable than trial and error methods and much easier to use than analytical or stress analysis based methods. An empirical design would bring together the extensive experience with strata reinforcement that now exists and provide a platform for future improvement.

4 MAJOR ACHIEVEMENTS AND CHALLENGES

The most significant achievements in strata reinforcement since its introduction in the 1950's have been:

1. The change away from load only and suspension concepts to the approach that relates load in the reinforcement with displacement of the strata.
2. Development of two part polyester based resin packaging to allow relatively straightforward and rapid roof bolt installation.
3. The understanding of how resin bonded bolts differ in their operation from point anchored bolts and the improvement in both bolt steel quality and external surface to improve bonding efficiency.
4. Flexible cable bolting systems to allow relatively long cables to be rapidly installed as both primary and secondary support.
5. Roof bolt and strata instrumentation to allow bolt load and downward strata movement to be measured in-situ.
6. The re-emergence of roof bolt and cable pre-tensioning to provide improved control of roof deformation during development and when subjected to longwall induced stress changes.
7. Miner mounted bolting rigs that allow simultaneous roof bolt installation and coal cutting.

In spite of the many achievements made, challenges still remain in strata reinforcement to provide safe working conditions while allowing roadway development rate productivity to rise. Major challenges are:

1. To create a reinforcement design approach which is straightforward to use, reliable and incorporates an appropriate level of monitoring to allow secondary support to be installed to maintain stable roof conditions.
2. To further quantify and better understand the benefits of pre-tensioning bolts and cables and roof conditions that will benefit most from pre-tensioning.
3. To improve the placement methods cable type reinforcement products to allow highly pre-tensioned cables to be fully bonded on installation.
4. To communicate the improved understanding of strata reinforcement principles to mine crew members and ensure that reinforcement installation is of consistent high quality.
5. To access mineable coal with more productive methods of roadway development and strata reinforcement such that safety, particularly against roof falls is never compromised.

5 REFERENCES

Bjurström, S. 1974. Shear strength of hard rock joints reinforced by grouted untensioned bolts. Proc. 3rd ISRM Congress. Denver : 1194-1199.

Dight, P.M. 1982. Improvements to the stability of rock walls in open pit mines. PhD thesis, Monash Univ.: 4.3.1.

Fabjanczyk, M.W. & G.C. Tarrant 1992. Load transfer mechanisms in reinforcing tendons. Proc. 11th Int. Conf. On Ground Control in Mining. Wollongong: 212-218.

Fairhurst, C. & B. Singh 1973. A theoretical study of the stability of bolted - laminated mine roof. Proc. Symp on Protection against Rock Fall. Katowice, Poland, I-5: 1-24.

Fairhurst, C. & B. Singh 1974. Roof bolting in horizontally laminated rock. Eng. & Min. J: 80-90.

Frith, R. & R. Thomas 1994. Roofbolt pretensioning and its effect on roof stability. Proc. Aust. Coal Assoc. Underground Roadway Workshop. Brisbane: 1-5.

Frith, R. & R. Thomas 1995. Roadway roof stability and its attainment through pretensioned bolting. ACARP Proj. Rep. 3032:1-15.

Frith, R. & R. Thomas 1998. The pretensioning pedigree. Australia's Mining Monthly Supplement-Underground Equipment and Technology. July: 68-72.

Fuller, P.G. & R.H.T. Cox 1978. Rock reinforcement design based on control of joint displacement-a new concept. Proc. 3rd Aust. Tunnelling Conf. Sydney : 28-35.

Fuller, P.G. & P. O'Grady 1993. Flexibolt flexible roof bolts. Proc. 12th Conf. on Ground Control in Mining. Morgantown, West Virginia: 24-34.

Fuller, P.G., P. O'Grady, B. McCowan & R.K. Butcher 1994. Operational experience with Flexibolt systems in Australian coal mines. Proc. 13th Int. Conf. On Ground control in Mining. Morgantown, West Virginia: 18-31.

Fuller, P.G., B.G. Hume & R.G. Hume 1996. Bolt load simulation and its practical application. Proc. North Amer. Rock Mech. Symp., Montreal. 1:187-193.

Fuller, P.G., M. Bijelac, I.R. Hulls & T. Rochford 1998. Prestressing of strands to improve cable performance. ACARP Proj. Rep. 4025: 1-38.

Gale, W., M. Fabjanczyk & G. Tarrant 1994. Design, configuration and control of roadway reinforcement systems. Proc. Aust. Coal Assoc. Underground Roadway Workshop. Brisbane: 1-4.

Gale, W.J., M.W. Fabjanczyk, J.A. Nemcik & S.M. Matthews 1998. Strata control considerations for improved mine roadway and panel driveage rates. Proc. AusIMM Symp.-21st Century higher production coal mining systems. Wollongong: 98-102.

Goris, J.M., L.A. Martin & R.P. Curtin 1996. Shear behaviour of cable bolt supports in horizontal bedded deposits. Proc. 15th Int. Conf. On Ground Control in Mining. Golden, Colorado: 511-521.

Grant, G. 1980. The effect of roof supporting on productivity. Proc. AusIMM Roof Support Colloquium: 7.1-7.9.

Gray, P. 1998. Bearing plates: New developments in the unsung heroes of ground support. Proc. 1st Aust. Coal Operators Conf. Wollongong: 167-178.

Gray, P. 1998. The pre-tensioning placebo. Australia's Mining Monthly Supplement-Underground Equipment & Technology. July: 64-68.

Haas, C.J. 1976. Shear resistance of rock bolts. Trans. Soc. Min. Eng., AIME. 260: 32-41.

Hutchins, W.B., R.R. Van Laeren & J. Hetherington 1997. A new high capacity cablebolt. Proc. Int. Symp. On Rock Support. Lillehammer, Norway: 528-539.

Iannacchione, A.T., D.R. Dolinar, L.J. Prosser, T.E. Marshall, D.C. Oyler & C.S. Compton. 1998. Controlling roof beam failures from high horizontal stresses in underground stone mines. Proc. Int. Conf. On Geomechanics/Ground Control in Mining & Underground Construction. Wollongong: 525-541.

Karabin, G.J. & W.L. Debevec 1976. Comparative evaluation of conventional and resin bolting systems. US Dept of the Interior, MESA: IR1033.

McCowan, B. 1994. New roof and rib support technology; a major breakthrough for Ellalong Colliery. Proc, Aust,. Coal Assoc. Underground roadway Workshop. Brisbane: 1-5.

Melrose, R. 1994. The application of the Alpine Bolter miners in comparison to standard wide head continuous miners under various roadway conditions. Proc. Aust. Coal Assoc. Underground Roadway Workshop. Brisbane: 1-6.

Panek, L.A. 1956. Principles of reinforcing bedded mine roof with bolts. US Bureau of Mines: RI 5156.

Panek, L.A. 1963. The combined effects of friction and suspension in bolting bedded mine roof. US Bureau of Mines: RI6139.

Panek, L.A. 1964. Design for bolting stratified roof Trans. Soc. Min. Eng., AIME. 229: 113-119.

Pearce, H.L. 1976. Standard type supports and costs. AMIRA Symposium on Influence of Excavation Design and Ground Support on Underground Mining Efficiency and Costs, Univ. of Wollongong: 87-101.

Rataj, M.,C. Hanson & R. Frith 1998. Innovative roof bolting technologies in Australian coal mines. Proc. Int. Conf. On Geomechanics/Ground Control in Mining and Underground Construction. Wollongong: 453-470.

Samuel, J. 1980. Development of heading support at Liddell Colliery. Proc. AusIMM Roof support Colloquim:12.1-12.7.

Seedsman, R. 1994. An engineering design approach to roadway support. Proc. Aust. Coal Assoc. Underground Roadway Workshop. Brisbane: 1-6.

Seedsman, R.W. 1997. Reinforcement design for laminated strata under high horizontal stresses. Proc. Int. Symp, on Rock Support. Lillehammer, Norway: 341-347.

Sleeman, J. 1980. Development and operation of on-the-miner bolting rigs at Harrow Creek trial colliery. Proc. AusIMM Roof support colloquium: 11.1-11.10.

Smith, M. & R. Seedsman 1998. A safe way to reduce roof support costs and improve safety and productivity. Proc. 1st Australasian Coal Operators Conf. Wollongong: 188-194.

Terzaghi, K. & R.B. Peck, 1948. Soil Mechanics in Engineering Practice. New York: John Wiley & Sons: 199-200.

Wade, L.V., A.J. Kwitowski & J.F. Judeikis 1976. Investigation of full column resin bolt reinforcement mechanisms. USBM rep.

Windsor, C.R. 1992. Cable bolting for underground and surface excavations. Proc. Int. Symp. On Rock Support. Sudbury, Ontario: 349-366.

Windsor, C.R. & A.G. Thompson 1993. Rock Reinforcement-Technology, Testing, Design and Evaluation. Comp. Rock Eng.: Principles, Practice and Projects. Pergamon Press. Oxford. 4: 451-484.

Wright, F.D. 1974. Lateral thrust, bedding and jointing in roof stability calculations. US Bureau of Mines. IC 8630: 86-91.

Rock Support and Reinforcement Practice in Mining, Villaescusa, Windsor & Thompson (eds)
© 1999 Balkema, Rotterdam, ISBN 90 5809 045 0

Innovations on stability control of roadways within soft strata in underground coal mines in China

C.Wang
Department of Mining Engineering and Mine Surveying, Western Australian School of Mines, Kalgoorlie, W.A., Australia

ABSTRACT: This paper presents three different and innovative methodologies for stability control of underground roadways within soft rocks in coal mines. The mechanism of destressing above a roadway is analyzed. The injection through rockbolts with cement grout for excavation stabilization is also described. The results are compared to those obtained using steel arches. Finally, a two-dimensional yieldable floor heave control mechanism is described.

1 INTRODUCTION

In general, rocks with unconfined compressive strength less that 30Mpa or 20Mpa can be classified as soft rocks. Criteria used to identify a rockmass as soft rock also include high density of discontinuities and fractures, high void ratio and water absorptivity along with high expansivity. Being argillously cemented, prone to deterioration when water is met or having a high softening coefficient factor are often recognized as the common characteristics used to specify a rock or rockmass under the soft rock category.

The behaviour of roadways excavated in soft rocks usually share the following features.

- Rapid reaction with a high rate of deformation in response to ground pressure redistribution.
- Evident rheological deformation.
- High sensitivity to stress disturbance and water invasion.
- Deformation occurring in all directions of a roadway cross section.
- Predominant floor heave in roof-to-floor convergence.

Considering the previous characteristics and the behavior of roadways in soft rocks, roadway support and stability maintenance should be designed in accordance with the following basic principles.

- Where possible, the first consideration should be to change the stress state or improving the mechanical properties of surrounding rocks where the roadway is accommodated, so that rock's capacity of self-stabilization is able to be brought into play to the full extent.
- To minimize the impact of abutments produced by a mining operation and roadway excavation by optimizing the sequence and rationalizing the entire underground roadway layout.
- To seal the surface of surrounding rock in order to protect the rock mass from deterioration as a result of water invasion.
- Preference of support pattern should be put on yieldable support instead of rigid support.
- In case of roadway within extremely soft strata, enclosed heavy support is necessary to provide the surrounding rock with a sufficient supporting resistance to control the rheological deformation of the surrounding rock.
- Special attention should be put on floor heave control.

However, it is hard to say that a prevailingly suitable approach has been developed to deal with issues associated with soft roadway support. The challenges facing researchers and engineers do not come from the variation of rock properties from one mine to another, or from one location to another in the same mine and same stratum, but instead from the lack of knowledge on the constitutive relations between rock stress and strain.

Roadway deformation of soft rock usually consists of multiple components including elastic deformation of rock, rock creep, rock expansion because of water invasion and volume expansion.

These issues make roadways developed and constructed in soft rocks even more complicated to tackle. Therefore, if a roadway is to be driven in soft strata, or a roadway has already been excavated in soft rocks and is likely to be influenced by huge stress changes, then a specific measures and design must be considered. The followings innovations regarding stability control and support of roadways in either soft or extremely soft rocks were successfully practiced in China in the last a few years.

2 DESTRESSING UPPER STRATA TO PROTECT AN UNDERLYING CHAMBER

Stress existing in strata is a fundamental cause resulting in the deformation of an underground opening excavated in a rock mass. Reducing stresses in the surrounding rocks of an opening therefore can reduce the rock deformation. If the stress can be reduced to within the elastic capacity of a rock mass, plastic deformation may be completely avoided.

Destressing is a method that can be employed in practice to reduce stresses of rocks around an underground opening. A number of methods can be used to perform destressing. These methods apply to their specific geotechnical conditions and mining sequence. This section is focused on destressing performed in the upper strata above the top of a chamber in an underground coal mine.

2.1 *Background of destressing upper strata*

The main belt haulage roadway of Baodie Colliery, Shandong Province, was driven in either mudstone or claystone located 20 to 60m below No. 3 coal seam (see Figure 1 below).

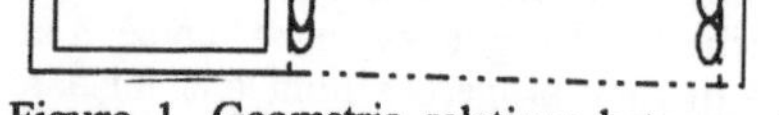

Figure 1. Geometric relations between the haulage roadway and longwall faces.

Original support pattern of bolts and yieldable steel sets of U-shape were installed to control its stability. Due to the impact of active abutment created by longwall face 1304 advancing in the No. 3 coal seam, a total amount of 200 to 400mm roof subsidence, 1200 to 1300mm of floor heave and 1250mm of wall-to-wall closure accumulated in the section of haulage roadway below longwall face No. 1304 while it was crossing over the roadway. It was predictable that a same amount of roadway deformation would take place in the section of roadway below longwall face No. 1306 while it was advancing across this section of haulage roadway. More importantly, a concern was raised regarding an inevitable damage to the convey chamber which was 70m aside longwall face No. 1306 due to the side abutment. Collapse of this chamber could occur if no specific measures were taken.

In view of the limitation of the arched stone support, which was originally provided to support the chamber, destressing was initiated in a bid to provide the convey chamber with an environment of relatively low stress state. It was assumed that the stability and safe use of the convey chamber would be secured if the stress state of the rocks immediately surrounding it could be artificially controlled to a level of no more than the stress state available before mining started. A concept of destressing the upper strata on the top of the chamber was therefore undertaken.

2.2 *Mechanism of destressing upper strata on the top of a chamber*

Numerical modelling of destressing the upper strata on the top of a chamber was carried out using the software ADINA. A comparison of the stress in the surrounding rock of the chamber under the condition of destressing and that without destressing was undertaken. A coefficient factor of 4 was applied to the numerical model to simulate the maximum front abutment created by longwall mining. Destressing in numerical modelling was realized by creating a slot in the upper strata on the top of the chamber.

Figure 2 illustrates the numerical modelling results and presents the stress contours of the rocks surrounding the chamber with and without the destressing slot.

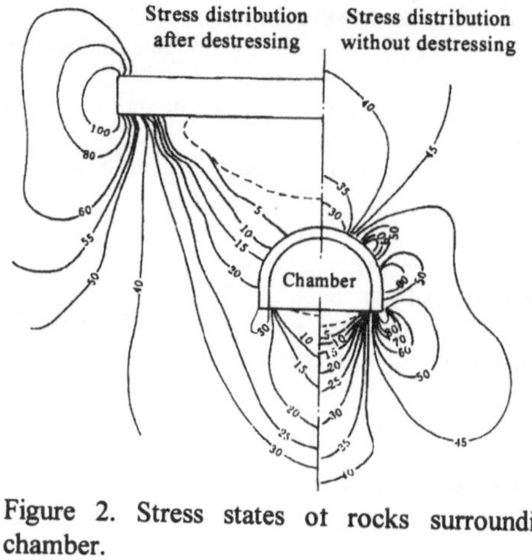

Figure 2. Stress states of rocks surrounding a chamber.

It was clearly shown that the rocks immediately surrounding the chamber, especially the roof of the chamber was completely destressed.

Physical modelling was simultaneously conducted at the Central Laboratory of Rock Mechanics and Ground Control, at China University of Mining and Technology by two-dimensional simulated tests on a scale of 1:50. The results from laboratory testing indicated that the roof-to-floor convergence of the chamber for the destressing case under the impact of front abutment of longwall face No. 1308 was only 6% of that for the non-destressed case.

2.3 *Destressing*

Destressing practice was implemented after optimizations of parameters was conducted through numerical modelling. Figure 3 shows two roadways on 14m centers, parallel to the convey chamber. The roadways were firstly excavated in the upper strata 8m above the convey chamber. Bore holes were drilled into two walls of each roadway and charged with a limited amount of explosive. Consequently, controlled blast was initiated to create a 19m wide, 2m high blasted rock area. This area prevented the immediate surrounding rock of the convey chamber from suffering excess stress caused by longwall face advance, and therefore from excess deformation and damage.

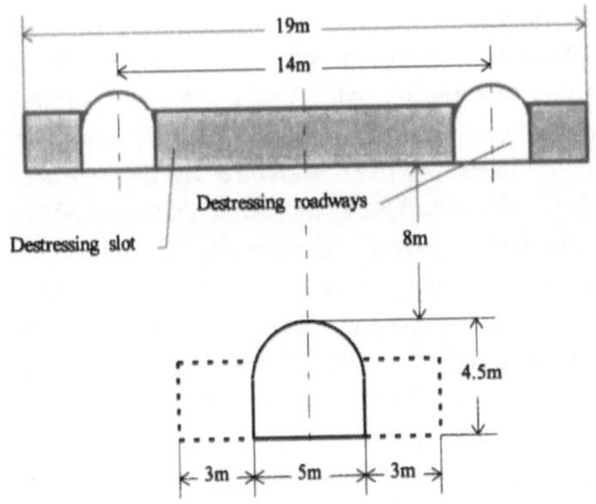

Figure 3. Schematic design of upper strata destress plan.

2.4 *Monitoring results*

Deformation monitoring carried out on both the destressed convey chamber and non-destressed haulage roadway close to the convey chamber showed that destressing reduced the wall-to-wall or roof-to-floor closure by 90% to 98%.

Supersonic detection carried out indicated that the thickness of plastic zone immediately surrounding the destressed chamber was only 0.4 to 0.7m. Monitoring of non-destressed sections of haulage roadway (under normal conditions of abutment) indicated a plastic zone of 5.0 to 5.5m thick.

3 HIGH PRESSURE INJECTION THROUGH BOLTS FOR ROCK REINFORCEMENT

Rock stability can also be maintained if the mechanical properties of a rockmass can be artificially improved to such an extent that the reinforced rock is capable of sustaining large ground pressures such as those caused by an active abutment. According to this concept, high pressure injection into roadway surrounding rocks through pumpable bolts was developed and introduced to the haulage roadway in Qisan Colliery.

3.1 *Haulage roadway reinforcement at Qisan Colliery*

The haulage roadway at the Qishan Colliery in the Jiangsu Province was excavated in sandy shale at 600m below surface. This roadway had a designed cross section area of 11.08 m^2 and was initially supported by split sets in conjunction with shotcrete. The roadway was kept in a good condition for more than a decade before it started to experience an active abutment pressure created by the mining activity of longwall faces located 25m above in the upper seam. The geometric relations between longwall faces and the haulage roadway were exactly the same as that in Baodian Colliery shown in Figure 1. The influence of abutment pressure on the behavior of the rocks surrounding the roadway became noticeable when the longwall face was within 150m of the roadway. The effect diminished gradually after the longwall face passed through and it was 100m away.

Framed steel support set was used as an expedient measure to maintain the safe use of this roadway when it experienced the impact of the first overlying longwall face. Unfortunately, this temporary support was not adequate to resist the large abutment pressures. Large amounts of deformation of the surrounding rock were experienced. Consequently, damage of support sets occurred and coal transportation was suspended during the period of roadway rehabilitation. Additionally, the installation of framed set in the roadway reduced its effective cross section area and caused an excessive rate of the wind within the roadway. The condition of roadway maintenance and deformation during the course of the longwall face advancing from the right side of the

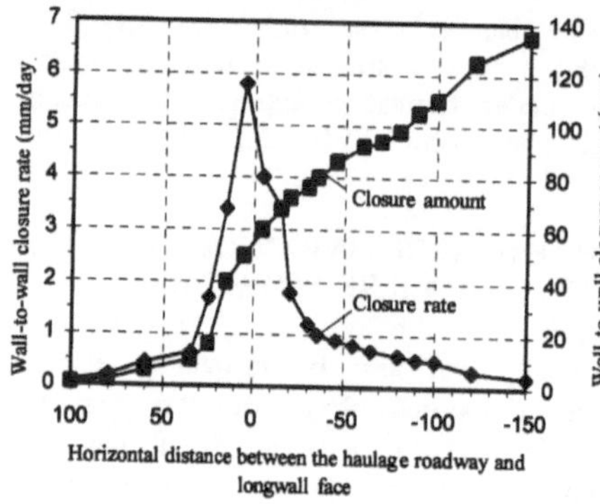

Figure 4. Deformation of the surrounding rock.

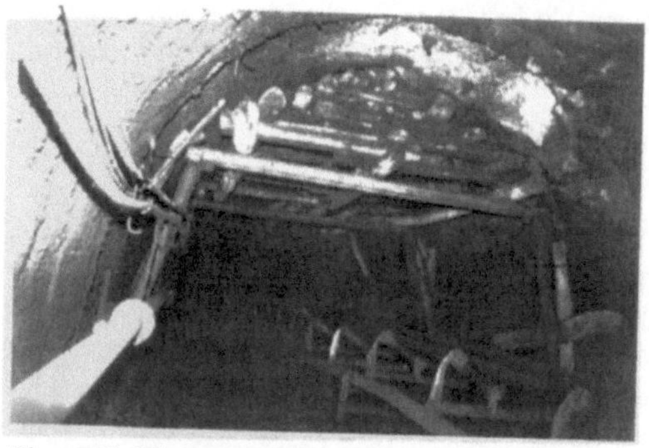

Figure 5. The roadway condition under additional steel set support.

roadway to its left side is shown in Figures 4 and 5. The peak roadway deformation rate was 6 mm/day and just occurred when the longwall face was 5m away from the roadway. Figure 5 reflects the serious roof scaling along with a big reduction of roadway cross section area taking place as a result of the large deformation of the rocks surrounding the roadway during the period of an active abutment pressure.

High pressure injection and bolting techniques were applied to the remaining section of the roadway which was expected to experience the impact of an active abutment caused by a second longwall face.

3.2 *Structure of the pumpable bolt*

Other than normal reinforcement, the unique design of a pumpable bolt enable itself to function as packer and injection pipe. High pressure injection can be performed by using the same borehole. Determination of bolt length depends on geotechnical conditions such as the dimension of the roadway, the width of the broken zone and the properties of the surrounding rock in which the bolt

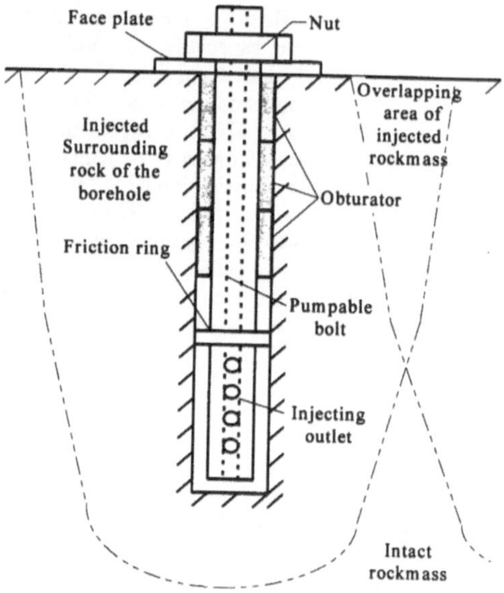

Figure 6. The structure of a pumpable bolt.

is installed. The injecting section is located at the end of the bolt. Several small diameter holes are needed for injecting. The exposed end of the pumpable bolt has a thread sector for the installation of a face plate and nut. This thread is also used to link the couple, cock and injecting slurry intake.

The obturator used to seal the borehole is a specially designed roll made from Rapidly Hardening and Swelling Cement (RHSC). Its outer and inner diameters are controlled by the selected diameters of borehole and the outer diameter of the pumpable bolt. Its length is normally 250 mm. Two to four RHSC rolls are employed depending upon the degree to which the surrounding rock of the roadway cracked, the value of the injection pressure and other parameters such as the viscosity or density of the injecting slurry.

The spacing of the individual borehole fans usually range from 1.0 to 1.5 m. In each fan, the boreholes are 0.8 to 1.2 m apart at the collars and 2.0 to 2.5 m apart at the toes.

3.3 *Implementation of pressure injection through a pumpable bolt*

Four steps were included in the procedure of implementing pressure injection through the pumpable bolts. Figure 7 demonstrates bolt injection.

Figure 7. Operating procedure for injection.

3.3.1 *Sealing the surface of the surrounding rock*

The surface of the rock surrounding a roadway always has more joints than the inner part of the surrounding rock and must be sealed properly around one week in advance before the implementation of injection. Otherwise, the injecting slurry will flow into the roadway instead of penetrating into the joints of the broken surrounding rock. Shotcrete with a thickness of 30 to 50mm is therefore usually employed for this purpose.

3.3.2 *Drilling and flushing the borehole*

Boreholes can be made by rotary drilling or rotary percussive drilling depending on the hardness of the rock. Water flushing or gas flushing has to be conducted to flush all the drillings out of the boreholes so that the injecting efficiency and adhesive force among the bolt, cement and the wall of the borehole can be achieved to their best.

3.3.3 *Installing bolt and obturators*

Two to four RHSC rolls to be used as obturators need to be soaked in water for about half minute before being inserted on the bolt. Subsequently, the bolt is put into the flushed borehole with the soaked RHSC being located as close to the hole collar as possible. A proper match among the diameters of bolt, RHSC roll and borehole along with the rapid expansion and hardening of RHSC rolls fixes the bolt immediately.

3.3.4 *High pressure injection*

As shown in Figure 7(b), the high pressure injection can be implemented simultaneously on multiple injecting bolts with each bolt being controlled individuallyby its own manometer.

3.4 *Performance of the haulage roadway at Qisan Colliery*

Boreholes having a diameter of 42mm and a length of 2.5m long were drilled with a fan spacing of 1.5m. The neighbouring boreholes on each fan were 1.0m apart. Shotcrete had been sprayed to seal the surface of the surrounding rock 10 days earlier. Four RHSC rolls were employed to seal the borehole and fix the pumpable bolt in place. The injecting material used was Portland Cement with a small amount of additives. Four pumpable bolts were injected simultaneously by the same pump. The designed injection pressure was 4 Mpa which was monitored by a manometer placed on each pumpable bolt.

3.4.1 *The behaviour of the injected and bolted rock surrounding the roadway*

The deformation of the rock surrounding the roadway was measured and presented in Figure 8. It can be found that, in contrast with the situation of a roadway section supported by framed steel sets (see Figure 4), the injected and bolted section was maintained in extremely good condition (see Figure 9) with its highest wall-to-wall closure rate of 1.25 mm/day occurring when the work face above was 5m away from the roadway. The wall-to-wall closure experienced while a longwall face was advancing 100m away from the roadway was only 25% of that experienced by the section supported by steel sets.

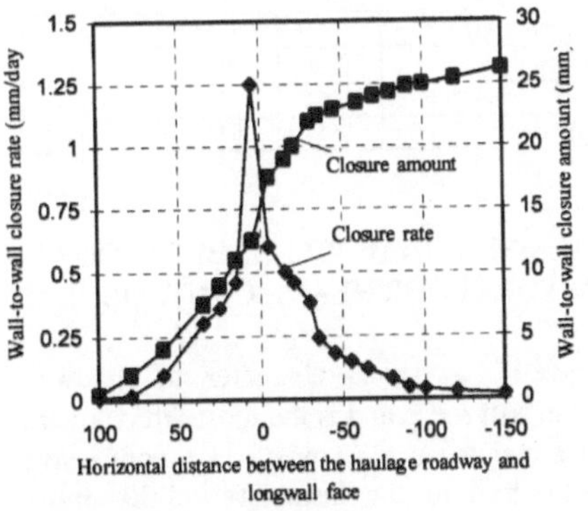

Figure 8. Deformation of the surrounding rock reinforced by bolt injection.

Figure 9. The roadway reinforced by bolt injection.

3.4.2 *The reinforcing effect of bolt injection on the properties of the surrounding rock*

Drill cores were collected before and after the performance of bolt injection. Properties of the core sample and the results of supersonic test carried out in the core holes are listed in Table 1. It indicates that the mechanical properties of surrounding rock after bolt injection were greatly improved. The enhancement of supersonic speed reveals that the integrity of the surrounding rock was greatly upgraded. That is why the bolted and injected section of the roadway was kept in a much better condition than the steel set supported section.

Table 1. Improvement of the properties of rocks after high pressure injection

Items		Steel set supported section	Bolted and injected section	Percentage Increase (%)
UCS (MPa)		29.8	69.7	134
Tensile strength (MPa)		1.83	2.38	30
Cohesion (MPa)		2.89	10.67	270
Friction angle(°)		26.9	36.8	37
Elastic modulus (MPa)		2.35×10^4	3.41×10^4	45
Poisson's ratio		0.26	0.2	-23
Sound wave speed (m/s)	Depth 0.7~1.1m	3502	4587	39
	Depth 0.5~1.7m	3547	4441	26

4 TWO DIMENSIONAL YIELDABLE FLOOR BEAM FOR FLOOR HEAVE CONTROL

Floor heave is a typical character of roadway behaviour in soft rocks and is the most difficult area of soft rock maintenance. In most cases, floor heave can directly lead to the instability of the entire roadway. Figure 10 illustrates a case where a roadway collapse resulted from floor heave.

Figure 10. Floor heave of the main air roadway in soft rocks in Xieqiao Colliery, Anhui Province.

A rule of thumb is that floor heave control is the key to control the entire stability of a roadway excavated in soft rock. Conventional floor bolts are usually impractical for two reasons. One is the particular difficulty in drilling downward bore holes because of the frequent holding up of the drill rod by the sticky borings and soft surrounding rocks. The other one is the trouble in dealing with the protruding bolts after floor failure eventually occurs. Therefore, application of floor bolts to floor heave control in roadways excavated in extremely soft rocks is not preferred by engineers and practitioners.

Lessons from previously unsuccessful practice indicated that steel sets with conventional floor beams yieldable in the direction of roadway width (see Figure 10) was not an effective approach. Very large floor heave and roadway failure occurred only 5 to 6 months following the installation of enclosed steel sets. In this case, the effort was focused on seeking an approach which could function as a floor surface sealing while applying a constant resistant force against floor heave.

4.1 *Pattern of innovative yieldable floor beams*

A well accepted practice to control soft ground in coal mines is to place backfill between the steel sets and the surrounding rock of a roadway. Although this is an expensive approach, the steel sets are recyclable and this might be the only choice in some particular circumstances. Experience shows that a round yieldable steel set with backfill behind is the best format to support a roadway with the deformation pattern shown in Figure 10. However, a low ratio of effective cross section area of a round

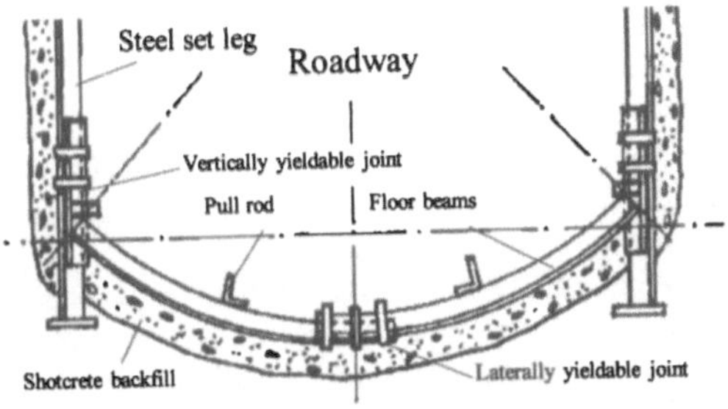

Figure 11. Details of the two-dimensional yieldable floor beams.

steel set to its excavated cross section area can make this uneconomical. In view of that, a two-dimensional yieldable floor beams was developed and put in practice in the rehabilitation of the main air roadways at Xieqiao Colliery (see Figure 11).

4.2 *Features of the two dimensional yieldable floor beams*

The following features characterize the two-dimensional yieldable floor beam.

- Yieldablility in both lateral and vertical directions in the cross section of the roadway allowed both floor heave and wall-to-wall closure to take place without damaging both floor beams and legs of the steel sets.
- Three pairs of check clamps connecting two floor beams secured a constant outward resistance to be generated during these two floor beams yielding under the action of wall-to-wall closure. A sliding U-type steel component held on each of the two legs of the steel set ensured a downward resistance to be produced while floor heave occurred.
- A layer of 200mm thick shotcrete with a mesh inside was sprayed under the floor beams. This shotcrete lining was designed to serve as surface sealing to protect the floor from water invasion and deterioration as well as a medium to even the load distribution applied to the floor beams.
- The pull rods between neighboring floor beams and the shotcrete lining jointly made the neighboring floor beams a whole unit, which was capable of producing resistance 3 to 4 times the sum of that produced by all individual floor beams according to laboratory testing results.

This two-dimensional yieldable floor beams along with shotcrete lining and pull rods between neighboring floor beams were capable of applying a resistance of 0.4 Mpa intensity. Meanwhile an amount of 300mm floor heave was allowed to be experienced during a designed service life of 10 years without compromising the normal use of the roadway.

4.3 *Applications*

Applications of this two-dimensional yieldable floor beams for roadway rehabilitation and development were consequently carried out in Xieqiao Colliery, Anhui Province, and Xiaokuang Colliery, Liaoning Province. The properties of the rocks surrounding the roadways are presented in Table 2. Figure 12 shows an application to the main air roadway in Xieqiao Colliery.

Long term monitoring the performance of this two-dimensional yieldable floor beams in conjunction with shotcrete lining and pull rods was carried out immediately after installation and continued for two years. Recording data revealed that the roof-to-floor rheological convergence rate was steadily controlled at 0.03~0.04 mm/day with the roadway being maintained in an absolutely good condition as shown

Table 2. Properties of extremely soft rock encountered at Xieqiao and Xialkuang Collieries.

Rock sampling site	Type of rock	Strength under the condition of natural water content		Strength under the condition of being saturated
		UCS	UTS	
Xieqiao Colliery	Mud stone	16.7 MPa	0.91 MPa	Dissolve in water in 3 minutes
Xiaokuang Colliery	Silt	17.6 MPa	0.79 MPa	UCS 12.3 MPa

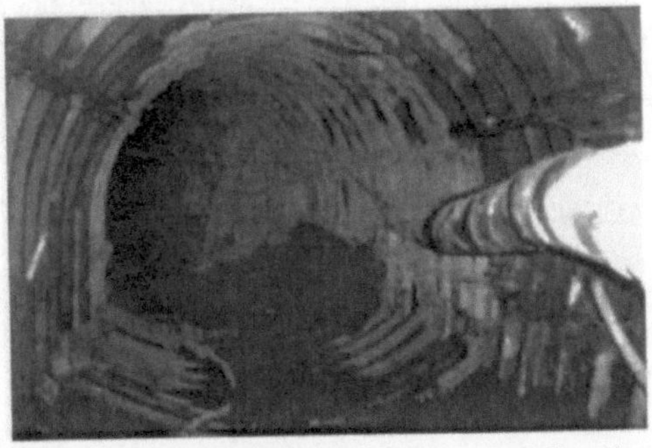

Figure 12. Implementation of the two-directional yieldable floor beams.

Figure 13. Condition of the main air roadway support by two-dimensional yieldable floor beams.

in Figure 13 which was photographed 18 and 15 months after the installation of the joint support at Xieqiao Colliery.

In situ instrumentation to determine the actual supporting intensity applied by the floor beams to the floor was also conducted by placing hydraulic flat jacks on the interface between rock surface of the floor and the shotcrete lining. A maximum pressure of 0.34 Mpa to 0.47 Mpa was recorded for flat jacks located beneath floor beams and between floor beams respectively 7 weeks after installation of the support. The order of magnitude of floor resistance thereafter generated by floor beams was maintained with a 10 to 18% deviation in response to the yielding of the joints of the steel sets.

5 CONCLUSION

Upper strata destressing can provide a very safe stress environment to an underground opening or roadway.

High pressure injection through pumpable bolts can be a flexible and cheap rock reinforcement alternative. The suitability of this technique depends upon the rock properties that control pumpability.

Steel sets with two-dimensional yieldable floor beams in conjunction with backfill might be the only option to control roadway stability in some extreme circumstances.

6 ACKNOWLEDGMENT

The author would like to thank his colleagues and work fellows for their close cooperation in research projects related to the innovations described in this paper.

REFERENCES

Y. Wang and C. Wang 1990. Effect of destressing the upper strata of a roadway on the stability of the roadway in soft rocks. *Journal of ground pressure and support*, 28-32.

C. Wang 1995. Study on the supporting mechanism and technology for roadway in soft rock. *Ph.D. thesis*. China University of Mining and Technology.

S. Lu, Y. Wang and C. Wang 1995. Relationship between support and surrounding rock of roadway in soft strata. *Proceedings of the International Symposium on Mining Technology and Science*, China Central Institute of Mining Technology. Beijing.

S. Lu and C. Wang 1993. Strata pressure law for roadway in soft rock and relationship between support and surrounding rock, *Proceedings of the Australian Conference on Geotechnical instrumentation and Monitoring in open pit and underground mining*, Kalgoorlie, WA, Australia.

Y. Chen and S. Lu 1994. *Strata control around coal mine roadways in China*. Express of China University of Mining and Technology.

Rock Support and Reinforcement Practice in Mining, Villaescusa, Windsor & Thompson (eds)
© 1999 Balkema, Rotterdam, ISBN 90 5809 045 0

Development of roofbolting in Australian coal mining

M.Rataj & M.Yearby
ANI Arnall, Newcastle, N.S.W., Australia

ABSTRACT: Roof bolting technology in Australian coal mines has advanced quite considerably during the 40 years plus since the introduction. The main area where the majority of improvements have taken place is in the stiffness of the support. This has been achieved by the introduction of chemical anchors and full encapsulation of bolts, higher grades of steel with more efficient deformations and the introduction of cable bolts and resin grouted strand bolts. Another important development occurred in the area of direct application of confinement to the rock mass by installing bolts in pre-failed rock conditions and application of high pre-tensioning to the support. All the changes introduced to the roof bolting technology which resulted in better rock reinforcement are analysed and the directions for further future development are discussed.

1 INTRODUCTION

Roof bolting continues to be improved despite the fact roof bolts have been used for well over 40 years in Australian coal mining. In fact, in recent years the rate of improvement is probably greater than ever, which can be attributed to such factors as better understanding of the strata/bolt interaction, availability of better materials and development of more efficient installation techniques.As a result, the rock bolts today are the most cost effective method of strata reinforcement but their development is not completed as yet.

The biggest improvement in the development of rock and roof bolts has been achieved in the area of stiffness of the support. Recent years have also seen significant development in installation techniques in Australian mines that are looking for more efficient confinement of the strata. One of the more substantial techniques is the high pre-tensioning of bolts. The individual roof bolting features evolved during the years will be discussed in order to form a view on the directions for future development.

2 HISTORICAL OVERVIEW OF ROOFBOLTING DEVELOPMENT

The concept of roof bolts was introduced into Australia from the USA in the late 1940's. Experiments with roof bolts were initially conducted at Elrington Colliery, NSW in 1949 (Gardener, 1971) as a possible means of roof support and to supplement timber. These first roof bolts were quite successful and BHP commenced manufacture of bolts in 1951 (Scotford, 1960)at their Port Kembla steelworks for use in their coal mines. Shortly after, the manufacture of roof bolts was transferred to The Titan Manufacturing Company Pty. Ltd. at an existing building at Newcastle. The ongoing success of roof bolts then led to a special purpose plant being built at Mayfield in 1961 to manufacture and distribute the new strata reinforcement product.

2.1 *The first bolts*

The first bolts used at Elrington Colliery were threaded 1" BSW x 7'6" (2250 mm) long slot & wedge bolts, and had a 9" (230 mm) slot on one end and screwed 6" (150 mm) on the other end. The bolt material conformed to ASA 1/1940. Installation was with pneumatic stopers.

At another pioneer mine, John Darling Colliery, the holes were drilled with an electric hand held borer which had been mounted on a hydraulic cylinder. The wedge was set by the manual use of a sledge hammer onto the end of the bolt. The nuts on the bolts were tightened by hand using a spanner about 600 mm long.

In the mid 1950's the first mobile bolting machines started to appear which created a change in demand from slot and wedge to expansion shell bolts. All of these mobile machines were of the rotary drilling type that allowed the machine to also tension the roof bolts. Local manufacture of expansion shells was also introduced which coincided with the introduction of mobile bolters.

The expansion shell type bolts were initially 5/8" diameter (14.2 mm bar), followed shortly after by ¾" diameter (17.1 mm bar) and 1" (23.1 mm bar). Hole diameters required were 1 7/16" (37 mm) for the 5/8" and ¾ " bolts and 1 5/8" (42 mm) for the 1" bolts.

A longer expansion shell for use with the 5/8" and ¾ "bolts was also introduced to improve the bearing area onto the rock and hence increase anchorage capacity. In addition to the 1" mild steel expansion bolt an 'extra strength' material was introduced to give two bolt grades in the expansion bolt range.

Although it was obvious that these expansion shell roof bolts were considered effective in supporting the roof there were a number of problems that had to be overcome if the product was to be truly viable. These were:

1. An improvement in anchorage capability, i.e. ability to provide anchorage in "softer" type rock. This had been overcome to a reasonable degree with longer expansion shells.

2. As seams became deeper, higher pressures became evident and strata control again started to become difficult.

3. Installation at the face in preference to out-bye. Note many of the mines had 'Fletcher' mobile bolters. They were dry drilling rotary machines which installed the roof bolts out-bye the coal cutting machines.

4. Loss of bolt tension from anchor creep particularly if explosives were used to win coal.

5. A need for reduced installation costs.

2.2 Slot & wedge bolts

The problems associated with expansion shell bolts were solved in the main by the higher use of hand held rotary impact, pneumatic rock drills (stopers). These machines allowed roof bolt installation at the face by withdrawal of mining machines. A gradual reduction in the use of expansion shell bolts followed as there was relatively little extra cost for the stronger/stiffer bolt.

The average slot & wedge bolt length in coal mines also increased towards 1800 mm over the next 10 years through to the mid 1970's.These bolts used a smaller diameter hole, 32 mm, and were usually wet drilled and installed with pneumatic, hand held, rotary impact stopers.

The material quality used to manufacture the slot and wedge bolt was restricted to "mild steel". As the slot in the bolt was oxy-cut, the flame hardening effect caused fracturing of the slot during installation on higher grades of steels.

The acceptance, and dependence on the slot and wedge bolt to support the roof became quite high. However the continuous demand from production personnel to increase installation rates led to angled placement of bolts with the hand held stopers. These rock drill, stopers, were used to drill and install several bolts from the one position. This, in turn, reduced the induced tension into the bolt from a nominal 4½ tonnes towards 1 tonne due to the nut corners fouling on the plate washer and resisting rotation. Effective installation hence suffered (Yearby & Shaw 1986). The available torque from these rotary impact stopers was 250 ft. lbs. (340 Nm). Cast iron deflection balls with domed plates initially overcame this problem. These were ultimately replaced with the hot forged steel product used today.

As development of coal mines were into greater depths, additional problems were becoming more evident.

1. The anchorage capability of slot and wedge bolts was not always regarded as adequate.

2. The bolt elongation of the point anchored slot & wedge bolt was excessive under load. Note that the mild steel slot & wedge bolt's elongation at maximum bolt strength is > 25%.

3. Hearing losses from the noise emitted from the rotary impact pneumatic drilling machines was becoming more evident.

2.3 Chemical bolts

In the early 1960's chemical anchors as an alternate to mechanical point anchors were being trialed with imported cartridges from Germany and subsequently from Australian manufacture.

Curing times after the two minutes mixing times with the bolt could be great as two hours subject to mixing and rock temperature.

Although some of the early tests were considered as 'commercial' failures, considerable technical and practical information became evident to give direction to development of both the anchor and future roof bolts, i.e.

(a) The early underground and laboratory testing of chemical anchors found that chemical anchorages were considerable stiffer than mechanical anchors (Barnes & Howe, 1964)

(b) The polyester anchor mixing efficiency could be improved by reduced annulus area and the external shape of the bolt. (Barns & Howe, 1964; and Wright, 1964).

By the mid 1970's the first commercially effective chemical anchors were being trialed. These anchors were manufactured from a polyester outer skin (like a sausage), were full of resin and a glass tube which contained a catalyst. Initial anchor lengths were 250 mm which gave about 400 mm of encapsulation to a 24 mm bolt installed in a 27 mm diameter hole. Subsequent anchor lengths were 330 and 440 mm. that gave longer point anchors of 500 mm and 700 mm respectively.

The first commercially economic chemical bolt was manufactured from mild steel round bar (21.7mm diameter) and had a 150 mm rolled M24 thread on each end, one of which was fitted with a nut.

It was also recognised that for effective mixing of the chemical anchor it was important that smaller hole diameters had to be drilled. The main difficulty at this time was to drill the 27 mm diameter hole that was preferred. In fact the pneumatic rotary impact stopers had difficulty in both drilling this hole diameter and rotating fast enough to effectively mix the chemical anchors. This was overcome by the introduction and development of lightweight, portable, pneumatic rotary drill legs and efficient wet drilling bits. Rotation speed under load was typically 500 RPM and chemical mixing efficiency dramatically improved.

From past experience with both the expansion shell and slot & wedge bolts the need for tensioning had been recognised and was beneficial to the effectiveness of the roof bolt installation.

Prior to the subsequent improvements in the portable rotary machines three methods were effective in providing bolt tension. These were rotary impact wrenches and torque multipliers (both heavy and expensive to maintain) and the 'booster' washer. This washer was manufactured from polypropylene and was assembled onto the bolt in the factory between the nut and domed ball. This washer acted as a lubricant and effectively allowed a doubling of the induced tension into the bolt from the available torque.

The early chemical bolts ultimately proved to be of considerable value in mines with variable types of roof that could be classed as medium to soft. In fact a number of new mines were opened using only chemical bolts and some excluded the use of timber completely.

However, it soon became apparent that anchorage capability could be improved by increasing the overall bolt length as well as extending the 150 mm thread length on the upper end of the bolt. This thread length ultimately became standard at 700 mm which was the effective encapsulation with a 440 x 25 mm diameter chemical cartridge (Norris & Yearby, 1980). The gel time of the 440 x 25 mm chemical anchor was 22 seconds and had an additional hold time (curing) before tensioning of 8 seconds being a total of 30 seconds from commencement of roof bolt installation.

During the 1970's a high number of different roof bolt configurations were trialed and rejected. The overall effect of this was the development, in the early 1980's, of a roof bolt bar feed which had a special left hand configuration deformations and was available in 2 steel grades being mild steel (AS) and a high strength (AH). In addition an extra high strength steel grade (AX) was introduced in the mid 1980's.

It is now known that a number of mines reduced their intersection failure dramatically during the 10 year period to the mid 1980's. This was primarily due to the collective improvement in roof bolts during that time including improved steel grades with a purpose designed bolt form, chemical anchors, increased encapsulations and longer bolts.

The average chemical bolt length also increased from 1800 mm towards 2400 mm from the mid to late 1980's. This bolt length growth was basically restricted by limitations in seam height.

In the early 1990's a new form of cold working was developed. This cold working improved the bolt yield. This new 24 mm AVH chemical bolt was designed to replace the mid range high strength (AH) bolt.

2.4 Longer than seam height bolts

In the mid 1960's cable bolts were starting to be used in coal mines. These cables were basically 7 wire plain cables of 15.2 mm diameter, commonly up to 10 metres long, and were fully cemented grouted and un-tensioned.

Subsequent cable and grout developments ("bird caging" or debonding of the cable wire plus non shrink high performance grouts with pumping equipment) led to considerable usage in coalmines. All were placed out bye the face in highly stressed areas, i.e. longwall gate roads and intersections. The ultimate tensile strength of a 7 wire x 15.2 mm diameter cable is 26 tonnes. These cables are often 'doubled up' giving a capacity of 52 tonnes.

2.4.1 Introduction of strand type bolts

In the early 1990's a new long tendon was introduced. This bolt was manufactured from 21 wire strand and had an ultimate capacity of 58 tonnes.

Important features of these Flexibolts were:

- They were stiff enough for installation similar to solid bolts using chemical anchors.

- Flexible enough to allow installation of long lengths.

- Could be installed at the face with existing drilling equipment in a 28mm diameter hole.

- Could be pre-tensioned up to 5 tonnes using a nut.

The success of the Flexibolt lead to the development and introduction in 1997 of the Hi-Ten Strand Bolt. This bolt is manufactured from the same strand, as above, but does not have a thread for tensioning. Tensioning at the face to 25 tonnes can now be readily accomplished with special end fittings and a pneumatic powered hydraulic jack.

Note that the term "Cable Bolt" generally is considered as being cement grouted whereas the term "Strand Bolt" refers to the bolt being resin/chemical grouted.

2.5 Other developments

There were a number of other roof bolt developments which have all been directed towards improving the overall efficiency or proficiency of the bolting system. These developments include:

1. Round shaped and stiffer plate washers.

2. Improved drive nuts which are available in multiple torque settings and which are "free" running for bolt tensioning as opposed to the early resin nuts.

3. Introduction of lower viscosity anchors together with two speed anchors which allows faster installation of fully encapsulated bolts and ability to pre-tensioned/load the bottom part of the roof.

4. High performance nuts which effectively doubles the induced bolt tension as given from the plastic 'booster washer'.

5. The shape and form of bar deformations that give a significant increase in the load transfer properties from the rock to the bolt

6. Introduction of higher tensile, multi-hole W straps.

3 ANAYLYSIS OF CHANGES RELATED TO SUPPORT STIFFNESS

All the changes that have been introduced to the roof bolting system can be divided into category groupings. The majority of them however, fall into a category effecting the stiffness of the roof bolting system.

It has been recognised in recent years that the effectiveness of the strata reinforcement can be improved by increased stiffness of the support. This view is also going to be confirmed by the analysis of all the changes that have provided the incremental improvements of the stiffness.

3.1 From mechanical to resin anchor

The change from mechanical anchors (expansion shell and slot & wedge) to the resin anchor resulted in a significant increase of stiffness, particularly in weak shale's or mud stone's. In the soft and weak sedimentary rock types the mechanical point anchor bolts rapidly lose efficiency (Howe 1968).

The mechanical anchor stiffness measured by the anchor displacement against applied load is very low in weak rock because of the rock

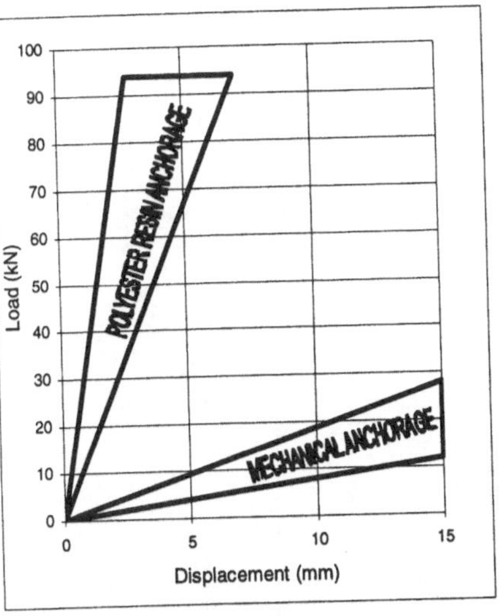

Figure1. Range of performance of mechanical and polyester resin anchors in coal at Hebburn No. 2 Colliery (Barnes and Howe 1964).

weathering or fracturing with the steel anchor/rock contact. Often the load drops to zero if the expansion shell plug pulls through the shell leaves.

Introduction of the resin anchor bolts dramatically improved the support stiffness as illustrated in Figure 1 by Barnes and Howe (1964).

3.2 Chemical anchor bolt deformation profile

To provide efficient bond between the bolt and the resin the bolt must have some type of deformations. The first type of deformations achieved by threading the end of the bar was not very efficient when compared to the deformations introduced later in form of the ribs protruding from the bar. These ribs move against the resin column as the bolt is loaded by the strata movement and develop radial (wedging) forces which compresses the resin between the rock and the bolt ribs. This then resists the bolt/rock relative movement.

The distance between the protruding bolt ribs is in the region of 10mm and is large enough to prevent shearing of the resin between them. This resistance to shear was not as great when bolt ends were threaded. The relatively small pitch of the thread form (2 or 4mm) only generated the marginal radial force required to wedge the resin between the bolt and the rock. The resistance, which prevented the bolt from being pulled out of the hole, relied mainly on the shear strength of the resin between the threads. Since the shear strength of the resin is about 3 times less than its compressive strength the anchors could fail at a relatively low loads.

The lack of radial forces generated in the resin caused failure of the threaded bolt anchor on the resin/rock interface particularly when the anchors were short and the rock was weak.

Subsequent extension of the anchor length by threading longer sections of the bar (up to 700mm) significantly improved its performance but the real breakthrough was the introduction of the ribbed LHA (Left Hand Anchor) bar or "T-bar". The average rib height of the LHA bar was 0.7-0.8mm.

Further improvements in the anchorage stiffness have been achieved by the recent introduction of bolts with significantly higher ribs when compared to the LHA bar. The rib height, which varies between 1.3 to 1.6 mm, provides a 40 to 60% increase in the anchorage peak shear stress. Anchorage stiffness has also improved significantly.

3.3 From point anchor to full encapsulation

Point anchored roof bolts, in areas of strata where a number of dilation zones occur, are loaded in a very fast and undesirable way. The entire free length of the bolt is loaded by the sum of all loads coming from individual dilation zones. Therefore, it is relatively easy to exceed the yield strength of the bar and once this happens, the bar becomes very soft (plastic mode). For example, the 1.5m free length of a point anchored bolt of 21.7 mm diameter bar in AS 3679.1 Gr.250 steel (which has yield strength of about 10 tonnes) will elongate about 150mm when loaded with 13 tonnes. This significant elongation of the bar may then allow for an escalation of roof deformation.

Implementation of full encapsulation of bolts dramatically improved the stiffness of the support. First of all, individual loading affects only the corresponding local sections of the fully grouted bolt. This eliminates, or at least delays, the building up of loads exceeding the yield strength of the bar.

Full encapsulation also increased the support stiffness due to longer anchors that provided a considerably higher anchorage strength which was directly proportioned to the length of the anchor.

Roof bolt stiffness also increased relative to the lateral movement of the rock because the fully grouted bolt provides an immediate

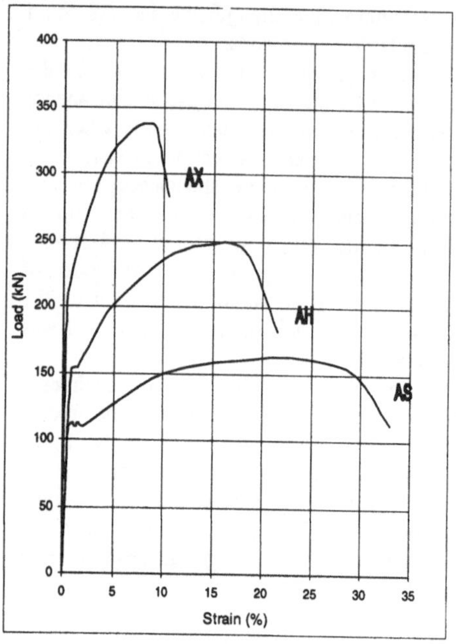

Figure 2. Mechanical characteristics of 24 mm bolts (21.7 mm bar diameter).

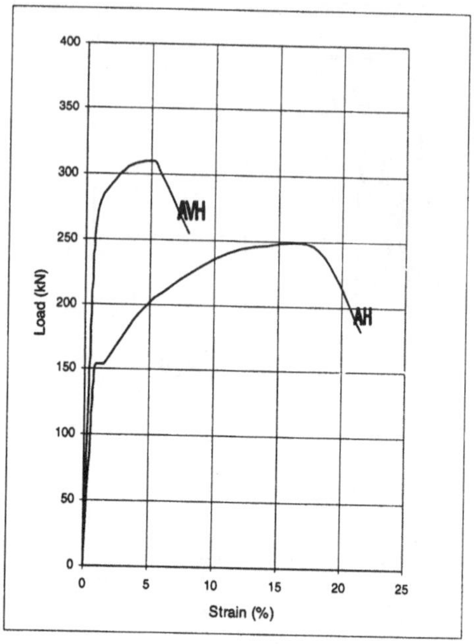

Figure 3. Mechanical Characteristics of AVH and AH bolts.

resistance to such movements as there is no void surrounding the bolt.

3.4 Introduction of higher strength steel

In high stress and high strata deformation areas roof bolts, even fully encapsulated, are sometimes subjected to loads exceeding their yield strength. This not only affects the localized areas but a quite significant length along the bolt which is due to the reduction of the bolt diameter when the bar is in the plastic mode. The bolt then reduces its diameter and de-bonds itself from the resin and as a result, a progressively longer section of the bar becomes 'free' and ready to elongate. The lower the yield strength of the bar the softer the system becomes under load conditions. Therefore introduction of bolts having higher yield strength again improves the stiffness.

Figure 2 shows the mechanical characteristics of the standard strength steel bolt (AS), the subsequently introduced high strength bolt (AH) and the extra height strength bolt (AX).

The introduction of the AX type bolt, having the significantly higher yield strength when compared to the other bolts (as shown in Figure 2), allowed significantly more efficient control of the strata in difficult conditions. It must be remembered that the graphs shown in the Fig 2, show the characteristics of bolts all having a 21.7 mm bar diameter, whereas the early bolts of smaller (14.4 mm and 18.0 mm) diameters bars, had mechanical characteristics even softer than the standard strength (AS) 21.7 mm bar.

In recent times two other bolt types (AVH and HPC) having high yield strength were also introduced to further improve the stiffness of the support. The AVH type bolt was introduced to improve performance of the AH type bolt as an alternate to the more expensive AX type bolt. The AVH bolt is manufactured by a cold working process which significantly increases the yield strength of the bar. The characteristics of the AVH and AH bolts are compared in Figure 3.

The recently introduced HPC type bolt is even stronger than the AX bolt. The higher strength was achieved by using a higher strength steel and making the bolt of slightly larger diameter.

It must be remembered that the higher strength steel bolts provide higher stiffness not only in regards to axial elongation but also where shear resistance is concerned. The higher the tensile strength of the bar, the higher its bending and shear resistance.

The reinforcing tendons currently in use having the highest yield strength are the double cable bolts and strand type bolts. Their performance will be discussed in paragraph 3.6.

3.5 Introduction of longer bolts

Improvements in strata reinforcement using longer bolts has been clearly observed. It is appreciated that these longer bolts create a thicker beam which gives an increased stability to the strata.

One of the major changes achieved by installing longer bolts was increased anchorage efficiency within the upper section of the bolt. In areas where the strata movement causes a tensile load in the top section of the bolt, the end of the bolt maybe pulled out from the hole at a relatively small load. For example, given the current regular bolting system in average rock of say 40 Mpa UCS conditions, a loading (dilation) zone which is 300mm below the end of the bolt will move that section of the bolt at a rate shown in Figure 4.

This Figure shows that in such conditions, load as little as say 8 tonnes causes as much as 2 mm bolt displacement. Therefore, increased bolt length bolts will increase the stiffness at the upper end of the bolt's anchorage.

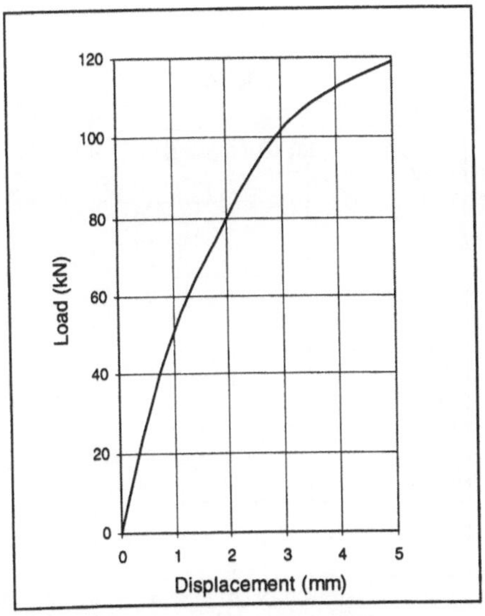

Figure 4. Typical bolt displacement versus load ratio for 300mm long anchor.

3.6 Introduction of cable and strand type bolts

The length of rigid bolts is normally limited by the height of the opening. Therefore in areas where rigid bolts of maximum lengths do not stabilize the roof cable bolts, typically 8 to 10m long, were introduced. The high effectiveness of the cable bolts in strata control may be explained for two most typical mechanisms of roof behaviour.

The first one applies to areas where the rigid bolts provide the expected reinforcement and form a non dilating beam but the beam is not strong enough to stop dilation of the roof above the bolted horizon and the roof is unstable.

Installation of cable bolts in such areas increases the beam thickness by securing the cables in long holes providing very efficient and stiff anchors.The second, and more typical, roof behavior mechanism is when the roof bolting system is not efficient enough to build a stable beam.

The lack of efficiency of the roof bolting support may be due to a lack of stiffness of the anchorage (bond) system that may be caused by a number of different factors. The most common being weak rock conditions, inefficient bolt profile, limited encapsulation length caused by resin losses in strata tissues or too large a hole diameter.

When the cable or strand bolts are used in such cases, the roof is being stabilised because the tendons again provide an additional stiffness together with increased anchorage given by the long holes.

In the early 1990's Flexibolts (strand bolts) were introduced. Their most typical length varies between 4 and 6m. These type bolts provide new possibility in the strata reinforcement techniques (Fuller et al, 1994). Because they can be installed in much less complicated way than the cable bolts they can be used close to the face and provide instantaneous and stiff support.

Apart from the increased support stiffness provided by the cable and strand bolts due to their length there is another significant factor which also positively affects the support stiffness. This is the high yield strength of these bolts.

Comparison between the mechanical characteristics of the Flexibolt (23.5mm diameter) and the solid bar AX bolt (21,7 mm diameter) is shown in Figure 5.

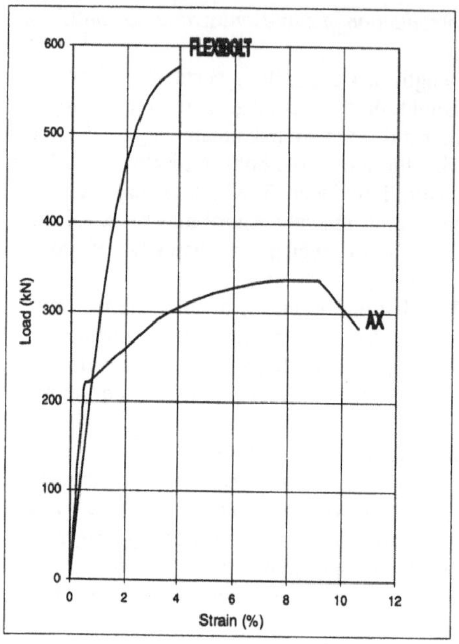

Figure 5. Mechanical characteristics of strand type bolt and the AX type bolt.

The previous figure shows that the AX bolt is slightly stiffer initially but once it goes to yield it becomes very soft, whereas the strand bolts (and cable bolts which have characteristics similar to the strand bolts) are very stiff up to much higher loads.

3.7 Angled Bolt Installation

The fully encapsulated bolt provides quite efficient resistance to the shear movement of the strata. Their effectiveness depends however on the angle between the bolt and shear plain. By angling the bolt in a way that its ends are inclined in the direction of the moving planes as shown in Figure 6, bolt A will prove to be more efficient providing stiffer shear resistance when compared to the bolt B which is angled less.

Similar mechanisms exist in the roof close to the rib side. It has been observed that roof bolts that are angled over the rib are more effective than those installed vertically. This proves that the stiffer shear restriction provides better roof conditions.

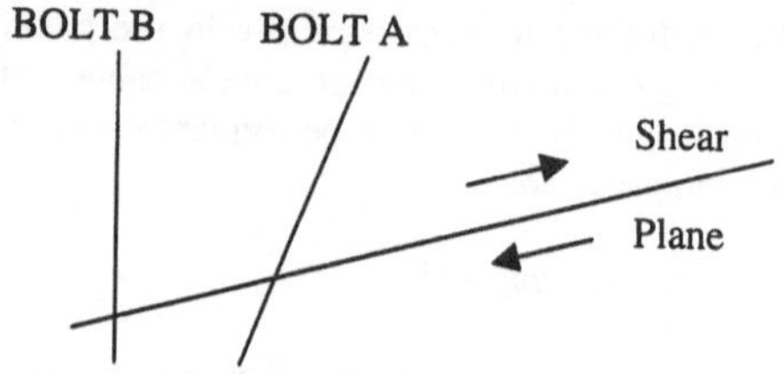

Figure 6. Bolt inclination in relation to shear plane movement.

3.8 Introduction of stiffer roof plates

The stiffness of the roof plate is also considered important as in high deformation areas plates are often heavily loaded, especially due to roof bolts being seldom fully encapsulated right up to the collar. The stiffness of the reinforcement system is therefore increased by the introduction of a new type of roof plate called the Turtle Plate. These plates have been formed in such a way that their stiffness is significantly higher when compared to conventional domed plates. As an example Figure 7. compares the behavioral characteristics of the conventional dome plate and the much stiffer Turtle Plate.

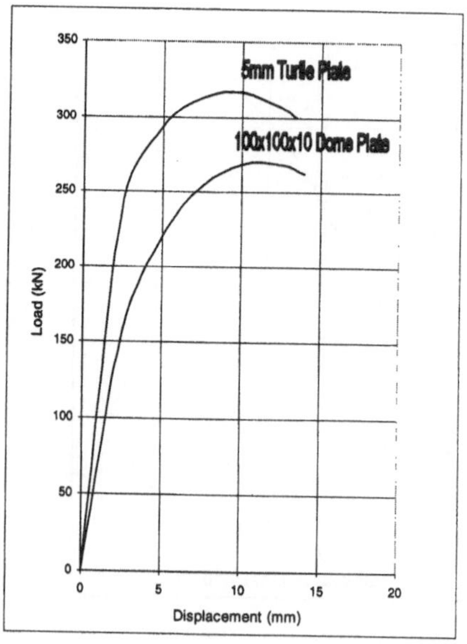

Figure 7. Loading Characteristics of 100 x 100 x 10mm Domed and 5mm thick Turtle Plates.

4 ANALYSIS OF CHANGES RELATED TO DIRECT ROCK MASS CONFINEMENT

All the introduced changes, as previously discussed, affect the stiffness of the roof bolting system. This improved stiffness provides a better confinement to the rock mass but only as a result
of the improved stiffness given from all components of the system. The following paragraphs discuss the techniques that improve the rock confinement in a direct way.

4.1 Installation close to face

It has been recognized that where strata softening occurs very quickly after roof exposure, which may be due to combinations of high stresses and weak rock conditions, installation of bolts as close to the face as possible results in an improving of roof stability.

The level of confinement provided by the bolts, in terms of load generated over the affected area, can be regarded as negligible in comparison to the level of stresses acting in the rock mass which may be as high as 40 MPa. However the net result of this is very significant in terms of improved roof behaviour. The early confinement of the strata provides not only improved stability of the roof but may also lead to the reduction of density of roof support.

Introduction of the resin grouted strand type bolts also provides opportunity for application of the early confinement to the strata as these bolts can be installed close to the face in the pre-failed roof conditions and provide the support immediately. In practical terms, this can not be done using cement grouted cable bolts as installation requires special equipment and crew and takes a considerably longer grout curing time.

4.2 Pre-tensioning

Pre-tensioning of roof bolts provides a more effective confinement to the immediate strata when compared to bolts which are un-tensioned. The level of pre-tension of the rigid bolts have been increased recently from typical 5 tonnes to 10 to 12 tonnes when special, low friction nuts are being used.

Data obtained from a number of collieries show that following the introduction of roof bolts with increased pre-tension the roof stability was improved allowing for reduction of amount of either the primary or the secondary reinforcement (Rataj & Thomas 1997).

Very encouraging results are being achieved by the introduction of strand type bolts tensioned to a level of 25 tonnes. A recent study at Central Colliery clearly identified the significant cost benefits achieved by introduction of pre-tensioned rigid and strand type bolts (Rataj, Hanson & Frith, 1998).

The active confinement to the immediate roof by applying a high pre-tension is particularly beneficial in areas where rock mass failure initially occurs within the bottom or lower zones in the roof. The extra load applied to the roof surface prevents separations along planes of weakness. On some occasions discontinuities which developed prior to application of the pre-tensioning, are being 'tightened up' as the upward movement of the roof is clearly noticeable when the 20 or 25 tonnes load is being applied to the bolt plate.

This applied extra load to the roof surface appears to provide confinement which is able to control or stabilize roof which is already 'on the move'. A number of examples illustrating the positive effect of pre-tensioning on roof behavior are provided by Frith & Thomas (1998).

5 DISCUSSION

There has been a large number of modifications introduced to rock reinforcement hardware and installation techniques. As a result, significant improvement in the strata control has been achieved, even though mining conditions have become more difficult due to greater depths, the number of roof falls have diminished significantly.

The improvements achieved over the years in strata control has been a result of developing roof bolting technology which has provided a more efficient confinement of the strata.

More efficient confinement has also been achieved through making the bolts react in a much faster and more efficient way to any discrete strata displacement (improved response stiffness) and through introduction of techniques which confine the strata in a direct way i.e. by timely placement of and by high pre-tensioning of roof support.

The above statement is a conclusion from the analysis of all the changes which have been introduced to roofbolting technology over the years. It is clear that the improved system support stiffness results in more efficient rock reinforcement. There has not been a single incident where support failure was due to its having too great a stiffness characteristic in any of its components.

It is hard to quantify the effects of individual changes on the support stiffness. These individual product changes and developments include:

- the introduction of chemical anchors
- full encapsulation of bolts
- improved steel quality
- more efficient deformations or design
- introduction of stiffer roof plates
- angled bolt installation
- introduction of cable bolts and
- strand bolts.

It can only be concluded that these were all very important individually as well as collectively.

The use of the cable and strand bolts is nowadays the most efficient form of strata reinforcement. These types of support supply a very high system stiffness and, in addition, they provide rock confinement in the areas above the normal solid bolt horizon.

The direct confinement of strata (where required) by timely installation of support and by high pre-tensioning has proven to be successful.

The confinement mechanism of the strata by applying high loads to the roof plates is not yet fully understood with regards to the distance of influence within the rock mass. It is not uncommon to see the immediate roof being pulled up while tensioning a strand bolt to 20 or 25 tonnes. This can be readily observed by checking the nuts on adjacent roof bolts which become loose. This also proves that defects in the immediate rock mass are being closed.

It is also clear that the extent of the deflection in the immediate roof is very site specific and not every roof will be pulled up. At this stage however, it is not clear to what extent the deflection or its magnitude and distribution is occurring. Currently there is no practical method of assessment as extensometers have to be installed only after exposure of the roof. A way to learn a little more may be by measurement of the rock displacement around the high pre-tensioned strand bolt at various sites.

As already stated, installation of bolts in pre-failed rock ie close to the face, results in improved stability of the roof. This happens in spite of the fact that the bolts provide relatively small resistance (and in localised areas only) when compared to the high horizontal stress, say 20 MPa, which may be present in the rock mass. (The 20 MPa stress equals a load of 2000 tonnes per square metre of the roof cross section area).

If we accept that the above is correct, i.e that when the limited and localised extra confinement achieved in the strata by the close to face bolt installation results in a significant improvement in roof stability in spite of the high rock mass stresses, perhaps this will explain why the extra confinement provided by the highly tensioned support is so successful in spite of being also very low in relation to the rock stresses.

5.1 Coal rib supporting technology

Development of rib support technology was not analysed in this paper as it followed a somewhat different course due to both the requirements of a cuttable reinforcement and understanding of the need for yielding type support in some mining and geological conditions.

6 CONCLUSIONS FOR FUTURE DEVELOPMENTS

It is clear that increased stiffness of the roof bolt support system provides better strata control. It is not very clear at this time if there is, and what is the limit of the stiffness level. It appears however, that there is room for further improvement since the roof dilation measured within the bolted horizon is still considerably large in some areas. Therefore the future development should continue to further increase the support stiffness.

Long tendon support has proved to be a very efficient way of rock reinforcement. Recent developments of the resin grouted strand bolts overcame a lot of installation difficulties associated with cable bolts. Work in this area is still required to make the installations even more economical.

High pre-tensions of roof bolts and strand bolts have also proved to be successful. Further studies are required to determine the best levels of pre-tensioning in given conditions.

The highly tensioned, high load capacity strand bolts, which can be installed in pre-failed conditions if required, gives the mining engineer a new and powerful tool not used in the past. Full encapsulation of these bolts which is available using special installation equipment further enhances capacity of this product. There is however, need for development of a simple post-grouting technique.

More work is also required in areas of the support design in terms of location of the secondary reinforcement, both tensioned and un-tensioned, with particular attention to angledbolts regarding assessment of their effect in various geological conditions. The advantage of the angled tendons in soft rock conditions is indicated by results from computer modelling of the strata failure around the roadway by Tarrant & Gale (1998).

REFERENCES

Barnes, E.L.S. & Howe, B.J. (1964). Surface and underground testing of polyester resin roof bolt anchorage. Australian Coal Association (Research) Ltd. P.R. 64-6.

Frith, R. & Thomas, R. (1998). The pre-tensioning pedigree. Australia's Mining Monthly – July 1998.

Fuller, P.G, O'Grady, P., McCowan, B., Butcher, T.K. (1994), 13th Conf. On Ground Control in Mining, Morgantown, USA.

Gardener, F.J. (1971). History of rock bolting. Symp. on Rock Bolting. Australian Inst. Of Mining and Metallurgy, Illawarra Branch Paper No.2.

Howe, B.J. (1968) Roof Bolt anchorage in mudstone. Aust. Coal Industry Res. Lab. Ltd., Published Report No. P.R. 68-1.

Moss, K.J. (1971) Roofbolts in current use in Australia AIMM, Illawarra Branch, Symp. on Rock Bolting Paper No. 5.

Norris, C. & Yearby, M. (1980). The choice of support available, Proc. Symp. On Support in Coal Mines (A.J. Hargraves, ed.) Aus. Inst. Metall., Wollongong, NSW, pp. 4.1-4.5.

Rataj, M. Hanson, C. & Frith, R. (1998). Innovative roofbolting technologies in Australian coal mines. Int. Conf. Geomechanics/Ground Control in Mining and U/G Construction, Wollongong, Australia, 14-17 July 1998.

Rataj, M. & Thomas, R. (1997). New methods and technologies of roofbolting in Australian coal mines. 16th Conf. on Ground Control, Morgantown, USA.

Scotford, R.F. (1960). Commercial literature. The Titan Manufacturing Company Pty Ltd.

Tarrant, G. & Gale, W.J. (1998). Experience in the application of computer modelling to coal mine roadway design in weak rock. Proc. Int. Conf. Geomechanical/Ground Control in Mining and U/G construction, Wollongong, Australia, 14-17 July 1998.

Wright, A.J. (1964). Pull out test on Titan rock bolts with "Airo" resin anchors. Snowy Mountains Hydro Electric Authority Scientific Services Div. Report No. S.M. 1394

Yearby, M. & Shaw, R. (1986). Materials used in roadway support development headings. In Australasian Coal Mining Practice, (C.H. Martin, ed.) Aus. Inst. Min. Metall.

Rock Support and Reinforcement Practice in Mining, Villaescusa, Windsor & Thompson (eds)
© 1999 Balkema, Rotterdam, ISBN 90 5809 045 0

Author index